Lecture Notes in Physics

Volume 863

For further volumes:
www.springer.com/series/5304

The Lecture Notes in Physics

The series Lecture Notes in Physics (LNP), founded in 1969, reports new developments in physics research and teaching—quickly and informally, but with a high quality and the explicit aim to summarize and communicate current knowledge in an accessible way. Books published in this series are conceived as bridging material between advanced graduate textbooks and the forefront of research and to serve three purposes:

- to be a compact and modern up-to-date source of reference on a well-defined topic
- to serve as an accessible introduction to the field to postgraduate students and nonspecialist researchers from related areas
- to be a source of advanced teaching material for specialized seminars, courses and schools

Both monographs and multi-author volumes will be considered for publication. Edited volumes should, however, consist of a very limited number of contributions only. Proceedings will not be considered for LNP.

Volumes published in LNP are disseminated both in print and in electronic formats, the electronic archive being available at springerlink.com. The series content is indexed, abstracted and referenced by many abstracting and information services, bibliographic networks, subscription agencies, library networks, and consortia.

Proposals should be sent to a member of the Editorial Board, or directly to the managing editor at Springer:

Christian Caron
Springer Heidelberg
Physics Editorial Department I
Tiergartenstrasse 17
69121 Heidelberg/Germany
christian.caron@springer.com

Gianluca Calcagni · Lefteris Papantonopoulos ·
George Siopsis · Nikos Tsamis

Editors

Quantum Gravity
and Quantum
Cosmology

 Springer

Editors
Gianluca Calcagni
Instituto de Estructura de la Materia
CSIC
Madrid, Spain

Lefteris Papantonopoulos
Department of Physics
National Technical University of Athens
Athens, Greece

George Siopsis
Department of Physics and Astronomy
The University of Tennessee
Knoxville, TN, USA

Nikos Tsamis
Crete Center for Theoretical Physics
Department of Physics
University of Crete
Heraklion, Greece

ISSN 0075-8450   ISSN 1616-6361 (electronic)
Lecture Notes in Physics
ISBN 978-3-642-33035-3   ISBN 978-3-642-33036-0 (eBook)
DOI 10.1007/978-3-642-33036-0
Springer Heidelberg New York Dordrecht London

Library of Congress Control Number: 2012952147

Printed on acid-free paper

Springer is part of Springer Science+Business Media (www.springer.com)

Preface

This book is an edited version of the review talks given in the Sixth Aegean School on *Quantum Gravity and Quantum Cosmology*, held in Chora on Naxos Island, Greece, from 12th to 17th of September 2011. The aim is to present an advanced multiauthored textbook meeting the needs of both postgraduate students and young researchers, in the fields of gravity, relativity, cosmology and quantum field theory.

Quantum gravity in a broad sense is a fast-growing subject in physics and its study is expected to give answers on the short-distance behaviour of the gravitational interaction. Probing the high-energy and high-curvature regimes of gravitating systems can shed some light on the ways to achieve an ultraviolet complete quantum theory of gravity, giving us information about fundamental problems of classical gravity such as the initial big-bang singularity, the cosmological constant problem and the physics at and beyond the Planck scale. On the other hand, it can give vital information on the early-time inflationary evolution of our Universe.

The selected contributions to this volume discuss quantum gravity theories in connection with cosmological models and observations, and explore what type of signature modern and mathematically rigorous frameworks can be detected by experiments.

In the first part of the book, the idea of quantum gravity is introduced and approached from different angles. In the article by Kelly Stelle, an overview is given of the way in which the unification program of particle physics has evolved into the proposal of superstring theory as a prime candidate for unifying gravity with the other forces and particles of nature. A key concern with quantum gravity has been the problem of ultraviolet divergences, which is naturally solved in string theory by replacing particles with spatially extended states as the fundamental excitations. Next, Abhay Ashtekar is presenting a broad perspective on loop quantum gravity and cosmology, while the article by Carlo Rovelli summarizes the present state of the covariant formulation of the loop quantum gravity dynamics. A lattice spinor gravity is formulated in the next article by Christof Wetterich, explaining why the key ingredient for lattice regularized quantum gravity is diffeomorphism symmetry. Andrzei Görlich describes the method of causal dynamical triangulations, a nonperturbative and background independent approach to quantum theory of gravity.

The first part of the book ends with the article by E. Bergshoeff, M. Kovacevic, J. Rosseel and Y. Yin who review the recent developments in massive gravity.

The second part of the book deals with quantum cosmology. Martin Bojoward presents loop quantum cosmology as an attempt to understand the dynamics of loop quantum gravity by realizing crucial effects in simpler, usually symmetric settings. The next article by Martin Reuter and Frank Saueressig, after introducing the basic ideas of the asymptotic safety approach to quantum Einstein gravity, discusses the implications of asymptotic safety for the cosmology of the early Universe. The last article is by Paul McFadden, about the recent developments in holographic cosmology which enables four-dimensional inflationary universes to be described in terms of three-dimensional dual quantum field theories.

In the third part of the book, the observational status of dark matter (the article by Joe Silk) and the observational status of dark energy (overviewed by Shinji Tsujikawa) are presented. The contribution by Robert Brandenberger describes two alternatives to the current cosmological scenario, the matter bounce and the string gas cosmology scenarios. The last article, by M. Romania, N. Tsamis and R. Woodard, presents a class of non-local, gravitational models obtained in quantum gravity in an accelerating cosmological background.

The Sixth Aegean School and the present book became possible with the kind support of many people and organizations. The School was organized and sponsored by the Albert Einstein Institute in Potsdam, the Physics Department of the University of Crete, the Physics Department of the University of Tennessee and the Physics Department of National Technical University of Athens, and it was cosponsored by the Municipality of Naxos and the General Secretariat of Aegean and Island Policy. We specially thank the Municipality of Naxos for making available to us all the excellent facilities of the Cultural Center in the former Ursuline School and all the staff of the center for helping us to run smoothly the school. We also thank Katerina Chiou-Lahanas for her valuable help in organizing the school in Naxos. The administrative support of the Sixth Aegean School was taken up with great care by Fani Siatra and Katerina Papantonopoulou. We acknowledge the help of Vassilis Zamarias who designed and maintained the webside of the School. We also thank Petros Skamagoulis for helping us in editing this book.

Last, but not least, we are grateful to the staff of Springer-Verlag, responsible for the Lecture Notes in Physics, whose abilities and help contributed greatly to the appearance of this book.

Gianluca Calcagni
Lefteris Papantonopoulos
George Siopsis
Nikos Tsamis

Contents

Part I
Quantum Gravity

Chapter 1
String Theory, Unification and Quantum Gravity

K.S. Stelle

Abstract An overview is given of the way in which the unification program of particle physics has evolved into the proposal of superstring theory as a prime candidate for unifying quantum gravity with the other forces and particles of nature. A key concern with quantum gravity has been the problem of ultraviolet divergences, which is naturally solved in string theory by replacing particles with spatially extended states as the fundamental excitations. String theory turns out, however, to contain many more extended-object states than just strings. Combining all this into an integrated picture, called M-theory, requires recognition of the rôle played by a web of nonperturbative duality symmetries suggested by the nonlinear structures of the field-theoretic supergravity limits of string theory.

1.1 Introduction: The Ultraviolet Problems of Gravity

Our currently agreed picture of fundamental physics involves four principal forces: strong, weak, electromagnetic; and gravitational. The first three are well described by the Standard Model, based on the nonabelian gauge group $SU(3)^{\text{strong}} \times (SU(2) \times U(1))^{\text{electroweak}}$. In the process of unifying these forces, one necessarily had to postulate new physical phenomena going beyond the specifically desired unification. Thus, in order to make the $SU(2) \times U(1)$ electroweak unification work, one had also to accept also the neutral Z^0 field in addition to the desired charged $W^\pm$ intermediate vector fields (needed to resolve the interactions of the nonrenormalizable 4-fermion Fermi theory). The experimental discovery of the corresponding Z^0 particle was a great triumph of the Standard Model.

Another key ingredient of our current perspective is the notion of spontaneous symmetry breaking: symmetries of the field equations may be broken by the vacuum, thus becoming non-linearly realized and at the same time allowing for the generation of masses for gauge fields—known as the Higgs effect. The Standard Model is moreover renormalizable: although ultraviolet infinities exist, they can be

K.S. Stelle (✉)
Imperial College London, London SW7 2AZ, UK
e-mail: k.stelle@imperial.ac.uk

G. Calcagni et al. (eds.), *Quantum Gravity and Quantum Cosmology*,
Lecture Notes in Physics 863, DOI 10.1007/978-3-642-33036-0_1,
© Springer-Verlag Berlin Heidelberg 2013

corralled into renormalizations of a finite set of parameters, thus allowing for consistent perturbative analysis of the rest of the theory. And most importantly, the Standard Model is now confirmed to very high precision by experiments at CERN, Fermilab and other laboratories.

Einstein's General Theory of Relativity, on the other hand, is nonrenormalizable, causing it to break down when interpreted as a quantum theory. One immediate indication of this is the dimensional character of the gravitational coupling constant $\kappa = \sqrt{8\pi G}$, which has dimensions of length (in units where $\hbar = c = 1$). Einstein gravity's uncontrolled divergences go on to corrupt otherwise well-behaved "matter" theories.

Consider, for example, a radiative correction to the Higgs mass caused by a gauge-particle emission and reabsorption:

In the Standard Model, with gauge coupling constant g, incoming momentum p and loop momentum k, the corresponding integral with a cutoff Λ has the form

$$g^2 \int^\Lambda d^4k \frac{k^2}{k^2((p+k)^2+m^2)} \tag{1.1}$$

which has logarithmic divergences $\sim g^2 \ln \Lambda \, p^2$, requiring a counterterm $(\partial\phi)^2$ and also another $\sim g^2 \ln \Lambda \, m^2$, requiring a counterterm $m^2\phi^2$. Since both of these counterterm operators are present in the Standard Model Lagrangian from the start, they can be accounted for by standard wavefunction and mass renormalizations.

When the system is coupled to gravity, however, the ultraviolet divergent integrals get much worse:

$$\kappa^2 \int^\Lambda d^4k \frac{k^4}{k^2((p+k)^2+m^2)} \tag{1.2}$$

producing now logarithmic divergences $\sim \kappa^2 \ln \Lambda(p^4, m^2 p^2, m^4)$ in addition to the flat-space SM divergences. The p^4 divergence would require a counterterm $(\partial^2\phi)^2$, which is an operator not present in the original theory. Moreover, this bad ultraviolet behavior gets worse and worse as the loop-order increases. At two loops, one encounters divergences $\sim \kappa^4 \ln \Lambda p^6 + \cdots$, requiring a counterterm like $(\partial^3\phi)^2$. Each new loop adds 2 to the divergence count. Thus, Einstein gravity is not only uncontrolled in its own divergence structure; it also renders otherwise well-behaved matter theories such as the Standard Model uncontrollable when coupled to gravity.

Pure General Relativity has a naïve degree of divergence at L loops in spacetime dimension D given by $\Delta = (D-2)L + 2$. When confronting the ultraviolet problem of quantum gravity, one wants to focus on the most serious divergent structures, whose elimination would require the introduction of genuinely new operators not present in the classical Lagrangian. For this purpose, candidate counterterms that vanish subject to the classical field equations can be handled by a more standard procedure, by making field-redefinition renormalizations, which generalize the wavefunction renormalizations of renormalizable theories. Leaving these more easily handled divergence structures to one side, one searches for counterterm structures that do *not* vanish subject to the classical equations of motion.

Using dimensional regularization to ensure a manifestly generally-coordinate-invariant quantization, one captures only the logarithmic divergences of a straight momentum-cutoff procedure. To balance engineering dimensions, this requires a number of factors of external momentum to be present on the external lines of a divergent diagram, in order to pick out just the logarithmically divergent part. Accordingly, at $L = 2$ loops in $D = 4$ dimensions, one expects $\Delta = 6$, which could be achieved by counterterms like $\int d^4x \sqrt{-g}(R_{\mu\nu\rho\sigma} R^{\rho\sigma\lambda\tau} R_{\lambda\tau}{}^{\mu\nu})$ or $\int d^4x \sqrt{-g}(R_{\mu\nu\rho\sigma} \Box R^{\rho\sigma\mu\nu})$ where $\Box = g^{\mu\nu}\nabla_\mu\nabla_\nu$ is a covariant d'Alembertian. However, use of the Bianchi identities shows that the second of these types vanishes subject to the classical equations of motion, so it may be dealt with by field-redefinition renormalizations. Only the first is a truly dangerous type. And indeed, in pure GR, such a (curvature)^3 counterterm does occur at the 2-loop order in $D = 4$. [1, 2].

In supergravity theories, local supersymmetry places additional constraints on counterterms. This has the consequence that the 2-loop divergence of pure GR is absent. In pure supergravities, the first counterterm that does not vanish subject to the classical equations of motion ("on-shell" in the jargon) then occurs at the 3-loop level:

The corresponding $D = 4$ counterterm has $\Delta = 8$ and starts with a purely gravitational part that is quadratic in the Bel-Robinson tensor, *i.e.* quartic in curvatures [3]

$$\int d^4x \sqrt{-g} T_{\mu\nu\rho\sigma} T^{\mu\nu\rho\sigma}, \quad T_{\mu\nu\rho\sigma} = R_\mu{}^\alpha{}_\nu{}^\beta R_{\rho\alpha\sigma\beta} + {}^*R_\mu{}^\alpha{}_\nu{}^\beta {}^*R_{\rho\alpha\sigma\beta}. \quad (1.3)$$

For lesser supergravities (with $N \leq 4$ independent gravitini), extensions of this structure remain as candidates for the first anticipated serious nonrenormalizable divergence.

1.2 String Theory Basics

The fundamental excitations of String Theory are not point particles, as in ordinary quantum field theories, but extended objects. Thus, point-particle worldline interactions such as in Fig. 1.1 become smoothed out to string worldsheet interactions like

Fig. 1.1 3-point
field-theoretic particle vertex

Fig. 1.2 3-closed-string
vertex: the splitting point is
determined by the choice of
time slicing, unlike the sharp
identification of the
interaction point in particle
theory

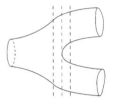

as in Fig. 1.2 with a consequent loss of sharpness in the spacetime localization of
the interaction.

The field-theory propagator

which has the usual overall momentum-space $\frac{1}{k^2}$ structure becomes in closed-string
theory that for a cylinder

with characteristic string length scale ℓ_s and momentum-space structure $\frac{e^{-\alpha' k^2}}{k^2}$
where α' is the *string slope parameter*, related to the characteristic string length
scale by $\alpha' = \frac{\ell_s^2}{2\hbar^2 c^2}$. The decreasing exponentials arising from string propagators
give rise to *convergent* loop diagrams for quantum corrections, yielding effectively
a cutoff to the field-theory divergences at a scale $\Lambda \sim (\ell_s)^{-1}$.

1.2.1 Reparametrization Invariance

An essential feature of all relativistic systems is the freedom to choose arbitrary
parametrizations for their histories. Begin with the analog of a relativistic particle,
whose action is obtained geometrically from the invariant proper length of its worldline as shown in Fig. 1.3.

This yields a worldline reparametrization-invariant action

$$I_{\text{particle}} = -m \int d\tau \left(-\frac{dx^\mu}{d\tau} \frac{dx^\nu}{d\tau} g_{\mu\nu}(x) \right)^{\frac{1}{2}}, \tag{1.4}$$

which has the following manifest local invariances:

Fig. 1.3 Particle worldline

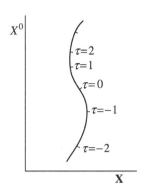

1. Spacetime general covariance:

$$x^\mu \to x^{\mu'}(x) \quad g'_{\mu\nu}(x') = \frac{\partial x^\rho}{\partial x^{\mu'}} \frac{\partial x^\sigma}{\partial x^{\nu'}} g_{\rho\sigma}(x) \tag{1.5}$$

2. Worldline reparametrization invariance:

$$
\begin{array}{ll}
& x'^\mu(\tau') = x^\mu(\tau) & \text{(worldline scalar)} \\
\tau \to \tau' & \dfrac{dx^\mu}{d\tau} \to \dfrac{dx^\mu}{d\tau'} = \dfrac{dx^\mu}{d\tau} \dfrac{d\tau}{d\tau'} & \text{(worldline vector)}
\end{array}
\tag{1.6}
$$

The worldline reparametrization invariance is physically important because it removes a negative-energy mode: for a metric $g_{\mu\nu}$ of Minkowski signature $(-+++\cdots)$, the $x^0(\tau)$ "scalar field" along the $d = 1$ worldline has the wrong sign of kinetic energy. However, this potential ghost mode is precisely removed from the theory by the worldline reparametrization invariance.

As is generally the case for gauge theories, the worldline reparametrization invariance gives rise, in the Hamiltonian formalism, to a constraint on the conjugate momenta:

$$p_\mu p_\nu g^{\mu\nu}(x) = -m^2, \quad \text{where } p_\mu = \frac{\partial \mathscr{L}}{\partial \left(\frac{\partial x^\mu}{\partial \tau}\right)}. \tag{1.7}$$

Thus, for a particle in D dimensional spacetime, $(D-1)$ degrees of freedom remain after taking into account the worldline reparametrization invariance and the corresponding Hamiltonian constraint. The constraint (1.7) is recognized as the mass-shell condition for the relativistic particle.

1.2.2 The String Action

Now generalize the relativistic particle action to that of a relativistic extended object with intrinsic spatial dimensionality $p = 1$. Instead of a worldline, one now has a

Fig. 1.4 Open string
worldsheet

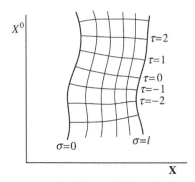

2-dimensional worldsheet as illustrated in Fig. 1.4 for an open string; for a closed string, one needs to identify $\sigma = 0$ and $\sigma = \pi$.

The string worldsheet action is then the reparametrization-invariant area of the worldsheet $\mathscr{W}$

$$I_{\text{string}} = -T \int_{\mathscr{W}} d^2\xi \left(-\det\left(\partial_i x^\mu(\xi)\partial_j x^\nu(\xi)g_{\mu\nu}\left(x(\xi)\right)\right)\right)^{\frac{1}{2}}. \tag{1.8}$$

As in the particle case, one has a number of local worldsheet invariances:

1. Spacetime general covariance $x^\mu(\tau,\sigma) \to x^{\mu'}(\tau,\sigma)$.
2. $d = 2$ worldsheet reparametrization invariance $x'^\mu(\tau',\sigma') = x^\mu(\tau,\sigma)$.
3. Exceptionally for the $d = 2 \leftrightarrow p = 1$ case among the general class of "p-branes", one has an additional local worldsheet invariance: *Weyl invariance*.

   Weyl invariance is crucial to the ability to carry out quantization of the string. Jealously preserving it leads to the notion of a *critical dimension* for string theory.

To see the Weyl invariance, reformulate the string action with an independent worldsheet metric $\gamma_{ij}(\xi)$ [4–6]:

$$I_{\text{string DZBdVHP}} = -\frac{1}{2}T \int d^2\xi \sqrt{-\det\gamma}\left(\gamma^{ij}(\xi)M_{ij}\right) \tag{1.9}$$

where $M_{ij} = \partial_i x^\mu \partial_j x^\nu g_{\mu\nu}(x)$ is the induced metric on the worldsheet and γ^{ij} is the matrix inverse of γ_{jk}. Varying $\gamma_{ij}(\xi)$ as an independent field, obtain its field equation $(\gamma^{ik}\gamma^{jl} - \frac{1}{2}\gamma^{ij}\gamma^{kl})M_{kl} = 0$. Note that for $d = 2$ worldsheet dimensions, the trace of this equation vanishes identically: $\gamma^{kl}M_{kl} - \frac{1}{2}\gamma^{ij}\gamma_{ij}\gamma^{kl}M_{kl} \equiv 0$.

This weakening of the set of algebraic equations for γ_{ij} corresponds to the local Weyl invariance of the DZBdVHP action:

$$\gamma_{ij} \to \Omega(\xi)\gamma_{ij} \tag{1.10}$$

$$\xi^i \to \xi^i \tag{1.11}$$

where $\Omega(\xi)$ is an arbitrary positive local scale factor. Varying the string action, one obtains the algebraic equation determining $\gamma_{ij}(\xi) = \Omega(\xi)M_{ij}$, with $\Omega(\xi)$ left undetermined, and the $d = 2$ covariant wave equation for $x^\mu(\xi)$:

$$\nabla^i_{(\gamma,g)}\partial_i x^\mu = 0. \tag{1.12}$$

For closed bosonic strings, the wave equation (1.12), plus periodicity in the spatial worldsheet coordinate σ (conventionally taken to identify $\sigma = 0$ with $\sigma = \pi$), give the full classical dynamical system of closed-string equations.

For open strings, the σ coordinate is conventionally considered to take its values in the closed interval $\sigma \in [0, \pi]$. Then, considering also the surface term arising in the variation of I_{DZBdVHP} upon integration by parts, one finds in addition the following Neumann boundary conditions:

$$M_{0i}\epsilon^{ik}\partial_k x^\mu = 0 \quad \text{at } \sigma = 0, \pi. \tag{1.13}$$

Considering strings in a flat spacetime background, $g_{\mu\nu} = \eta_{\mu\nu}$, and picking conformal gauge for the worldsheet reparametrization symmetries, $\gamma_{ij} = \Omega(\xi)\,\text{diag}(-1, 1)$, the $x^\mu(\xi)$ wave equation and open-string boundary conditions become

$$\Box x^\mu = 0 \quad \text{where } \Box = \eta^{ij}\partial_i\partial_j \text{ is the flat-space } d = 2 \text{ d'Alembertian} \tag{1.14}$$

$$\frac{\partial}{\partial\sigma}x^\mu = 0 \quad \text{at } \sigma = 0, \pi. \tag{1.15}$$

These may be interpreted classically as requiring waves to travel back and forth along the string at speed $c = 1$, while the boundary conditions imply that the endpoints of the open string travel through the embedding spacetime at speed $c = 1$.

For closed strings, there are periodicity conditions instead of reflective boundary conditions. In that case, there can be independent left- and right-moving waves travelling around the string at speed $c = 1$.

A simple solution to the open-string equations of motion and boundary conditions is

$$x^0 = \frac{1}{2}\left(p^+ + \frac{A^2}{p^+}\right)\tau \qquad x^3 = \frac{1}{2}\left(p^+ - \frac{A^2}{p^+}\right)\tau \tag{1.16}$$

$$x^1 = A\cos\sigma\cos\tau \qquad x^2 = A\cos\sigma\sin\tau. \tag{1.17}$$

Boosting to a Minkowski reference frame where $x^3 = 0$, find $p^+ = \frac{A^2}{p^+} = \pm A$; in this frame, the center-of-mass of the open string at $\sigma = \pi/2$ remains stationary while the string profile at any time τ describes a straight line of length $2A$ rotating with period $2\pi A$ (with respect to the background Minkowski time $t = x^0 = A\tau$).

The total string energy for this solution is $E = \frac{\pi}{2}\ell T$, where $\ell = 2A$ is the string length. Thus, the parameter T should be interpreted as the string tension.

The angular momentum for this solution is $J^3 = \frac{\pi}{8}\ell^2 T = \frac{E^2}{2\pi T}$. This linear relationship between angular momentum and (energy)^2 is known as *Regge behavior*.

Fig. 1.5 Linear Regge
trajectories relating spin and
(mass)^2 of particle states

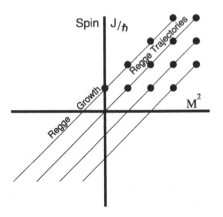

One can now make a rough Bohr-Sommerfeld estimate of the quantum spectrum,
requiring $|J| = n\hbar$, $n \in \mathbb{Z}$ and considering the excitations in their rest frames where
$E = M$. Then $n = \frac{|J|}{\hbar} = \alpha' M^2$ where $\alpha' = \frac{1}{2\pi \hbar c T}$ is the string slope parameter. The
quantized states lie on linear Regge trajectories making an angle α' in a $J/\hbar$ versus
M^2 plot (see Fig. 1.5).

Of course, finding such Regge trajectories in the physical particle spectrum
would be a spectacular confirmation of string theory.

The above semiclassical analysis makes the lowest-lying string state a massless
scalar. However, a more careful quantum analysis reveals a feature missed by the
Bohr-Sommerfeld analysis: the intercept at $n = 0$ is shifted down: $\alpha' E^2 = n - 1 \leftrightarrow$
$M^2 = \frac{n-1}{\alpha' c^4}$. Thus, the $n = 0$ lowest-lying state of the bosonic string becomes a neg-
ative M^2 tachyon, while the $n = 1$ first excited state with $|J| = \hbar$ becomes mass-
less. Accordingly, the open-string quantum spectrum contains massless spin 1 gauge
fields.

The closed string dispenses with the reflective open-string boundary conditions
and accordingly has twice as many modes: independent left- and right-moving excitations. It turns out that the closed-string spectrum is a tensor product of open-string
spectra in the R & L sectors, together with a level-matching condition: the R and L
level numbers must be equal. The closed-string $(n_L, n_R) = (1, 1)$ states thus contain the tensor product of (spin 1)$_L$ × (spin 1)$_R$ states: the closed-string spectrum
contains *massless spin 2*.

1.3 Effective Field Equations

The spin 2 mode identified in the closed-string spectrum is not merely a hint that
closed-string theory has something to do with gravity. The full Einstein action also
emerges when one considers string theory from an effective-field-theory point of
view. The key to understanding this is the requirement that anomalies in the local
Weyl symmetry cancel.

Analysis of the spectrum of any string theory shows the presence of at least three types of massless field: the graviton $g_{\mu\nu}(x)$, a 2-form antisymmetric tensor gauge field $B_{\mu\nu}(x)$ and a "dilatonic" scalar $\phi(x)$. In supersymmetric theories, the infamous tachyon of bosonic string theory is absent. In non-supersymmetric contexts, the tachyon is interpreted as indicating that the presumed "vacuum" around which one is trying to quantize is unstable and so one should shift instead to a stable vacuum background. This shift is made explicit in string field theory.

To begin with, consider just the massless backgrounds $(g_{\mu\nu}(x), B_{\mu\nu}(x), \phi(x))$. The string action on this effective-field background is then

$$I_{\text{gen. back.}}$$

$$= -\frac{1}{4\pi\alpha'} \int d^2\xi \sqrt{-\gamma} \left[\left(\gamma^{ij} g_{\mu\nu}(x) - \epsilon^{ij} B_{\mu\nu}(x) \right) \partial_i x^\mu \partial_j x^\nu + \alpha' R(\gamma)\phi(x) \right].$$

$$(1.18)$$

Note that the 2-form background gauge field $B_{\mu\nu}(x)$ has rank needed to pull back using $\partial_i x^\mu$ to a 2-form on the worldsheet, precisely as needed to contract with the $d = 2$ Levi-Civita tensor ϵ^{ij}. Note also that the coupling to the dilaton $\phi(x)$ involves the worldsheet Ricci scalar $R(\gamma)$ and enters with an additional factor of α', as is appropriate if $g_{\mu\nu}$, $B_{\mu\nu}$, ϕ and γ_{ij} are all taken to be dimensionless.

The worldsheet Weyl symmetry $\gamma_{ij}(\xi) \rightarrow \Omega(\xi)\gamma_{ij}(\xi)$ is respected by the $B_{\mu\nu}$ coupling (since $\sqrt{-\gamma}\epsilon^{ij}_{\text{tensor}} = \epsilon^{ij}_{\text{density}}$ is γ_{ij} independent), but it is violated by the dilaton coupling $\phi R(\gamma)$. This is intentional: the dilaton coupling is introduced precisely to complete the cancellation of Weyl-symmetry anomalies arising in the perturbative α' expansion.

$I_{\text{gen. back.}}$ is manifestly invariant under spacetime general coordinate transformations $x^\mu \rightarrow x^{\mu'}$ provided $g_{\mu\nu}$ and $B_{\mu\nu}$ transform as tensors and the dilaton ϕ is a scalar. It is also invariant under the $B_{\mu\nu}$ gauge transformation $B_{\mu\nu} \rightarrow B_{\mu\nu} + \partial_\mu \zeta_\nu - \partial_\nu \zeta_\mu$, which causes the integrand of $I_{\text{gen. back.}}$ to vary by a total derivative.

The general-coordinate and 2-form gauge invariances are precisely what are needed to give agreement with the expected degree-of-freedom counts for these massless backgrounds: $(\underbrace{\frac{1}{2}D(D-3)}_{\text{metric}}, \underbrace{\frac{1}{2}(D-2)(D-3)}_{\text{2-form}})$.

Imposing on this background-coupled string system the requirement that the Weyl symmetry anomalies cancel gives differential-equation restrictions on the background fields $(g_{\mu\nu}, B_{\mu\nu}, \phi)$; these may be viewed as *effective field equations* for these massless modes.

The system of effective field equations for $(g_{\mu\nu}, B_{\mu\nu}, \phi)$ is, remarkably, derivable from an *effective action* for the D dimensional massless modes [7].

$$I_{\text{eff}} = \int d^D x \sqrt{-g} e^{-2\phi} \left[(D - 26) - \frac{3}{2}\alpha' \left(R + 4\nabla^2\phi - 4(\nabla\phi)^2 \right. \right.$$

$$\left. \left. - \frac{1}{12} F_{\mu\nu\rho} F^{\mu\nu\rho} \right) + \mathcal{O}(\alpha')^2 \right].$$

$$(1.19)$$

Note the appearance of a *critical dimension*: the "cosmological term" vanishes only for $D = 26$, showing that, for a flat background, the Weyl anomalies can be cancelled in this way only in 26 dimensional spacetime.

In superstring theories, there are additional anomaly contributions from the fermionic modes which change the critical dimension to 10. Moreover, in supergravity theories, the tachyon is absent, so $D = 10$ flat space becomes a stable background of the massless modes. Aside from the change of the critical dimension to 10, however, the above effective action remains valid for a subset of the bosonic background of the theory, known as the Neveu-Schwarz sector.

Now specialize to $D = 10$ for the superstring and accordingly drop the cosmological term. Moreover, the unfamiliar $e^{-2\phi}$ factor in front of the Ricci scalar R may be eliminated together with the $4e^{-2\phi}\nabla^2\phi$ term by redefining the metric: $g_{\mu\nu}^{(e)} = e^{-\phi/2}g_{\mu\nu}^{(s)}$ where $g^{(s)}$ is the previous string-frame metric and $g^{(e)}$ is the new Einstein-frame metric.

In the Einstein frame, the Neveu-Schwarz sector effective action then becomes

$$I_{\text{Einstein}} = \int d^{10}x\sqrt{-g^{(e)}}\left[R\left(g^{(e)}\right) - \frac{1}{2}\nabla_\mu\phi\nabla^\mu\phi - \frac{1}{12}e^{-\phi}F_{\mu\nu\rho}F^{\mu\nu\rho}\right]. \quad (1.20)$$

Including effective-action contributions for the other (Ramond sector) bosonic backgrounds and also for fermionic backgrounds, one obtains thus a correspondence between superstring theories and related supergravity theories: a supergravity theory describes the massless field-theory sector of the corresponding superstring theory. One obtains in this way effective supergravity theories for the following superstring theory variants: type IIA, type IIB, type I with gauge group SO(32), heterotic SO(32) and heterotic $E_8 \times E_8$.

1.4 Dimensional Reduction and T-Duality

In order to extract a more realistic physical scenario from the higher-dimensional contexts native to string theory, one needs to reduce the effective theory down to $D = 4$ one way or another. The most straightforward way to do this is by a traditional Kaluza-Klein reduction.

The basic idea can be explained in terms of a massless scalar field in $D = 5$ on a spacetime with the 5th direction periodically identified: $y \sim y + 2\pi\mathcal{R}$. Periodicity requirements on the de Broglie waves $e^{ipy/\mathcal{R}}$ then require the momenta in the y direction to be quantized, $p_n = \frac{n\hbar}{\mathcal{R}}$. Thus, expand the $D = 5$ field $\phi(x^\mu, y)$, $\mu = 0, 1, 2, 3$, using a complete set of eigenfunctions of the Laplace operator on a circle, *i.e.* in terms of plane waves with quantized momenta:

$$\phi\left(x^\mu, y\right) = \sum_{n\in\mathbb{Z}}\phi_n\left(x^\mu\right)e^{iny/\mathcal{R}}. \quad (1.21)$$

Inserting this expansion into the $D = 5$ Klein-Gordon field equation gives an infinite number of $D = 4$ equations for the independent modes $\phi_n(x^\mu)$:

$$\frac{1}{c^2}\frac{\partial^2 \phi_n}{\partial t^2} - \nabla^2 \phi_n + \frac{n^2}{\mathscr{R}^2}\phi_n = 0. \tag{1.22}$$

Thus, the $n \neq 0$ modes ϕ_n are massive, with masses $m_n = \frac{n}{\mathscr{R}}$.

The basic physical picture is that at energies low compared to $\frac{\hbar}{c\mathscr{R}}$, the massive modes $\phi_{n>0}$ are frozen out, so the theory effectively reduces to just ϕ_0. Dimensional reduction of the supergravity theories associated to the various $D = 10$ string theories produces the family of supergravity theories existing in lower spacetime dimensions, including the maximally extended $N = 8$ supergravity in $D = 4$.

1.4.1 Dimensional Reduction of Strings and T-Duality

Consider now string theory in a background spacetime with a compactified direction, $x^M \rightarrow (x^\mu, y)$, $\mu = 0, \ldots, (D-2)$. The Regge towers of string states can be individually treated as particle fields; massless string states give rise to massless states in the $(D-1)$ lower dimensions plus Kaluza-Klein towers of states with masses $\frac{n}{\mathscr{R}}$, just like in Kaluza-Klein field theory.

Strings, however, can do something different from particles in that they can *wrap* *around* the compactified dimension. Consider a closed-string mode expansion

$$x^M(\tau, \sigma) = q^M(\tau) + p^M \ell^2 \tau + 2\tilde{n}\mathscr{R}\sigma\delta_y^M$$
$$+ \frac{i\ell}{2}\sum_{k \neq 0}\left(\frac{\alpha_k^M}{k}e^{-2ik(\tau-\sigma)} + \frac{\tilde{\alpha}_k^M}{k}e^{-2ik(\tau+\sigma)}\right) \tag{1.23}$$

where $n, \tilde{n} \in \mathbb{Z}$ and $\ell^2 = 2\alpha'$, the (string length)^2.

As expected, the momentum in the compactified direction is quantized, $p^y = \frac{n}{\mathscr{R}}$. However, owing to the fact that the string can wind around the compactified y dimension a number $\tilde{n}$ times (Fig. 1.6), the energy (*i.e.* mass) formula for the string spectrum considered from the viewpoint of the dimensionally reduced theory has a generalized form:

$$M^2 = \frac{\hbar^2}{c^2}\left(\frac{n^2}{\mathscr{R}^2} + \frac{\tilde{n}^2\mathscr{R}^2}{\alpha'^2}\right) + \text{contributions from ordinary oscillator modes.} \tag{1.24}$$

This mass formula suggests a striking symmetry of string theory that is not present for particle theories: interchanging $n \leftrightarrow \tilde{n}$ and simultaneously inverting the compactification radius, $\mathscr{R} \rightarrow \alpha'/\mathscr{R}$ leaves the spectrum invariant.

This symmetry is *T-duality:* a string propagating on a compact direction of radius $\mathscr{R}$ with momentum mode n and winding mode $\tilde{n}$ is equivalent to a string propagating on a compact direction of radius $\alpha'/\mathscr{R}$ with interchanged mode numbers: momentum $\tilde{n}$ and winding n.

Fig. 1.6 Winding modes
with various $\tilde{n}$ values

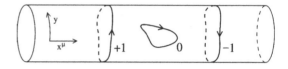

Because string and background configurations related by a T-duality transformation are *identified*, this symmetry, although discrete, extends the notion of local symmetry in string theory beyond the ordinary context of general coordinate and gauge invariances.

T-duality has a dramatic effect on curved background geometries. Start from a simplified closed-string action without the dilaton:

$$I_{g,B} = -\frac{1}{2}\int d^2\xi \sqrt{-\gamma}\left(\gamma^{ij}\partial_i x^M \partial_j x^N g_{MN} - \epsilon^{ij}\partial_i x^M \partial_j x^N B_{MN}\right). \qquad (1.25)$$

Now suppose that there is an *isometry* in the y direction, *i.e.* that g_{MN} and B_{MN} don't depend on y. Of course, $y(\tau,\sigma)$ is still a string variable—the string is not prevented from moving in the y direction of spacetime. But the background functional dependence on y is trivial owing to the isometry. Accordingly, the string variable $y(\tau,\sigma)$ appears only through its derivative $\partial_i y$.

Now replace $\partial_i y$ everywhere in the action by v_i, a worldsheet vector. Enforce the curl-free nature of v_i by a Lagrange multiplier term $\int d^2\xi \sqrt{-\gamma}\epsilon^{ij}\partial_i z v_j$. Then eliminate v_i by its algebraic equation of motion. The result is the T-dualized version of the string action written in terms of $\tilde{x}^M(\tau,\sigma) = (x^\mu(\tau,\sigma), z(\tau,\sigma))$.

The net effect of a T-duality transformation may be seen by reassembling the results into an action $I_{\tilde{g},\tilde{B}}$ of the same general form as $I_{g,B}$ but now for string variables $\tilde{x}^M(\tau,\sigma)$ and with dualized backgrounds $\tilde{g}_{\tilde{M}\tilde{N}}(\tilde{x})$, $\tilde{B}_{\tilde{M}\tilde{N}}(\tilde{x})$ given by [8]

$$\tilde{g}_{\mu\nu} = g_{\mu\nu} + g_{yy}^{-1}(B_{\mu y}B_{\nu y} - g_{\mu y}g_{\nu y})$$

$$\tilde{g}_{\mu z} = g_{yy}^{-1}B_{\mu y} \qquad \tilde{g}_{zz} = g_{yy}^{-1}$$

$$\tilde{B}_{\mu\nu} = B_{\mu\nu} + g_{yy}^{-1}(g_{\mu y}B_{\nu y} - g_{\nu y}B_{\mu y})$$

$$\tilde{B}_{\mu z} = g_{yy}^{-1}g_{\mu y}.$$

Careful attention to the effect of T-duality transformations reveals that they can map not only between different solutions of a given string theory, but they can even map between solutions of *different* string theories. In particular, paying careful attention to the effect on spinor backgrounds shows [9–11] that T-duality maps between type IIA and type IIB closed-string theories:

Type IIA on S^1 of radius $\mathscr{R} \overset{T}{\longleftrightarrow}$ Type IIB on S^1 of radius $\alpha'/\mathscr{R}$

1.5 M-Theory and the Web of Dualities

Another essential duality symmetry of string theory is strong-weak coupling duality, or S-duality. The *dilaton* field plays a crucial rôle in this, as its expectation value serves as the *coupling constant* for string interactions. String theory has no other à priori determined parameters (except for the scale-setting slope parameter α').

All the essential coupling constants are determined by vacuum expectation values of scalar fields present in the theory, with coupling constants typically given by the VEVs of exponentials like e^ϕ. Since, in a dimensional-reduction context, massless scalar fields derive from the moduli of the reduction manifold (*e.g.* torus circumferences, twist parameters, *etc.*), scalar fields with undetermined vacuum expectation values are generically called *moduli* fields.

The most accessible illustration of the geometry of such moduli and the symmetries acting upon them is to be found in the massless sector of Type IIB theory, whose effective action is Type IIB supergravity. The bosonic part of the action for Type IIB supergravity is

$$I_{10}^{\text{IIB}} = \int d^{10}x \left[e\mathscr{R} + \frac{1}{4}e\, \text{tr}\big(\nabla_\mu \mathscr{M}^{-1}\nabla^\mu \mathscr{M}\big) - \frac{1}{12}e H_{[3]}^T \mathscr{M} H_{[3]} \right.$$
$$\left. - \frac{1}{240}e H_{[5]}^2 - \frac{1}{2\sqrt{2}}\epsilon_{ij}{}^* \big(B_{[4]} \wedge dA_{[2]}^{(i)} \wedge dA_{[2]}^{(j)}\big) \right], \qquad (1.26)$$

subject to the further constraint of self-duality for the 5-form field strength $H_{\mu_1...\mu_5} = \frac{1}{5!}\epsilon_{\mu_1...\mu_5\mu_6...\mu_{10}}H^{\mu_1...\mu_{10}}$. The 3-form field strengths $H_{[3]} = \binom{dB_{[2]}^1}{dB_{[2]}^2}$ contract into the 2×2 matrix built from the scalars ϕ and χ

$$\mathscr{M} = \begin{pmatrix} e^{-\phi} + \chi^2 e^\phi & \chi e^\phi \\ \chi e^\phi & e^\phi \end{pmatrix}. \qquad (1.27)$$

Multiplying out the scalar kinetic terms, one finds a more familiar form:

$$-\frac{1}{2}\int d^{10}x \sqrt{-g}\big(\partial_\mu \phi \partial_\nu \phi g^{\mu\nu} + e^{2\phi}\partial_\mu \chi \partial_\nu \chi g^{\mu\nu}\big). \qquad (1.28)$$

From the above form of the IIB action, one can see that it has an $SL(2, \mathbb{R})$ symmetry $\mathscr{M} \to \Lambda \mathscr{M} \Lambda^T$, $H_{[3]} \to (\Lambda^T)^{-1}H_{[3]}$, $H_{[5]} \to H_{[5]}$ where $\Lambda = \binom{a\ b}{c\ d}$ with $\det \Lambda = 1$ is an $SL(2, \mathbb{R})$ matrix. While the action of $SL(2, \mathbb{R})$ on $\mathscr{M}$ is linear, the action on (ϕ, χ) is nonlinear: these fields form an $SL(2, \mathbb{R})/U(1)$ nonlinear sigma model. The action of $SL(2, \mathbb{R})$ on the scalars may be reformulated in terms of its action on the modular field $\tau = \chi + ie^{-\phi}$, which transforms in a fractional linear fashion as $\tau \to \frac{a\tau+b}{c\tau+d}$.

At the nonperturbative quantum level, the $SL(2, \mathbb{R})$ symmetry gets reduced to its discrete subgroup $SL(2, \mathbb{Z})$. This is necessary in order for the Gauss's law charges associated to $H_{[3]}^i$ to obey a Dirac quantization condition; $SL(2, \mathbb{Z})$ is the subgroup that preserves the resulting charge lattice. The surviving $SL(2, \mathbb{Z})$ may be considered to be generated by two elementary transformations, $\tau \to \tau + 1$ and $\tau \to -\frac{1}{\tau}$. For $\chi = 0$, the second of these inverts the v.e.v. of e^ϕ, hence the string coupling constant g_s. So this is called *S-duality* because it exchanges strong and weak string coupling.

Given that apparently different string theories can be related by T-duality transformations and that different coupling-constant regimes can be related by S-duality transformations, one naturally searches for the full interrelated set of theories and coupling regimes related by duality transformations, known as the "web of dualities".

A key link in this web of dualities concerns the strong-coupling limit of type IIA theory. There is no known duality that gives this limit purely within the type IIA theory, but the relation between string-theory dualities and supergravity dualities does suggest what the strong-coupling regime of type IIA string theory might become. There is one more maximal supergravity theory which had not yet been integrated into the general picture of string & supergravity theories: *supergravity in 11-dimensional spacetime*. This theory has as bosonic fields just the metric g_{MN} and a 3-form gauge field C_{MNP}, and as fermionic field the gravitino ψ_M^α ($\alpha = 1, \ldots, 32$). Overall there are 128 bosonic and 128 fermionic physical degrees of freedom per spacetime point.

$D = 11$ supergravity contains no scalar fields, but when it is dimensionally reduced to $D = 10$ on a circle S^1, straightforward Kaluza-Klein reduction generates one scalar, basically from the $g_{11\,11}$ component of the $D = 11$ metric. The reduced theory precisely reproduces $D = 10$ type IIA theory at the classical level, with the Kaluza-Klein scalar ϕ becoming the dilaton of the type IIA theory and $g_s = \langle e^\phi \rangle$ being the supergravity realization of the type IIA string coupling constant. Since $g_{11\,11}$ gives the metric on the reduction circle S^1, the modulus field ϕ controls the circumference of that circle. Thus, strong coupling, $g_s \to \infty$, corresponds to the limit where the S^1 reduction circle circumference tends to infinity.

Now consider just *compactification* of $D = 11$ supergravity instead of *dimensional reduction* down to $D = 10$, *i.e.* define the theory on a circle S^1 but don't discard the Kaluza-Klein towers of massive states. Taking the limit $g_s \to \infty$ now corresponds to returning the theory to uncompactified $D = 11$ supergravity.

If there is to be a $D = 11$ picture of $D = 10$ Type IIA theory, where can the Kaluza-Klein towers of states come from? Well, the dimensional reduction of massless $D = 11$ states produces massive states that also carry a $U(1)$ charge corresponding to the Kaluza-Klein vector, derived from $g_{11,\mu}$: they are $\frac{1}{2}$ BPS states originating in the Ramond sector of the theory. And, in fact, Type IIA theory does have just such states: the tower of $\frac{1}{2}$ BPS *black hole* states, carrying charges under the vector gauge field A_μ of the Type IIA theory [12, 13].

For increasing $g_s = \langle e^\phi \rangle$, the spacing between the BPS mass levels decreases, approaching a continuum as one approaches the decompactification limit of infinite S^1 circumference, where the full $D = 11$ nature of the theory becomes more and more manifest. Accordingly, the strong g_s coupling limit of Type IIA string theory is hypothesized to be described by a phase whose full quantum properties remain incompletely known, but which has $D = 11$ supergravity as a field-theory limit. This phase of the overall picture has been called *M-Theory* (Fig. 1.7).

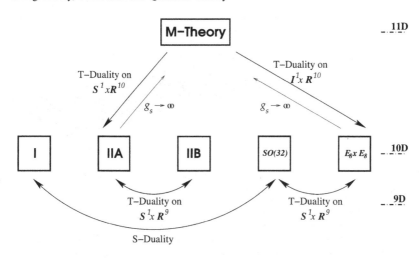

Fig. 1.7 "Not to take this web of dualities as a sign that we are on the right track would be a bit like believing that God had put fossils into the rocks in order to mislead Darwin about the evolution of life." Steven Hawking

1.6 Branes and Duality

Consider a T-duality transformation on the worldsheet variable $y(\tau, \sigma)$ now in a bit more detail, specializing to a flat background spacetime in which $g^{yy} = k$. The relevant part of the string action is $-\frac{k}{2} \int d^2\xi \sqrt{-\gamma} \gamma^{ij} \partial_i y \partial_j y$. Now replace $\partial_i y \rightarrow v_i$ and include as before a Lagrange multiplier $z(\tau, \sigma)$ in order to enforce the vanishing of $\partial_i v_j - \partial_j v_i$: $\int d^2\xi(-\frac{k}{2}\sqrt{-\gamma}\gamma^{ij} v_i v_j + \epsilon^{ij} v_i \partial_j z)$. For v^i, find the algebraic equation $v^i = \frac{1}{k}\epsilon^{ij}\partial_j z$. Substituting back into the action then gives the T-dualized result $-\frac{1}{2k} \int d^2\xi \sqrt{-\gamma} \gamma^{ij} \partial_i z \partial_j z$.

Now consider, however, the effect of the above procedure on the usual open-string Neumann boundary condition $\partial_\sigma y = 0$ at the endpoints (endpoint worldline normal derivative vanishes). After T-dualization, this becomes $\partial_\tau z = 0$ at the endpoints (endpoint worldline tangential derivative vanishes). Thus, for the T-dualized coordinate $z(\tau, \sigma)$, one obtains a *Dirichlet boundary condition* $z =$ constant at the endpoints (Fig. 1.8):

$$\text{Neumann b.c.} \overset{T}{\longleftrightarrow} \text{Dirichlet b.c.}$$

The surfaces on which Dirichlet boundary conditions are imposed obviously would break Lorentz invariance if they were considered to be imposed externally to the theory. However, considering them to be dynamical objects similar to solitons in the theory restores Lorentz symmetry.

Analysis of the open-string modes in which p background spatial dimensions are treated with Dirichlet and the remaining $(10 - (p + 1))$ spatial dimensions with Neumann boundary conditions reveals modes associated to the $(p + 1)$ dimensional "worldvolume" of the Dirichlet surface (p spatial dimensions plus time). These are a massless $U(1)$ gauge field A_i together with $(9 - p)$ massless scalar modes.

Fig. 1.8 Open strings
starting and ending with
Dirichlet boundary conditions
on a p-dimensional D-brane
hyperplane in the target
spacetime

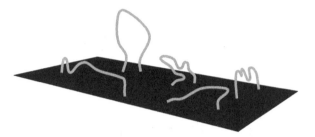

The massless worldvolume modes can be interpreted as Goldstone modes for the broken antisymmetric tensor gauge symmetry and for the Poincaré translation symmetries broken by the choice of Dirichlet boundary-condition integration constants. In other words, these massless scalar worldvolume modes may be seen to describe motions of the Dirichlet surface transverse to the worldvolume. This dynamical object is called a D_p brane. The motions of a D_p brane are described by an effective action of Dirac-Born-Infeld type [14]:

$$I_{D_p} = -T_p \int d^{p+1}\xi\, e^{-\phi(x(\xi))} \left[-\det\left(M_{ij} + B_{ij} + 2\pi\alpha' F_{ij} \right) \right]^{\frac{1}{2}}$$

$$M_{ij} = \partial_i x^\mu \partial_j x^\nu g_{\mu\nu}(x),\ \ B_{ij} = \partial_i x^\mu \partial_j x^\nu B_{\mu\nu}(x),\ \ F_{ij} = \partial_i A_j(\xi) - \partial_j A_i(\xi).$$

$$(1.29)$$

The dynamical extended-object hypersurfaces encountered as D_p-branes in string theory have natural analogue p-brane solutions in the associated supergravity theories. In fact, the supergravity solutions extend the brane family beyond those seen directly as D_p branes in perturbative string theory, indicating a yet richer family of nonperturbative extended-object solutions.

A representative example is the string itself, viewed now as an extended-object solution to the effective theory's field equations. In the various $D = 10$ supergravities associated to superstring theories, one always has a Neveu-Schwarz sector

$$I_{NS} = \int d^{10}x \sqrt{-g} \left[R - \frac{1}{2}\nabla_\mu \phi \nabla^\mu \phi - \frac{1}{12} H_{\mu\nu\rho} H^{\mu\nu\rho} \right] \qquad (1.30)$$

where $H_{\mu\nu\rho} = \partial_\mu B_{\nu\rho} + \partial_\nu B_{\rho\mu} + \partial_\rho B_{\nu\mu}$.

This effective action has an explicit solution:

$$ds^2 = \mathscr{H}^{-\frac{3}{4}}(y)dx^i dx^j \eta_{ij} + \mathscr{H}^{\frac{1}{4}}(y)dy^m dy^m$$

$$B_{ij} = \epsilon_{ij}\mathscr{H}^{-1}(y) \qquad\qquad\qquad\qquad (1.31)$$

$$e^\phi = \mathscr{H}^{-\frac{1}{2}}(y) \qquad \mathscr{H}(y) = 1 + \frac{k}{(y^m y^m)^3}.$$

The singular surface at $y = \sqrt{y^m y^m} = 0$, parametrized by x^i, $i = 0, 1$, corresponds to the static worldsheet of an infinite string extending from $x^1 = -\infty$ to $x^1 = +\infty$,

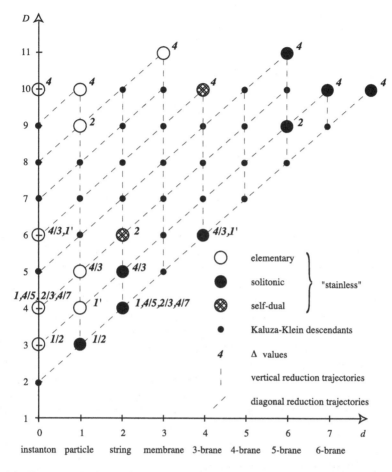

Fig. 1.9 "Brane-scan" of supergravity p-brane solutions, linked by worldvolume (diagonal) and transverse-space (vertical) dimensional reductions

with an 8-dimensional transverse space $\mathcal{M}_8$ within which the solution is spherically symmetric.

The solution has a charge as well, given by Gauss's law: $U = \int_{\partial \mathcal{M}_8} d^7 \Sigma^m H_{m01} = 6k\Omega_7$, where Ω_7 is the volume of the unit 7-sphere corresponding to the infinite boundary of $\mathcal{M}_8$. This charge is equal to the ADM *tension* (energy/unit x^1 length) of the solution, so this string solution is an analogue of the extremal Reissner-Nordstrom solution of Einstein-Maxwell theory.

There is a great variety of p-brane solutions in supergravity theories, of diverse worldvolume and transverse dimensionalities, as shown in Fig. 1.9. The supersymmetric p-brane spectrum naturally generalizes the extremal black holes of Einstein-Maxwell theory, which may be viewed as 0-branes. In a given dimension of spacetime, the brane spectrum also naturally carries a representation of the corresponding supergravity *duality group*.

Table 1.1 Supergravity $E_{11-D(11-D)}(\mathbb{R})$ duality symmetries, K_D maximal compact subgroups and the superstring $E_{11-D(11-D)}(\mathbb{Z})$ discretizations

D	$E_{11-D(11-D)}(\mathbb{R})$	K_D	$E_{11-D(11-D)}(\mathbb{Z})$
10A	$\mathbb{R}^+$	1	1
10B	$Sl(2,\mathbb{R})$	$SO(2)$	$Sl(2,\mathbb{Z})$
9	$Sl(2,\mathbb{R}) \times \mathbb{R}^+$	$SO(2)$	$Sl(2,\mathbb{Z})$
8	$Sl(3,\mathbb{R}) \times Sl(2,\mathbb{R})$	$SO(3) \times SO(2)$	$Sl(3,\mathbb{Z}) \times Sl(2,\mathbb{Z})$
7	$Sl(5,\mathbb{R})$	$SO(5)$	$Sl(5,\mathbb{Z})$
6	$SO(5,5,\mathbb{R})$	$SO(5) \times SO(5)$	$SO(5,5,\mathbb{Z})$
5	$E_{6(6)}(\mathbb{R})$	$USp(8)$	$E_{6(6)}(\mathbb{Z})$
4	$E_{7(7)}(\mathbb{R})$	$SU(8)/\mathbb{Z}_2$	$E_{7(7)}(\mathbb{Z})$
3	$E_{8(8)}(\mathbb{R})$	$SO(16)$	$E_{8(8)}(\mathbb{Z})$

Dimensional reduction of the maximal theory down from $D = 11$ automatically generates a $GL(11 - D, \mathbb{R})$ nonlinearly realized symmetry of the D dimensional reduced supergravity. However, special features of supergravity theories lead to an enhancement of this anticipated duality group. These features include the combination of vectors and scalars coming from the $D = 11$ metric and the $D = 11$ 3-form gauge field, and also the dualization of higher-rank form fields to lower-rank fields by Hodge dualization of the corresponding field strengths.

The resulting duality groups for maximal supergravity are shown in Table 1.1, where $E_{11-D(11-D)}$ is the nonlinearly realized duality symmetry in spacetime dimension D, K_D is its linearly realized maximal compact subgroup and $E_{11-D(11-D)}(\mathbb{Z})$ is the discretized "U-duality" [12, 13] form consistent with the Dirac quantization condition, which is conjectured to survive in superstring theory.

1.7 The Onset of Supergravity Divergences

Now we shall return to the initial question of field-theoretic gravity theories and their quantum problems. We have seen that there is a rich tapestry of supergravity limits, with surprizing additional duality symmetries, which emerge as "zero-slope" $\alpha' \to 0$ limits of superstring/M-theory. The question remains whether links to theories based upon extended objects as the fundamental excitations have a bearing on the original ultraviolet problems of field-theoretic gravity and supergravity theories. The dimensional character of Newton's constant and the related nonlinearity of the Einstein-Hilbert action leads to a general expectation of nonrenormalizability in gravity and supergravity theories. Resolving the ultraviolet problem of quantum gravity has consequently been one of the main aims of superstring theory. However, it is important to understand precisely how the quantum properties of superstring theories differ from those of the corresponding supergravities when the latter are subject to standard field-theoretic quantization.

In order to understand this relation, the precise order of onset of nonrenormalizable divergences in supergravity theories has remained an intensely studied question. Local supersymmetry brings about at least significant delays in the onset of ultraviolet divergences, but the full reach of the corresponding nonrenormalization theorems is still not fully clear. What is clear, however, is that links to superstring and M-theory have led to some genuinely surprizing ultraviolet cancellations.

Explicit calculations of ultraviolet divergence coefficients have been carried out using traditional Feynman diagram techniques up to the 2-loop level [1, 2]. Continuing on this way to higher loop orders, however, quickly becomes prohibitive: for the important 3-loop level at which the first dangerous counterterms occur, an estimate of the number of terms in a standard Feynman diagram calculation is of the order of 10^{20}, owing to the complexity of the vertices and propagators.

Nonetheless, important progress using new techniques developed since 1998 has been made in the calculation of loop-diagram divergences in maximal supergravity and maximal super Yang-Mills theories. These new methods use heavily the unitarity properties of Feynman diagrams, which generalize the optical theorem $\mathrm{Im}\, T = T^* T$ of ordinary quantum mechanics [15–17].

Normally, one might think that one can only learn about the imaginary parts of quantum amplitudes using unitarity. However when the unitarity diagram cutting rules are combined with an expanded use of dimensional regularization, much more can be learned. In dimensional regularization, one analytically continues the dimension of spacetime in Feynman integrals away from the dimension of interest, *e.g.* replacing $\int d^4 k$ loop integrals by $\int d^{4+\epsilon} k$.

The ordinary use of dimensional regularization focuses simply on the $\frac{1}{\epsilon}$ poles in quantum amplitudes, corresponding to logarithmic divergences in a straightforward high momentum cutoff regularization. However, one gets useful information by retaining the full $(4 + \epsilon)$ dimensional amplitude. In such an analytically continued integral, an integrand $f(s)$ (where, *e.g.*, s is a Mandelstam momentum invariant, quadratic in loop momenta) will become deformed to $f(s)s^{-\epsilon/2}$ in order to balance dimensions. Then, since $s^{-\epsilon/2} = 1 - (\epsilon/2)\ln s + \cdots$ and since $\ln(s) = \ln(|s|) + i\pi\theta(s)$, one can learn about the real parts of an amplitude by retaining imaginary terms at order ϵ.

The unitarity-based techniques allow for large classes of diagrams to become *cut constructible*, and allow for the eventual reduction of a higher-loop amplitude to integrals over products of tree amplitudes. At that point, other recent progress in the understanding of tree amplitudes comes into play. Although the individual Feynman diagrams at tree level are very complicated, sums of diagrams representing complete amplitudes can have striking simplicity, and in particular can satisfy powerful recursion relations [18].

To date, these techniques have allowed the explicit calculation of maximal supergravity divergences to proceed up to the 4-loop level (Fig. 1.10), something that would have been unthinkable using traditional Feynman diagram techniques [19].

The current status of the first possible maximal supergravity ultraviolet divergences is summarized in Table 1.2, with the currently known divergences shown in gray. The BPS degree represents the degree of supersymmetry invariance of the integrand prior to superspace integration. The surprizing feature of these results is the tardiness of maximal supergravity in getting around to revealing its ultraviolet

Fig. 1.10 Four out of the 50 diagrams arising in the calculation of the 4-loop divergences in maximal supergravity

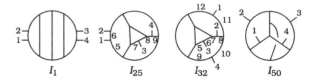

I_1 $\quad$ I_{25} $\quad$ I_{32} $\quad$ I_{50}

Table 1.2 Maximal supergravity first possible divergences & BPS degree from unitarity-based calculations. Known divergences are shown in gray

Dimension D	11	10	8	7	6	5	4
Loop order L	2	2	1	2	3	6?	5?
BPS degree	0	0	$\frac{1}{2}$	$\frac{1}{4}$	$\frac{1}{8}$	0	$\frac{1}{4}$
Gen. form	$\partial^{12}R^4$	$\partial^{10}R^4$	R^4	$\partial^6 R^4$	$\partial^6 R^4$	$\partial^{12}R^4$	$\partial^4 R^4$

divergences. At $D = 8$ dimensions and $L = 1$, one does indeed encounter the expected R^4 counterterm, but in lower spacetime dimensions this $\frac{1}{2}$ BPS divergence does not occur, leaving it to significantly less constrained counterterms to be the first UV divergence candidates.

We will next see that careful analysis of the available counterterms reveals the reasons for the unanticipated divergence cancellations, and will push the anticipated first divergences even farther out than the calculational front shown in Table 1.2.

1.7.1 Supergravity Counterterm Analysis

The surprizing resilience of maximal $N = 8$ supergravity to the threat of anticipated ultraviolet divergences has led to some speculation that perhaps superstring theory isn't actually necessary after all. Certainly, it has led to more than a decade of discussion between the unitarity-based calculators and supersymmetry practitioners in trying to understand what is going on. The current state of affairs reflects a significant deepening in understanding of the consequences of local supersymmetry and also of the rôle of duality symmetries of the maximal theory.

In the 1980's, the understanding was that allowable counterterms would be those subject to nonrenormalization theorems based upon linearly realized supersymmetry, generalizing the famous nonrenormalization theorem that disallows as counterterms chiral superspace integrals (also called "F terms") like $\int d^4x\, d^2\theta\, W(\phi)$ in $N = 1$, $D = 4$ supersymmetry, where $W(\phi)$ is a holomorphic function and ϕ is a chiral superfield satisfying $\bar{D}\phi = 0$. Although such terms are fully allowed in a theory's classical action (and, indeed, play a critical rôle in supersymmetric extensions of the Standard Model), only full superspace integrals like $\int d^4x\, d^2\theta d^2\bar{\theta}\, K(\phi, \bar{\phi})$ are allowed to occur as counterterms.

It was known by the mid 1980's that maximal supergravity and maximal super Yang-Mills theory could be quantized with at least half of their full supersymmetry

$$\Delta I_{SYM} = \int (d^4\theta d^4\bar\theta)_{105} \, \text{tr}(\phi^4)_{105} \qquad \boxplus \; 105 \qquad \phi_{ij} \quad \boxminus \; 6 \text{ of } SU(4)$$

$$\Delta I_{SG} = \int (d^8\theta d^8\bar\theta)_{232848} (W^4)_{232848} \quad \boxplus \; 232848 \qquad W_{ijkl} \quad \boxminus \; 70 \text{ of } SU(8)$$

Fig. 1.11 Half-BPS candidate counterterms for maximal super Yang-Mills and maximal supergravity theories

manifestly linearly realized—in a so-called "off-shell" formalism. This was explicitly constructed for the full maximal $N = 4$ super Yang-Mills theory, but only for the linearized theory in the case of maximal $N = 8$ supergravity. The resulting expectation was that the first allowed counterterms would have a $\int d^8\theta$ superspace integral structure in the case of maximal super Yang-Mills and a $\int d^{16}\theta$ structure in the case of maximal supergravity—*i.e.* full-superspace integrals for the linearly-realizable half supersymmetry.

Accordingly, the first allowed counterterms in maximal super Yang-Mills and maximal supergravity were considered to be [20, 21] of the structures shown in Fig. 1.11.

Since $\int d^{16}\theta$ integration has dimension $16/2 = 8$, one finds terms with 8 derivatives among the many terms produced by superspace integration in the maximal supergravity counterterm; general covariance requires these to be of the general form $\int R^4$. So the above understanding would just allow the dangerous 3-loop anticipated counterterm for the first $D = 4$ divergence.

This expectation of $L = 3$ loop first divergences has clearly been upset by the unitarity-based calculations. This has led to a more detailed investigation of the nonrenormalization theorems, and in particular to use of the Ward identities for the full supersymmetry using a Batalin-Vilkovisky version of the BRST quantum formalism—even though the transformations for the maximal theories are highly nonlinear and close to form an expected supersymmetry algebra only subject to the classical equations of motion.

To do this requires adding source fields for a whole range of additional operators as needed to formulate the Ward identities. The resulting identities then take the form of a *cohomology problem* for a generalized exterior derivative, acting on triples of forms of adjacent rank. In addition, instead of computing beta functions for the coefficients of the expected counterterms, it is advantageous to consider operator insertions of the classical Lagrangian and its corresponding cocycle into its own quantum amplitudes, and then to calculate gamma functions for the allowed operator mixings of the Lagrangian cocycle with candidate counterterms and their cocycles, the latter being required to be consistent in structure with that of the Lagrangian cocycle [22, 23].

In the case of maximal super Yang-Mills, such considerations are sufficient to explain all of the previously unanticipated single-trace operator cancellations. In the case of maximal supergravity, similar arguments show why the R^4 counterterm is in the end ruled out. But the unanticipated maximal supergravity cancellations found by the unitarity methods go much further than that: $D = 5$, $L = 2 \& 4$ and $D = 4$, $L = 3 \& 4$ divergences have all failed to occur.

Table 1.3 Maximal supergravity current first divergence expectations & BPS degree. Known divergences are shown in gray and the first anticipated $D = 4$ and $D = 5$ divergences are shown in black

Dimension D	11	10	8	7	6	5	4
Loop order L	2	2	1	2	3	6	7
BPS degree	0	0	$\frac{1}{2}$	$\frac{1}{4}$	$\frac{1}{8}$	0	$\frac{1}{8}$
Gen. form	$\partial^{12} R^4$	$\partial^{10} R^4$	R^4	$\partial^6 R^4$	$\partial^6 R^4$	$\partial^{12} R^4$	$\partial^8 R^4$

In order to understand these further divergence cancellations, one needs to turn an apparent bug of the supergravity Ward identities into a feature. Their analysis is complicated by the density character of integrands in a locally supersymmetric theory. Counterterm cocycle component forms need to be pulled back to a bosonic, or "body" coordinate frame, and in local supersymmetry this involves not only the gravitational vielbeins, but also the fermionic gravitino fields.

To turn this density character to our advantage, one needs to combine it with the requirements of maximal supergravity's continuous duality invariance: $E_{7(7)}(\mathbb{R})$ in the $D = 4$ case. For this purpose, one needs to know how to quantize while maintaining manifest duality symmetry, and this can be done in the case of the maximal $N = 8$, $D = 4$ theory by sacrificing manifest Lorentz invariance so as to handle the fact that only the $F^i_{\mu\nu}$ field strengths of the 28 vector fields can form an $E_{7(7)}(\mathbb{R})$ representation, separating them into self-dual and anti-self-dual parts in order to form a **56** of $E_{7(7)}(\mathbb{R})$. Sacrificing manifest Lorentz symmetry in this way, one can double the number of spacelike components of the vector fields, but reduce the number by a factor of $1/2$ by a duality constraint. In this formalism, one needs to check for the absence of anomalies in the now non-manifest Lorentz symmetry and also check the $SU(8)$ divisor subgroup of the duality group. Happily, these anomaly checks succeed, and as a result one may require that the perturbative field-theory counterterms preserve the continuous duality symmetry [24].

Combining the requirements of the full local supersymmetry Ward identities with the requirement of continuous duality symmetry, all of the outstanding divergence cancellations currently found by the unitarity methods have been explained, and new cancellations are now predicted at $D = 4$, $L = 5$ & 6. The resulting pattern of anticipated[1] first divergences is shown in Table 1.3.

---

[1]The $D = 4$, $L = 7$ situation requires special care [25]. At the linearized level, it would seem that the $\partial^8 R^4$ candidate could be the first full-superspace non-BPS counterterm in $D = 4$. The volume of superspace, $\int d^4x\, d^{32}\theta \det(E^A_M)$ would seem to be the obvious candidate. However, rather surprisingly, it turns out that this superspace volume vanishes subject to the classical field equations, so there is no need for such a counterterm in the renormalized action. Instead, what looks like a non-BPS $\partial^8 R^4$ counterterm at the linearized level turns into a $\frac{1}{8}$-BPS counterterm at the full nonlinear level. This illustrates the important rôle that nonlinear structure can play in quantum gravity divergence analysis.

1.7.2 Supergravity Divergences from Superstrings

A satisfying aspect of the current understanding of the relation between superstring theory and maximal supergravity is that one now obtains exactly the above predictions for the onset of supergravity divergences from a superstring perspective as well. In this regard, one may view superstring theory as a rather elaborate "regulator" of the supergravity quantum amplitudes.

In the critical dimension $D = 10$ for superstrings, the R^4 correction to the effective field-theory Lagrangian occurs with a coefficient α'^3, as can be seen on dimensional grounds, since α' has dimensions of (length)^2, so $\alpha'^3 R^4$ has the same dimensions as R. How then can there be divergences in the effective field theory, which is obtained by taking the limit $\alpha' \to 0$?

Recall that in order to compare $D = 4$ maximal supergravity to the string-theory effective action, one must dimensionally reduce on a torus T^6. Start from the string frame, in which the gravitational Lagrangian $e^{-2\phi} R$ has a scalar prefactor whose v.e.v. g_s^{-2} gives a $D = 10$ Newton's constant $G_{10} \sim g_s^2 \ell_s^8$, where ℓ_s is the string scale, needed on dimensional grounds and related to α' by $\alpha' \sim \frac{\ell_s^2}{2\hbar^2 c^2}$; g_s is the string coupling constant. If the typical scale of one of the compact toroidal dimensions is $\mathscr{R}$, then reduction of the effective action on T^6 produces an extra prefactor of $\mathscr{R}^6$ in the $D = 4$ effective action, giving a $D = 4$ Newton's constant $G_4 \sim \frac{g_s^2 \ell_s^8}{\mathscr{R}^6}$. Thus, the $D = 4$ Plank length $\ell_4 \sim (G_4)^{\frac{1}{2}}$ is related to the string coupling constant by $g_s \sim \frac{\mathscr{R}^3}{\ell_s^4} \ell_4$.

In order to compare the dimensionally reduced string effective action to quantized $D = 4$ maximal supergravity, one needs to ensure that the $D = 4$ Newton's constant remains finite, while the towers of excited string states and also the Kaluza-Klein excitations from the dimensional reduction all have masses that are infinitely large compared to the $D = 4$ Planck scale ℓ_4. This is achieved by taking $\frac{1}{\mathscr{R}}, \frac{1}{\ell_s}$ & $\frac{\mathscr{R}}{\ell_s^2} \gg \frac{1}{\ell_4}$, which is compatible with holding g_s and ℓ_4 fixed while taking $\ell_s \sim \mathscr{R}^{\frac{3}{4}} (\frac{\ell_4}{g_s})^{\frac{1}{4}} \to 0$.

This analysis would seem to indicate that the effective field theory could be ultraviolet finite. However, this string analysis can be misleading because string nonperturbative effects can conspire to swamp what one might otherwise want to identify as the field-theoretic supergravity contribution [26]. Analysis of this "decoupling" problem in the above supergravity limit shows that decoupling can be carried out for loop orders $L \leq 6$, but that beyond that order the decoupling issues prevent conclusions about field-theoretic finiteness from being made [27]. Moreover, analysis of the superstring effective action contributions also shows they can be continuously $E_{7(7)}(\mathbb{R})$ invariant to the same order [28, 29].

1.8 Other Aspects of String Theory

1.8.1 The String Scale

Let us now consider the physically relevant energy or inverse-length scales characterizing the higher dimensions inherent in string theory. The analysis will be similar
to that of Sect. 1.7.2, except that instead of looking at the field-theory limit $\ell_s \to 0$,
we now consider the *finite* values of ℓ_s that are compatible with perturbatively realistic particle physics. First consider the context of a traditional Kaluza-Klein reduc-
tion, starting in $D = 10$ with $I_{10} = \ell_s^{-8} \int d^{10}x \, e^{-2\phi}(R + \ell_s^2 F^2)$, where F represents
the Yang-Mills field strength. Next, dimensionally reduce down to $D = 4$ on a man-
ifold of volume V_6 while replacing ϕ by $\phi_0 = \langle \phi \rangle$. The $D = 4$ reduced action then
becomes

$$I_4 = \frac{V_6}{\ell_s^8} e^{-2\phi_0} \int d^4x \left(R + \ell_s^2 F^2\right)$$

so in $D = 4$ we can identify

$$M_{\mathrm{Pl}} = \frac{V_6^{\frac{1}{2}}}{\ell_s^4} e^{-\phi_0} \quad \text{and} \quad g_{\mathrm{YM}} = \frac{e^{\phi_0} \ell_s^3}{V_6^{\frac{1}{2}}}.$$

Now, to avoid strong coupling in the $D = 10$ string theory, one requires $e^{\phi_0} < 1$
while in $D = 4$, $g_{\mathrm{YM}}^2 \sim \frac{1}{30}$. Hence $V_6 \ell_s^{-6} = e^{2\phi_0} g_{\mathrm{YM}}^{-2} \le 30$ and so for $V_6 \sim \mathscr{R}^6$,
one finds $\ell_s \sim \mathscr{R}$. Moreover, substituting for $e^{-\phi_0}$ in terms of g_{YM}, one finds
$M_{\mathrm{Pl}} = (\ell_s g_{\mathrm{YM}})^{-1}$, requiring $\ell_s \sim \frac{10}{M_{\mathrm{Pl}}}$. This is as one might expect: in the standard dimensional reduction scenario, intrinsic string-theory effects cannot occur too
far below the Planck scale.

Consider now how the analysis changes when one switches from the standard
Kaluza-Klein reduction to a braneworld scenario. In a braneworld scenario, one
proposes that the observable lower-dimensional universe is concentrated on a sub-
surface of the higher-dimensional spacetime, instead of being smeared evenly over
the extra dimensions, *i.e.* instead of assuming there is no dependence on the extra
coordinates.

To see how the string-scale analysis changes, consider a D_p-brane with $p > 3$ and
with gauge fields defined only on the $d = p + 1$ dimensional D-brane worldvolume.
Of the p spatial D-brane dimensions, $p - 3$ are compactified and the remaining 3
coincide with the spatial directions of the reduced $D = 4$ spacetime. The starting
action is now $\ell_s^{-8} \int d^{10}x \, e^{-2\phi} R + \ell_s^{3-p} \int d^{p+1}\xi \, e^{-\phi} F^2$. As before, one has $M_{\mathrm{Pl}} =$
$V_6^{\frac{1}{2}} \ell_s^{-4} e^{-\phi_0}$ but now $g_{\mathrm{YM}} = e^{\frac{1}{2}\phi_0} \ell_s^{\frac{1}{2}(p-3)} V_{p-3}^{-\frac{1}{2}}$.

Limiting again the string coupling to perturbative values $e^{\phi_0} \le 1$, find $V_{p-3} \ell_s^{3-p}$
$\le g_{\mathrm{YM}}^{-2} \sim 30$. Now, however, for $V_6 = V_{p-3} V_{9-p}$ one can have *independent* $\ell_p =$
$V_{p-3}^{\frac{1}{p-3}}$ and $\mathscr{R} = V_{9-p}^{\frac{1}{9-p}}$. One may write $M_{\mathrm{Pl}}^2 = \ell_s^{-2}(V_{p-3} \ell_s^{3-p})(V_{9-p} \ell_s^{p-9}) e^{-2\phi_0}$.

Fig. 1.12 Topology change
at a boundary of moduli space

Then from $V_{p-3}\ell_s^{3-p} \leq 30$, one learns $\ell_s \sim \ell_p(30)^{-\frac{1}{p-3}}$, while if $\mathscr{R} \gg \ell_p$ (giving a highly asymmetric V_6), one has $V_{9-p}\ell_s^{p-9} \gg 1$. Thus $\ell_s^2 \gg M_{pl}^{-2}$ is now possible, *i.e.* one can have a string mass scale *significantly below* the Planck scale [30].

Such brane-world considerations lie at the base of current experimental protocols searching for string or gravitational phenomena at the Large Hadron Collider at CERN. A low string mass scale corresponds also to a low scale for inherently quantum-gravitational effects such as the threshold production of black holes. These would immediately then decay by Hawking radiation, but the resulting resonance could have a striking new experimental signature.

1.8.2 Boundaries of Moduli Space

The infinities of perturbative field theory are tamed by string theory. But the story of infinities does not end there. In string theory, the most important singularities occur at the *boundaries of moduli space, e.g.* in amplitudes where moduli are about to pinch off so that topology change can take place in a Riemann surface (Fig. 1.12).

These boundary configurations are the places where one really has to check whether string theory is free of infinities. And often, when divergences seem apparent, one realizes that the singularities should be "blown up", with new modes particular to the singularity structure of the Riemann surface appearing. These can contribute to blown-up singularity sectors with new gauge fields and other massless modes in the effective field theory, so such special points in moduli space can also give rise to important physical effects at the same time as posing a challenge to the analysis of divergences.

The calculation of higher-loop string amplitudes requires detailed control of the Riemann surface moduli in the quantum path integral. This has been comprehensively done up through the two-loop order [31, 33, 34]. This has enabled proofs of the finiteness of string theory up to this level to be given [35, 36]. More recent developments have employed a pure-spinor approach to the string worldsheet [37, 38] are expected to lead to an all-orders resolution of this key question.

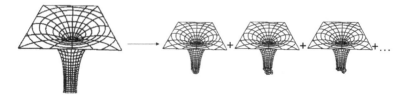

Fig. 1.13 "Fuzzball" effective geometry from an average over nonsingular geometries

1.8.3 String and Gravity Thermodynamics

One of the most famous results of string theory has been the derivation of the Bekenstein-Hawking formula $S = \frac{A}{4G}$ for the entropy of a black hole in terms of the area A of the horizon. This derivation employed nearly supersymmetric (*i.e.* nearly BPS) configurations in order to enable a detailed microstate counting agreeing with the Bekenstein-Hawking entropy formula [39]. Current work includes study of the deviation from a blackbody to a "greybody" spectrum in the emitted Hawking radiation.

Related work is aimed at understanding whether string theory evades Hawking's prediction that black holes lead to a loss of quantum information. This has led to "fuzzball" formulations of macroscopic blackholes (Fig. 1.13) in terms of an average over BPS coherent states describing individual nonsingular geometries, in order to give an account of the thermodynamics of non-supersymmetric black holes.

1.9 Conclusion

With the expansion of string theory to the M-theory picture of a web of perturbatively-defined theories (plus $D = 11$ supergravity) linked by duality symmetries, a breathtaking unification vista has opened. And yet, this unification remains largely an aspiration: despite many striking confirmations of duality relations, a proper definition of the fundamental states of M-theory and a corresponding derivation of the duality symmetries acting on them remain to be given. Considering the history of nonperturbative relations in field theory, this should not be so surprising. Many of the currently most important theoretical constructions (Yang-Mills theory, supersymmetry, even ϕ^4 theory) are not rigorously formulated outside perturbation theory. Perhaps M-theory will upend this situation and in the end become better formally grounded than phenomenological field theory currently is.

The theme of duality, meanwhile, continues to make further surprising connections between previously distinct subjects. The picture of stacks of $(3 + 1)$-dimensional D-branes in type IIB theory, with braneworld Goldstone modes constituting vector supermultiplets and interactions via open superstrings linking the D-brane stacks to form nonabelian gauge theories, forms a "holomorphic" construction of supersymmetric Yang-Mills theory on the boundary of a $(4 + 1)$-dimensional spacetime. This "gauge-gravity duality" provides powerful nonperturbative relations on both the "gravity" (or superstring) and "gauge" sides of the correspondence (for a review, see [40]).

The gauge-gravity correspondence has evolved to cover many situations where nonperturbative gauge-theory phenomena on boundaries are important. Key applications currently include the dynamics of quark-gluon plasmas, Navier-Stokes fluid hydrodynamics, and applications in high-temperature superconductivity. The gauge-gravity duality constructions can provide information about phases where the usual field-theoretic partition function is neither convergent not Borel-summable: in other words, in deeply nonperturbative regions of phase space.

Historically, string theory was born out of the study of resonances in strong-interaction physics. With some of the recent developments, it may in a sense be returning to its birth context. But it also remains the strongest current approach towards solving the key remaining problem of theoretical physics, which is the formulation of a quantum theory of gravity.

References

1. M.H. Goroff, A. Sagnotti, The ultraviolet behavior of Einstein gravity. Nucl. Phys. B **266**, 709 (1986)
2. A.E.M. van de Ven, Two loop quantum gravity. Nucl. Phys. B **378**, 309 (1992)
3. S. Deser, J.H. Kay, K.S. Stelle, Renormalizability properties of supergravity. Phys. Rev. Lett. **38**, 527 (1977)
4. S. Deser, B. Zumino, A complete action for the spinning string. Phys. Lett. B **65**, 369 (1976)
5. L. Brink, P. Di Vecchia, P.S. Howe, A locally supersymmetric and reparametrization invariant action for the spinning string. Phys. Lett. B **65**, 471 (1976)
6. A.M. Polyakov, Quantum geometry of bosonic strings. Phys. Lett. B **103**, 207 (1981)
7. C.G. Callan Jr., E.J. Martinec, M.J. Perry, D. Friedan, Strings in background fields. Nucl. Phys. B **262**, 593 (1985)
8. T.H. Buscher, A symmetry of the string background field equations. Phys. Lett. B **194**, 59 (1987)
9. J. Dai, R.G. Leigh, J. Polchinski, New connections between string theories. Mod. Phys. Lett. A **4**, 2073 (1989)
10. M. Dine, P.Y. Huet, N. Seiberg, Large and small radius in string theory. Nucl. Phys. B **322**, 301 (1989)
11. M. Cvetic, H. Lu, C.N. Pope, K.S. Stelle, T duality in the Green-Schwarz formalism, and the massless / massive IIA duality map. Nucl. Phys. B **573**, 149 (2000). hep-th/9907202
12. C.M. Hull, P.K. Townsend, Unity of superstring dualities. Nucl. Phys. B **438**, 109 (1995). hep-th/9410167
13. E. Witten, String theory dynamics in various dimensions. Nucl. Phys. B **443**, 85 (1995). hep-th/9503124
14. E.S. Fradkin, A.A. Tseytlin, Quantum string theory effective action. Nucl. Phys. B **261**, 1 (1985)
15. Z. Bern, L.J. Dixon, D.C. Dunbar, M. Perelstein, J.S. Rozowsky, On the relationship between Yang-Mills theory and gravity and its implication for ultraviolet divergences. Nucl. Phys. B **530**, 401 (1998). hep-th/9802162
16. Z. Bern, J.J.M. Carrasco, H. Johansson, Perturbative quantum gravity as a double copy of gauge theory. Phys. Rev. Lett. **105**, 061602 (2010). arXiv:1004.0476 [hep-th]

17. Z. Bern, J.J.M. Carrasco, L.J. Dixon, H. Johansson, R. Roiban, Simplifying multiloop integrands and ultraviolet divergences of gauge theory and gravity amplitudes. arXiv:1201.5366 [hep-th]
18. R. Britto, F. Cachazo, B. Feng, E. Witten, Direct proof of tree-level recursion relation in Yang-Mills theory. Phys. Rev. Lett. **94**, 181602 (2005). hep-th/0501052
19. Z. Bern, J.J. Carrasco, L.J. Dixon, H. Johansson, R. Roiban, The ultraviolet behavior of $N = 8$ supergravity at four loops. Phys. Rev. Lett. **103**, 081301 (2009). arXiv:0905.2326 [hep-th]
20. P.S. Howe, K.S. Stelle, P.K. Townsend, Superactions. Nucl. Phys. B **191**, 445 (1981)
21. R.E. Kallosh, Counterterms in extended supergravities. Phys. Lett. B **99**, 122 (1981)
22. G. Bossard, P.S. Howe, K.S. Stelle, The ultra-violet question in maximally supersymmetric field theories. Gen. Relativ. Gravit. **41**, 919 (2009). arXiv:0901.4661 [hep-th]
23. G. Bossard, P.S. Howe, K.S. Stelle, A note on the UV behaviour of maximally supersymmetric Yang-Mills theories. Phys. Lett. B **682**, 137 (2009). arXiv:0908.3883 [hep-th]
24. G. Bossard, C. Hillmann, H. Nicolai, E7(7) symmetry in perturbatively quantised $N = 8$ supergravity. J. High Energy Phys. **1012**, 052 (2010). arXiv:1007.5472 [hep-th]
25. G. Bossard, P.S. Howe, K.S. Stelle, P. Vanhove, The vanishing volume of $D = 4$ superspace. Class. Quantum Gravity **28**, 215005 (2011). arXiv:1105.6087 [hep-th]
26. M.B. Green, H. Ooguri, J.H. Schwarz, Nondecoupling of maximal supergravity from the superstring. Phys. Rev. Lett. **99**, 041601 (2007). arXiv:0704.0777 [hep-th]
27. M.B. Green, J.G. Russo, P. Vanhove, String theory dualities and supergravity divergences. J. High Energy Phys. **1006**, 075 (2010). arXiv:1002.3805 [hep-th]
28. H. Elvang, M. Kiermaier, Stringy KLT relations, global symmetries, and $E_{7(7)}$ violation. J. High Energy Phys. **1010**, 108 (2010). arXiv:1007.4813 [hep-th]
29. N. Beisert, H. Elvang, D.Z. Freedman, M. Kiermaier, A. Morales, S. Stieberger, E7(7) constraints on counterterms in $N = 8$ supergravity. Phys. Lett. B **694**, 265 (2010). arXiv:1009.1643 [hep-th]
30. I. Antoniadis, N. Arkani-Hamed, S. Dimopoulos, G.R. Dvali, New dimensions at a millimeter to a Fermi and superstrings at a TeV. Phys. Lett. B **436**, 257 (1998). hep-ph/9804398
31. E. D'Hoker, D.H. Phong, Two loop superstrings. 1. Main formulas. Phys. Lett. B **529**, 241 (2002). hep-th/0110247
32. E. D'Hoker, D.H. Phong, Two loop superstrings. 2. The chiral measure on moduli space. Nucl. Phys. B **636**, 3 (2002). hep-th/0110283
33. E. D'Hoker, D.H. Phong, Two loop superstrings. 3. Slice independence and absence of ambiguities. Nucl. Phys. B **636**, 61 (2002). hep-th/0111016
34. E. D'Hoker, D.H. Phong, Two loop superstrings 4: the cosmological constant and modular forms. Nucl. Phys. B **639**, 129 (2002). hep-th/0111040
35. S. Mandelstam, The n loop string amplitude: explicit formulas, finiteness and absence of ambiguities. Phys. Lett. B **277**, 82 (1992)
36. N. Berkovits, Finiteness and unitarity of Lorentz covariant Green-Schwarz superstring amplitudes. Nucl. Phys. B **408**, 43 (1993). hep-th/9303122
37. N. Berkovits, Covariant quantization of the Green-Schwarz superstring in a Calabi-Yau background. Nucl. Phys. B **431**, 258 (1994). hep-th/9404162
38. N. Berkovits, Quantization of the superstring with manifest U(5) superPoincare invariance. Phys. Lett. B **457**, 94 (1999). hep-th/9902099
39. A. Strominger, C. Vafa, Microscopic origin of the Bekenstein-Hawking entropy. Phys. Lett. B **379**, 99 (1996). hep-th/9601029
40. O. Aharony, S.S. Gubser, J.M. Maldacena, H. Ooguri, Y. Oz, Large N field theories, string theory and gravity. Phys. Rep. **323**, 183 (2000). hep-th/9905111

Chapter 2
Introduction to Loop Quantum Gravity and Cosmology

Abhay Ashtekar

Abstract The goal of the lecture is to present a broad perspective on loop quantum gravity and cosmology for young researchers which would serve as an introduction to lectures by Rovelli and Bojowald. The first part is addressed to beginning students and the second to young researchers who are already working in quantum gravity.

2.1 Introduction

This section, addressed to beginning researchers, is divided into two parts. The first provides a historical perspective and the second illustrates key physical and conceptual problems of quantum gravity. Researchers who are already quite familiar with quantum gravity can/should go directly to Sect. 2.2; there will be no loss of continuity.

2.1.1 Development of Quantum Gravity: A Bird's Eye View

The necessity of a quantum theory of gravity was pointed out by Einstein already in a 1916 paper in the Sitzungsberichte der Preussischen Akademie. He wrote:

> Nevertheless, due to the inneratomic movement of electrons, atoms would have to radiate not only electromagnetic but also gravitational energy, if only in tiny amounts. As this is hardly true in Nature, it appears that quantum theory would have to modify not only Maxwellian electrodynamics but also the new theory of gravitation.

These words appeared nearly a hundred years ago. While there have been notable advances in the field especially over the past 25 years, we do not yet have a fully satisfactory quantum theory of gravity. Why is the problem so hard? While there are many difficulties, I believe that the central ones arise from the fact that in

A. Ashtekar (✉)
Physics Department, Institute for Gravitational Physics and Geometry, 104 Davey, Penn State, University Park, PA 16802, USA
e-mail: ashtekar@gravity.psu.edu

G. Calcagni et al. (eds.), *Quantum Gravity and Quantum Cosmology*,
Lecture Notes in Physics 863, DOI 10.1007/978-3-642-33036-0_2,
© Springer-Verlag Berlin Heidelberg 2013

general relativity *gravity is encoded in the very geometry of space-time.* Therefore, in a quantum theory of gravity, space-time geometry itself must be described by quantum physics, with all its fuzziness. How do we do physics if there is no sharply defined space-time in the background, serving as a grand arena for all dynamics?

To appreciate this point, let us begin with field theories in Minkowski space-time, say Maxwell's theory to be specific. Here, the basic dynamical field is represented by a tensor field $F_{\mu\nu}$ on Minkowski space. The space-time geometry provides the kinematical arena on which the field propagates. The background, Minkowskian metric provides light cones and the notion of causality. We can foliate this space-time by a one-parameter family of space-like three-planes, and analyze how the values of electric and magnetic fields on one of these surfaces determine those on any other surface. The isometries of the Minkowski metric let us construct physical quantities such as fluxes of energy, momentum, and angular momentum carried by electromagnetic waves.

In general relativity, by contrast, there is no background geometry. The space-time metric itself is the fundamental dynamical variable. On the one hand it is analogous to the Minkowski metric in Maxwell's theory; it determines space-time geometry, provides light cones, defines causality, and dictates the propagation of all physical fields (including itself). On the other hand it is the analog of the Newtonian gravitational potential and therefore the basic dynamical entity of the theory, similar in this respect to the $F_{\mu\nu}$ of the Maxwell theory. This dual role of the metric is in effect a precise statement of the equivalence principle that is at the heart of general relativity. It is this feature that is largely responsible for the powerful conceptual economy of general relativity, its elegance and its aesthetic beauty, its strangeness in proportion. However, this feature also brings with it a host of problems. We see already in the classical theory several manifestations of these difficulties. It is because there is no background geometry, for example, that it is so difficult to analyze singularities of the theory and to define the energy and momentum carried by gravitational waves. Since there is no a priori space-time, to introduce notions as basic as causality, time, and evolution, one must first solve the dynamical equations and *construct* a space-time. As an extreme example, consider black holes, whose traditional definition requires the knowledge of the causal structure of the entire space-time. To find if the given initial conditions lead to the formation of a black hole, one must first obtain their maximal evolution and, using the causal structure determined by that solution, ask if the causal past $J^-(\mathscr{I}^+)$ of its future infinity $\mathscr{I}^+$ is the entire space-time. If not, space-time contains a black hole and the future boundary of $J^-(\mathscr{I}^+)$ within that space-time is its event horizon. Thus, because there is no longer a clean separation between the kinematical arena and dynamics, in the classical theory substantial care and effort is needed even in the formulation of basic physical questions.

In quantum theory the problems become significantly more serious. To see this, recall first that, because of the uncertainty principle, already in non-relativistic quantum mechanics, particles do not have well-defined trajectories; time-evolution only produces a probability amplitude, $\Psi(x, t)$, rather than a specific trajectory, $x(t)$. Similarly, in quantum gravity, even after evolving an initial state, one would not be left with a specific space-time. In the absence of a space-time geometry, how is one to introduce even habitual physical notions such as causality, time, scattering states, and black holes?

Loop quantum gravity provides a background approach to address such deep conceptual problems. In its initial formulation it was based on the canonical approach. Therefore, let us begin with a summary of ideas underlying canonical gravity. Here one notices that, in spite of the absence of a background space-time geometry, the Hamiltonian formulation of general relativity is well-defined and attempts to use it as a stepping stone to quantization. The fundamental canonical commutation relations are to lead us to the basic uncertainty principle. The motion generated by the Hamiltonian is to be thought of as time evolution. The fact that certain operators on the fixed ('spatial') three-manifold commute is supposed to capture the appropriate notion of causality. The emphasis is on preserving the geometrical character of general relativity, on retaining the compelling fusion of gravity and geometry that Einstein created. In the first stage of the program, completed in the early 1960s, the Hamiltonian formulation of the classical theory was worked out in detail by Dirac, Bergmann, Arnowitt, Deser and Misner and others [1–5].[1] The basic canonical variable was the 3-metric on a spatial slice. The ten Einstein's equations naturally decompose into two sets: four constraints on the metric and its conjugate momentum (analogous to the equation DivE = 0 of electrodynamics) and six evolution equations. Thus, in the Hamiltonian formulation, general relativity could be interpreted as the dynamical theory of 3-geometries. Wheeler therefore baptized it *geometrodynamics* [6, 7].

In the second stage, this framework was used as a point of departure for quantum theory by Bergmann, Komar, Wheeler, DeWitt and others. The basic equations of the quantum theory were written down and several important questions were addressed [5, 7]. Wheeler also launched an ambitious program in which the internal quantum numbers of elementary particles were to arise from non-trivial, microscopic topological configurations and particle physics was to be recast as 'chemistry of geometry'. However, most of the work in quantum geometrodynamics continued to remain formal; indeed, even today the field theoretic difficulties associated with the presence of an *infinite number of degrees of freedom* in the Wheeler–DeWitt equation remain unresolved. Furthermore, even at the formal level, is has been difficult to solve the quantum Einstein's equations. Therefore, after an initial burst of activity, the quantum geometrodynamics program became stagnant. Interesting results have been obtained by Misner, Wheeler, DeWitt and others in the limited context of quantum cosmology where one freezes all but a finite number of degrees of freedom. However, even in this special case, the initial singularity could not be resolved without additional 'external' inputs into the theory, such as the use of matter violating energy conditions.

---

[1] Since this introduction is addressed to non-experts, I will generally refer to books and review articles which summarize the state of the art at various stages of development of quantum gravity. References to original papers can be found in these reviews.

The third stage in the canonical approach began with the following observation: the geometrodynamics program laid out by Dirac, Bergmann, Wheeler and others simplifies significantly if we regard a spatial connection (rather than the 3-metric) as the basic object. In fact we now know that, among others, Einstein and Schrödinger had recast general relativity as a theory of connections already in the fifties. (For a brief account of this fascinating history, see [8].) However, they used the 'Levi-Civita connection' that features in the parallel transport of vectors and found that the theory becomes rather complicated. This episode had been forgotten and connections were re-introduced afresh in the mid 1980s [9].[2] However, now these are 'spin-connections' required to parallel propagate spinors, and they turn out to *simplify* Einstein's equations considerably. For example, the dynamical evolution dictated by Einstein's equations can now be visualized simply as a *geodesic motion* on the space of spin-connections (with respect to a natural metric extracted from the constraint equations). Since general relativity is now regarded as a dynamical theory of connections, this reincarnation of the canonical approach is called 'connection-dynamics'.

Perhaps the most important advantage of the passage from metrics to connections is that the phase-space of general relativity is now the same as that of gauge theories [9, 10]. The 'wedge between general relativity and the theory of elementary particles' that Steve Weinberg famously referred to in his monograph on general relativity largely disappears *without having to sacrifice the geometrical essence of the theory.* One could now import into general relativity techniques that have been highly successful in the quantization of gauge theories. At the kinematic level, then, there is a unified framework to describe all four fundamental interactions. The dynamics, of course, depends on the interaction. In particular, while there is a background space-time geometry in electroweak and strong interactions, there is none in general relativity. Therefore, qualitatively new features arise. These were exploited in the late eighties and early nineties to solve simpler models—general relativity in $2 + 1$ dimensions [9, 11, 12], linearized gravity clothed as a gauge theory [9], and certain cosmological models. To explore the physical, $3 + 1$ dimensional theory, a 'loop representation' was introduced by Rovelli and Smolin. Here, quantum states

---

[2]This reformulation used (anti-)self-dual connections which are complex. These have a direct interpretation in terms space-time geometry and also render the constraint equations polynomial in the basic variables. This simplicity was initially regarded as crucial for passage to quantum theory. However, one is then faced with the task of imposing appropriate quantum 'reality conditions' to ensure that the classical limit is real general relativity. Barbero introduced real connection variables by replacing the $\pm i$ in the expression of the (anti-)self-dual connections with a real parameter β. However, now the connection does not have a natural space-time interpretation and the constraints are no longer polynomial in the basic variables. But the strategy became viable after Thiemann introduced novel ideas to handle quantization of the specific non-polynomial terms that now feature in the constraints. Since then, this strategy has become crucial because the rigorous functional calculus on the space of connections has so far been developed only for real connections. Immirzi suggested that the value of β could be chosen so that the leading term in black hole entropy is precisely (area/$4\ell_{\mathrm{Pl}}^2$). That is why β (which is often denoted by γ in later papers) is referred to as the *Barbero–Immirzi* parameter.

were taken to be suitable functions of loops on the 3-manifold.[3] This led to a number of interesting and intriguing results, particularly by Gambini, Pullin and their collaborators, relating knot theory and quantum gravity [13]. Thus, there was rapid and unanticipated progress in a number of directions which rejuvenated the canonical quantization program. Since the canonical approach does not require the introduction of a background geometry or use of perturbation theory, and because one now has access to fresh, non-perturbative techniques from gauge theories, in relativity circles there is a hope that this approach may lead to well-defined, *non-perturbative* quantum general relativity, or its supersymmetric version, supergravity.

However, a number of these considerations remained rather formal until midnineties. Passage to the loop representation required an integration over the infinite-dimensional space of connections and the formal methods were insensitive to possible infinities lurking in the procedure. Indeed, such integrals are notoriously difficult to perform in interacting field theories. To pay due respect to the general covariance of Einstein's theory, one needed diffeomorphism invariant measures and there were folk-theorems to the effect that such measures did not exist!

Fortunately, the folk-theorems turned out to be incorrect. To construct a well-defined theory capable of handling field theoretic issues, a *quantum theory of Riemannian geometry* was systematically constructed in the mid-nineties [14]. This launched the fourth stage in the canonical approach. Just as differential geometry provides the basic mathematical framework to formulate modern gravitational theories in the classical domain, quantum geometry provides the necessary concepts and techniques in the quantum domain. It is a rigorous mathematical theory which enables one to perform integration on the space of connections for constructing Hilbert spaces of states and to define geometric operators corresponding, e.g., to areas of surfaces and volumes of regions, even though the classical expressions of these quantities involve non-polynomial functions of the Riemannian metric. There are no infinities. One finds that, at the Planck scale, geometry has a definite discrete structure. Its fundamental excitations are one-dimensional, rather like polymers, and the space-time continuum arises only as a coarse-grained approximation. The fact that the structure of space-time at Planck scale is qualitatively different from Minkowski background used in perturbative treatments reinforced the idea that quantum general relativity (or supergravity) may well be non-perturbatively finite (see, e.g., [14–16]).

Over the last six years, another frontier has advanced in loop quantum gravity: *spin foams* (and the associated development of group field theory) which provide a sum-over-histories formulation [15, 17, 18]. The new element here is that the histories that enter the sum are *quantum* geometries of a specific type; they can be regarded as the 'time evolution' of the polymer-like quantum 3 geometries that emerged in the canonical approach. So far the sum has not been systematically derived starting from the classical theory as one generally does in, say, gauge theories.

---

[3]This is the origin of the name 'loop quantum gravity'. The loop representation played an important role in the initial stages. Although this is no longer the case in the current, fourth phase, the name is still used to distinguish this approach from others.

Rather, one uses semi-heuristic considerations to arrive at a definition of the 'transition amplitudes' and then explores physical properties of the resulting quantum theory. There are detailed arguments to the effect that one recovers the Einstein–Hilbert action in an appropriate limit. Furthermore, although the underlying theory is diffeomorphism invariant, given a suitable 'boundary state', there is a conceptual framework to calculate n-point functions normally used in perturbative treatments. Information about the background space-time on which these n-point functions live is encoded in the chosen 'boundary state'. However, a number of important problems still remain. The status is described in some detail in the lectures by Rovelli at this school.

As of now, neither the Hamiltonian theory nor the spin foam program is complete. Therefore we do not have a complete quantum gravity theory incorporating full quantum dynamics. Nonetheless, by using suitable truncations of the theory, very significant progress has occurred over the last decade. These truncations have enabled researchers to obtain a number of results of direct physical interest. I will conclude with three main illustrations. By restricting oneself to the sector of general relativity consisting of space-times that admit an isolated horizon [19] as the inner boundary, the very large entropy of black holes has been accounted for, using microstates of the quantum horizon geometry. By restricting oneself to truncations generally used in cosmology, it has been possible to show that quantum geometry effects naturally lead to the resolution of the big-bang and big-crunch singularities that had eluded the Wheeler–DeWitt theory. Similarly, by restricting oneself to a truncation that captures small fluctuations around Minkowski space-time, it has been possible to derive the standard spin-2 graviton propagator starting with quantum space-time geometries that feature in spin-foams. In the next section, I will discuss these and other advances further. Some of them are also discussed in the contributions by Bojowald and Rovelli in this volume.

Finally, various researchers have also built bridges between loop quantum gravity and other approaches which also focus on problems of quantum gravity in its own right, without tying it to the issue of unification of all interactions. These include the Euclidean path integral approach [20], Regge calculus [21], asymptotic safety scenarios [22] (see the contribution by Reuter and Saueressig to the present volume), discrete approaches [23], causal dynamical triangulations [24, 25], twistor theory [26–28] and the theory of H-spaces [29], asymptotic quantization [30], non-commutative geometry [31], causal sets [32, 33] and Topos theory [34, 35].

2.1.2 Physical Questions of Quantum Gravity

Approaches to quantum gravity face two types of issues: Problems that are 'internal' to individual programs and physical and conceptual questions that underlie the whole subject. Examples of the former are: Incorporation of physical (rather than half flat) gravitational fields in the twistor program, mechanisms for breaking of supersymmetry and dimensional reduction in string theory, and issues of space-time covariance in the canonical approach. In this sub-section, I will focus on the second type of issues by recalling some of the long-standing issues that *any* satisfactory quantum theory of gravity should address.

• *The big bang and other singularities*: It is widely believed that the prediction of a singularity, such as the big bang of classical general relativity, is primarily a signal that the physical theory has been pushed beyond the domain of its validity. A key question to any quantum gravity theory, then, is: What replaces the big bang? Are the classical geometry and the continuum picture only approximations, analogous to the 'mean (magnetization) field' of ferro-magnets? If so, what are the microscopic constituents? What is the space-time analog of a Heisenberg quantum model of a ferro-magnet? When formulated in terms of these fundamental constituents, is the evolution of the *quantum* state of the universe free of singularities? General relativity predicts that the space-time curvature must grow unboundedly as we approach the big bang or the big crunch but we expect the quantum effects, ignored by general relativity, to intervene, making quantum gravity indispensable before infinite curvatures are reached. If so, what is the upper bound on curvature? How close to the singularity can we 'trust' classical general relativity? What can we say about the 'initial conditions', i.e., the quantum state of geometry and matter that correctly describes the big bang? If they have to be imposed externally, is there a *physical* guiding principle?

• *Black holes.* In the early seventies, using imaginative thought experiments, Bekenstein argued that black holes must carry an entropy proportional to their area [19, 20, 36]. About the same time, Bardeen, Carter and Hawking (BCH) showed that black holes in equilibrium obey two basic laws, which have the same form as the zeroth and the first laws of thermodynamics, provided one equates the black hole surface gravity κ to some multiple of the temperature T in thermodynamics and the horizon area a_{hor} to a corresponding multiple of the entropy S [19, 20, 36]. However, at first this similarity was thought to be only a formal analogy because the BCH analysis was based on *classical* general relativity and simple dimensional considerations show that the proportionality factors must involve Planck's constant $\hbar$. Two years later, using quantum field theory on a black hole background space-time, Hawking showed that black holes in fact radiate quantum mechanically as though they were black bodies at temperature $T = \hbar\kappa/2\pi$ [20, 37]. Using the analogy with the first law, one can then conclude that the black hole entropy should be given by $S_{BH} = a_{hor}/4G\hbar$. This conclusion is striking and deep because it brings together the three pillars of fundamental physics—general relativity, quantum theory and statistical mechanics. However, the argument itself is a rather hodge-podge mixture of classical and semi-classical ideas, reminiscent of the Bohr theory of atom. A natural question then is: what is the analog of the more fundamental, Pauli–Schrödinger theory of the hydrogen atom? More precisely, what is the statistical mechanical origin of black hole entropy? What is the nature of a quantum black hole and what is the interplay between the quantum degrees of freedom responsible for entropy and the exterior curved geometry? Can one derive the Hawking effect from first principles of quantum gravity? Is there an imprint of the classical singularity on the final quantum description, e.g., through 'information loss'?

• *Planck scale physics and the low-energy world*. In general relativity, there is no background metric, no inert stage on which dynamics unfolds. Geometry itself is dynamical. Therefore, as indicated above, one expects that a fully satisfactory quantum gravity theory would also be free of a background space-time geometry. However, of necessity, a background-independent description must use physical concepts and mathematical tools that are quite different from those of the familiar, low-energy physics. A major challenge then is to show that this low-energy description does arise from the pristine, Planckian world in an appropriate sense, bridging the vast gap of some 16 orders of magnitude in the energy scale. In this 'top-down' approach, does the fundamental theory admit a 'sufficient number' of semi-classical states? Do these semi-classical sectors provide enough of a background geometry to anchor low-energy physics? Can one recover the familiar description? If the answers to these questions are in the affirmative, can one pin point why the standard 'bottom-up' perturbative approach fails? That is, what is the essential feature which makes the fundamental description mathematically coherent but is absent in the standard perturbative quantum gravity?

There are of course many more challenges: adequacy of standard quantum mechanics, the issue of time, of measurement theory and the associated questions of interpretation of the quantum framework, the issue of diffeomorphism invariant observables and practical methods of computing their properties, convenient ways of computing time evolution and S-matrices, exploration of the role of topology and topology change, and so on. In loop quantum gravity described in the rest of this chapter, one adopts the view that the three issues discussed in detail are more basic from a physical viewpoint because they are rooted in general conceptual questions that are largely independent of the specific approach being pursued. Indeed they have been with us longer than any of the current leading approaches.

2.2 Loop Quantum Gravity and Cosmology

In this section, I will summarize the overall viewpoint, achievements, challenges and opportunities underlying loop quantum gravity. The emphasis is on structural and conceptual issues. For details, see [14–16, 38] and references therein. The development of the subject can be seen by following older monographs [9, 10, 13]. For a treatment at a more elementary (i.e., advanced undergraduate) level, see [39].

2.2.1 Viewpoint

In loop quantum gravity, one takes the central lesson of general relativity seriously: gravity *is* geometry whence, in a fundamental quantum gravity theory, there should be no background metric. Geometry and matter should *both* be 'born quantum mechanically'. Thus, in contrast to approaches developed by particle physicists, one does not begin with quantum matter on a background geometry and use perturbation theory to incorporate quantum effects of gravity. There *is* a manifold but no metric, or indeed any other physical fields, in the background.[4]

In classical gravity, Riemannian geometry provides the appropriate mathematical language to formulate the physical, kinematical notions as well as the final dynamical equations. This role is now taken by *quantum* Riemannian geometry. In the classical domain, general relativity stands out as the best available theory of gravity, some of whose predictions have been tested to an amazing degree of accuracy, surpassing even the legendary tests of quantum electrodynamics. Therefore, it is natural to ask: *Does quantum general relativity, coupled to suitable matter* (or supergravity, its supersymmetric generalization), *exist as consistent theories non-perturbatively?* There is no implication that such a theory would be the final, complete description of Nature. Nonetheless, this is a fascinating and important open question in its own right.

As explained in Sect. 2.1.1, in particle-physics circles the answer is often assumed to be in the negative, not because there is concrete evidence against non-perturbative quantum gravity, but because of the analogy to the theory of weak interactions. There, one first had a 4-point interaction model due to Fermi which works quite well at low energies but which fails to be renormalizable. Progress occurred not by looking for non-perturbative formulations of the Fermi model but by replacing the model by the Glashow–Salam–Weinberg renormalizable theory of electro-weak interactions, in which the 4-point interaction is replaced by $W^\pm$ and Z propagators. Therefore, it is often assumed that perturbative non-renormalizability of quantum general relativity points in a similar direction. However, this argument overlooks the crucial fact that, in the case of general relativity, there is a qualitatively new element. Perturbative treatments pre-suppose that the space-time can be assumed to be a continuum *at all scales* of interest to physics under consideration. This assumption is safe for weak interactions. In the gravitational case, on the other hand, the scale of interest is *the Planck length* ℓ_{Pl} and there is no physical basis to pre-suppose that the continuum picture should be valid down to that scale. The failure of the standard perturbative treatments may largely be due to this grossly incorrect assumption and a non-perturbative treatment which correctly incorporates the physical micro-structure of geometry may well be free of these inconsistencies.

Are there any situations, outside loop quantum gravity, where such physical expectations are borne out in detail mathematically? The answer is in the affirmative. There exist quantum field theories (such as the Gross–Neveu model in three dimensions) in which the standard perturbation expansion is not renormalizable although the theory is *exactly soluble*! Failure of the standard perturbation expansion can occur because one insists on perturbing around the trivial, Gaussian point rather than

---

[4]In 2 + 1 dimensions, although one begins in a completely analogous fashion, in the final picture one can get rid of the background manifold as well. Thus, the fundamental theory can be formulated combinatorially [9, 11]. While some steps have been taken to achieve this in 3 + 1 dimensions, by considering 'abstract' spin networks in the canonical approach and 2-complexes in spin foams, one still needs a more complete handle on the underlying mathematics.

the more physical, non-trivial fixed point of the renormalization group flow. Interestingly, thanks to recent work by Reuter, Lauscher, Percacci, Perini and others there is now non-trivial and growing evidence that situation may be similar in Euclidean quantum gravity. Impressive calculations have shown that pure Einstein theory may also admit a non-trivial fixed point [22, 40, 41]. Furthermore, the requirement that the fixed point should continue to exist in presence of matter constrains the couplings in non-trivial and interesting ways [42].

However, as indicated in the Introduction, even if quantum general relativity did exist as a mathematically consistent theory, there is no *a priori* reason to assume that it would be the 'final' theory of all known physics. In particular, as is the case with classical general relativity, while requirements of background independence and general covariance do restrict the form of interactions between gravity and matter fields and among matter fields themselves, the theory would not have a built-in principle which *determines* these interactions. Put differently, such a theory would not be a satisfactory candidate for unification of all known forces. However, just as general relativity has had powerful implications in spite of this limitation in the classical domain, quantum general relativity should have qualitatively new predictions, pushing further the existing frontiers of physics. Indeed, unification does not appear to be an essential criterion for usefulness of a theory even in other interactions. QCD, for example, is a powerful theory even though it does not unify strong interactions with electro-weak ones. Furthermore, the fact that we do not yet have a viable candidate for the grand unified theory does not make QCD any less useful.

2.2.2 Advances

From the historical and conceptual perspectives of Sect. 2.1, loop quantum gravity has had several successes. Thanks to the systematic development of quantum geometry, several of the roadblocks encountered by quantum geometrodynamics were removed. Functional analytic issues related to the presence of an infinite number of degrees of freedom are now faced squarely. Integrals on infinite-dimensional spaces are rigorously defined and the required operators have been systematically constructed. Thanks to this high level of mathematical precision, the Hamiltonian and the spin foam programs in loop quantum gravity have leaped past the 'formal' stage of development. More importantly, although key issues related to quantum dynamics still remain, it has been possible to use the parts of the program that are already well established to extract useful and highly non-trivial physical predictions. In particular, some of the long standing issues about the nature of the big bang, physics of the very early universe, properties of quantum black holes, giving meaning to the n-point functions in a background independent framework have been resolved. In this sub-section, I will further clarify some conceptual issues and discuss some recent advances.

- *Quantum geometry.* The specific quantum Riemannian geometry underlying loop quantum gravity predicts that eigenvalues of geometric operators (such as areas of 2-surfaces and volumes of 3-dimensional regions) are discrete. Thus, continuum underlying general relativity is only a coarse-grained approximation. What is the direct *physical* significance of this specific discreteness? Recall first that, in the classical theory, differential geometry simply provides us with formulas to compute areas of surfaces and volumes of regions in a Riemannian manifold. To turn these quantities into physical observables of general relativity, one has to define the surfaces and regions *operationally*, e.g., by focusing on surfaces of black holes or regions in which matter fields are non-zero. Once this is done, one can simply use the formulas supplied by differential geometry to calculate values of these observable. The situation is rather similar in loop quantum gravity. For instance, the area of the isolated horizon is a Dirac observable in the classical theory and the application of the quantum geometry area formula to *this* surface leads to physical results. In $2 + 1$ dimensions, Freidel, Noui and Perez have introduced point particles coupled to gravity. The physical distance between these particles is again a Dirac observable. When used in this context, the spectrum of the length operator has direct physical meaning. In all these situations, the operators and their eigenvalues correspond to the 'proper' lengths, areas and volumes of physical objects, measured in the rest frames. Finally, sometimes questions are raised about compatibility between discreteness of these eigenvalues and Lorentz invariance. As was emphasized by Rovelli and Speziale, there is no tension whatsoever: it suffices to recall that discreteness of eigenvalues of the angular momentum operator $\hat{J}_z$ of non-relativistic quantum mechanics is perfectly compatible with the rotational invariance of that theory.

• *Quantum cosmology.* In Friedmann–Lemaître–Robertson–Walker (FLRW) models, loop quantum gravity has resolved the long-standing physical problem of the fate of the big bang in quantum gravity [38]. Work by Bojowald, Ashtekar, Pawłowski, Singh and others has shown that non-perturbative effects originating in quantum geometry create an effective repulsive force which is negligible when the curvature falls significantly below the Planck scale but rises very quickly and dramatically in the deep Planck regime to overcome the classical gravitational attraction, thereby replacing the big bang by a quantum bounce. The same is true with the big-crunch singularity in the closed models. More generally, using effective equations, Singh has shown that these quantum geometry effects also resolve *all* strong curvature singularities in homogeneous isotropic models where matter sources have an equation of state of the type $p = p(\rho)$, including the exotic singularities such as the big rip. (These can occur with non-standard matter, still described by an equation of state $p = p(\rho)$.)

A proper treatment of anisotropies (i.e., Bianchi models) has long been a highly non-trivial issue in general bouncing scenarios because the anisotropic shears dominate in Einstein's equations in the contracting phase before the bounce, diverging (as $1/a^6$ which is) faster than, say, the dust or radiation matter density. Therefore, if anisotropies are added even as a perturbation to a FLRW model, they tend to grow unboundedly. What is the situation in loop quantum cosmology? The issue turned out to be quite subtle and there were some oversights at first. But a careful examination by Ashtekar, Wilson-Ewing and others has shown that the singularity is again resolved: any time a shear scalar (a potential for the Weyl curvature) or matter density approaches the Planck regime, the repulsive force of quantum geometry grows to dilute it. As in the isotropic case, effective equations can again be used to gain physical insights. In particular, they show that the matter density is again bounded above. Singularity resolution in these Bianchi models is also important from a more general consideration. There is a conjecture due to Belinskii, Khalatnikov and Lifshitz (BKL) that says that as one approaches a space-like singularity in classical general relativity, 'the terms containing time derivatives dominate over those containing spatial derivatives, so that the dynamics of the gravitational field at any one spatial point are better and better approximated by the dynamics of Bianchi models'. By now considerable evidence has accumulated in support of the BKL conjecture and it is widely believed to be essentially correct. One might therefore hope that the singularity resolution in the Bianchi models in loop quantum cosmology has opened a door to showing that all strong curvature, space-like singularities are resolved by the quantum geometry effects underlying loop quantum gravity.

Finally, the simplest type of (non-linear) inhomogeneous models (the 1-polarization Gowdy space-times) have also been analyzed in detail. These models were studied extensively in the early quantum gravity literature, prior to the advent of loop quantum cosmology (LQC). In all cases the singularity had persisted. A systematic study in the context of loop quantum cosmology was initiated by Mena, Martín-Benito, Pawłowski and others by making an astute use of the fact that the homogeneous modes of the model correspond to a Bianchi I space-time. Once again, the underlying quantum geometry resolves the big-bang singularity.

I will conclude with the discussion of a conceptual point. In general relativity, non-singular bouncing models can be and have been constructed by using matter fields that violate energy conditions. In loop quantum cosmology, by contrast, matter fields satisfy all energy conditions. How can the theory then evade singularity theorems of Penrose, Hawking and others? It does so because the quantum geometry effects modify the geometric, left-hand side of Einstein's equations, whence these theorems are inapplicable. However, there are more recent singularity theorems due to Borde, Guth and Vilenkin which do *not* refer to field equations at all. How are these evaded? These theorems were motivated by the inflationary scenario and therefore assume that the universe has been eternally undergoing an expansion. In loop quantum cosmology, even with an inflationary potential, the pre-bounce branch is contracting. Thus again the singularity is avoided because the solutions violate a key assumption of these theorems as well.

• *Quantum horizons.* Loop quantum cosmology illuminates dynamical ramifications of quantum geometry but within the context of mini- and midi-superspaces where an infinite number of degrees of freedom are frozen. The application to the black hole entropy problem is complementary in that one considers the full theory but probes consequences of quantum geometry which are not sensitive to full quantum dynamics. I will discuss this topic in a little more detail because it was not covered in any of the main lectures at this school.

As explained in the Introduction, since mid-seventies, a key question in the subject has been: What is the statistical mechanical origin of the entropy $S_{BH} = (a_{hor}/4\ell_{Pl}^2)$ of large black holes? What are the microscopic degrees of freedom that account for this entropy? This relation implies that a solar-mass black hole must have $\exp 10^{77}$ quantum states, a number that is *huge* even by the standards of statistical mechanics. Where do all these states reside? To answer these questions, in the early 1990s Wheeler had suggested the following heuristic picture, which he christened 'It from Bit'. Divide the black hole horizon into elementary cells, each with one Planck unit of area, ℓ_{Pl}^2 and assign to each cell two microstates, or one 'Bit'. Then the total number of states $\mathcal{N}$ is given by $\mathcal{N} = 2^n$ where $n = (a_{\mathrm{hor}}/\ell_{\mathrm{Pl}}^2)$ is the number of elementary cells, whence entropy is given by $S = \ln \mathcal{N} \sim a_{\mathrm{hor}}$. Thus, apart from a numerical coefficient, the entropy ('It') is accounted for by assigning two states ('Bit') to each elementary cell. This qualitative picture is simple and attractive. But can these heuristic ideas be supported by a systematic analysis from first principles?

Ashtekar, Baez, Corichi and Krasnov used quantum geometry to provide such an analysis. The first step was to analyze the structure of 'isolated horizons' in general relativity [19] and use it in conjunction to quantum geometry to define an isolated *quantum horizon*. To probe its properties, one has to combine the isolated horizon boundary conditions from classical general relativity and quantum Riemannian geometry of loop quantum gravity with the Chern–Simons theory on a punctured sphere, the theory of a non-commutative torus and subtle considerations involving mapping class groups. This detailed analysis showed that, while qualitative features of Wheeler's picture are borne out, geometry of a quantum horizon is much more subtle. First, while Wheeler's ideas hold for any 2-surface, the loop quantum gravity calculation requires a quantum horizon. Second, basic features of both of Wheeler's arguments undergo a change: (i) the elementary cells do not have Planck area; values of their area are dictated by the spectrum, $\sim \sqrt{j(j+1)}$, of the area operator in loop quantum gravity, where j is a half integer; (ii) individual cells carry much more than just one 'bit' of information; the number of states associated with any one cell is $2j + 1$.

Nonetheless, a careful counting of states by Lewandowski, Domagała, Meissner and others has shown that the number of microstates is again proportional to the area of the isolated horizon. To get the exact numerical factor of $1/4$, one has to fix the Barbero–Immirzi parameter of loop quantum gravity to a specific value. One can use a specific type of isolated horizon for this—e.g., the spherically symmetric one with zero charge, or the cosmological one in the de Sitter space-time. Once the value of the parameter is fixed, one gets the correct numerical coefficient in the leading order contribution for isolated horizons with arbitrary mass and angular momentum moments, charge, and so on. (One also obtains a precise logarithmic sub-leading correction, whose coefficient does not depend on the Barbero–Immirzi parameter.) The final result has two significant differences with respect to the string theory calculations: (i) one does not require near-extremality; one can handle ordinary 4-dimensional black holes of direct astrophysical interest which may be distorted and/or rotating; and, (ii) one can simultaneously incorporate cosmological horizons for which thermodynamics considerations also apply [20].

Why does this value of the Barbero–Immirzi parameter not depend on non-gravitational charges? This important property can be traced back to a key consequence of the isolated horizon boundary conditions: detailed calculations show that only the gravitational part of the symplectic structure has a surface term at the horizon; the matter symplectic structures have only volume terms. (Furthermore, the gravitational surface term is insensitive to the value of the cosmological constant.) Consequently, there are no independent surface quantum states associated with matter. This provides a natural explanation of the fact that the Hawking–Bekenstein entropy depends only on the horizon area and is independent of electro-magnetic (or other) charges. (For more detailed accounts of these results, see [14, 19].)

Over the last three years there has been a resurgence of interest in the subject, thanks to the impressive use of number theory techniques by Barbero, Villasenor, Agullo, Borja, Díaz-Polo and to sharpen and very significantly extend the counting of horizon states. These techniques have opened new avenues to further explore the microstates of the quantum horizon geometry through contributions by Perez, Engle, Noui, Pranzetti, Ghosh, Mitra, Kaul, Majumdar and others.

To summarize, as in other approaches to black hole entropy, concrete progress could be made in loop quantum gravity because: (i) the analysis does not require detailed knowledge of how quantum dynamics is implemented in *full* theory, and, (ii) restriction to large black holes implies that the Hawking radiation is negligible, whence the black hole surface can be modeled by an isolated horizon [19]. The states responsible for entropy have a direct interpretation in *space-time* terms: they refer to the geometry of the quantum, isolated horizon.

• *Quantum Einstein's equations in the canonical framework.* The challenge of quantum dynamics in the full theory is to find solutions to the quantum constraint equations and endow these physical states with the structure of an appropriate Hilbert space. The general consensus in the loop quantum gravity community is that while the situation is well-understood for the Gauss and diffeomorphism constraints, it is far from being definitive for the Hamiltonian constraint. Non-trivial development due to Thiemann is that well-defined candidate operators representing the Hamiltonian constraint exist on the space of solutions to the Gauss and diffeomorphism constraints [16]. However there are several ambiguities [14] and, unfortunately, we do not understand the physical meaning of choices made to resolve them. Detailed analysis in the limited context of loop quantum cosmology has shown that choices which appear to be mathematically natural can nonetheless lead to unacceptable physical consequences such as departures from general relativity in completely tame situations with low curvature [38]. Therefore, much more work is needed in the full theory.

The current status can be summarized as follows. Four main avenues have been pursued to construct and solve the quantum Hamiltonian constraint. The first is the 'Master Constraint program' introduced by Thiemann [16]. The idea here is to avoid using an infinite number of Hamiltonian constraints $\mathscr{S}(N) = \int N(x)\mathscr{S}(x)d^3x$, each smeared by a so-called 'lapse function' N. Instead, one squares the integrand $\mathscr{S}(x)$ itself in an appropriate sense and then integrates it on the 3-manifold M. In simple examples, this procedure leads to physically viable quantum theories. However, in loop quantum gravity the procedure does not remove any of the ambiguities in the definition of the Hamiltonian constraint. Rather, if the ambiguities are resolved, the principal strength of the strategy lies in its potential to complete the last step in quantum dynamics: finding the physically appropriate scalar product on physical states. The general philosophy is similar to that advocated by John Klauder over the years in his approach to quantum gravity based on coherent states [43]. A second strategy to solve the quantum Hamiltonian constraint is due to Gambini, Pullin and their collaborators. It builds on their extensive work on the interplay between quantum gravity and knot theory [13]. The more recent of these developments use the relatively new invariants of *intersecting* knots discovered by Vassiliev. This is a novel approach which furthermore has a potential of enhancing the relation between topological field theories and quantum gravity. As our knowledge of invariants of intersecting knots deepens, this approach could provide increasingly significant insights. In particular, it has the potential of leading to a formulation of quantum gravity which does not refer even to a background manifold.

The third approach comes from spin-foam models [17, 44] which, as discussed below, provide a path integral approach to quantum gravity. Over the last four years, there has been extensive work in this area, discussed in articles by Rovelli, Speziale, Baratin, Perini, Fairbairn, Bianchi, and Kaminski. Transition amplitudes from path integrals can be used to restrict the choice of the Hamiltonian constraint operator in the canonical theory. This is a very promising direction and Freidel, Noui, Perez, Rovelli and others have analyzed this issue especially in $2 + 1$ dimensions. The idea in the fourth approach, due to Varadarajan, Laddha, Henderson, Tomlin and others, is to use insights gained from the analysis of parameterized field theories. Now the emphasis is on drastically reducing the large freedom in the choice of the definition of the Hamiltonian constraint by requiring that the quantum constraint algebra closes, so that one is assured that there is no obstruction to obtaining a large number of *simultaneous* solutions to all constraints. Because the Poisson bracket between two Hamiltonian constraints is a diffeomorphism constraint, one has to find a viable expression of the operator generating *infinitesimal* diffeomorphisms. (Until this work, the focus was on the action only of *finite* diffeomorphisms in the kinematical setup.) Very recently, this program has witnessed promising advances. The Hamiltonian constraint one is led to define shares qualitative features of 'improved dynamics' of loop quantum cosmology that lies at the foundation of the most significant advances in that area.

In this discussion I have focused primarily on pure gravity. In the mid 1990s Brown, Kuchař and Romano had introduced frameworks in which matter fields can be used as 'rods and clocks' thereby providing a natural 'de-parametrization' of the constraints in the classical theory. Giesel, Thiemann, Tambornino, Domagała, Kaminski, Lewandowski, Husain and Pawłowski have used these considerations as the point of departure to construct loop quantum gravity theories for these systems. Deparametrization greatly facilitates the task of finding Dirac observables and makes it easier to interpret the quantum theory. However, as in the Master Constraint program, issues associated with quantization ambiguities still remain and the domain on which matter fields serve as good clocks and rods still needs to be clarified.

• *Spin foams.* Four different avenues to quantum gravity have been used to arrive at spin-foam models (SFMs). The fact that ideas from seemingly unrelated directions converge to the same type of structures and models has provided a strong impetus to the spin foam program. Indeed, currently this is the most active area on the mathematical physics side of loop quantum gravity [44].

The first avenue is the Hamiltonian approach to loop quantum gravity [14–16]. By mimicking the procedure that led Feynman [45] to a sum-over-histories formulation of quantum mechanics, Rovelli and Reisenberger proposed a space-time formulation of this theory. This work launched the spin-foam program. The second route stems from the fact that the starting point in canonical loop quantum gravity is a rewriting of classical general relativity that emphasizes connections over metrics [14]. Therefore in the passage to quantum theory it is natural to begin with the path integral formulation of appropriate gauge theories. A particularly natural candidate is the topological BF theory because in three space-time dimensions it is equivalent to Einstein gravity, and in higher dimensions general relativity can be regarded as a constrained BF theory [17, 46]. The well-controlled path integral formulation of the BF theory provided the second avenue and led to the SFM of Barrett and Crane.

The third route comes from the Ponzano–Regge model of 3-dimensional gravity that inspired Regge calculus in higher dimensions. Here one begins with a simplicial decomposition of the space-time manifold, describes its discrete Riemannian geometry using edge lengths and deficit angles and constructs a path integral in terms of them. If one uses holonomies and discrete areas of loop quantum gravity in place of edge lengths, one is again led to a spin foam. These three routes are inspired by various aspects of general relativity. The fourth avenue starts from approaches to quantum gravity in which gravity is to emerge from a more fundamental theory based on abstract structures that, to begin with, have nothing to do with space-time geometry. Examples are matrix models for 2-dimensional gravity and their extension to 3-dimensions (the Boulatov model) where the basic object is a field on a group manifold rather than a matrix. The Boulatov model was further generalized to a group field theory tailored to 4-dimensional gravity [15, 18]. The perturbative expansion of this group field theory turned out be very closely related to 'vertex expansions' in SFMs. Thus the SFMs lie at a junction where four apparently distinct paths to quantum gravity meet. Through contributions of many researchers it has now become an active research area (see, e.g., [15, 17]).

Four years ago, two groups, Engle–Livine–Pereira–Rovelli (EPRL), and Freidel–Krasnov (FK), put forward precise proposals for the sum over quantum geometries that could provide detailed dynamics in loop quantum gravity. The motivations were different but for the physically interesting values of the Barbero–Immirzi parameter (selected, e.g., by the black hole entropy considerations), the two proposals agree. This is an improvement over the earlier Barrett–Crane model which cured some of the problems faced by that model. Perhaps more importantly, thanks to the generalizations by Kaminski, Kisielowski and Lewandowski, the canonical and path integral approaches have been brought closer to one another: they use the same kinematics. However, there does not yet exist a systematic 'derivation' leading to this proposal starting from classical general relativity, say, along the lines used in textbooks to arrive at the path integral formulation of gauge theories. Nonetheless the program has attracted a large number of researchers because: (i) there do exist semi-heuristic considerations motivating the passage; (ii) as I indicated above, it can be arrived at from four different avenues; and (iii) detailed asymptotic analysis by Barrett, Hellmann, Dowdall, Fairbairn, Pereira and others strongly indicates that these models have the correct classical limit; and, (iv) because of the use of quantum geometry (more precisely, because there is a non-zero area gap) this sum over quantum geometries has no ultraviolet divergences. More recently, Fairbairn, Meusberger, Han and others have extended these considerations to include a cosmological constant by a natural use of quantum groups. It is then argued that, for a given 2-simplex, the sum is also infrared finite.

However, the issue of whether to sum over distinct 2-complexes or to take an appropriate 'continuum limit' is still debated and it is not known whether the final result would be finite in either case.[5] In the cosmological mini-superspaces, the situation is well-controlled: under a single assumption that a sum and an integral can be interchanged, the analog of the sum over 2-complexes (called the vertex expansion in the spin foam literature) has been shown to converge, and furthermore, converge to the 'correct' result that is already known from a well-established Hamiltonian theory [38].

A much more detailed discussion of spin foams can be found in [17, 44]

2.2.3 Challenges and Opportunities

Developments summarized so far should suffice to provide a sense of the extent to which advances in loop quantum gravity already provide an avenue to a non-perturbative and background independent formulation of quantum gravity. I will conclude by providing an illustrative list of the open issues. Some of these are currently driving the field while others provide challenges and opportunities for further work. This discussion assumes that the reader is familiar with basic ideas behind current research in loop quantum gravity.

2.2.3.1 Foundations

• *Hamiltonian theory.* In Sect. 2.2.2 I outlined four strategies that are being used to extract quantum dynamics. I will now sketch another avenue, inspired in part by the success of loop quantum cosmology, that has been proposed by Domagała, Giesel, Kaminski and Lewandowski. In loop quantum cosmology, a massless scalar field often serves as an 'internal clock' with respect to which observables of physical interest evolve [38]. The idea is to take over this strategy to full quantum gravity by focusing on general relativity coupled with a massless scalar field. This is a particularly interesting system because, already in the 1990s, Kuchař and Romano showed that one can rearrange the constraints of this system so that they form a

---

[5]Mathematically, this situation is somewhat reminiscent to perturbative super-string theory, where there is evidence that each term in the expansion is finite but the sum is not controlled.

true Lie algebra, where the *Hamiltonian constraints Poisson commute with each other on the entire phase space*. Interestingly, under seemingly mild assumptions one can show that solutions of this system admit space-like foliations on which ϕ is constant. Consequently, even though the system has infinitely many degrees of freedom, *as in LQC* one can use ϕ as a relational time variable. With $\mathbb{T}^3$ spatial topology for definiteness, one can decompose all fields into homogeneous and *purely* inhomogeneous modes. If one were to truncate the system by setting the inhomogeneous modes to zero, the resulting quantum theory would be precisely the loop quantum cosmology of Bianchi I models that has been analyzed in detail by Ashtekar and Wilson-Ewing. One might imagine incorporating the inhomogeneous modes using the 'hybrid' quantization scheme that has been successfully used in the Gowdy models by Mena, Martín-Benito, Pawłowski and others, although it will have to be non-trivially generalized to handle the fact that there are no Killing fields. As in the Gowdy models, this will likely involve some gauge fixing of the diffeomorphism and Gauss constraints. Even if these gauge fixing strategies do not work globally on the full phase space, one should still obtain a quantum theory tailored to a 'non-linear neighborhood' of FLRW or Bianchi I space-times. Finally, effective equations for this system would also provide valuable insights into the singularity resolution (which we expect to persist in an appropriate, well-defined sense). In particular, one would be able to compare and contrast their prediction with the simple BKL behavior near the general relativistic singularity, found by Andersson and Rendall for this system. More generally, this analysis will enable one to place loop quantum cosmology in the setting of full loop quantum gravity.

A second important open issue is to find restrictions on matter fields and their couplings to gravity for which this non-perturbative quantization can be carried out to a satisfactory conclusion. Supersymmetry, for example, is known to allow only very specific matter content. Recent work by Bodendorfer, Thiemann and Thurn has opened a fresh window for this analysis. A second possibility is suggested by the analysis of the closure of the constraint algebra in quantum theory. When it is extended to allow for matter couplings, the recent work by Varadarajan, Laddha, and Tomlin referred to in Sect. 2.2.2 could provide a promising approach to explore this issue in detail. Finally, as mentioned in Sect. 2.1.1, the renormalization group approach has provided interesting hints. Specifically, Reuter et al. have presented significant evidence for a non-trivial fixed point for vacuum general relativity in 4 dimensions [22]. When matter sources are included, it continues to exist only when the matter content and couplings are suitably restricted. For scalar fields, in particular, Percacci and Perini have found that polynomial couplings (beyond the quadratic term in the action) are ruled out, an intriguing result that may 'explain' the triviality of such theories in Minkowski space-times [42]. Are there similar constraints coming from loop quantum gravity?

• *Spin foams*. As discussed in Sect. 2.2.2, the spin foam program has made significant advances over the last four years. Results on the classical limit and finiteness of the sum over histories for a fixed 2-complex are especially encouraging. Therefore it is now appropriate to invest time and effort on key foundational issues.

First, we need a better understanding of the physical meaning of the 'vertex expansion' that results when one sums over arbitrary 2-complexes. In particular, is there a systematic physical approximation that lets us terminate the sum after a finite number of terms? In group field theory each term is multiplied by a power of the coupling constant [15, 18] but the physical meaning of this coupling constant in space-time terms is not known. Analysis by Ashtekar, Campiglia and Henderson in the cosmological context bears out an early suggestion of Oriti that the coupling constant is related to the cosmological constant. In the full theory, Fairbairn, Meusberger, Han and others have shown that the cosmological constant can be incorporated using quantum groups (which also makes the spin foam sum infrared finite for a fixed 2-complex). It is then natural to ask if there is a precise sense in interpreting the vertex expansion as a perturbation series in a parameter physically related to the cosmological constant also in the full theory.

Second, as I mentioned in Sect. 2.2.2, the issue of whether one should actually sum over various 2-complexes (i.e., add up all terms in the vertex expansion), or take an appropriately defined continuum limit is still open. Rovelli and Smerlak have argued that there is a precise sense in which the two procedures coincide. But so far there is no control over the sum and experts in rigorous field theory have expressed the concern that, unless a new principle is invoked, the number of terms may grow uncontrollably as one increases the number of vertices. Recent work on group field theory by Oriti, Rivasseau, Gurau, Krajewski and others may help streamline this analysis and provide the necessary mathematical control.

Finally, because the EPRL and FK models are motivated from the BF theory, they inherit certain ('Plebanski') sectors which classically do not correspond to general relativity. In addition, analysis of cosmological spin foams re-enforces an early idea due to Oriti that one should only sum over 'time oriented' quantum geometries. Some of these issues are now being analyzed in detail by Engle and others. But more work is needed on these basic conceptual issues.

• *Low-energy physics.* In low-energy physics one uses quantum field theory on given background space-times. Therefore one is naturally led to ask if this theory can be arrived at by starting from loop quantum gravity and making systematic approximations. Here, a number of interesting challenges appear to be within reach. Fock states have been isolated in the polymer framework [14] and elements of quantum field theory on quantum geometry have been introduced [16]. These developments lead to concrete questions. For example, in quantum field theory in flat space-times, the Hamiltonian and other operators are regularized through normal ordering. For quantum field theory on quantum geometry, on the other hand, the Hamiltonians are expected to be manifestly finite [14, 16]. Can one then show that, in a suitable approximation, normal ordered operators in the Minkowski continuum arise naturally from these finite operators? Can one 'explain' why the so-called Hadamard states of quantum field theory in curved space-times are special? These considerations could also provide valuable hints for the construction of viable semi-classical states of quantum geometry.

Since quantum field theory in FLRW space-times plays such an important role in the physics of the early universe, it is especially important to know if it can be systematically derived from loop quantum gravity. A number of obstacles immediately come to mind. In the standard treatment of quantum fields on cosmological space-times, one typically works with conformal or proper time, makes a heavy use of the causal structure made available by the fixed background space-time, and discusses dynamics as an unitary evolution in the chosen time variable. In quantum geometry state of loop quantum cosmology, none of these structures are available. Even in the 'deparameterized picture' it is a *scalar field* that plays the role of internal time; proper and conformal times are at best operators. Even when the quantum state is sharply peaked on an effective solution, we have only a probability distribution for various space-time geometries; we do not have a single, well-defined, classical causal structure. Finally, in loop quantum gravity, dynamics is teased out of the constraint while in quantum field theory in curved space-times it is dictated by a Hamiltonian. These obstacles seem formidable at first; Ashtekar, Kaminski and Lewandowski have shown that they can be overcome if one works with spatially compact topology and focuses just on a finite number of modes of the test field. The first assumption frees one from infrared issues which can be faced later, while the second restriction was motivated by the fact that, in the inflationary scenario, only a finite number of modes of perturbations are relevant to observations. It is important to remove these restrictions and use the resulting framework to analyze the questions on the quantum gravity origin of Hadamard states and of the adiabatic regularization procedure routinely used in cosmology.

2.2.3.2 Applications

• *The very early universe.* Because the initial motivations for inflation are not as strong as they are often portrayed to be, several prominent relativists were put off by the idea. As a consequence, recent developments in the inflationary paradigm have not drawn due attention in general relativity circles. In my view, there is a compelling case to take the paradigm seriously: it predicted the main features of inhomogeneities in the cosmic microwave background (CMB) which were subsequently observed and which serve as seeds for structure formation.

Let me first explain this point in some detail. Note first that one analyzes CMB inhomogeneities in terms of their Fourier modes and observationally relevant wave numbers are in a finite range, say Δk. Using this fact, we can write down the four assumptions on which the inflationary scenario is based:

1. Some time in its very early history, the universe underwent a phase of accelerated expansion during which the Hubble parameter H was nearly constant.
2. During this phase the universe is well-described by a FLRW background space-time together with linear perturbations.
3. A few e-foldings before the longest wavelength mode in the family Δk under consideration exited the Hubble radius, these Fourier modes of quantum fields describing perturbations were in the Bunch–Davis vacuum.
4. Soon after a mode exited the Hubble radius, its quantum fluctuation can be regarded as a classical perturbation and evolved via linearized Einstein's equations.

Then quantum field theory on FLRW space-times and classical general relativity imply the existence of tiny inhomogeneities in CMB which have been seen by the 7-year WMAP data. Numerical simulations show that these seeds grow to yield the large-scale structure that is observed today. Although the assumptions are by no means compelling, the overall economy of thought is nonetheless impressive. In particular, in this paradigm, the origin of the large-scale structure of the universe lies just in vacuum fluctuations! Therefore, it is of considerable interest to attempt to provide a quantum gravity completion of this paradigm.

The issues that are left open by this standard paradigm are of two types: Particle Physics issues and Quantum Gravity issues. Let me focus on the second for now:

1. Initial singularity: The paradigm assumes classical general relativity and theorems due to Borde, Guth and Vilenkin then imply that space-time had an initial big bang singularity. For reasons discussed in Sect. 2.1, this is an artifact of using general relativity in domains where it is not applicable. Therefore, one needs a viable treatment of the Planck regime and the corresponding extension of the inflationary paradigm.
2. Probability of inflation: In loop quantum cosmology, the big bang is replaced by a quantum bounce. So it is natural to introduce initial conditions there. Will a generic homogeneous, isotropic initial state for the background, when evolved, encounter a phase of slow roll inflation compatible with the 7-year WMAP data?
3. Trans-Planckian issues: In classical general relativity, if we evolve the Fourier modes of interest back in time, they become trans-Planckian. We need a quantum field theory on *quantum cosmological space-times* to adequately handle them.
4. Observations: The question then is whether the initial quantum state at the bounce, when evolved forward in time agrees sufficiently with the Bunch–Davis vacuum at the onset of inflation so as not to contradict the observations. More importantly, are there small deviations which could be observed in future missions?

Recent work by Ashtekar, Sloan, Agullo, Nelson and Barreau, Cailleteau, Grain and Mielczarek has made notable advances in facing these questions [38] but there are ample opportunities for other research that will provide both a viable quantum gravity completion of the inflationary paradigm and potentially observable predictions.

Finally, even if loop quantum gravity does offer a viable quantum gravity completion of the inflationary paradigm, open issues related to particle physics will still remain. In particular: What is the physical origin of the inflaton field? Of the potential one must use to get a sufficiently long slow roll? Is there only one inflaton or many? If many, what are their interactions? What are the couplings that produce the known particles as the inflaton oscillates around the minimum of the potential at the end of inflation (the so-called 'reheating')? Therefore, it would be healthy to look also for alternatives to inflation. Indeed, the alternatives that have been advocated by Brandenberger and others involve bouncing models and therefore have similarities with the general loop quantum cosmology paradigm. Because the expansion of the universe from the bounce to the surface of last scattering in the post-bounce branch is *much* smaller than that in the inflationary scenario, at the bounce, modes of direct interest to the CMB observations now have physical frequencies *much* below the Planck scale. Therefore, the trans-Planckian issue is avoided and quantum field theory in curved space-times should be viable for these modes. This fact, coupled with the absence of singularity, enables one to calculate a transfer matrix relating modes in the pre-bounce epoch to those in the post-bounce epoch. Under suitable assumptions, Brandenberger and others have shown that this relation gives rise to a nearly scale invariant spectrum of scalar and tensor modes in the post-bounce phase. But the underlying premise in these calculations is that perturbations originate in the distant past of the contracting branch where the geometry is nearly flat and quantum fields representing perturbations are taken to be in their vacuum state. This idea that the entire evolution from the distant past in the contracting phase to the bounce is well described by a homogeneous model with small perturbations is not at all realistic. But since the general paradigm has several attractive features, it would be of considerable interest to investigate whether loop quantum cosmology bounces allow similar alternatives to inflation without having to assume that non-linearities can be neglected throughout the pre-bounce phase.

• *Black hole evaporation: The issue of the final state.* Black hole thermodynamics was initially developed in the context of stationary black holes. Indeed, until relatively recently, there were very few analytical results on dynamical black holes in classical general relativity. This changed with the advent of dynamical horizons which provide the necessary analytical tools to extract physics from numerical simulations of black hole formation and evaporation. It also led to some new insights on the fundamental side. In particular, it was also shown that the first law can be extended to these time-dependent situations and the leading term in the expression of the entropy is again given by $a_{\mathrm{hor}}/4\ell_{\mathrm{Pl}}^2$ [19]. Hawking radiation will cause the horizon of a large black hole to shrink *very* slowly, whence it is reasonable to expect that the description of the quantum horizon geometry can be extended from isolated to dynamical horizons in this phase of the evaporation. The natural question then is: Can one describe in detail the black hole evaporation process and shed light on the issue of information loss?

The space-time diagram of the evaporating black hole, conjectured by Hawking, is shown in the left-hand drawing in Fig. 2.1. It is based on two ingredients: (i) Hawking's original calculation of black hole radiance, in the framework of quantum field theory on a *fixed* background space-time; and (ii) heuristics of back-reaction effects which suggest that the radius of the event horizon must shrink to zero. It is generally argued that the semi-classical process depicted in this figure should be reliable until the very late stages of evaporation when the black hole has shrunk to Planck size and quantum gravity effects become important. Since it takes a very long time for a large black hole to shrink to this size, one then argues that the quantum gravity effects during the last stages of evaporation will not be sufficient to restore the correlations that have been lost due to thermal radiation over such a long period. Thus there is loss of information. Intuitively, the lost information is 'absorbed' by the 'left-over piece' of the final singularity which serves as a new boundary to space-time.

However, loop quantum gravity considerations suggest that this argument is incorrect in two respects. First, the semi-classical picture breaks down not just at the

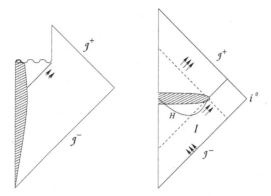

Fig. 2.1 Conjectured space-time diagrams of evaporating black holes in full quantum theory. (**a**) *Left figure*: Standard paradigm, originally proposed by Hawking. Information is lost because part of the incoming state on $\mathscr{I}^-$ falls into the part of the future singularity that is assumed to persist in the full quantum gravity theory. (**b**) *Right figure*: New paradigm motivated by the singularity resolution in LQC. What forms and evaporates is a dynamical horizon H. Quantum space-time is larger and the incoming information on the full $\mathscr{I}^-$ is adequately recovered on the $\mathscr{I}^-$ of this larger space-time

end point of evaporation but in fact *all along what is depicted as the final singularity*. Using ideas from quantum cosmology, the interior of the Schwarzschild horizon was analyzed in the context of loop quantum gravity by Ashtekar, Bojowald, Modesto, Vandersloot and others. This analysis is not as complete or refined as that in the cosmological context. But the qualitative conclusion that the singularity is resolved due to quantum geometry effects is likely to be robust. If so, the space-time does *not* have a singularity as its final boundary. The second limitation of this semi-classical picture is its depiction of the event horizon. The notion of an event horizon is teleological and refers to the *global* structure of space-time. Resolution of the singularity introduces a domain in which there is no classical space-time, whence the notion ceases to be meaningful; it is simply 'transcended' in quantum theory. Using these considerations Ashtekar and Bojowald introduced a new paradigm for black hole evaporation in loop quantum gravity, depicted in the right hand drawing of Fig. 2.1: Now, it is the dynamical horizon that evaporates with emission of quantum radiation, and the initial pure state evolves to a final pure state on the future null infinity of the extended space-time. Thus, there is no information loss. In this paradigm, the semi-classical considerations would not be simply dismissed; they would be valid in certain space-time regions and under certain approximations. But for fundamental conceptual issues, they would not be inadequate.

However, this is still only a paradigm and the main challenge is to develop it into a detailed theory. Just as Wheeler's 'It from Bit' ideas were transformed into a detailed theory of quantum horizon geometry, it should be possible to construct a detailed theory of black hole evaporation based on this paradigm. More recently, this paradigm was put on a firm footing by Ashtekar, Taveras and Varadarajan in the case of 2-dimensional black holes first introduced by Callen, Giddings, Harvey and Strominger. The model is interesting especially because its action and equations of motion closely mimic those governing 4-dimensional, spherically symmetric black holes formed by the gravitational collapse of a scalar field. Ashtekar, Pretorius and Ramazanoglu have used a combination of analytical and numerical methods to analyze the mean field approximation in complete detail. It explicitly shows that some of the common assumptions regarding effects of including back reaction, discussed in the last paragraph, are incorrect. This analysis further reinforces the paradigm of the figure on the right. It is therefore of considerable interest to extend all this analysis to four dimensions in the loop quantum gravity setting. The very considerable work on spherically symmetric midi-superspaces by Gambini, Pullin, Bojowald and others will serve as a point of departure for this analysis.

• *Contact with low-energy physics.* Spin foam models provide a convenient arena to discuss issues such as the graviton propagator, n-point functions and scattering, that lie at the heart of perturbative treatments. At first, it seems impossible to have non-trivial n-point functions in a diffeomorphism invariant theory. Indeed, how could one say that the 2-point function falls off as $1/r^n$ when the distance r between the two points has no diffeomorphism invariant meaning? Thanks to a careful conceptual set-up by Oeckl, Colosi, Rovelli and others, this issue has been satisfactorily resolved. To speak of n-point functions, one needs to introduce a boundary state (in which the expectation values are taken) and the notion of distance r descends from the boundary state. Interestingly, a detailed calculation of the 2-point function brought out some limitations of the Barrett–Crane model and provided new impetus for the EPRL and FK models. These calculations by Bianchi, Ding, Magliaro and Perini strongly indicate that, to the leading order, a graviton propagator with the correct functional form and tensorial structure will arise from these models.

However, these calculations can be improved in a number of respects and their full implications have yet to be properly digested. In particular, one needs a better handle on contributions from 2-complexes with large numbers of vertices and the physics of the sub-leading terms. These terms seem to be sensitive to the choice of the boundary state and there is not a canonical one representing Minkowski space. Therefore, comparison with the standard perturbation theory in Minkowski space is difficult. This is a fertile and important area for further research. Indeed the key challenge in this area is to 'explain' why perturbative quantum general relativity fails if the theory exists non-perturbatively. As mentioned in Sect. 2.1, heuristically the failure can be traced back to the insistence that the continuum space-time geometry is a good approximation even below the Planck scale. But a more detailed answer is needed. For example, is it because, as developments in the asymptotically safe scenarios indicate [22, 40, 41], the renormalization group has a *non-Gaussian* fixed point?

• *Unification.* Finally, there is the issue of unification. At a kinematical level, there is already an unification because the quantum configuration space of general relativity is the same as in gauge theories which govern the strong and electroweak interactions. But the non-trivial issue is that of dynamics. To conclude, let us consider a speculation. One possibility is to use the 'emergent phenomena' scenario where new degrees of freedom or particles, which were not present in the initial Lagrangian, emerge when one considers excitations of a non-trivial vacuum.

For example, one can begin with solids and arrive at phonons; start with superfluids and find rotons; consider superconductors and discover Cooper pairs. In loop quantum gravity, the micro-state representing Minkowski space-time will have a highly non-trivial Planck-scale structure. The basic entities will be 1-dimensional and polymer-like. One can argue that, even in absence of a detailed theory, the fluctuations of these 1-dimensional entities should correspond not only to gravitons but also to other particles, including a spin-1 particle, a scalar and an anti-symmetric tensor. These 'emergent states' are likely to play an important role in Minkowskian physics derived from loop quantum gravity. A detailed study of these excitations may well lead to interesting dynamics that includes not only gravity but also a selected family of non-gravitational fields. It may also serve as a bridge between loop quantum gravity and string theory. For, string theory has two *a priori* elements: unexcited strings which carry no quantum numbers and a background space-time. Loop quantum gravity suggests that both could arise from the quantum state of geometry, peaked at Minkowski (or, de Sitter) space. The polymer-like quantum threads which must be woven to create the classical ground state geometries could be interpreted as unexcited strings. Excitations of these strings, in turn, may provide interesting matter couplings for loop quantum gravity.

Acknowledgements My understanding of classical and quantum gravity has deepened through discussions with a large number of colleagues. Much of this overview is taken from the somewhat more detailed review article in the Proceedings of the Third School on Quantum Geometry and Quantum Gravity, Pos/QGQG2011-001. I would like to thank Miguel Campiglia, Kristina Giesel, Alok Laddha and especially Carlo Rovelli for their comments on the manuscript. This work was supported in part by the European Science Foundation through its network *Quantum Geometry and Quantum Gravity*, NSF grant PHY 1205388 and the Eberly research funds of The Pennsylvania State University.

References

1. R. Arnowitt, S. Deser, C.W. Misner, in *Gravitation: An Introduction to Current Research*, ed. by L. Witten (Wiley, New York, 1962), pp. 227–265. gr-qc/0405109
2. A. Komar, in *Relativity*, ed. by M. Carmeli, S.I. Fickler, L. Witten (Plenum Press, New York, 1970), pp. 19–29
3. P.G. Bergmann, A. Komar, in *General Relativity and Gravitation*, vol. 1, ed. by A. Held (Plenum Press, New York, 1980), pp. 227–254
4. A. Ashtekar, R. Geroch, Rep. Prog. Phys. **37**, 1211 (1974)
5. K. Kuchař, in *Quantum Gravity 2, a Second Oxford Symposium*, ed. by C.J. Isham, R. Penrose, D.W. Sciama (Clarendon Press, Oxford, 1981), pp. 329–374
6. J.A. Wheeler, *Geometrodynamics* (Academic Press, New York, 1962)
7. J.A. Wheeler, in *Relativity, Groups and Topology*, ed. by C.M. DeWitt, B.S. DeWitt (Gordon and Breach, New York, 1964), pp. 467–500
8. A. Ashtekar, in *The Universe: Visions and Perspectives*, ed. by N. Dadhich, A. Kembhavi (Kluwer, Dordrecht, 2000), pp. 13–34. gr-qc/9901023
9. A. Ashtekar, *Lectures on Non-perturbative Canonical Gravity* (World Scientific, Singapore, 1991)
10. J. Baez, J.P. Muniain, *Gauge Fields, Knots and Gravity* (World Scientific, Singapore, 1994)
11. A. Ashtekar, in *Gravitation and Quantizations: Proceedings of the 1992 Les Houches Summer School*, ed. by B. Julia, J. Zinn-Justin (Elsevier, Amsterdam, 1995), pp. 181–284. gr-qc/9302024

12. S. Carlip, *Quantum Gravity in 2 + 1 Dimensions* (Cambridge University Press, Cambridge, 1998)
13. R. Gambini, G. Pullin, *Loops, Knots, Gauge Theories and Quantum Gravity* (Cambridge University Press, Cambridge, 1996)
14. A. Ashtekar, J. Lewandowski, Class. Quantum Gravity **21**, R53 (2004). gr-qc/0404018
15. C. Rovelli, *Quantum Gravity* (Cambridge University Press, Cambridge, 2004)
16. T. Thiemann, *Introduction to Modern Canonical Quantum General Relativity* (Cambridge University Press, Cambridge, 2005)
17. A. Perez, Class. Quantum Gravity **20**, R43 (2003). gr-qc/0301113
18. D. Oriti, in *Approaches to Quantum Gravity* (Cambridge University Press, Cambridge, 2009), pp. 310–331. gr-qc/0607032
19. A. Ashtekar, B. Krishnan, Living Rev. Relativ. **7**, 10 (2004). http://www.livingreviews.org/lrr-2004-10. gr-qc/0407042
20. W. Israel, S.W. Hawking (eds.), *General Relativity, an Einstein Centenary Survey* (Cambridge University Press, Cambridge, 1980)
21. R.W. Williams, P.M. Tuckey, Class. Quantum Gravity **9**, 1409 (1992)
22. M. Niedermaier, M. Reuter, Living Rev. Relativ. **7**, 9 (2006). http://www.livingreviews.org/lrr-2006-5
23. R. Gambini, J. Pullin, in *Approaches to Quantum Gravity* (Cambridge University Press, Cambridge, 2009), pp. 378–392. gr-qc/0512065
24. R. Loll, Living Rev. Relativ. **1**, 13 (1998). http://www.livingreviews.org/lrr-1998-13. gr-qc/9805049
25. J. Ambjørn, J. Jurkiewicz, R. Loll, Phys. Rev. Lett. **93**, 131301 (2004). hep-th/0404156
26. R. Penrose, in *Quantum Gravity, an Oxford Symposium*, ed. by C.J. Isham, R. Penrose D. W. Sciama (Clarendon Press, Oxford, 1975), pp. 268–407
27. R. Penrose, W. Rindler, *Spinors and Space-Times*, vol. 2 (Cambridge University Press, Cambridge, 1988)
28. R. Penrose, in *100 Years of Relativity*, ed. by A. Ashtekar (World Scientific, Singapore, 2005), pp. 465–505
29. M. Ko, M. Ludvigsen, E.T. Newman, P. Tod, Phys. Rep. **71**, 51 (1981)
30. A. Ashtekar, *Asymptotic Quantization* (Bibliopolis, Naples, 1984)
31. A. Connes, *Non-commutative Geometry* (Academic Press, New York, 1994)
32. R. Sorkin, in *Lectures on Quantum Gravity*, ed. by A. Gomberoff, D. Marolf (Plenum Press, New York, 2005), pp. 305–327. gr-qc/0309009
33. F. Dowker, in *100 Years of Relativity*, ed. by A. Ashtekar (World Scientific, Singapore, 2005), pp. 445–464. gr-qc/0508109
34. C.J. Isham, in *Deep Beauty*, ed. by H. Halvorson (Cambridge University Press, Cambridge, 2011), pp. 187–205. arXiv:1004.3564
35. T. Dahlen, arXiv:1111.5685
36. R.M. Wald, Living Rev. Relativ. **4**, 6 (2001). http://www.livingreviews.org/lrr-2001-6. gr-qc/9912119
37. R.M. Wald, *Quantum Field Theory in Curved Spacetime and Black Hole Thermodynamics* (Chicago University Press, Chicago, 1994)
38. A. Ashtekar, P. Singh, Class. Quantum Gravity **28**, 213001 (2011). arXiv:1108.0893
39. R. Gambini, J. Pullin, *A First Course in Loop Quantum Gravity* (Oxford University Press, Oxford, 2011)
40. J.M. Daum, M. Reuter, arXiv:1012.4280
41. J.M. Daum, M. Reuter, arXiv:1111.1000
42. D. Perini, Ph.D. Dissertation, SISSA, 2004
43. J. Klauder, Int. J. Mod. Phys. D **12**, 1769 (2003)
44. C. Rovelli, arXiv:1102.3660
45. R.P. Feynman, Rev. Mod. Phys. **20**, 367 (1948)
46. J.C. Baez, Lect. Notes Phys. **543**, 25 (2000). gr-qc/9905087

Chapter 3
Covariant Loop Gravity

Carlo Rovelli

Abstract I summarize and illustrate the present state of the covariant formulation of the loop quantum gravity dynamics.

3.1 The Definition of the Theory

A quantum theory of gravity is defined by the family of transition amplitudes [1–5].

$$W_{\mathscr{C}}(h_l) = \int_{SU(2)} dh_{vf} \prod_f \delta(h_f) \prod_v A(h_{vf}), \qquad (3.1)$$

where the vertex amplitude is given by

$$A(h_{ab}) = \sum_{j_{ab}} \int_{SL(2C)} dg_a \prod_{ab} tr_{j_{ab}} [h_{ab} Y_\gamma^\dagger g_a g_b^{-1} Y_\gamma]. \qquad (3.2)$$

$\mathscr{C}$ is a two-complex with faces f, edges e and vertices v, bounded by a graph $\Gamma = \partial\mathscr{C}$ with links l and nodes n, and Y_γ is the linear map

$$Y_\gamma : |j; m\rangle \mapsto |(\gamma(j+1), j); j, m\rangle \qquad (3.3)$$

from the $SU(2)$ spin-j representation space $\mathscr{H}_j$ to the *unitary $SL(2, C)$* representation space $\mathscr{H}_{p,k}$ with quantum numbers $p = \gamma(j+1)$ and $k = j$. See [2] for the full details on the notation. These transition amplitudes are ultraviolet finite and admits a q-deformed version [6–8] where they are finite. The theory can be coupled to fermions and Yang–Mills fields [9].

The transition amplitude depends on L $SU(2)$ elements h_l associated with the L links l of the graph that bounds $\mathscr{C}$ and is therefore an element W of the boundary Hilbert space $H_{\partial\mathscr{C}}$ [2]. Semi-classical states ψ in $H_{\partial\mathscr{C}}$ describe discretized 3d geometries q_ψ formed by glued polyhedra [10, 11]. In particular, Regge geometries.

Evidence has piled up [12–15] that (3.1) converges to the Hamilton function $S_\Delta[q]$ of a Regge discretization of gravity in the classical limit, if $\mathscr{C}$ is the two-skeleton of the dual of the triangulation Δ. The structure of the theory is therefore

C. Rovelli (✉)
Centre de Physique Théorique, Case 907, Luminy, 13288 Marseille, France
e-mail: rovelli@cpt.univ-mrs.fr

G. Calcagni et al. (eds.), *Quantum Gravity and Quantum Cosmology*,
Lecture Notes in Physics 863, DOI 10.1007/978-3-642-33036-0_3,
© Springer-Verlag Berlin Heidelberg 2013

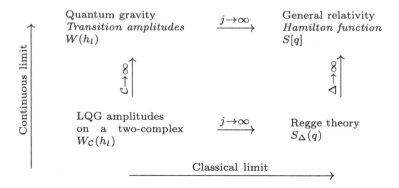

Thus, the LQG transition amplitudes on a two-complex define a family of approximations to the full theory, and the classical limit of each of these (for $\mathscr{C}$ the two-skeleton of Δ^*) is given by Regge gravity on Δ.

In the rest of this article, I illustrate properties and various aspects of these transition amplitudes. This article does not cover the quantum geometry defined by the boundary states of the theory, which is amply discussed in many reviews. (See for instance [2].)

3.2 Properties

The structure described above can be compared with that of lattice QCD, where $\mathscr{C}$ corresponds to a given lattice. Here Eqs. (3.1) and (3.2) can be shown to provide a discretization of the formal "sum over 4-geometries" of the exponential of the GR action [16]

$$Z \sim \int Dg \, e^{\frac{i}{\hbar} \int R \sqrt{g} \, d^4 x}, \tag{3.4}$$

on a manifold with boundaries with fixed geometry. There is a crucial difference, however. In QCD, the continuous limit requires the refinement of the lattice as well as the tuning of a coupling constant to a critical value. In gravity, instead, there is no parameter to be tuned. The difference can be traced directly to the general covariance of the continuous theory. As shown in detail in [17], a general covariant system can always be discretized without introducing a lattice spacing, and therefore a scale. This difference is important and is the source of confusion.

A second important difference is that the three-metric q cannot be fully diagonalized in $H_{\partial\mathscr{C}}$. A maximal set of commuting operators in $H_{\partial\mathscr{C}}$ is formed by areas and volumes of the polyhedra, but these quantities do not suffice to determine the geometry of the glued polyhedra. From this perspective the metric is like angular momentum: the full set of data that determine the geometry is formed by operators that do not commute. Therefore, classical 3-metrics can only be associated with semi-classical states in $H_{\partial\mathscr{C}}$. To have semi-classical states (where fluctuations are small with respect to expectation values) we need large quantum numbers. In particular, we need large spins, and therefore large distances compared to the Planck scale. Physically, this means that at small scale the geometry of space cannot be Riemannian: it is a quantized geometry, and there is a Heisenberg uncertainty preventing the full 3-geometry to be sharp. In other words, the $\hbar \to 0$ limit of the theory is also necessarily a large-distance limit. At short scale, quantum gravity does not have a proper continuous limit taking it to classical general relativity (GR). This was of course expected on physical grounds.

Individual amplitudes (3.1)–(3.2) can be obtained as Feynman amplitudes of a proper quantum field theory (QFT), using the group-field-theory formalism [18]. By separating the terms with vanishing spins and reinterpreting them as defined on sub-two-complexes, it can be possible to re-express the limit as a series [19]. This observation, and the analogy with the standard Feynman expansion provide a second interpretation to the expansion in $\mathscr{C}$, as a perturbative expansion. States in $\mathscr{H}_\Gamma$ can be viewed as formed by N quanta, where N is the number of nodes of the graph. Each node of the graph is like a particle in QED, namely a quantum of electromagnetic field. Here, each node represents a quantum of gravitational field. But there is a key difference. QED Fock quanta carry quantum numbers coding where they are located in the background space-manifold. Here, since in general relativity the gravitational field is also physical space, individual quanta of gravity are also quanta of space. Therefore they do not carry information about their localization in space, but only information about the *relative* location with respect to one another. This information is coded by the graph structure. Thus, the quanta of gravity form themselves the texture of physical space. Therefore the graph can *also* be seen as a generalization of the lattices of lattice QCD.

This convergence between the perturbative-QED picture and the lattice-QCD picture follows directly from the key physics of general relativity: the fact that the gravitational field is physical space itself. Indeed, the lattice sites of lattice QCD are small regions of space; according to general relativity, these are excitations of the gravitational field, therefore they are themselves quanta of a (generally covariant) quantum field theory. An N-quanta state of gravity has therefore the same structure as a Yang–Mills state on a lattice with N sites. This convergence between the perturbative-QED and the lattice-QCD pictures is a beautiful feature of loop gravity.

3.3 The Discretization of Parametrized Systems

To understand the particular features of the transition amplitudes of loop quantum gravity, it is useful to compare them with the following definition of the amplitudes of a finite-dimensional 'general covariant' system. Consider a dynamical system with configuration variable $q \in \mathscr{C}$, and Lagrangian $L(q, \dot{q})$. Given an initial configuration q at time t and a final configuration q' at time t', let $q_{q,t,q',t'} : R \to \mathscr{C}$ be a solution of the equations of motion such that $q_{q,t,q',t'}(t) = q$ and $q_{q,t,q',t'}(t') = q'$.

Assume for the moment this exists and is unique. The Hamilton function is the function on $(\mathscr{C} \times R)^2$ defined by

$$S(q, t, q', t') = \int_t^{t'} dt\, L(q_{q,t,q',t'}, \dot{q}_{q,t,q',t'}),\qquad(3.5)$$

namely, the value of the action on the solution of the equation of motion determined by given initial and final data. This function, introduced by Hamilton in 1834 [20] codes the solution of the dynamics of the system, has remarkable properties and is a powerful tool that remains meaningful in background-independent physics. Let H be the quantum Hamiltonian operator of the system and $|q\rangle$ the eigenstates of its q observables. The transition amplitude

$$W(q, t, q', t') = \langle q' | e^{-\frac{i}{\hbar} H(t'-t)} | q \rangle\qquad(3.6)$$

codes all the quantum dynamics. Now, consider a parametrized formulation of this same system. That is, consider a new action which depends on $n + 1$ variables $x = (q, t) \in \mathscr{C} \times \mathscr{R} \equiv \mathscr{C}_{ex}$ evolving in τ

$$I[q] = \int dt\, L(q, \dot{q}) \rightarrow \int d\tau\, \dot{t} L(q, \dot{q}/\dot{t}) = \int d\tau\, \mathscr{L}(x, \dot{x}) \equiv I[x],\qquad(3.7)$$

where on the left-hand side of the arrow the dot indicates a derivative with respect to t, while on the right-hand side of the arrow (and from now on) it indicates a derivative with respect to τ. For instance, a single-particle Lagrangian $L = m\dot{q}^2/2 - V(q)$ gives $\mathscr{L} = m\dot{q}^2/(2\dot{t}) - \dot{t}V(q)$. This 'parametrized' system describes the same physics as the original one, but in a different logic. The physical time variable t is now treated on the same footing as the other dynamical variables[1] and the system has a local gauge invariance under reparametrizations of τ. As was recognized in the early days of the canonical analysis of general relativity [24], this is precisely the structure of general relativity. The general relativistic coordinates x^μ play the same role as τ above and reparametrization invariance is the invariance under general coordinate transformations. Above, the 'parametrized' form of the dynamics has been derived from an original 'un-parametrized' form, while GR is directly written in the parametrized form, with a local gauge-invariance. 'De-parametrizing' GR is possible in principle, but spoils many of its formal properties, making the formalism far more cumbersome and intractable. In GR we better live with reparametrization invariance and, in fact, take advantage of it.

What is the Hamilton function of the parametrized system? From the definition,

$$S(x, \tau, x', \tau') = \int_\tau^{\tau'} d\tau\, \mathscr{L}(x_{x,x'}, \dot{x}_{x,x'}),\qquad(3.8)$$

where $x_{x,x'}(\tau)$ is a solution of the equations of motion such that $x_{x,x'}(\tau) = x$ and $x_{x,x'}(\tau') = x'$. But a moment of reflection will convince the reader that

$$S(x, \tau, x', \tau') = S(x, x'),\qquad(3.9)$$

---

[1]That is, dynamics is not anymore interpreted as the description of the evolution in t of the n configuration variables q, but rather as a description of the possible *relations* between the $n + 1$ variables $x = (q, t)$ (see Sect. 3.2.4 in [21] and [22, 23]).

namely the Hamilton function of a parametrized system is independent from τ and τ'.[2] Second, because of the gauge there are many solutions of the classical equations with the same boundary data, but $S(x, x')$ is independent from the one chosen, that is, it is gauge invariant. Both facts are immediate consequence of the invariance of the action under reparametrizations of τ. Furthermore, since the new action is just the old one in new variables, $S(x, x')$ is precisely nothing else than (3.5).

That is, remarkably, the Hamilton function of the parametrized system is the same object as the Hamilton function of the original system. In fact, notice that the original Hamilton function was already a function of (two copies of) the *extended* configuration space $\mathscr{C}_{\text{ex}} = \mathscr{C} \times \mathscr{R}$. The Hamilton function was already born as if it knew it had to work in a parametrized language.

Let me know give a definition of the path integral that defines the transitions amplitudes of this system, as a limit of multiple integrals, as in the original Feyman's derivation of path integral, or as in lattice QCD. This can be defines as

$$W_N(x, x') = \int \frac{dq_n dt_n}{\mu(q_n, t_n)} \, e^{\frac{i}{\hbar} \sum_{n=1}^{N} (t_n - t_{n-1}) L(q_n, q_{n-1}, t_n, t_{n-1})}. \tag{3.11}$$

The essential point is that the discretization of the parametrized action does not involve a lattice spacing. This can be seen explicitly, for instance, by noticing that the discretized form of the general-covariant action of a one-dimensional system with kinetic and potential energy

$$S_N[q_n, t_n] = \sum_{n=1}^{N} a \left[\frac{m}{2} \frac{\frac{(q_{n+1} - q_n)^2}{a^2}}{\frac{t_{n+1} - t_n}{a}} - \frac{t_{n+1} - t_n}{a} V(q_n) \right]$$

$$= \sum_{n=1}^{N} \left[\frac{m}{2} \frac{(q_{n+1} - q_n)^2}{t_{n+1} - t_n} - (t_{n+1} - t_n) V(q_n) \right]$$

cancels exactly! Notice that the 'lattice spacing' a is replaced by the variable quantity $(t_n - t_{n-1})$ [17].

We can therefore recover the structure illustrated in the diagram in Sect. 3.1. The discretized quantum transition amplitude (3.11) yields the full quantum transition amplitude in the $N \to \infty$ limit and the Hamilton function of a discretization of the classical parametrized theory in the $\hbar \to 0$ limit. A concrete example is illustrated in detail in [17].

---

[2] S is a solution of the Hamilton–Jacobi equation

$$\partial S / \partial \tau = H(x, \partial S / \partial x). \tag{3.10}$$

But the canonical Hamiltonian $H(x, p_x)$ vanishes because of the gauge invariance and therefore $\partial S / \partial \tau = 0$.

3.4 The Discretization of Classical General Relativity

Consider a finite 4d region $\mathscr{B}$ surrounded by a 3d surface Σ.[3] For simplicity, I begin with metric variables—later on I will use other variables to describe the gravitational field. We can fix pure gravity boundary data on Σ by giving the three-metric q of Σ. Let g_q be a 4d metric on $\mathscr{B}$ which satisfies the Einstein's equations and induces the 3-metric q on Σ. The possibility that such g_q might be non-existent or non-unique does not concern us here, as discussed in the previous section. Consider the action of GR, including the boundary term

$$I[g] = \frac{1}{2} \int_{\mathscr{B}} d^4x \sqrt{g} R[g] + \int_{\Sigma} d^3x \sqrt{q} k, \tag{3.12}$$

where $k = k^{ab} q_{ab}$ is the trace of the extrinsic curvature k^{ab} on the boundary and q is the induced 3-metric. If g is a solution of the equations of motion, the first term vanishes. Thus, the Hamilton function of q is the action of g_q, that is

$$S[q] = I[g_q] = \int_{\Sigma} d^3x \sqrt{q} k^{ab}[q](x) q_{ab}(x). \tag{3.13}$$

The non-trivial part of this expression is the dependence of the extrinsic curvature $k^{ab}[q](x)$ on the 3-metric. This dependence is non-local: in general the extrinsic curvature in a point x depends on the value of the metric on the entire surface Σ.[4]

Now, consider a discretization of GR. As a warm-up, consider the best-known discretization, which is Regge calculus. Fix a triangulation Δ of $\mathscr{B}$ and consider Regge geometries on $\mathscr{B}$. A Regge geometry is a geometry which is everywhere flat except on the triangles t of Δ, where the curvature is distributional and fully determined by the deficit angle θ_t at the triangle, determined by the sum of the dihedral angles of the flat 4-simplices meeting at t. This geometry is uniquely determined by giving the lengths l_i of the segments i of Δ, which determine the deficit angles $\theta_t = \theta_t(l_i)$ and the areas $A_t = A_t(l_i)$ of these triangles. The Regge action is a local function of these lengths and reads

$$I_\Delta[l_i] = \sum_{t \in \Delta} \theta_t(l_i) A_t(l_i) + \sum_{t \in \partial\Delta} \theta_t(l_{i_b}) A_t(l_{i_b}). \tag{3.14}$$

The triangulation of $\mathscr{B}$ indices a boundary triangulation $\partial\Delta$ on the boundary surface Σ. The boundary 3-metric is fully determined by the lengths of the boundary

---

[3]In the applications, we are particularly interested in the cases where $\mathscr{B}$ is a ball or a segment of a cylinder, and therefore Σ has the topology of S_3, or $S_3 \times S_3$, which are respectively relevant to discuss scattering and cosmology.

[4]For a simple example of this dependence consider the Euclidean theory, and say that the metric q is that of a metric 3-sphere with radius a. Such a sphere can be imbedded in many curved 4d manifolds, and the extrinsic curvature of the imbedding is *not* determined by q. But if the 4d Riemannian manifold is a solution of the Euclidean Einstein equations, then this freedom is drastically reduced. A solution of the Einstein equations is flat space. A metric 3-sphere can be imbedded in a flat 4d space as the surface of the ball with radius a and this fixes the extrinsic curvature to be $k^{ab} = a^{-1}q^{ab}$, so that $H[q(a)] = 6\pi^2 a^2$.

links i_b, which I denote as $q = \{l_{i_b}\}$. The Hamilton function of the Regge theory is the discrete analog of (3.13):

$$S_\Delta[q] = \sum_{t \in \partial \Delta} \theta_t[l_{i_b}] A_t(l_{i_b}), \tag{3.15}$$

where the sum is here over the triangles in the boundary and θ_t is the discrete extrinsic curvature at the boundary triangles, namely, the angle between the 4-normals of the two boundary tetrahedra separated by the triangle t. While the area of a boundary triangle depends only on the length of its three sides, the extrinsic curvature $\theta_t[l_{i_b}]$ is a non-trivial function of all the boundary lengths, determined by solving the bulk Regge equations of motion. (I have used the square-brackets notation to emphasize this non-local character of the dependence.)

Regge theory approximates GR in the following sense. First, say that a sequence of Regge geometries g_n converges to a Riemannian geometry g if there is map f from one to the other such that for any two points x and y, $|d_g(f(x), f(y)) - d_{g_n}(x, y)| < \epsilon$, where $d_g(x, y)$ is the distance between x and y in the geometry g. Then consider a sequence Δ_n of refinements of a triangulations. Let a Regge geometry q_n be on the boundary of Δ_n, such that the sequence q_n converges to q. A discretization is good if

$$\lim_{n \to \infty} S_{\Delta_n}[q_n] = S[q]. \tag{3.16}$$

The truncated transition amplitudes given at the opening of this article gave therefore two relevant limits. The $\hbar \to 0$ limit gives the Regge Hamilton function.

$$\lim_{\hbar \to 0} (-i\hbar) \log W_\Delta[q] = S_\Delta[q]. \tag{3.17}$$

If we restore physical units, we must add the Newton constant $1/8\pi G$ in front of the action and (3.17) reads

$$\lim_{l_P \to 0} \left(-i8\pi l_P^2\right) \log W_\Delta[q_i] = S_\Delta[q_{\partial\Delta}], \tag{3.18}$$

where $l_P = \sqrt{\hbar G}$ is the Planck length. Then, taking (3.16) into account, it is reasonable to *define* the quantum gravity transition amplitudes

$$W[q] = \lim_{n \to \infty} W_{\Delta_n}[q_n] \tag{3.19}$$

if $\lim_{n \to \infty} q_n = q$. (The choice of a particular sequence Δ_n can be avoided by defining a stricter limit, for $\Delta \to \infty$ in the sense of nets [19].) The limit itself, however, is not of great significance from the perspective of a physicist, since the quantum theory is already defined by its family of approximations (3.1).

It is reasonable to expect these approximation to be good in the regime where the corresponding classical approximations are good, namely where (3.16) converges fast. This is the regime where the bulk deficit angles are small, namely the regime around flat space.[5]

---

[5]The expansion in n should not be confused with the standard perturbative expansion used in QFT. The latter is an expansion in the amplitude of the field, keeping all its modes. The former is an expansion in the number of modes.

But should we expect the limit (3.19) to actually converge, and converge fast? A rigorous answer to this equation is missing, but there are circumstantial arguments that indicate that this could be the case in some regimes. The Regge approximation becomes exact on flat space. That is, the Regge action and the continuous Einstein–Hilbert action are equal for a flat geometry. It follows that if the geometry is flat, the Regge discretization is 'topological' in the sense that it is invariant under a refinement of the triangulation. This very peculiar property was called Ditt-invariance in [17] from Bianca Dittrich, who has emphasized it in her work with Benjamin Bahr [25–27]. When the boundary data are near flatness, in the sense mentioned, a refinement of the triangulation becomes therefore irrelevant. More precisely, let q_n be a sequence of boundary Regge metrics converging to a 3-geometry q such that g_q is close to a *flat* geometry. Then one may expect that in this regime $W_{\Delta_n}[q_n]$ converges fast, and therefore a small n is sufficient to give a good approximation to the physical amplitudes.

3.5 Conclusion

Quantum gravity is often confusing. Field operator insertions in the path integral, which are the main tool for analyzing conventional QFT, are uninteresting in quantum gravity, due to diff-invariance. The discretization used to define boundary amplitudes is often confused with the quantum discreteness of space. The general structure of a background-independent quantum theory is different from a conventional QFT. What are good observables in quantum gravity, and how do we describe evolution? See for example [23, 28–41] and [42] for an overview.

I have given a tentative overall picture of the structure of the theory, the observables, and the form of the continuous and classical limits. The truncated boundary transition amplitudes (3.1)–(3.2) are the tool for extracting physics from the theory. From these quantities one can derive standard observables, for instance for analyzing cosmological evolution [43], or particle scattering [44], where the old idea that the gravitons live on the non-perturbative quantum states (e.g., [45]) is realized concretely by having the quantum excitations on the nodes of the boundary state.

These quantities admit two distinct limits. In the classical limit where quantum effects are disregarded, they converge to the Hamilton function of a truncation of GR. In the continuum limit, we expect them to converge to the transition amplitudes of the full theory. If the present indications are confirmed, there should be a regime in q where the convergence is rapid and therefore the truncation can work as an effective expansion. The expansion parameter is g_q's deviation from flatness.

The truncation introduced by $\mathscr{C}$ (or Δ) should not be confused with the physical quantum discreteness of the geometry. The quantum discreteness of the geometry is the fact that the geometrical *size* of the cells of the complex takes discrete values. It disappears in the semi-classical limit, where the theory is studied at distances large with respect to the Planck scale, while it persists in the continuum limit, with an arbitrary large two-complex. In other words, no refinement of the cellular complex can make the size of the cells go smoothly to zero, because geometry is physically discrete at the Planck scale. This is the most characteristic aspect of quantum gravity.

Finally, not much is known about the effect of the radiative corrections on this structure (for partial results, see [18, 46–48]). These are finite in the deformed version of (3.1) [6, 7] but this does not make them irrelevant. The main open problem in quantum gravity, I think, is to study their effect on the convergence of the continuous limit.

References

1. C. Rovelli, arXiv:1004.1780
2. C. Rovelli, arXiv:1102.3660
3. J. Engle, E. Livine, R. Pereira, C. Rovelli, Nucl. Phys. B **799**, 136 (2008). arXiv:0711.0146
4. L. Freidel, K. Krasnov, Class. Quantum Gravity **25**, 125018 (2008). arXiv:0708.1595
5. W. Kaminski, M. Kisielowski, J. Lewandowski, Class. Quantum Gravity **27**, 095006 (2010). arXiv:0909.0939
6. M. Han, J. Math. Phys. **52**, 072501 (2011). arXiv:1012.4216
7. W.J. Fairbairn, C. Meusburger, J. Math. Phys. **53**, 022501 (2012). arXiv:1012.4784
8. M. Han, Phys. Rev. D **84**, 064010 (2011). arXiv:1105.2212
9. E. Bianchi, M. Han, E. Magliaro, C. Perini, W. Wieland, arXiv:1012.4719
10. L. Freidel, S. Speziale, Phys. Rev. D **82**, 084040 (2010). arXiv:1001.2748
11. E. Bianchi, P. Donà, S. Speziale, Phys. Rev. D **83**, 044035 (2011). arXiv:1009.3402
12. F. Conrady, L. Freidel, Class. Quantum Gravity **25**, 245010 (2008). arXiv:0806.4640
13. J.W. Barrett, R.J. Dowdall, W.J. Fairbairn, F. Hellmann, R. Pereira, Class. Quantum Gravity **27**, 165009 (2010). arXiv:0907.2440
14. J.W. Barrett, R.J. Dowdall, W.J. Fairbairn, H. Gomes, F. Hellmann, J. Math. Phys. **50**, 112504 (2009). arXiv:0902.1170
15. E. Magliaro, C. Perini, arXiv:1105.0216
16. C.W. Misner, Rev. Mod. Phys. **29**, 497 (1957)
17. C. Rovelli, arXiv:1107.2310
18. J. Ben Geloun, R. Gurau, V. Rivasseau, Europhys. Lett. **92**, 60008 (2010). arXiv:1008.0354
19. C. Rovelli, M. Smerlak, Class. Quantum Gravity **29**, 055004 (2012). arXiv:1010.5437
20. W. Hamilton, British Association Report, 1834, 513
21. C. Rovelli, *Quantum Gravity* (Cambridge University Press, Cambridge, 2004)
22. C. Rovelli, Found. Phys. **41**, 1475 (2011). arXiv:0903.3832
23. C. Rovelli, Phys. Rev. D **65**, 124013 (2002). arXiv:gr-qc/0110035
24. S. Deser, R. Arnowitt, C. Misner, Nuovo Cimento **19**, 668 (1961)
25. B. Dittrich, Adv. Sci. Lett. **2**, 151 (2009). arXiv:0810.3594
26. B. Bahr, B. Dittrich, S. Steinhaus, Phys. Rev. D **83**, 105026 (2011). arXiv:1101.4775
27. B. Bahr, B. Dittrich, Phys. Rev. D **80**, 124030 (2009). arXiv:0907.4323
28. S.B. Giddings, D. Marolf, J.B. Hartle, Phys. Rev. D **74**, 064018 (2006). hep-th/0512200
29. N. Tsamis, R. Woodard, Ann. Phys. **215**, 96 (1992)
30. J.B. Hartle, in *Gravitation and Quantizations: Proceedings of the 1992 Les Houches Summer School*, ed. by B. Julia, J. Zinn-Justin (Elsevier, Amsterdam, 1995), pp. 285–480. gr-qc/9304006
31. E. Witten, hep-th/0106109
32. C.E. Dolby, gr-qc/0406034
33. T. Banks, W. Fischler, S. Paban, J. High Energy Phys. **0212**, 062 (2002). hep-th/0210160
34. C. Rovelli, Class. Quantum Gravity **8**, 297 (1991)
35. H.W. Hamber, Phys. Rev. D **50**, 3932 (1994). hep-th/9311024

36. A. Ashtekar, R. Tate, C. Uggla, Int. J. Mod. Phys. D **2**, 15 (1993). gr-qc/9302027
37. D.N. Page, W.K. Wootters, Phys. Rev. D **27**, 2885 (1983)
38. A. Strominger, J. High Energy Phys. **0110**, 034 (2001). hep-th/0106113
39. D. Marolf, Class. Quantum Gravity **12**, 2469 (1995). gr-qc/9412016
40. R.M. Wald, Phys. Rev. D **48**, 2377 (1993). gr-qc/9305024
41. K. Kuchař, Int. J. Mod. Phys. Proc. Suppl. D **20**, 3 (2011)
42. E. Anderson, arXiv:1009.2157
43. E. Bianchi, C. Rovelli, F. Vidotto, Phys. Rev. D **82**, 084035 (2010). arXiv:1003.3483
44. E. Bianchi, L. Modesto, C. Rovelli, S. Speziale, Class. Quantum Gravity **23**, 6989 (2006). gr-qc/0604044
45. J. Iwasaki, C. Rovelli, Int. J. Mod. Phys. D **1**, 533 (1993)
46. C. Perini, C. Rovelli, S. Speziale, Phys. Lett. B **682**, 78 (2009). arXiv:0810.1714
47. T. Krajewski, J. Magnen, V. Rivasseau, A. Tanasa, P. Vitale, Phys. Rev. D **82**, 124069 (2010). arXiv:1007.3150
48. V. Bonzom, M. Smerlak, Ann. Henri Poincaré **13**, 185 (2012). arXiv:1103.3961

Chapter 4
Spinor Gravity and Diffeomorphism Invariance on the Lattice

C. Wetterich

Abstract The key ingredient for lattice regularized quantum gravity is diffeomorphism symmetry. We formulate a lattice functional integral for quantum gravity in terms of fermions. This allows for a diffeomorphism invariant functional measure and avoids problems of boundedness of the action. We discuss the concept of lattice diffeomorphism invariance. This is realized if the action does not depend on the positioning of abstract lattice points on a continuous manifold. Our formulation of lattice spinor gravity also realizes local Lorentz symmetry. Furthermore, the Lorentz transformations are generalized such that the functional integral describes simultaneously euclidean and Minkowski signature. The difference between space and time arises as a dynamical effect due to the expectation value of a collective metric field. The quantum effective action for the metric is diffeomorphism invariant. Realistic gravity can be obtained if this effective action admits a derivative expansion for long wavelengths.

4.1 Introduction

The conceptual unification of general relativity and quantum theory is one of the central goals of theoretical physics. The aim of the present work is an approach to a consistent and mathematically well defined formulation of quantum gravity within the general framework of quantum field theory. It is based on the formulation of a functional integral which is regularized on a lattice of space-time points. As long as the number of space-time points remains finite we deal with a finite number of degrees of freedom. Then all operations for the functional integral are well defined. The continuum limit of infinite volume or vanishing lattice distance is taken at the end.

The central ingredient for general relativity is diffeomorphism symmetry which accounts for the invariance of the formulation under general coordinate transformations. In the presence of fermions this has to be supplemented by local Lorentz sym-

C. Wetterich (✉)
Institut für Theoretische Physik, Universität Heidelberg, Philosophenweg 16, 69120 Heidelberg, Germany
e-mail: c.wetterich@thphys.uni-heidelberg.de

G. Calcagni et al. (eds.), *Quantum Gravity and Quantum Cosmology*,
Lecture Notes in Physics 863, DOI 10.1007/978-3-642-33036-0_4,
© Springer-Verlag Berlin Heidelberg 2013

metry, with Lorentz transformations acting on spinors and the vierbein. In our view the implementation of these symmetries is crucial, much more basic than the choice of particular degrees of freedom as the metric, the vierbein, spinors, or geometrical objects. In a diffeomorphism invariant quantum field theory a metric will be induced as a collective degree of freedom even if it is not present as a fundamental degree of freedom in the formulation of the functional integral. If the quantum effective action for the metric admits a derivative expansion for long wavelengths only a few terms matter for the long distance physics. The possible terms are strongly restricted by diffeomorphism symmetry. The first two are a cosmological constant term with zero derivatives and an Einstein-Hilbert term involving the curvature scalar R with two derivatives. For a small enough value of the cosmological constant this yields a realistic theory for gravity. This holds even if the strict derivative expansion breaks down in higher orders, for example by terms involving $R^2 \ln R$.

We require the following six criteria for a quantum field theory for quantum gravity:

(1) For a finite number of lattice points the functional integral is well defined.
(2) The lattice action and functional measure are invariant under lattice diffeomorphisms.
(3) A continuum limit exists where lattice diffeomorphism invariance turns into the continuous diffeomorphism symmetry.
(4) The lattice theory is invariant under local Lorentz transformations. This symmetry is then preserved in the continuum limit.
(5) The continuum limit describes some massless (or very light) degrees of freedom. It comprises gravitational interactions.
(6) A derivative expansion gives a reasonable approximation for the gravitational degrees of freedom at long wavelengths.

A model obeying these criteria realizes a consistent quantum gravity. It describes realistic gravitational interactions if the cosmological constant is small enough, and if there are no additional massless gravitational degrees of freedom beyond the metric or vierbein which induce observable effects like torsion.

4.2 Spinors as Fundamental Degrees of Freedom

For the formulation of a functional integral we first have to decide for the "fundamental degrees of freedom" which appear in the action and the functional measure. There are various proposals for lattice formulations, based on spinors [1–3], nonlinear σ-models [2], vierbein and spin connection, the length of edges of simplices in Regge gravity [4] or other objects of lattice geometry [5, 6].

The metric is a central object for general relativity. It would therefore seem natural to use directly the metric field variables for the formulation of the functional integral. Such an approach encounters, however, two major difficulties. The first is the difficulty to find a functional measure on the lattice that respects diffeomorphism invariance. The second is the problem of boundedness of the action. Depending on the metric configuration the curvature scalar can be positive or negative. Also the determinant of the metric has no definite sign unless one imposes a non-linear signature-constraint. These problems carry over to a formulation with the vierbein as fundamental degree of freedom. Now it is the determinant e of the vierbein that is not positive definite. This determinant appears as a multiplicative factor in the Lagrange density $\mathscr{L}$, as required by the transformation of $\mathscr{L}$ as a scalar density under diffeomorphisms. With action $S = \int_x \mathscr{L} = \int_x eL$ and L a diffeomorphism invariant one can construct configurations of the vierbein with arbitrarily large positive or negative values of S.

In Regge gravity the metric as basic degree of freedom is replaced by the lengths of edges of simplices from which a metric can be computed. One can formulate an invariant lattice functional measure. Despite some remarkable achievements it remains unclear at the present stage if the problem of "boundedness of the action" can be overcome such that a satisfactory continuum limit is reached.

Spinors as basic degrees of freedom avoid both problems. They transform as scalars under general coordinate transformations and the formulation of an invariant functional measure therefore poses no problem. Since spinors are Grassmann variables the functional integral becomes a Grassmann functional integral. Grassmann integrals are always well defined for a finite number of Grassmann variables such that the problem of boundedness of the action is completely absent. Scalar fields with a non-linear constraint (non-linear σ-models) are also a possibility. The functional measure is trivially diffeomorphism invariant, and the action can become bounded since the constraint forbids arbitrarily large values of the fields.

In this work we concentrate on spinors as basic degrees of freedom. We will discuss a lattice formulation which is close in spirit to spinor gravity as formulated in Refs. [7–9]. The metric and the vierbein arise then as collective fields involving an even number of spinors. First observations that a diffeomorphism invariant action for fermions can be formulated without the use of a metric, and the conjecture that the metric is a composite field, have been made long ago [10–13]. (The actual implementation in these early approaches is not always fully consistent—for example the inverse of products of Grassmann variables does not exist.) We build on these ideas, but we propose a different action that implements local Lorentz symmetry.

For continuous spacetime the action of spinor gravity has to be diffeomorphism invariant. A loop expansion for a simple model [7, 8] has indeed shown that the quantum effective action for the metric can indeed contain Einstein's curvature scalar. Early formulations of spinor gravity as in [7, 8] exhibit, however, only global and not local Lorentz symmetry.

A simple geometrical quantity that can be constructed from two spinors is the vierbein bilinear [10, 11],

$$\tilde{E}_\mu^m = i\bar{\psi}\gamma^m \partial_\mu \psi, \tag{4.1}$$

with γ^m the Dirac matrices. The spinors ψ are scalars with respect to diffeomorphisms, such that $\tilde{E}_\mu^m$ is a vector. With respect to global Lorentz transformations the vierbein bilinear also transforms as a vector. We may consider $\tilde{E}_\mu^m$ as a matrix with first index μ and second index m. Then the determinant $\tilde{E} = \det(\tilde{E}^m_\mu)$ transforms as a scalar density under general coordinate transformations and is invariant under global Lorentz transformations. An action

$$S = \sim \int d^4x \tilde{E} \tag{4.2}$$

is diffeomorphism symmetric and invariant under global Lorentz transformations. One may try to identify the expectation value of $\tilde{E}^m_\mu$ with the vierbein. Doing so, one encounters the problem that the action (4.2) is invariant under global but not under local Lorentz transformations [7, 8]. The lack of local Lorentz symmetry leads to additional massless degrees of freedom that are contained in the vierbein, beyond the ones corresponding to the metric. (In case of local Lorentz symmetry these would be gauge degrees of freedom.) The resulting torsion terms in the effective action have been discussed in detail in Ref. [8]. It was found that one of the torsion invariants— the only one generated at one loop order—is actually compatible with all present observations, while a second possible invariant is excluded by the tests of general relativity. In the present work we avoid this difficulty by formulating a model with local Lorentz symmetry, with analogies to the higher dimensional model in Ref. [9].

The formulation of a basic theory only involving fermions can be viewed as a possible path towards a unified theory of all interactions. In this case all bosons arise as collective fields—in distinction to supersymmetry where fermions and bosons are on equal footing. Gravitons, photons, gluons, W- and Z-bosons and the Higgs scalar are all composite. Only at length scales large compared to the Planck length they look fundamental, similar to the bosonic hydrogen atom at length scales large compared to Bohr radius. Realizing massless bosonic bound states in a purely fermionic theory is quite common in other physical systems. For example, a Nambu-Jona-Lasinio model [14] with spontaneous symmetry breaking of the global chiral symmetry leads to massless pions if the chiral symmetry is exact. Many "massless" bosonic excitations are known in condensed matter physics, as spin waves for an antiferromagnet in case of spontaneous breaking of the continuous spin-rotation symmetry. Usually, there is some physical reason for the presence of massless bosons, as the Goldstone theorem for spontaneously broken continuous symmetries. In our case a massless graviton is related to the spontaneous breaking of diffeomorphism symmetry by the selection of a particular metric for the "ground state" or cosmological solution.

In the present work we concentrate on gravitational degrees of freedom for the composite bosons. Gauge bosons or scalars as the Higgs-doublet can arise either directly in a four-dimensional formulation as suitable fermion bilinears, or by dimensional reduction of a higher-dimensional theory of gravity. In our present formulation we find indeed additional symmetries. For two flavors of fermions the continuum limit exhibits an $SU(2)_L \times SU(2)_R$ gauge symmetry.

Most important, the local Lorentz-transformations of the group $SO(1, 3)$ are extended to complex transformation parameters realizing the group $SO(4, \mathbb{C})$, which also includes the euclidean rotation symmetry $SO(4)$. No signature for space and time are singled out in the basic formulation—both appear on completely equal footing. The difference in signature between space and time arises as a dynamical effect through expectation values of composite fields [9].

In the first part of this work we will discuss the continuum action for our model of spinor gravity. It exhibits diffeomorphism symmetry and local Lorentz symmetry. We proceed to the lattice formulation in the second part. The third part investigates the issue of lattice diffeomorphism invariance. The fourth part addresses the emergence of geometry from lattice spinor gravity. We describe the diffeomorphism invariant quantum effective action for the collective metric field. Our treatment will be largely based on Refs. [1, 2].

4.3 Action and Functional Integral

Let us explore a setting with 16 Grassmann variables ψ_γ^a at every spacetime point x, $\gamma = 1 \dots 8$, $a = 1, 2$. Here γ denotes the eight real variables of a complex four-component Dirac spinor and a is a flavor index for two species of Dirac fermions. The coordinates x parametrize four real numbers, i.e. $x^\mu = (x^0, x^1, x^2, x^3)$. In the lattice formulation these numbers are discrete. We will later associate $t = x^0$ with a time coordinate, and x^k, $k = 1, 2, 3$, with space coordinates. There is, however, a priori no difference between time and space coordinates. The partition function Z is defined as

$$Z = \int \mathscr{D}\psi\, g_f \exp(-S) g_{in},$$

$$\int \mathscr{D}\psi = \prod_x \prod_{a=1}^2 \left\{ \int d\psi_1^a(x) \dots \int d\psi_8^a(x) \right\}. \tag{4.3}$$

For discrete spacetime points on a lattice the Grassmann functional integral (4.3) is well defined mathematically. We assume that the time coordinate $x^0 = t$ obeys $t_{in} \le t \le t_f$. The boundary term g_{in} is a Grassmann element constructed from $\psi_\gamma^a(t_{in}, \mathbf{x})$, while g_f involves terms with powers of $\psi_\gamma^a(t_f, \mathbf{x})$, were $\mathbf{x} = (x^1, x^2, x^3)$. If S as well as g_{in} and g_f are elements of a real Grassmann algebra the partition function is real. We may restrict the range of the space coordinates or use a torus T^3 instead of $\mathbb{R}^3$. For a discrete spacetime lattice the number of Grassmann variables is then finite.

Observables $\mathscr{A}$ will be represented as Grassmann elements constructed from $\psi_\gamma^a(x)$. We will consider only bosonic observables that involve an even number of Grassmann variables. Their expectation value is defined as

$$\langle \mathscr{A} \rangle = Z^{-1} \int \mathscr{D}\psi\, g_f \mathscr{A} \exp(-S) g_{in}. \tag{4.4}$$

"Real observables" are elements of a real Grassmann algebra, i.e. they are sums of powers of $\psi_\gamma^a(x)$ with real coefficients. For real S, g_{in} and g_f all real observables have real expectation values. We will take the continuum limit of vanishing lattice distance at the end. Physical observables are those that have a finite continuum limit.

For the formulation of the action we will first investigate the continuum limit with $x \in \mathbb{R}^4$. For this purpose we will work with complex Grassmann variables $\varphi_\alpha^a, \alpha = 1 \ldots 4$,

$$\varphi_\alpha^a(x) = \psi_\alpha^a(x) + i\psi_{\alpha+4}^a(x), \tag{4.5}$$

with α the "Dirac index" and a the "flavor index". We propose an action which involves twelve Grassmann variables and realizes diffeomorphism symmetry and local $SO(4, \mathbb{C})$ symmetry

$$S = \alpha \int d^4x \varphi_{\alpha_1}^{a_1} \ldots \varphi_{\alpha_8}^{a_8} \varepsilon^{\mu_1\mu_2\mu_3\mu_4}$$

$$\times J_{\alpha_1\ldots\alpha_8\beta_1\ldots\beta_4}^{a_1\ldots a_8 b_1\ldots b_4} \partial_{\mu_1}\varphi_{\beta_1}^{b_1} \partial_{\mu_2}\varphi_{\beta_2}^{b_2} \partial_{\mu_3}\varphi_{\beta_3}^{b_3} \partial_{\mu_4}\varphi_{\beta_4}^{b_4} + c.c., \tag{4.6}$$

where we sum over repeated indices. The choice of J is dictated by the requirement of Lorentz symmetry and will be discussed in the following chapters. The complex conjugation $c.c.$ replaces $\alpha \to \alpha^*, J \to J^*$ and $\varphi_\alpha(x) \to \varphi_\alpha^*(x) = \psi_\alpha(x) - i\psi_{\alpha+4}(x)$, such that $S^* = S$. In terms of the Grassmann variables $\psi_\gamma^a(x)$ the action S as well as $\exp(-S)$ are elements of a real Grassmann algebra.

Invariance of the action under general coordinate transformations follows from the use of the totally antisymmetric product of four derivatives $\partial_\mu = \partial/\partial x^\mu$. Indeed, with respect to diffeomorphisms $\varphi(x)$ transforms as a scalar, and $\partial_\mu\varphi(x)$ as a vector. The particular contraction with the totally antisymmetric tensor $\varepsilon^{\mu_1\mu_2\mu_3\mu_4}$, $\varepsilon^{0123} = 1$, allows for a realization of diffeomorphism symmetry without the use of a metric.

4.4 Generalized Lorentz Transformations

We want to construct an action that is invariant under local generalized Lorentz transformations. Thus the tensor $J_{\alpha_1\ldots\alpha_8\beta_1\ldots\beta_4}^{a_1\ldots a_8 b_1\ldots b_4}$ must be invariant under global $SO(4, \mathbb{C})$ transformations. We will often use double indices $\varepsilon = (\alpha, a)$ or $\eta = (\beta, b)$, $\varepsilon, \eta = 1 \ldots 8$. The tensor $J_{\varepsilon_1\ldots\varepsilon_8\eta_1\ldots\eta_4}$ is totally antisymmetric in the first eight indices $\varepsilon_1 \ldots \varepsilon_8$, and totally symmetric in the last four indices $\eta_1 \ldots \eta_4$. This follows from the anticommuting properties of the Grassmann variables $\varphi_\varepsilon\varphi_\eta = -\varphi_\eta\varphi_\varepsilon$. We will see that for any J invariant under global $SO(4, \mathbb{C})$ transformations the action (4.6) is also invariant under local $SO(4, \mathbb{C})$ transformations.

Local $SO(4, \mathbb{C})$ transformations act infinitesimally as

$$\delta\varphi_\alpha^a(x) = -\frac{1}{2}\varepsilon_{mn}(x)\big(\Sigma_E^{mn}\big)_{\alpha\beta}\varphi_\beta^a(x), \tag{4.7}$$

with arbitrary complex parameters $\varepsilon_{mn}(x) = -\varepsilon_{nm}(x)$, $m = 0, 1, 2, 3$. The complex 4×4 matrices Σ_E^{mn} are associated to the generators of $SO(4)$ in the (reducible) four-component spinor representation. They can be obtained from the euclidean Dirac matrices

$$\Sigma_E^{mn} = -\frac{1}{4}[\gamma_E^m, \gamma_E^n], \quad \{\gamma_E^m, \gamma_E^n\} = 2\delta^{mn}. \tag{4.8}$$

Subgroups of $SO(4, \mathbb{C})$ with different signatures obtain by appropriate choices of ε_{mn}. Real parameters ε_{mn} correspond to euclidean rotations $SO(4)$. Taking ε_{kl}, $k, l = 1, 2, 3$ real, and $\varepsilon_{0k} = -i\varepsilon_{0k}^{(M)}$ with real $\varepsilon_{0k}^{(M)}$, realizes the Lorentz transformations $SO(1, 3)$. The Lorentz transformations can be written equivalently with six real transformation parameters $\varepsilon_{mn}^{(M)}$, $\varepsilon_{kl}^{(M)} = \varepsilon_{kl}$, using Lorentz-generators Σ_M^{mn} and signature $\eta^{mn} = \mathrm{diag}(-1, 1, 1, 1)$,

$$\delta\varphi = -\frac{1}{2}\varepsilon_{mn}^{(M)}\Sigma_M^{mn}\varphi, \tag{4.9}$$

with

$$\Sigma_M^{mn} = -\frac{1}{4}[\gamma_M^m, \gamma_M^n], \quad \{\gamma_M^m, \gamma_M^n\} = \eta^{mn}. \tag{4.10}$$

The euclidean and Minkowski Dirac matrices are related by $\gamma_M^0 = -i\gamma_E^0$, $\gamma_M^k = \gamma_E^k$. The transformation of a derivative involves an inhomogeneous part

$$\delta\partial_\mu\varphi_\beta = -\frac{1}{2}\varepsilon_{mn}\left(\Sigma^{mn}\partial_\mu\varphi\right)_\beta - \frac{1}{2}\partial_\mu\varepsilon_{mn}\left(\Sigma^{mn}\varphi\right)_\beta, \tag{4.11}$$

with $\Sigma^{mn} = \Sigma_E^{mn}$, $\gamma^m = \gamma_E^m$. The first "homogeneous term" $\sim \partial_\mu\varphi$ transforms as φ_β. Thus an invariant tensor J guarantees an invariant action if the second term in Eq. (4.11) does not contribute to δS. Contributions of the second "inhomogeneous term" to the variation of the action δS involve at least nine spinors at the same position x, i.e. $(\Sigma^{mn}\varphi)_\beta^b(x)\varphi_{\alpha_1}^{a_1}(x)\dots\varphi_{\alpha_8}^{a_8}(x)$. Therefore this inhomogeneous contribution to δS vanishes due to the identity $\varphi_\alpha(x)\varphi_\alpha(x) = 0$ (no sum here)—at most eight different complex spinors can be placed on a given position x. The invariance of S under global $SO(4, \mathbb{C})$ transformations entails the invariance under local $SO(4, \mathbb{C})$ transformations. We have constructed in Ref. [9] a model for sixteen dimensional spinor gravity with local $SO(16, \mathbb{C})$ symmetry. The present four-dimensional model shows analogies to this.

It is important that all invariants appearing in the action (4.6) involve either only factors of $\varphi_\alpha = \psi_\alpha + i\psi_{\alpha+4}$ or only factors of $\varphi_\alpha^* = \psi_\alpha - i\psi_{\alpha+4}$. It is possible to construct $SO(1, 3)$ invariants which involve both φ and φ^*. Those will not be invariant under $SO(4, \mathbb{C})$, however. We can also construct invariants involving φ and φ^* which are invariant under euclidean $SO(4)$ rotations. They will not be invariant under $SO(1, 3)$. The only terms which are invariant under both $SO(4)$ *and* $SO(1, 3)$, and more generally $SO(4, \mathbb{C})$, are those constructed from φ alone or φ^* alone, or products of such invariants. (Invariants involving both φ and φ^* can be constructed as products of invariants involving only φ with invariants involving only φ^*.) We conclude that for a suitable $SO(4, \mathbb{C})$-invariant tensor J the action has the symmetries required for a realistic theory of gravity for fermions, namely diffeomorphism symmetry and local $SO(1, 3)$ Lorentz symmetry. No signature and no metric are introduced at this stage, such that there is no difference between time and space [9].

4.5 Lorentz Invariant Spinor Bilinears

We next want to construct the $SO(4, \mathbb{C})$ invariant tensor J in Eq. (4.6). We do this in steps by discussing first simpler invariants, out of which we will compose J. Our model with two flavors allows us to construct two symmetric invariants with two Dirac indices

$$S^{\pm}_{\eta_1 \eta_2} = \left(S^{\pm}\right)^{b_1 b_2}_{\beta_1 \beta_2} = \mp (C_{\pm})_{\beta_1 \beta_2} (\tau_2)^{b_1 b_2}, \tag{4.12}$$

where τ_k denotes the Pauli matrices. The invariant tensors $C_{\pm}$ are antisymmetric [15]

$$(C_{\pm})_{\beta_2 \beta_1} = -(C_{\pm})_{\beta_1 \beta_2}, \tag{4.13}$$

such that $S^{\pm}$ is symmetric under the exchange $(\beta_1, b_1) \leftrightarrow (\beta_2, b_2)$, or, in terms of the double index $\eta = (\beta, b)$,

$$S^{\pm}_{\eta_2 \eta_1} = S^{\pm}_{\eta_1 \eta_2}. \tag{4.14}$$

The $SO(4, \mathbb{C})$-invariants $C_{\pm}$ can best be understood in terms of Weyl spinors. The matrix

$$\bar{\gamma} = -\gamma^0 \gamma^1 \gamma^2 \gamma^3 \tag{4.15}$$

commutes with Σ^{mn} such that the two doublets

$$\varphi_+ = \frac{1}{2}(1 + \bar{\gamma})\varphi, \qquad \varphi_- = \frac{1}{2}(1 - \bar{\gamma})\varphi \tag{4.16}$$

correspond to inequivalent two component complex spinor representations (Weyl spinors). We employ here a representation of the Dirac matrices γ^m where $\bar{\gamma} = \mathrm{diag}(1, 1, -1, -1)$, namely

$$\gamma^0 = \tau_1 \otimes 1, \qquad \gamma^k = \tau_2 \otimes \tau_k. \tag{4.17}$$

(The general structure is independent of this choice. Our representation corresponds to the Weyl basis of Ref. [16] where details of conventions can be found.) In this representation one has

$$
\begin{aligned}
C_+ &= \frac{1}{2}(C_1 + C_2) = \frac{1}{2}C_1(1 + \bar{\gamma}) = \begin{pmatrix} \tau_2 & 0 \\ 0 & 0 \end{pmatrix}, \\
C_- &= \frac{1}{2}(C_1 - C_2) = \frac{1}{2}C_1(1 - \bar{\gamma}) = \begin{pmatrix} 0 & 0 \\ 0 & -\tau_2 \end{pmatrix}
\end{aligned}
\tag{4.18}
$$

such that $\psi^T_{\pm} C_1 = \psi^T_{\pm} C_{\pm} = \psi^T C_{\pm}$.

It is straightforward to construct invariants only involving the two Weyl spinors φ_+^1 and φ_+^2 by combining C_+ with an appropriate flavor matrix. For this purpose we can restrict the index η to the values $1 \ldots 4$. The action of $SO(4, \mathbb{C})$ on φ_+ is given by the subgroup of complexified $SU(2, \mathbb{C})_+$ transformations. In our basis the generators of $SU(2, \mathbb{C})_+$ read

$$\Sigma^{0k} = -\frac{i}{2}\tau_k, \qquad \Sigma^{kl} = \varepsilon^{klm}\Sigma^{0m}, \tag{4.19}$$

such that Σ^{kl} is linearly dependent on Σ^{0k}. (For $SU(2, \mathbb{C})_-$ the generators Σ^{kl} are identical, while $\Sigma^{0k} = \frac{i}{2}\tau_k$. The subgroup of special unitary transformations $SU(2)$ obtains for real transformation parameters, while we consider here arbitrary complex transformation parameters.)

We observe that we can also consider a group $SU(2, \mathbb{C})_L$ acting on the flavor indices of φ_+. With respect to $SU(2, \mathbb{C})_+ \times SU(2, \mathbb{C})_L$ the four component spinor $\varphi_{+,\eta}$ ($\eta = 1 \ldots 4$) transforms as the $(2, 2)$ representation. Since the matrix $(\tau_2)^{ab}$ in Eq. (4.12) is invariant under $SU(2, \mathbb{C})_L$, the invariant S^+ is invariant under the group

$$SO(4, \mathbb{C})_+ \equiv SU(2, \mathbb{C})_+ \times SU(2, \mathbb{C})_L. \tag{4.20}$$

(Here $SO(4, \mathbb{C})_+$ should be distinguished from the generalized Lorentz transformation since it acts both in the space of Dirac and flavor indices.) With respect to $SO(4, \mathbb{C})_+$ the two-flavored spinor φ_+ transforms as a four component vector. The classification of tensors, invariants and symmetries can be directly inferred from the analysis of four-dimensional vectors. Invariants only involving φ_- can be constructed in a similar way with $SU(2, \mathbb{C})_R$ acting on the flavor indices of φ_- and $SO(4, \mathbb{C})_- = SU(2, \mathbb{C})_- \times SU(2, \mathbb{C})_R$.

4.6 Action with Local Lorentz Symmetry

A totally symmetric invariant four index object can be constructed as

$$L_{\eta_1\eta_2\eta_3\eta_4} = \frac{1}{6}\left(S_{\eta_1\eta_2}^+ S_{\eta_3\eta_4}^- + S_{\eta_1\eta_3}^+ S_{\eta_2\eta_4}^- + S_{\eta_1\eta_4}^+ S_{\eta_2\eta_3}^- \right.$$
$$\left. + S_{\eta_3\eta_4}^+ S_{\eta_1\eta_2}^- + S_{\eta_2\eta_4}^+ S_{\eta_1\eta_3}^- + S_{\eta_2\eta_3}^+ S_{\eta_1\eta_4}^- \right). \tag{4.21}$$

The global invariant with four derivatives

$$D = \varepsilon^{\mu_1\mu_2\mu_3\mu_4}\partial_{\mu_1}\varphi_{\eta_1}\partial_{\mu_2}\varphi_{\eta_2}\partial_{\mu_3}\varphi_{\eta_3}\partial_{\mu_4}\varphi_{\eta_4}L_{\eta_1\eta_2\eta_3\eta_4} \tag{4.22}$$

involves two Weyl spinors φ_+ and two Weyl spinors φ_-. Furthermore, an invariant with eight factors of φ involves the totally antisymmetric tensor for the eight values of the double-index ε

$$A^{(8)} = \frac{1}{8!}\varepsilon_{\varepsilon_1\varepsilon_2\ldots\varepsilon_8}\varphi_{\varepsilon_1}\cdots\varphi_{\varepsilon_8}$$

$$= \frac{1}{(24)^2} \varepsilon_{\alpha_1 \alpha_2 \alpha_3 \alpha_4} \varphi^1_{\alpha_1} \cdots \varphi^1_{\alpha_4} \varepsilon_{\beta_1 \beta_2 \beta_3 \beta_4} \varphi^2_{\beta_1} \cdots \varphi^2_{\beta_4}$$

$$= \varphi^1_1 \varphi^1_2 \varphi^1_3 \varphi^1_4 \varphi^2_1 \varphi^2_2 \varphi^2_3 \varphi^2_4. \tag{4.23}$$

An action with local $SO(4, \mathbb{C})$ symmetry takes the form

$$S = \alpha \int d^4x A^{(8)} D + c.c. \tag{4.24}$$

Indeed, the inhomogeneous contribution (4.11) to the variation of $D(x)$ contains factors $(\Sigma^{mn} \varphi^b)_\beta(x)$. As discussed before, it vanishes when multiplied with $A^{(8)}(x)$, since the Pauli principle $(\varphi^a_\alpha(x))^2 = 0$ admits at most eight factors φ for a given x. In consequence, the inhomogeneous variation of the action (4.24) vanishes and S is invariant under *local $SO(4, \mathbb{C})$* transformations. In contrast to $\int d^4x D(x)$ the action S in Eq. (4.24) is not a total derivative. Besides local $SO(4, \mathbb{C})$, it is also invariant under local $SO(4, \mathbb{C})_F$ gauge transformations, with $SO(4, \mathbb{C})_F = SU(2, \mathbb{C})_L \times SU(2, \mathbb{C})_R$.

The derivative-invariant D can be written in the form

$$D = \varepsilon^{\mu_1 \mu_2 \mu_3 \mu_4} D^+_{\mu_1 \mu_2} D^-_{\mu_3 \mu_4}, \tag{4.25}$$

with

$$D^\pm_{\mu_1 \mu_2} = \partial_{\mu_1} \varphi_{\eta_1} S^\pm_{\eta_1 \eta_2} \partial_{\mu_2} \varphi_{\eta_2}. \tag{4.26}$$

Inserting Eq. (4.25) into Eq. (4.24) we recognize the contraction of four derivatives with the totally antisymmetric ε-tensor which explains the invariance of S under diffeomorphisms. Equation (4.25) shows that D is invariant under the exchange $\varphi_{+,\eta} \leftrightarrow \varphi_{-,\eta}$. The transformation $\varphi \to \gamma^0 \varphi$ maps $S^+_{\eta_1 \eta_2} \leftrightarrow S^-_{\eta_1 \eta_2}$ and therefore $D^+_{\mu_1 \mu_2} \leftrightarrow D^-_{\mu_1 \mu_2}$, such that again D is invariant. (For our choice $\gamma^0 = \tau_1 \otimes 1$ the transformation $\varphi \to \gamma^0 \varphi$ actually corresponds to $\varphi_{+,\eta} \leftrightarrow \varphi_{-,\eta}$.) We can also decompose

$$A^{(8)} = A^+ A^-, \tag{4.27}$$

with

$$A^+ = \varphi^1_{+1} \varphi^1_{+2} \varphi^2_{+1} \varphi^2_{+2}, \tag{4.28}$$

and similarly for A^-. The combinations

$$F^\pm_{\mu_1 \mu_2} = A^\pm D^\pm_{\mu_1 \mu_2} \tag{4.29}$$

are invariant under local $SO(4, \mathbb{C}) \times SO(4, \mathbb{C})_F$ transformations. They involve six Weyl spinors φ_+ or six Weyl spinors φ_-, respectively. The action involves products of F^+ and F^-,

$$S = \alpha \int d^4x \varepsilon^{\mu_1 \mu_2 \mu_3 \mu_4} F^+_{\mu_1 \mu_2} F^-_{\mu_3 \mu_4} + c.c. \tag{4.30}$$

We define the Minkowski action by

$$S = -i S_M, \qquad e^{-S} = e^{i S_M}, \tag{4.31}$$

which yields the usual "phase factor" for the functional integral written in terms of S_M. We can define the operation of a transposition as a total reordering of all Grassmann variables. The result of transposition for a product of Grassmann variables depends only on the number of factors N_φ. For $N_\varphi = 2, 3 \bmod 4$ the transposition results in a minus sign, while for $N_\varphi = 4, 5 \bmod 4$ the product is invariant. In consequence, one finds that S_M is symmetric. With respect to the complex conjugation c.c. used in Eq. (4.5) the Minkowski action is antihermitean. This complex conjugation, which is defined for the Grassmann variables ψ_γ by the involution $\psi_{\alpha+4}^a \to -\psi_{\alpha+4}^a$ for $\alpha = 1 \ldots 4$, is, however, not unique. We may define a different conjugation by an involution where the Grassmann variables changing sign are $\psi_5^1, \psi_6^1, \psi_7^1, \psi_8^1, \psi_3^2, \psi_4^2, \psi_5^2$ and ψ_6^2. In this case we use the same definition as before for φ_α^1 and φ_1^2, φ_2^2, but we replace φ_3^2 and φ_4^2 by new complex Grassmann variables

$$\xi_3^2 = \psi_7^2 - i\psi_3^2, \qquad \xi_4^2 = \psi_8^2 - i\psi_4^2,$$
$$(\xi_3^2)^* = \psi_7^2 + i\psi_3^2, \qquad (\xi_4^2)^* = \psi_8^2 + i\psi_4^2. \tag{4.32}$$

The new complex conjugation can be interpreted as a multiplication of c.c. in Eq. (4.5) with the transformation $\varphi_-^2 \to -\varphi_-^2$. Expressing the euclidean action in terms of $\varphi_\pm^1$, φ_+^2 and ξ_-^2 it changes sign under the new complex conjugation. With respect to this conjugation the Minkowski action is real and symmetric and therefore hermitean. We can use the first complex conjugation in order to establish that we work with a real Grassmann algebra, and the second one to define hermiticity of S_M which is related to a unitary time evolution.

4.7 Gauge and Discrete Symmetries

Besides the generalized Lorentz transformations $SO(4, \mathbb{C})$ the action (4.24), (4.30) is also invariant under continuous gauge transformations. By the same argument as for local $SO(4, \mathbb{C})$ symmetry, any global continuous symmetry of the action is also a local symmetry due to the Pauli principle. We have already encountered the symmetry $SU(2, \mathbb{C})_L$ which transforms

$$\delta\varphi_{+\alpha}^a(x) = \frac{i}{2}\tilde\alpha_{+k}(x)(\tau_k)^{ab}\varphi_{+\alpha}^b(x), \tag{4.33}$$

with three complex parameters $\tilde\alpha_{+k}$, and similar for $SU(2, \mathbb{C})_R$ acting on φ_-. For real $\tilde\alpha_{+k}$ these are standard gauge transformations with compact gauge group $SU(2)$. Altogether, we have four $SU(2, \mathbb{C})$ factors. With respect to $G = SU(2, \mathbb{C})_+ \times SU(2, \mathbb{C})_- \times SU(2, \mathbb{C})_L \times SU(2, \mathbb{C})_R$ the Weyl spinors φ_+ and φ_- transform as $(2, 1, 2, 1)$ and $(1, 2, 1, 2)$, respectively, and the action is invariant.

Discrete symmetries are also a useful tool to characterize the properties of the model. Simple symmetries of the action are Z_{12} phase-transformations or multiplications with $\bar{\gamma}$ or γ^0, e.g.

$$\varphi \to \exp(2\pi i n/12)\varphi, \qquad \varphi \to \bar{\gamma}\varphi, \qquad \varphi \to \gamma^0 \varphi. \qquad (4.34)$$

The reflection of the three space coordinates

$$\psi_\gamma^a(x) \to \psi_\gamma^a(Px), \qquad P(x^0, x^1, x^2, x^3) = (x^0, -x^1, -x^2, -x^3), \qquad (4.35)$$

changes the sign of the action. If this transformation is accompanied by any other discrete transformation which inverts the sign of S the combined transformation amounts to a type of parity symmetry. As an example, we may consider

$$\varphi^1(x) \to \gamma^0 \varphi^1(x), \qquad \varphi^2(x) \to \gamma^0 \bar{\gamma}\varphi^2(x). \qquad (4.36)$$

Time reflection symmetry can be obtained in a similar way by combining $\psi_\gamma^a(x) \to \psi_\gamma^a(-Px)$ with a suitable transformation that changes the sign of S, as for Eq. (4.36). Reflections of an even number of coordinates, including the simultaneous space and time reflections, $\psi_\gamma^a(x) \to \psi_\gamma^a(-x)$, leave the action invariant.

4.8 Discretization

In the second part we formulate a regularized version of the functional integral (4.3). For this purpose we will use a lattice of space-time points. We recall that the action (4.24) is invariant under $SO(4)$ and $SO(1, 3)$ transformations and does not involve any metric. The regularization will therefore be valid simultaneously for a Minkowski and a euclidean theory.

Let us consider a four-dimensional hypercubic lattice with lattice distance Δ. We distinguish between the "even sublattice" of points $y^\mu = \tilde{y}^\mu \Delta$, $\tilde{y}^\mu$ integer, $\Sigma_\mu \tilde{y}^\mu$ even, and the "odd sublattice" $z^\mu = \tilde{z}^\mu \Delta$, $\tilde{z}^\mu$ integer, $\Sigma_\mu \tilde{z}^\mu$ odd. The odd sublattice is considered as the fundamental lattice, and we associate to each position z^μ the 16 ("real") Grassmann variables $\psi_\gamma^a(z)$, or their complex counterpart $\varphi_\alpha^a(z)$. (We use here z instead of x in order to make the difference between lattice coordinates and continuum coordinates more visible.) The functional measure (4.3) is invariant under local $SO(4, \mathbb{C})$ transformations since it can be written as a product of invariants of the type A_+, A_- in Eq. (4.28) and their complex conjugate for every z. It is also invariant under local $SU(2, \mathbb{C})_L \times SU(2, \mathbb{C})_R$ gauge transformations.

We write the action as a sum over local terms or Lagrangians $\mathscr{L}(y)$,

$$S = \tilde{\alpha} \sum_y \mathscr{L}(y) + c.c. \qquad (4.37)$$

Here y^μ denotes a position on the even sublattice or "dual lattice". It has eight nearest neighbors on the fundamental lattice, with distance Δ from y. To each point

y we associate a "cell" of those eight points $\tilde{x}_j$ whose $\tilde{z}$-coordinates are given by

$$\tilde{z}^\mu = \tilde{y}^\mu \pm (w_\nu)^\mu, \tag{4.38}$$

with $(w_\nu)^\mu = \delta_\nu^\mu$. The Lagrangian $\mathcal{L}(y)$ is given by a sum of "hyperloops". A hyperloop is a product of an even number of Grassmann variables located at positions $\tilde{x}_j(\tilde{y})$, $j = 1\ldots 8$, within the cell at $\tilde{y}$. In accordance with Eq. (4.6) we will consider hyperloops with twelve spinors. In a certain sense the hyperloops are a four-dimensional generalization of the plaquettes in lattice gauge theories.

We want to preserve the local $SO(4, \mathbb{C})$-symmetry for the lattice regularization of spinor gravity. We therefore employ hyperloops that are invariant under local $SO(4, \mathbb{C})$ transformations. Local $SO(4, \mathbb{C})$ symmetry can be implemented by constructing the hyperloops as products of invariant bilinears involving two spinors located at the same position $\tilde{z}$,

$$\tilde{\mathcal{H}}_\pm^k(\tilde{z}) = \varphi_\alpha^a(\tilde{z})(C_\pm)_{\alpha\beta}(\tau_2\tau_k)^{ab}\varphi_\beta^b(\tilde{z}). \tag{4.39}$$

Since the local $SO(4, \mathbb{C})$ transformations (4.7) involve the same $\varepsilon_{mn}(\tilde{z})$ for both spinors the six bilinears $\tilde{\mathcal{H}}_\pm^k$ are all invariant. The three matrices $\tilde{\tau}_k = \tau_2\tau_k$ are symmetric, such that $C_\pm \otimes \tau_2\tau_k$ is antisymmetric, as required by the Pauli principle. An $SO(4, \mathbb{C})$ invariant hyperloop can be written as a product of six factors $\tilde{\mathcal{H}}(\tilde{z})$, with $\tilde{z}$ belonging to the hypercube $\tilde{y}$ and obeying Eq. (4.38). We will take all six positions to be different. Furthermore, we will take three factors $\tilde{\mathcal{H}}_+$ and three factors $\tilde{\mathcal{H}}_-$ in order to realize the global symmetries of the continuum limit.

4.9 Lattice Action

The lattice action is a sum of local terms $\mathcal{L}(y)$ for all hypercubes $\tilde{y}$, where each $\mathcal{L}(y)$ is a combination of hyperloops. Using only the bilinears (4.39) the local Lorentz symmetry is guaranteed. We need a lattice implementation of the contraction of four derivatives with $\varepsilon^{\mu_1\mu_2\mu_3\mu_4}$ in order to realize diffeomorphism symmetry in the continuum limit. As basic building blocks we define

$$\begin{aligned}
\mathcal{F}_{\mu\nu}^\pm(\tilde{y}) = {}& \frac{1}{12}\varepsilon^{klm}\,\bar{\mathcal{H}}_k^\pm(\tilde{y}) \\
& \times \left[\mathcal{H}_l^\pm(\tilde{y}+w_\mu) - \mathcal{H}_l^\pm(\tilde{y}-w_\mu)\right] \\
& \times \left[\mathcal{H}_m^\pm(\tilde{y}+w_\nu) - \mathcal{H}_m^\pm(\tilde{y}-w_\nu)\right],
\end{aligned} \tag{4.40}$$

with $\bar{\mathcal{H}}(\tilde{y})$ the cell average

$$\bar{\mathcal{H}}_k^\pm(\tilde{y}) = \frac{1}{8}\sum_\nu\left(\mathcal{H}_k^\pm(\tilde{y}+w_\nu) + \mathcal{H}_k^\pm(\tilde{y}-w_\nu)\right). \tag{4.41}$$

A lattice diffeomorphism invariant action in four dimensions can be written as

$$S = \frac{\alpha}{128} \sum_{\tilde{y}} \varepsilon^{\mu\nu\rho\sigma} \mathscr{F}^+_{\mu\nu} \mathscr{F}^-_{\rho\sigma} + c.c. \tag{4.42}$$

We observe that the action is invariant under $\pi/2$-rotations in all six planes spanned by pairs of two coordinates z^μ. It is also odd under all four reflections of a single coordinate, $z^\mu \to -z^\mu$, as well as under diagonal reflections $z^\mu \leftrightarrow z^\nu$ or $z^\mu \to -z^\nu (\mu \neq \nu)$.

Finally, we note that the three components $\mathscr{H}^k_+$ in Eq. (4.39) transform as a three-component vector with respect to global $SU(2, \mathbb{C})_L$ gauge transformations. Thus the contraction (4.40) with the invariant tensor ε^{klm} yields a $SU(2, \mathbb{C})_L$-singlet, and $\mathscr{F}^+_{\mu\nu}(\tilde{y})$ is invariant under global $SU(2, \mathbb{C})_L$ transformations. The lattice action is invariant under global $SU(2, \mathbb{C})_L \times SU(2, \mathbb{C})_R$ gauge transformations. It is, however, not invariant under local gauge transformations of this kind. Local gauge transformations transform the factors $\mathscr{H}^k_\pm$ at different positions $\tilde{x}_j$ differently. If we would like to realize local $SU(2)$ gauge symmetry we would have to replace $(\tilde{\tau}_k)^{ab}$ in Eq. (4.39) by the invariant $\tilde{\tau}_0 = \tau_2$. This is not compatible with local Lorentz symmetry. The 4×4 matrices $C_\pm \otimes \tilde{\tau}_0$ are symmetric, such that $\mathscr{H}$ would vanish due to the Pauli principle. One could try to realize a local $U(1)$-symmetry by employing a different structure where only $\mathscr{H}^3_\pm$ appears. This is, however, not compatible with the required transformation properties of the lattice action with respect to reflections.

We define the lattice derivatives by the four relations

$$\mathscr{H}(\tilde{y} + w_\nu) - \mathscr{H}(\tilde{y} - w_\nu) = \left(x^+_\nu - x^-_\nu\right)^\mu \hat{\partial}_\mu \mathscr{H}(\tilde{y}), \tag{4.43}$$

where $x^\pm_\nu = x_p(\tilde{y} \pm w_\nu)$. Here we extend our discussion to a general assignment of points in a manifold $x_p(\tilde{z})$ for any discrete label $\tilde{z}$ of the lattice points. Our special case of a regular lattice corresponds to $x_p(\tilde{z}) = \Delta \tilde{z}$. With $\Delta_\nu = (x^+_\nu - x^-_\nu)/2$ the cell volume amounts to

$$V(\tilde{y}) = 2\varepsilon_{\mu\nu\rho\sigma} \Delta^\mu_0 \Delta^\nu_1 \Delta^\rho_2 \Delta^\sigma_3$$

$$= \frac{1}{12} \varepsilon_{\mu\nu\rho\sigma} \varepsilon^{\mu'\nu'\rho'\sigma'} \Delta^\mu_{\mu'} \Delta^\nu_{\nu'} \Delta^\rho_{\rho'} \Delta^\sigma_{\sigma'}. \tag{4.44}$$

The volume depends on the particular choice of positions $x_p(\tilde{z})$. (We only consider $V(\tilde{y}) > 0$.) Also the expressions for the lattice derivatives $\hat{\partial}_\mu \mathscr{H}$, which follow from solving Eq. (4.43), depend on this "positioning of the lattice points".

Using $\int d^4x = \sum_{\tilde{y}} V(\tilde{y})$ one finds that the action does not depend on the positioning of the lattice points, if it is expressed in terms of lattice derivatives and a continuous integral $\int d^4x$, i.e.

$$S = \frac{\alpha}{16} \int d^4x \, \varepsilon^{\mu\nu\rho\sigma} \hat{\mathscr{F}}^+_{\mu\nu} \hat{\mathscr{F}}^-_{\rho\sigma} + c.c., \tag{4.45}$$

with

$$\hat{\mathcal{F}}^{\pm}_{\mu\nu}(\tilde{y}) = \frac{1}{12}\varepsilon^{klm}\,\bar{\mathcal{H}}^{\pm}_k(\tilde{y})\hat{\partial}_\mu\mathcal{H}^{\pm}_l(\tilde{y})\hat{\partial}_\nu\mathcal{H}^{\pm}_m(\tilde{y}). \tag{4.46}$$

The positioning dependence of the derivatives is cancelled by the one of the volume. This will be crucial for the lattice diffeomorphism symmetry discussed in the next section.

The continuum limit $\bar{\mathcal{H}} \to \mathcal{H}$, $\hat{\partial}_\mu \to \partial_\mu$, is diffeomorphism invariant due to the contraction of the partial derivatives with the ε-tensor. It obtains formally as $\Delta \to 0$ at fixed y^μ—for details see Ref. [2]. We use Eq. (4.29) and find for the continuum limit

$$\hat{\mathcal{F}}^{\pm}_{\mu\nu} \to \pm 4i\,F^{\pm}_{\mu\nu}. \tag{4.47}$$

One recovers the diffeomorphism symmetric action (4.30).

4.10 Lattice Diffeomorphism Invariance

In the third part of this work we discuss the lattice equivalent of diffeomorphism symmetry of the continuum action. This "lattice diffeomorphism invariance" should be a special property of the lattice action that guarantees diffeomorphism symmetry for the continuum limit and the quantum effective action. We have no fundamental metric or vierbein at our disposal. Neither do we employ geometrical objects as simplices in order to perform a "functional integration over geometries." Our concept of lattice diffeomorphism invariance differs therefore substantially from the approach in Regge gravity [17–20]. The lattice points are associated to points in a coordinate manifold. The latter is simply a region in $\mathbb{R}^d$ and we have to formulate an invariance principle for this type of setting.

In the continuum, the invariance of the action under general coordinate transformations states that it should not matter if fields are placed at a point x or some neighboring point $x + \xi$, provided that all fields are transformed simultaneously according to suitable rules. In particular, scalar fields $\mathcal{H}(x)$ are simply replaced by $\mathcal{H}(x + \xi)$. After an infinitesimal transformation the new scalar field $\mathcal{H}'(x)$ at a given position x is related to the original scalar field $\mathcal{H}(x)$ by

$$\mathcal{H}'(x) = \mathcal{H}(x - \xi) = \mathcal{H}(x) - \xi^\mu \partial_\mu \mathcal{H}(x). \tag{4.48}$$

Diffeomorphism symmetry states that the action is the same for $\mathcal{H}(x)$ and $\mathcal{H}'(x)$. Implicitly the general coordinate transformations assume that the same rule for forming derivatives is used before and after the transformation.

We want to implement a similar principle for a lattice formulation. For this purpose we associate the abstract lattice points $\tilde{z} = (\tilde{z}^0, \tilde{z}^1, \tilde{z}^2, \tilde{z}^3)$, with integer $\tilde{z}^\mu$, with points on a manifold. We consider here a piece of $\mathbb{R}^d$ with cartesian coordinates $x^\mu = (x^0, x^1, \ldots, x^{d-1})$, but we do not specify any metric a priori, nor assume its existence. A map $\tilde{z} \to x^\mu_p(\tilde{z})$ defines the positioning of lattice points in the manifold.

We can now compare two different positionings, as a regular lattice $x_p^\mu(\tilde{z}) = \tilde{z}^\mu \Delta$, or some irregular one with different coordinates $x_p'^\mu(\tilde{z})$, for the same abstract lattice points $\tilde{z}$. In particular, we can compare two positionings related to each other by an arbitrary infinitesimal shift $x_p'^\mu = x_p^\mu + \xi_p^\mu(x)$. The notion of an *infinitesimal* neighborhood requires a continuous manifold and cannot be formulated for the discrete abstract lattice points $\tilde{z}$.

Positioning of the lattice points on a manifold is also required for the notion of a lattice derivative. One can define the meaning of two neighboring lattice points $\tilde{z}_1$ and $\tilde{z}_2$ in an abstract sense. A lattice derivative of a field will then be connected to the difference between field values at neighboring sites, $\mathcal{H}(\tilde{z}_1) - \mathcal{H}(\tilde{z}_2)$. For the definition of a lattice derivative $\hat{\partial}_\mu \mathcal{H}$ we need, in addition, some quantity with dimension of length. This is provided by the positioning on the manifold and follows from solving Eq. (4.43) for $\hat{\partial}_\mu \mathcal{H}$. Furthermore, the positioning of $\tilde{z}$ on a manifold is a crucial ingredient for the formulation of a continuum limit, where one switches from $\mathcal{H}(\tilde{z})$ to $\mathcal{H}(x)$ and derivatives thereof.

If the lattice action is originally formulated in terms of $\mathcal{H}(\tilde{z})$ only, its expression in terms of lattice derivatives will in general depend on the positioning, since the relation between $\mathcal{H}(\tilde{z})$ and lattice derivatives (4.43) depends on the positioning. We can now state the principle of "lattice diffeomorphism invariance". A lattice action is lattice diffeomorphism invariant if its expression in terms of lattice derivatives and a continuous integral does not depend on the positioning of the lattice points. For infinitesimally close positionings the lattice action is then independent of ξ_p. The lattice action (4.45) exhibits this property of lattice diffeomorphism invariance.

The usual discussion of lattice theories considers implicitly a given fixed positioning, for example a regular lattice. We investigate here a much wider class of positionings. Only the comparison of different positionings allows the formulation of lattice diffeomorphism invariance. One can show that the continuum limit of a lattice diffeomorphism invariant action exhibits diffeomorphism symmetry [2]. Also the quantum effective action is invariant under general coordinate transformations. This extends to the effective action for the metric which appears in our setting as the expectation value of a suitable collective field. The gravitational field equations are therefore covariant, with a similar general structure as in general relativity.

In order to show diffeomorphism symmetry of the continuum limit and the effective action a central ingredient is the observation that diffeomorphism transformations can be realized by repositionings of the lattice variables, without transforming the lattice variables themselves. One employs the concept of interpolating functions [2] and defines a version of partial derivatives of interpolating functions that takes into account the lack of knowledge of details of the interpolation. At the positions of lattice cells these derivatives equal the lattice derivatives. For smooth interpolating fields they coincide with the standard definition of partial derivatives. In this view, the lattice does not reflect a basic discreteness of space and time. It rather expresses the fact that only a finite amount of information is available in practice, and that arbitrarily accurate continuous functions are an idealization since they require an infinite amount of information. In a sense, we treat continuous functions similar to numerical simulations. In our formulation diffeomorphism transformations are nothing else than moving the lattice points, where the information about the function is given, within a manifold. Diffeomorphism symmetry is realized if the action in terms of fields and their derivatives does not notice this change in positions.

4.11 Lattice Diffeomorphism Invariance in Two Dimensions

Basic construction principles of a lattice diffeomorphism invariant action can be understood in two dimensions. We label abstract lattice points by two integers $\tilde{z} = (\tilde{z}^0, \tilde{z}^1)$, with $\tilde{z}^0 + \tilde{z}^1$ odd. For the discussion of lattice diffeomorphism invariance only the transformation of $\mathcal{H}_k$ as a scalar matters. Our discussion therefore also applies for fundamental scalars $\mathcal{H}_k$ [2]. For lattice spinor gravity $\mathcal{H}_k$ is again a fermion bilinear. We use for every lattice point two species, $a = 1, 2$, of two-component complex Grassmann variables $\varphi_\alpha^a(\tilde{z})$, $\alpha = 1, 2$, or equivalently eight real Grassmann variables $\psi_\gamma^a(\tilde{z})$, $\gamma = 1 \ldots 4$, with $\varphi_1^a(\tilde{z}) = \psi_1^a(\tilde{z}) + i\psi_3^a(\tilde{z})$, $\varphi_2^a(\tilde{z}) = \psi_2^a(\tilde{z}) + i\psi_4^a(\tilde{z})$. The functional measure (4.3) is replaced by

$$\int \mathcal{D}\psi = \prod_{\tilde{z}} \prod_{\gamma} \left(d\psi_\gamma^1(\tilde{z}) d\psi_\gamma^2(\tilde{z}) \right). \tag{4.49}$$

We introduce the bilinears $\mathcal{H}_k$ as in Eq. (4.39), with $\alpha, \beta = 1, 2$, and define the action as a sum over local cells located at $\tilde{y} = (\tilde{y}^0, \tilde{y}^1)$, with $\tilde{y}^\mu$ integer and $\tilde{y}^0 + \tilde{y}^1$ even, as in Eq. (4.37). Each cell consists of four lattice points that are nearest neighbors of $\tilde{y}$, denoted by $\tilde{x}_j(\tilde{y})$, $j = 1 \ldots 4$. Their lattice coordinates are $\tilde{z}(\tilde{x}_1(\tilde{y})) = (\tilde{y}^0 - 1, \tilde{y}^1)$, $\tilde{z}(\tilde{x}_2(\tilde{y})) = (\tilde{y}^0, \tilde{y}^1 - 1)$, $\tilde{z}(\tilde{x}_3(\tilde{y})) = (\tilde{y}^0, \tilde{y}^1 + 1)$, and $\tilde{z}(\tilde{x}_4(\tilde{y})) = (\tilde{y}^0 + 1, \tilde{y}^1)$. The local term $\mathcal{L}(\tilde{y})$ involves lattice fields on the four sites of the cell that we denote by $\mathcal{H}_k(\tilde{x}_j)$. We choose

$$\mathcal{L}(\tilde{y}) = \frac{1}{48} \varepsilon^{klm} \left[\mathcal{H}_k(\tilde{x}_1) + \mathcal{H}_k(\tilde{x}_2) + \mathcal{H}_k(\tilde{x}_3) + \mathcal{H}_k(\tilde{x}_4) \right]$$
$$\times \left[\mathcal{H}_l(\tilde{x}_4) - \mathcal{H}_l(\tilde{x}_1) \right] \left[\mathcal{H}_m(\tilde{x}_3) - \mathcal{H}_m(\tilde{x}_2) \right]. \tag{4.50}$$

At this point no notion of a manifold is introduced. We specify only the connectivity of the lattice by grouping lattice points $\tilde{z}$ into cells $\tilde{y}$ such that each cell has four points and each point belongs to four cells. This defines neighboring cells as those that have two common lattice points. Neighboring lattice points belong to at least one common cell.

We now proceed to an (almost) arbitrary positioning of the lattice points on a piece of $\mathbb{R}^2$ by specifying positions $x_p^\mu(\tilde{z})$. This associates to each cell a "volume" $V(\tilde{y})$,

$$V(\tilde{y}) = \frac{1}{2} \varepsilon_{\mu\nu} \left(x_4^\mu - x_1^\mu \right) \left(x_3^\nu - x_2^\nu \right), \tag{4.51}$$

with $\varepsilon_{01} = -\varepsilon_{10} = 1$ and x_j^μ shorthands for the positions of the sites $\tilde{x}_j$ of the cell $\tilde{y}$, i.e. $x_j^\mu = x_p^\mu(\tilde{z}(\tilde{x}_j(\tilde{y})))$. The volume corresponds to the surface inclosed by straight lines joining the four lattice points $\tilde{x}_j(\tilde{y})$ of the cell in the order $\tilde{x}_1, \tilde{x}_2, \tilde{x}_4, \tilde{x}_3$. For simplicity we restrict the discussion to "deformations" of the regular lattice, $x_p^\mu = \tilde{z}^\mu \Delta$, where $V(\tilde{y})$ remains always positive and the path of one point during the deformation never touches another point or a straight line between two other points at the boundary of the surface. We use the volume $V(\tilde{y})$ for the definition of an integral over the relevant region of the manifold

$$\int d^2x = \sum_{\tilde{y}} V(\tilde{y}), \tag{4.52}$$

where we define the region by the surface covered by the cells appearing in the sum.

We next express the action (4.37), (4.50) in terms of average fields in the cell

$$\mathscr{H}_k(\tilde{y}) = \frac{1}{4} \sum_j \mathscr{H}_k(\tilde{x}_j(\tilde{y})) \tag{4.53}$$

and lattice derivatives associated to the cell

$$\begin{aligned}
\hat{\partial}_0 \mathscr{H}_k(\tilde{y}) &= \frac{1}{2V(\tilde{y})} \{ (x_3^1 - x_2^1)(\mathscr{H}_k(\tilde{x}_4) - \mathscr{H}_k(\tilde{x}_1)) \\
&\quad - (x_4^1 - x_1^1)(\mathscr{H}_k(\tilde{x}_3) - \mathscr{H}_k(\tilde{x}_2)) \}, \\
\hat{\partial}_1 \mathscr{H}_k(\tilde{y}) &= \frac{1}{2V(\tilde{y})} \{ (x_4^0 - x_1^0)(\mathscr{H}_k(\tilde{x}_3) - \mathscr{H}_k(\tilde{x}_2)) \\
&\quad - (x_3^0 - x_2^0)(\mathscr{H}_k(\tilde{x}_4) - \mathscr{H}_k(\tilde{x}_1)) \}.
\end{aligned} \tag{4.54}$$

For the pairs $(\tilde{x}_{j_1}, \tilde{x}_{j_2}) = (\tilde{x}_4, \tilde{x}_1)$ and $(\tilde{x}_3, \tilde{x}_2)$ the lattice derivatives obey

$$\mathscr{H}_k(\tilde{x}_{j_1}) - \mathscr{H}_k(\tilde{x}_{j_2}) = (x_{j_1}^\mu - x_{j_2}^\mu) \hat{\partial}_\mu \mathscr{H}_k(\tilde{y}), \tag{4.55}$$

similar to Eq. (4.43). In terms of average and derivatives all quantities in $\mathscr{L}(\tilde{y})$ depend on the cell variable $\tilde{y}$ or the associated position of the cell $x_p^\mu(\tilde{y})$ that we take somewhere inside the surface of the cell, the precise assignment being unimportant at this stage. In this form we denote $\mathscr{L}(\tilde{y})$ by $\hat{\mathscr{L}}(\tilde{y}; x_p)$ or $\hat{\mathscr{L}}(x; x_p)$, where $\hat{\mathscr{L}}(x)$ only depends on quantities with support on discrete points in the manifold corresponding to the cell positions. We indicate explicitly the dependence on the choice of the positioning by the argument x_p.

The action appears now in a form referring to the positions on the manifold

$$S(x_p) = \tilde{\alpha} \int d^2x \bar{\mathscr{L}}(\tilde{y}; x_p) + c.c = \tilde{\alpha} \int d^2x \bar{\mathscr{L}}(x; x_p) + c.c., \tag{4.56}$$

with

$$\bar{\mathscr{L}}(\tilde{y}; x_p) = \bar{\mathscr{L}}(x; x_p) = \frac{\hat{\mathscr{L}}(\tilde{y}; x_p)}{V(\tilde{y}; x_p)}. \tag{4.57}$$

Lattice diffeomorphism invariance states that for fixed $\mathscr{H}(\tilde{y})$ and $\hat{\partial}_\mu \mathscr{H}(\tilde{y})$ the ratio $\mathscr{L}(\tilde{y}; x_p)$ is independent of the positioning, or independent of ξ_p for infinitesimal changes of positions $x'_p = x_p + \xi_p$,

$$\bar{\mathscr{L}}(\tilde{y}; x_p + \xi_p) = \bar{\mathscr{L}}(\tilde{y}; x_p), \qquad S(x_p + \xi_p) = S(x_p). \tag{4.58}$$

The ξ_p-independence of $\bar{\mathscr{L}}(\tilde{y}; x_p)$ means that the dependence of $V(\tilde{y}; x_p)$ and $\hat{\mathscr{L}}(\tilde{y}; x_p)$ on the positioning x_p must cancel. Inserting Eqs. (4.53), (4.55) in Eq. (4.50) yields

$$\hat{\mathscr{L}}(\tilde{y}) = \frac{1}{12} \varepsilon^{klm} V(\tilde{y}) \mathscr{H}_k(\tilde{y}) \varepsilon^{\mu\nu} \hat{\partial}_\mu \mathscr{H}_l(\tilde{y}) \hat{\partial}_\nu \mathscr{H}_m(\tilde{y}), \tag{4.59}$$

and we find indeed that the factor $V(\tilde{y})$ cancels in $\bar{\mathscr{L}}(\tilde{y}) = \hat{\mathscr{L}}(\tilde{y})/V(\tilde{y})$. Thus the action (4.37), (4.50) is lattice diffeomorphism invariant. This property is specific for a certain class of actions—for example adding to ε^{klm} a quantity s^{klm} which is symmetric in $l \leftrightarrow m$ would destroy lattice diffeomorphism symmetry. For all typical lattice theories the formulation of $\mathscr{L}(\tilde{y})$ only in terms of next neighbors and common cells (not using a distance) does not refer to any particular positioning. However, once one proceeds to a positioning of the lattice points and introduces the concept of lattice derivatives, the independence on the positioning of $\bar{\mathscr{L}}(\tilde{y}; x_p)$ for fixed $\mathscr{H}(\tilde{y})$ and $\hat{\partial}_\mu \mathscr{H}(\tilde{y})$ is not shared by many known lattice theories. For example, standard lattice gauge theories are not lattice diffeomorphism invariant.

Using the concept interpolating functions for fermion-bilinears [2] the continuum limit obtains by replacing lattice derivatives by partial derivatives and all average fields by local fields. This yields for the continuum action as a functional of the interpolating fields

$$S = \frac{\tilde{\alpha}}{12} \int d^2 x \varepsilon^{klm} \varepsilon^{\mu\nu} \mathscr{H}_k(x) \partial_\mu \mathscr{H}_l(x) \partial_\nu \mathscr{H}_m(x) + c.c. \tag{4.60}$$

The lattice derivatives for the Grassmann variables are defined similar to Eq. (4.55) by the two relations

$$\varphi^a_\alpha(\tilde{x}_{j_1}) - \tilde{\varphi}^a_\alpha(\tilde{x}_{j_2}) = (x^\mu_{j_1} - x^\mu_{j_2}) \hat{\partial}_\mu \varphi^a_\alpha(\tilde{y}) \tag{4.61}$$

for $(j_1, j_2) = (4, 1)$ and $(3, 2)$. With

$$\mathscr{H}_k(\tilde{x}_{j_1}) - \mathscr{H}_k(\tilde{x}_{j_2}) = (\varphi^a_\alpha(\tilde{x}_{j_1}) + \varphi^a_\alpha(\tilde{x}_{j_2}))(\tau_2)_{\alpha\beta}$$
$$\times (\tau_2 \tau_k)^{ab} (\varphi^b_\beta(\tilde{x}_{j_1}) - \varphi^b_\beta(\tilde{x}_{j_2})), \tag{4.62}$$

and using reordering of the Grassmann variables, one obtains from Eq. (4.50)

$$\mathscr{L}(y) = -8i\alpha A(\tilde{y})(\varphi^a_\alpha(\tilde{x}_4) - \varphi^a_\alpha(\tilde{x}_1))(\tau_2)_{\alpha\beta}(\tau_2)^{ab}$$
$$\times (\varphi^b_\beta(\tilde{x}_3) - \varphi^b_\beta(\tilde{x}_2)) + \cdots, \tag{4.63}$$

with

$$A(\tilde{y}) = \bar{\varphi}_1^1(\tilde{y})\bar{\varphi}_2^1(\tilde{y})\bar{\varphi}_1^2(\tilde{y})\bar{\varphi}_2^2(\tilde{y}), \tag{4.64}$$

and $\bar{\varphi}_\alpha^a(\tilde{y})$ the cell average. The dots indicate terms that do not contribute in the continuum limit. In terms of lattice derivatives (4.61) one finds the action

$$\bar{\mathscr{L}}(\tilde{y}) = -8i\tilde{\alpha} A(\tilde{y})\varepsilon^{\mu\nu}\hat{\partial}_\mu\varphi_\alpha^a(\tilde{y})(\tau_2)_{\alpha\beta}(\tau_2)^{ab}\hat{\partial}_\nu\varphi_\beta^b(\tilde{y}) + \cdots. \tag{4.65}$$

For fixed spinor lattice derivatives (4.61) the leading term (4.65) is again lattice diffeomorphism invariant.

The continuum limit (4.60) can be expressed in terms of spinors using $\partial_\mu\mathscr{H}_k(x) = 2\varphi(x)\tau_2 \otimes \tau_2\tau_k\partial_\mu\varphi(x)$, where the first 2×2 matrix E in $E \otimes F$ acts on spinor indices α, the second F on flavor indices a. With

$$F_{\mu\nu} = -A\partial_\mu\varphi\tau_2 \otimes \tau_2\partial_\nu\varphi \tag{4.66}$$

one obtains

$$S = 4i\tilde{\alpha} \int d^2x\varepsilon^{\mu\nu} F_{\mu\nu} + c.c., \tag{4.67}$$

in accordance with Eq. (4.65). Two comments are in order: (i) For obtaining a diffeomorphism invariant continuum action it is sufficient that the lattice action is lattice diffeomorphism invariant up to terms that vanish in the continuum limit. (ii) The definition of lattice diffeomorphism invariance is not unique, differing, for example, if we take fixed lattice derivatives (4.54) for spinor bilinears or the ones (4.61) for spinors. It is sufficient that the action is lattice diffeomorphism invariant for *one* of the possible definitions of lattice derivatives kept fixed.

We finally note that A and $F_{\mu\nu}$ are invariant under $SO(4, \mathbb{C})$ transformations. This symmetry group rotates among the four complex spinors φ_α^a, with complex infinitesimal rotation coefficients. For real coefficients, one has $SO(4)$, whereas other signatures as $SO(1, 3)$ are realized if some coefficients are imaginary. The continuum action (4.67) or (4.60) exhibits a local $SO(4, \mathbb{C})$ gauge symmetry. A subgroup of $SO(4, \mathbb{C})$ is the two-dimensional Lorentz group $SO(1, 1)$. The action (4.50) is therefore a realization of lattice spinor gravity [1] in two dimensions.

The extension of this discussion to four dimensions is straightforward. One verifies that the lattice action (4.42) is lattice diffeomorphism invariant. One can also define the concept of lattice diffeomorphism transformations [2] which is directly linked to the repositioning of lattice points within a continuous manifold.

4.12 Effective Action

The quantum effective action for fermions is introduced in the usual way by introducing Grassmann valued sources and making a Legendre transform of $\ln Z$. We can also introduce the effective action for bosonic collective fields. As an example, we discuss here first the fermion bilinear $\mathscr{H}_k$.

The generating functional for the connected Greens functions of collective lattice variables $\mathcal{H}_k$ is defined in the usual way

$$W[J(\tilde{y})] = \ln \int \mathcal{D}\psi \, \mathcal{H} \exp\left\{-S + \sum_{\tilde{y}}(\mathcal{H}_k(\tilde{y})J_k^*(\tilde{y}) + c.c)\right\}, \qquad (4.68)$$

with

$$\frac{\delta W}{\delta J_k^*(\tilde{y})} = \langle\mathcal{H}_k(\tilde{y})\rangle = h_k(\tilde{y}). \qquad (4.69)$$

(We don not write explicitly the boundary terms g_f and g_{in} in Eq. (4.3) for Z. They may be incorporated formally into $\int \mathcal{D}\psi$.) In the continuum limit the source term becomes

$$\sum_{\tilde{y}}\mathcal{H}_k(\tilde{y})J_k^*(\tilde{y}) = \int_x \mathcal{H}_k(\tilde{y})j_k^*(\tilde{y}) = \int_x \mathcal{H}_k(x)j_k^*(x), \qquad (4.70)$$

where the lattice source field $j(\tilde{y}) = J(\tilde{y})/V(\tilde{y})$ transforms as a scalar density under lattice diffeomorphisms. One also may define

$$\Gamma[h, J] = -W[J] + \sum_{\tilde{y}}(h_k(\tilde{y})J_k^*(\tilde{y}) + c.c.), \qquad (4.71)$$

which becomes the usual quantum effective action $\Gamma[h]$ (generating functional of 1PI-Greens functions) if we solve Eq. (4.69) for $J^*(\tilde{y})$ as a functional of $h(\tilde{y})$ and insert this solution into Eq. (4.71).

We have shown in ref. [2] that the effective action $\Gamma[h]$ is lattice diffeomorphism invariant. Its continuum limit exhibits the usual diffeomorphism symmetry if $h(x)$ transforms as a scalar. The proof relies on the observation that if the action does not "notice" the positioning of lattice points on the coordinate manifold, the same holds true for the effective action. No information about a specific positioning is introduced by the construction (4.68)–(4.71).

4.13 Metric

In the fourth part of this note we discuss the emergence of geometry from our formulation of lattice spinor gravity. So far we have used the coordinates x^μ only for the parametrization of a region of a continuous manifold. We have not used the notion of a metric and the associated "physical distance". (The physical distance differs from the coordinate distance $|x - y|$, except for the metric $g_{\mu\nu} = \delta_{\mu\nu}$.) The notion of a metric and the associated physical distance, topology and geometry can be inferred from the behavior of suitable correlation functions [21]. Roughly speaking, for a euclidean setting the distance between two points x and y gets larger if a suitable properly normalized connected two-point function $G(x, y)$ gets smaller.

This is how one world "measure" distances intuitively. In our case we may consider the two point function for collective fields $G(x, y) = \langle \mathcal{H}_k(x) \mathcal{H}_k(y) \rangle$.

We define the metric as

$$g_{\mu\nu}(x) = \frac{1}{2} (\langle G_{\mu\nu}(x) + G^*_{\mu\nu}(x) \rangle),$$

$$G_{\mu\nu}(x) = \mu_0^{-2} \sum_k \partial_\mu \mathcal{H}_k(x) \partial_\nu \mathcal{H}_k(x). \tag{4.72}$$

Here $\mathcal{H}_k(x)$ stands for the continuum limit or for a suitable interpolating field. The real normalization constant μ_0^{-1} has dimension of length such that $G_{\mu\nu}$ and $g_{\mu\nu}$ are dimensionless. In general, the elements $\langle G_{\mu\nu}(x) \rangle$ can be complex such that $g_{\mu\mu}$ can be positive or negative real numbers. The signature of the metric is not defined a priori. Points where $\det(g_{\mu\nu}(x)) = 0$ indicate singularities—either true singularities or coordinate singularities. More generally, the geometry and topology (e.g. singularities, identification of points etc.) of the space can be constructed from the metric [21]. The metric is the central object in general relativity and appears in our setting as the expectation value of a suitable collective field.

On the lattice we may use interpolating functions [2] for $\mathcal{H}_k(x)$. For x coinciding with the position of one of the cells $y_n = x_p(\tilde{y}_n)$ the derivative $\partial_\mu \mathcal{H}_k(x)$ is then given by the lattice derivative $\hat{\partial}_\mu \mathcal{H}_k(\tilde{y})$. For these values of x the field $G_{\mu\nu}(x)$ can be expressed by lattice quantities

$$G_{\mu\nu}(x) = \mu_0^{-2} \sum_k \hat{\partial}_\mu \mathcal{H}_k(\tilde{y}) \hat{\partial}_\nu \mathcal{H}_k(\tilde{y})$$

$$= \mu_0^{-2} a_\mu^{\tilde{\mu}}(x) a_\nu^{\tilde{\nu}}(x) G^{(L)}_{\tilde{\mu}\tilde{\nu}}, \tag{4.73}$$

with "lattice metric"

$$G^{(L)}_{\tilde{\mu}\tilde{\nu}} = p_{k,\tilde{\mu}} p_{k,\tilde{\nu}} \tag{4.74}$$

and

$$p_{k,\tilde{\mu}} = \mathcal{H}_k(\tilde{y} + v_{\tilde{\mu}}) - \mathcal{H}_k(\tilde{y} - v_{\tilde{\mu}}). \tag{4.75}$$

Similar to the lattice derivatives, the x-dependence of the metric arises only through the functions $a_\mu^{\tilde{\mu}}(x)$ which reflect the positioning of the lattice points. These functions obey

$$\hat{\partial}_\mu \mathcal{H}_k(\tilde{y}) = a_\mu^{\tilde{\mu}}(x) p_{k,\tilde{\mu}}, \tag{4.76}$$

and their explicit form can be extracted from Eq. (4.57). For interpolating functions $\mathcal{H}_k(x)$ transforming as scalars under general coordinate transformations the metric (4.72) transforms as a covariant second rank symmetric tensor. This is matched by the transformation properties of the expression (4.73) under lattice diffeomorphisms.

As a particular positioning we can use the regular lattice $x^\mu(\tilde{z}) = \Delta \tilde{z}^\mu$. This corresponds to a fixed choice of coordinates in general relativity. With this choice one has $d^{\tilde{\mu}}_\mu = 2\Delta \delta^{\tilde{\mu}}_\mu$, $V(\tilde{y}) = 2\Delta^2$ and therefore

$$a^{\tilde{\mu}}_\mu(x) = \frac{1}{2\Delta} \delta^{\tilde{\mu}}_\mu. \tag{4.77}$$

Choosing $\mu_0^{-2} = 4\Delta^2$, the collective field $G_{\mu\nu}$ in Eq. (4.73) coincides with the lattice metric $G^{(L)}_{\mu\nu}$ in Eq. (4.74).

Properties of the metric can often be extracted from symmetries. If the expectation values preserve lattice translation symmetry the metric $g_{\mu\nu}(x)$ will be independent of x. Invariance under a parity reflection implies $g_{0k} = g_{k0} = 0$. Symmetry of the expectation values under lattice rotations would imply a flat euclidean metric $g_{\mu\nu} \sim \delta_{\mu\nu}$. A Minkowski metric $g_{\mu\nu} = \eta_{\mu\nu}$ requires that the expectation values violate the euclidean rotation symmetry.

4.14 Effective Action for Gravity and Gravitational Field Equations

The quantum effective action for the metric, $\Gamma[g_{\mu\nu}]$, can be constructed in the usual way by introducing sources for the collective field,

$$W[\tilde{T}] = \ln \int \mathcal{D}\mathcal{H} \exp\left\{ -S + \int_x G^R_{\mu\nu}(x) \tilde{T}^{\mu\nu}(x) \right\},$$

$$G^R_{\mu\nu} = \frac{1}{2}(G_{\mu\nu} + G^*_{\mu\nu}), \qquad \frac{\delta W[\tilde{T}]}{\delta \tilde{T}^{\mu\nu}(x)} = g_{\mu\nu}(x). \tag{4.78}$$

Solving formally for $\tilde{T}^{\mu\nu}$ as a functional of $g_{\mu\nu}$, the quantum effective action for the metric obtains by a Legendre transform

$$\Gamma[g_{\mu\nu}] = -W + \int_x g_{\mu\nu}(x) \tilde{T}^{\mu\nu}(x). \tag{4.79}$$

The metric obeys the exact quantum field equation

$$\frac{\delta \Gamma}{\delta g_{\mu\nu}(x)} = \tilde{T}_{\mu\nu}(x), \tag{4.80}$$

and we realize that $\tilde{T}^{\mu\nu}$ can be associated to the energy momentum tensor $T^{\mu\nu}$ by $\tilde{T}^{\mu\nu} = \sqrt{g}T^{\mu\nu}$, $g = |\det g_{\mu\nu}|$.

Under a general coordinate transformation $\mathcal{H}_k(x)$ transforms as a scalar

$$\delta \mathcal{H}_k(x) = -\xi^\nu \partial_\nu \mathcal{H}_k(x). \tag{4.81}$$

This implies that $\partial_\mu \mathcal{H}_k$ and $G^R_{\mu\nu}$ transform as covariant vectors and second rank symmetric tensors, respectively. In consequence, $\tilde{T}^{\mu\nu}$ transforms as a contravariant tensor density, with $T^{\mu\nu}$ a symmetric second rank tensor. Thus $\int_x G^R_{\mu\nu} \tilde{T}^{\mu\nu}$ and $\int_x g_{\mu\nu} \tilde{T}^{\mu\nu}$ are diffeomorphism invariant, and $\Gamma[g_{\mu\nu}]$ is diffeomorphism invariant if $W[\tilde{T}]$ is diffeomorphism invariant. This is indeed the case for $\tilde{T}_{\mu\nu}$ transforming as a tensor density [2]—the argument is similar as for the diffeomorphism symmetry of $\Gamma[h(x)]$.

The functional integral (4.78) is well defined and regularized for a finite number of lattice points. Therefore also $\Gamma[g_{\mu\nu}(x)]$ is a well defined functional that is, in principle, unambiguously calculable. (More precisely, this holds for all metrics for which the third equation (4.78) is invertible.)

A key question concerns the general form of the effective action $\Gamma[g_{\mu\nu}]$. If Γ is diffeomorphism invariant and sufficiently local in the sense that an expansion in derivatives of $g_{\mu\nu}$ yields a good approximation for slowly varying metrics, then only a limited number of invariants as a cosmological constant or Einstein's curvature scalar R contribute at long distances. The signature of the metric is not fixed a priori. For $g \neq 0$ the inverse metric $g^{\mu\nu}$ is well defined—this contrasts with the Grassmann element $G_{\mu\nu}$ or $G^R_{\mu\nu}$ for which no inverse exists. The existence of $g^{\mu\nu}$ opens the possibility that $\Gamma[g_{\mu\nu}]$ also involves the inverse metric. Two dimensions are special for gravity since the graviton does not propagate. Our construction generalizes, however, in a straightforward way to four dimensions.

4.15 Conclusions and Discussion

We have constructed a lattice regularized functional integral for fermions with local Lorentz symmetry. The continuum limit of both the action and the quantum effective action exhibits invariance under general coordinate transformations. We thus have realized the first four of the six criteria for realistic quantum gravity that we have specified in the introduction. The remaining two criteria (4.5) and (4.6) depend on the form of the quantum effective action for the metric. The diffeomorphism invariance of the effective action suggests that it can describe a massless graviton if the cosmological constant vanishes. For a verification of this conjecture one needs, however, an explicit computation of the long wavelength limit of the effective action.

The symmetry properties of our model suggest that it can be used as a promising starting point for realistic quantum gravity. We have only sketched the way to geometry. Much remains to be done before the effective action for the composite metric can be computed explicitly. For our regularized model this issue is, at least, well defined. However, only an explicit calculation can settle the issue if diffeomorphism invariant terms involving explicit length scales, as a cosmological constant or Einstein's curvature scalar multiplied by the Planck mass, can be generated by fluctuations. The classical continuum action (4.24) is dilatation symmetric— the only coupling α is dimensionless. If the effective action for the graviton preserves dilatation symmetry no dimensional couplings can be present. In this case one would expect gravitational invariants involving two powers of the curvature tensor, as $R_{\mu\nu\rho\sigma}R^{\mu\nu\rho\sigma}$, $R_{\mu\nu}R^{\mu\nu}$ or R^2. Also composite scalar fields may play a role, such that terms $\sim \xi^2 R$ can induce an Einstein-Hilbert term in the effective action by spontaneous dilatation symmetry breaking through an expectation value of ξ [22–24]. As an alternative, an explicit mass scale could be generated by running couplings, which constitute a dilatation anomaly through quantum fluctuations.

For a regularized functional integral realizing lattice diffeomorphism invariance the lattice distance Δ does not introduce an explicit length scale. It neither appears in the lattice action nor in the continuum limit of the action. The parameter Δ only characterizes a particular regular positioning of the abstract lattice points on a manifold, and one can vary its value by repositioning. This absence of a length scale suggests that the ultraviolet limit of quantum gravity is characterized by an ultraviolet fixed point. Such a fixed point would realize the "asymptotic safety" scenario for non-perturbative renormalizable gravity [25]. Recent progress [26, 27] in computations of the flow of gravitational couplings, based on functional renormalization of the effective average action or flowing action [28, 29], give many hints in this direction.

Besides the metric, a consistent coupling of fermions to gravity also needs the vierbein. In our formulation of lattice spinor gravity we have several candidates of the type

$$\tilde{E}_\mu^m = \varphi^a C \gamma_M^m \partial_\mu \varphi^b V_{ab}, \tag{4.82}$$

with $C = C_1$ or C_2 defined in Eq. (4.18) and V_{ab} a suitable 2×2 matrix in flavor space. All objects (4.82) transform as vectors under general coordinate transformations, and as vectors under global generalized $SO(4, \mathbb{C})$-Lorentz transformations. (Further objects transforming as vectors under global $SO(1, 3)$-transformations can be constructed by replacing φ_α by a suitable linear combination of φ_β^*.) From this point of view the expectation value

$$e_\mu^m = \langle \tilde{E}_\mu^m \rangle / \mu_e, \tag{4.83}$$

with μ_e a suitable mass scale, resembles in many aspects the vierbein.

There are, however, also new unfamiliar features. The bilinear $\tilde{E}_\mu^m$ does not transform as a vector under local Lorentz transformations, but rather acquires an inhomogeneous piece [7–9]. By construction the quantum effective action for e_μ^m, which is formulated similarly to the effective action for the collective metric (4.79) or the scalar bilinear h_k (4.71), is invariant under local Lorentz transformations. In view of the inhomogeneous transformation property one may expect some differences to Cartan's formulation of gravity [30].

As another striking feature we observe that $\tilde{E}_\mu^m$ does not transform as a singlet with respect to the gauge transformations $SU(2)_L \times SU(2)_R$ which act in flavor space. (Exceptions are particular subgroups for particular choices of V_{ab}.) This hints to a more intrinsic entanglement between gauge transformations and Lorentz transformations. It remains to be seen if this new form of "gauge-gravity unification" could lead to observable effects.

References

1. C. Wetterich, Phys. Lett. B **704**, 612 (2011)
2. C. Wetterich, arXiv:1110.1539
3. D. Diakonov, arXiv:1109.0091 [hep-th]
4. H.W. Hamber, arXiv:0901.0964 [gr-qc]
5. J. Ambjorn, J. Jurkiewicz, R. Loll, arXiv:1105.5582 [hep-lat]
6. C. Rovelli, arXiv:1102.3660 [gr-qc]
7. A. Hebecker, C. Wetterich, Phys. Lett. B **57**, 269 (2003)
8. C. Wetterich, Phys. Rev. D **70**, 105004 (2004)
9. C. Wetterich, Phys. Rev. Lett. **94**, 011602 (2005)
10. K. Akama, Y. Chikashige, T. Matsuki, H. Terazawa, Prog. Theor. Phys. **60**, 868 (1978)
11. K. Akama, Prog. Theor. Phys. **60**, 1900 (1978)
12. D. Amati, G. Veneziano, Phys. Lett. B **105**, 358 (1981)
13. G. Denardo, E. Spallucci, Class. Quantum Gravity, 89 (1987)
14. Y. Nambu, G. Jona-Lasinio, Phys. Rev. **122**, 345 (1961)
15. C. Wetterich, Nucl. Phys. B **211**, 177 (1983)
16. C. Wetterich, Nucl. Phys. B **852**, 174 (2011)
17. C. Itzykson, Fields on a random lattice, in *Progess in Gauge Field Theory*, Cargèse (1983)
18. T.D. Lee, in *Discrete Mechanics*, 1983. Erice School of Subnuclear Physics, vol. 21 (Plenum Press, New York, 1985)
19. J.B. Hartle, J. Math. Phys. **26**, 804 (1985)
20. H.W. Hamber, R.M. Williams, Nucl. Phys. B **435**, 361 (1995)
21. C. Wetterich, Nucl. Phys. B **397**, 299 (1993)
22. C. Wetterich, Nucl. Phys. B **302**, 645 (1988)
23. C. Wetterich, Nucl. Phys. B **302**, 668 (1988)
24. Y. Fujii, Phys. Rev. D **26**, 2580 (1982)
25. S. Weinberg, in *General Relativity: An Einstein Centenary Survey*, ed. by S.W. Hawking, W. Israel (Cambridge University Press, Cambridge, 1979), p. 790
26. M. Reuter, Phys. Rev. D **57**, 971 (1998)
27. A. Codello, R. Percacci, C. Rahmede, Ann. Phys. **324**, 414 (2009)
28. C. Wetterich, Phys. Lett. B **301**, 90 (1993)
29. M. Reuter, C. Wetterich, Nucl. Phys. B **417**, 181 (1994)
30. E. Cartan, Ann. Sci. Ec. Norm. Super. **40**, 325 (1923)

Chapter 5
Introduction to Causal Dynamical Triangulations

Andrzej Görlich

Abstract The method of causal dynamical triangulations is a non-perturbative and background-independent approach to quantum theory of gravity. In this review we present recent results obtained within the four dimensional model of causal dynamical triangulations. We describe the phase structure of the model and demonstrate how a macroscopic four-dimensional de Sitter universe emerges dynamically from the full gravitational path integral. We show how to reconstruct the effective action describing scale factor fluctuations from Monte Carlo data.

5.1 Introduction

The model of causal dynamical triangulations (CDT) was proposed some years ago by J. Ambjørn, J. Jurkiewicz and R. Loll with the aim of defining a lattice formulation of quantum gravity from first principles [1–4]. The foundation of this model is the formalism of path integrals applied to quantize a theory of gravitation. The causal dynamical triangulations method is a natural generalization of discretization procedure, introduced in the definition of quantum mechanical Feynman's path integral, to higher dimensions. In the path integral formulation of quantum gravity, the role of a particle trajectory is played by the geometry of four-dimensional spacetime. CDT provide an explicit recipe for calculating the path integral and for specifying the class of virtual geometries which should be superimposed in the path integral. Let us emphasize that no *ad hoc* discreteness of space-time is assumed from the outset, and the discretization appears only as a regularization, which is intended to be removed in the continuum limit. The presented approach has the virtue that it allows quantum gravity to be relatively easily represented and studied by computer simulations.

Classical theory of gravitation, general relativity, in contrast with other known interactions describes the dynamics of space-time geometry where the considered degree of freedom is the geometry associated with the metric field $g_{\mu\nu}(x)$. The non-

A. Görlich (✉)

Niels Bohr Institute, University of Copenhagen, Blegdamsvej 17, 2100 Copenhagen, Denmark

e-mail: goerlich@nbi.dk

G. Calcagni et al. (eds.), *Quantum Gravity and Quantum Cosmology*,
Lecture Notes in Physics 863, DOI 10.1007/978-3-642-33036-0_5,

vanishing curvature of the underlying space-time geometry is interpreted as a gravitational field. The starting point for construction of the quantum theory of gravitation is the classical Einstein–Hilbert action ($\{-, +, +, +\}$ signature and sign convention as in [5, 6])

$$S_{EH}[g_{\mu\nu}] = \frac{1}{16\pi G} \int_{\mathcal{M}} d^4x \sqrt{-\det g}(R - 2\Lambda), \tag{5.1}$$

where G and Λ are respectively the Newton's gravitational constant and the cosmological constant, $\mathcal{M}$ is the space-time manifold equipped with a pseudo-Riemannian metric $g_{\mu\nu}$ with Minkowskian signature $\{-, +, +, +\}$ and R denotes the associated Ricci scalar curvature [7, 8]. We used the natural Planck units $c = \hbar = 1$. For simplicity, we assume that the topology of $\mathcal{M}$ is $S^1 \times S^3$.

Path-integrals are one of the most important tools used for the quantization of classical field theories. The path integral or partition function of quantum gravity is defined as a formal integral over all space-time geometries, i.e., equivalence classes of space-time metrics g with respect to the diffeomorphism group Diff$_{\mathcal{M}}$ on $\mathcal{M}$, also called histories,

$$Z = \int \mathcal{D}_{\mathcal{M}}[g] e^{i S_{EH}[g]}. \tag{5.2}$$

5.1.1 Causal Triangulations

To make sense of the formal gravitational path integral (5.2), the causal dynamical triangulations model uses a standard method of regularization, and replaces the path integral over geometries by a sum over a discrete set $\mathbb{T}$ of all causal triangulations $\mathcal{T}$. In other words, CDT serve as a regularization of smooth space-time histories present in the formal path integral (5.2) with piecewise linear manifolds.

The building blocks of four dimensional CDT are four-simplices. A simplex is a generalization of a triangle, which itself is a two-dimensional simplex, to higher dimensions. Each four-dimensional simplex is composed of five vertices connected to each other and is taken to be a subset of a four-dimensional Minkowski space-time together with its inherent light-cone structure. Thus the metric inside every simplex is flat. Figure 5.1 presents a visualization of four-simplices together with a light-cone sketch. A four-dimensional simplicial manifold, with a given topology, is obtained by properly gluing pairwise four-simplices along common tetrahedral faces. A simplicial manifold takes over a metric from simplices of which it is built. In general, such n-dimensional complex cannot be embedded in $\mathbb{R}^n$, which signifies a non-vanishing curvature. The curvature is singular and localized on the triangles.

The underlying assumption of CDT is the causality condition. It has a significant impact on desirable properties of the theory. As a consequence of the original Lorentzian signature of space-time, in a gravitational path integral one should sum over causal geometries only. We will consider only globally hyperbolic pseudo-Riemannian manifolds which allow introducing a global proper-time foliation. The

Fig. 5.1 A visualization of fundamental building blocks of four-dimensional causal dynamical triangulations: four-simplices. The simplices join two successive slices t and $t + 1$, and are divided into two types: $\{4, 1\}$ and $\{3, 2\}$. The simplices are equipped with the flat Minkowski metric imposing the light-cone structure

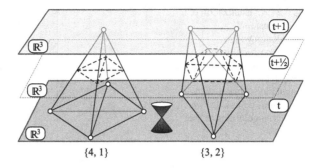

$\{4, 1\}$ $\{3, 2\}$

leaves of the foliation are spatial three-dimensional Cauchy surfaces Σ and are called *slices*. Because topology changes of the spatial slices are often associated with causality violation, we forbid the topology of the leaves to alter in time. Figure 5.2 illustrates a triangulation with imposed a foliation violating the causality condition. For simplicity, we choose the spatial slices to have a fixed topology $\Sigma = S^3$, that of a three-sphere, and establish periodic boundary conditions in the time direction. Therefore, we assume space-time topology to be $\mathscr{M} = S^1 \times S^3$, where S^1 corresponds to time and S^3 to space. The spatial slices are enumerated by a discrete *time* coordinate i. At each integer proper-time step i, a spatial slice itself forms a triangulation of S^3, made up of equilateral tetrahedra with a side length $a_s > 0$, with *an* induced metric which has a Euclidean signature. Each vertex lies in one spatial slice and is assigned the corresponding discrete time coordinate i.

Two successive slices, given respectively by triangulations $\mathscr{T}^{(3)}(t)$ and $\mathscr{T}^{(3)}(t + 1)$, are connected with four-simplices. The simplices are joined in such a way that they form a four-dimensional piecewise linear geometry. Such an object takes the form of a four-dimensional *slab* with the topology of $[0, 1] \times S^3$ and has $\mathscr{T}^{(3)}(t)$ and $\mathscr{T}^{(3)}(t + 1)$ as the three-dimensional boundaries. A set of slabs glued one after another builds the whole simplicial complex. Such connection of two consecutive slices, by interpolating the *space* between them with properly glued four-simplices, does not spoil the *causal structure*. The triangulation of the later slice wholly lies in the future of the earlier one.

Because each simplex connects two consecutive spatial slices and contains vertices lying in both of them, there are four kinds of simplices, namely $\{1, 4\}$, $\{2, 3\}$, $\{3, 2\}$ and $\{4, 1\}$. The first number denotes the number of vertices lying in slice $\mathscr{T}^{(3)}(t)$, and the second lying in slice $\mathscr{T}^{(3)}(t + 1)$. Figure 5.1 illustrates four-simplices of type $\{4, 1\}$ and $\{3, 2\}$ connecting slices t and $t + 1$.

Similarly, due to the causal structure, we distinguish two types of edges. The space-like links connecting two vertices in the same slice have length $a_s > 0$. The time-like links connecting two vertices in adjacent slices have length a_t. In causal dynamical triangulations, the lengths a_s and a_t are constant but not necessarily equal. Let us denote the asymmetry factor between the two lengths by α:

$$a_t^2 = \alpha\, a_s^2. \tag{5.3}$$

Fig. 5.2 A visualization of a
two-dimensional
triangulation with a
light-cone structure and a
branching point marked. In
causal dynamical
triangulations spatial slices
are not allowed to split, which
prevents singularities of the
time arrow

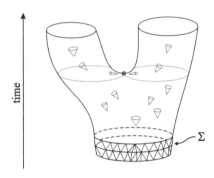

In the Lorentzian case $\alpha < 0$. The volumes and angles of simplices are functions of a_s and a_t and differ for the two types $\{4, 1\}$ and $\{3, 2\}$. Because no coordinates are introduced, the CDT model is manifestly diffeomorphism-invariant. Such a formulation involves only geometric invariants such as lengths and angles.

5.1.2 The Regge Action and the Wick Rotation

The Einstein–Hilbert action (5.1) has a natural realization on piecewise linear manifolds called the Regge action. Let N_{41} mean the number of simplices of type $\{1, 4\}$ or $\{4, 1\}$, and N_{32} the number of simplices of type $\{2, 3\}$ or $\{3, 2\}$. They sum up to the total number of simplices, $N_4 = N_{41} + N_{32}$. The total physical four-volume $\int_{\mathscr{T}} d^4x \sqrt{|\det g|}$ is given by a linear combination of N_{41} and N_{32}. Similarly, the global curvature $\int_{\mathscr{T}} d^4x \sqrt{|\det g|} R$ can be expressed using the angle deficits which are localized at triangles, and is a linear function of total volumes N_{32}, N_{41} and the total number of vertices N_0. The Regge action, i.e., action (5.1) calculated for a causal triangulation $\mathscr{T}$, can be written in a very simple form,

$$S[\mathscr{T}] \equiv -K_0 N_0[\mathscr{T}] + K_4 N_4[\mathscr{T}] + \Delta\big(N_{41}[\mathscr{T}] - 6N_0[\mathscr{T}]\big), \qquad (5.4)$$

where K_0, K_4 and Δ are bare coupling constants, and are nonlinear functions of parameters appearing in the continuous Einstein–Hilbert action, namely G and Λ, and the asymmetry factor $\alpha = a_t^2/a_s^2$ which is a regularization parameter [3]. K_4 plays a similar role as a cosmological constant, it controls the total volume. K_0 may be viewed as the inverse of the gravitational coupling constant G. Δ is related to the asymmetry factor α between lengths time-like and spatial-like links. It is zero when $a_t = a_s$ and does not occur in the Euclidean dynamical triangulations [9]. Δ will play an important role since a change in Δ will be associated with geometric phase transitions which might determine the ultraviolet limit of the lattice theory.

Causal dynamical triangulations provide a regularization of histories appearing in the formal gravitational path integral (5.2). The integral is now discretized by replacing it with a sum over the set of all causal triangulations $\mathbb{T}$ weighted with the Regge action (5.4), providing a meaningful definition of the partition function,

$$Z \equiv \sum_{\mathscr{T} \in \mathbb{T}} \frac{1}{C_{\mathscr{T}}} e^{i\,S[\mathscr{T}]}. \tag{5.5}$$

$C_{\mathscr{T}}$ is the order of the automorphism group of a triangulation $\mathscr{T}$, and might be viewed as the remnant of the division by the volume of the diffeomorphism group $\text{Diff}_{\mathscr{M}}$.

The advantage of the CDT approach is that for a fixed size of the triangulations, understood as the number of simplices N_4, the number of combinations is finite, which in general makes it possible to use numerical calculations. Nonetheless, this number grows exponentially with the size. Because of the oscillatory behavior of the integrand (5.5), we are still led into problems in defining the path integral, and in addition the mentioned numerical techniques are not useful. We may evade this problem by applying a trick called Wick rotation, which, roughly, is based on the analytical continuation of the time coordinate to imaginary values, and results in the change of the space-time signature from Lorentzian to Euclidean and a substitution of the complex amplitudes by real probabilities,

$$e^{i\,S^{\text{Lor}}} \rightarrow e^{-S^{\text{Euc}}}. \tag{5.6}$$

Due to the global proper-time foliation, the Wick rotation is well defined. It can be simply implemented by analytical continuation of the lengths of all time-like edges, $a_t \rightarrow i a_t$,

$$a_t^2 = \alpha\, a_s^2, \quad \alpha > 0.$$

This procedure is possible, because we have a distinction between time-like and space-like links. The Regge action rotated to the Euclidean sector, after the redefinition applied in (5.6), $S^{\text{Euc}} = -i\,S^{\text{Lor}}$, has exactly the same simple form as its original Lorentzian version (5.4). An exact derivation of the Wick-rotated Regge action can be found in [3].

As a consequence of the regularization procedure and Wick rotation to the Euclidean signature, the partition function (5.2) is finally written as a real sum over the set of all causal triangulations $\mathbb{T}$,

$$Z = \sum_{\mathscr{T} \in \mathbb{T}} \frac{1}{C_{\mathscr{T}}} e^{-S[\mathscr{T}]}. \tag{5.7}$$

We should keep in mind that the Euclidean Regge action $S[\mathscr{T}]$ and the partition function Z depend on bare coupling constants K_0, K_4 and Δ. With the partition function (5.7) there is associated a probability distribution on the space of triangulations $P[\mathscr{T}]$ which defines the quantum expectation value

$$\langle \mathcal{O} \rangle \equiv \sum_{\mathscr{T} \in \mathbb{T}} \mathcal{O}[\mathscr{T}] P[\mathscr{T}], \qquad P[\mathscr{T}] \equiv \frac{1}{Z} \frac{1}{C_{\mathscr{T}}} e^{-S[\mathscr{T}]}, \tag{5.8}$$

where $\mathcal{O}[\mathscr{T}]$ denotes some observable defined on $\mathbb{T}$. The above partition function defines a statistical mechanical problem which is free of oscillations and may be

Fig. 5.3 A sketch of the
phase diagram of the
four-dimensional causal
dynamical triangulations. The
phases correspond to regions
on the bare coupling constant
K_0–Δ plane. We observe
three phases: a *crumpled*
phase A, a *branched polymer*
phase B and the most
interesting genuinely
four-dimensional de Sitter
phase C

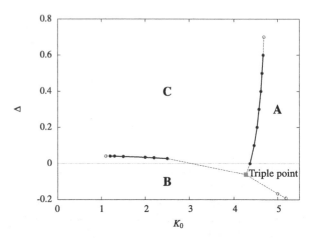

tackled in an approximate manner using Monte Carlo methods. Equation (5.7) is
the starting point for computer simulations which further allow us to measure expectation values defined by (5.8) and to obtain physically relevant information.

5.2 Phase Diagram

The standard version of the causal dynamical triangulations model uses the Regge
action (5.4), which depends on a set of three bare coupling constants K_0, Δ and K_4.
For simulation-related technical reasons it is preferable to keep the total four-volume
fluctuating around some finite prescribed value during Monte Carlo simulations. The
number of configurations grows exponentially with the size, but the contribution to
the partition function coming from extremely large configurations is suppressed by
the term involving K_4. A value of K_4 below the critical value would make the partition function ill defined. Thus K_4, acting as Lagrange multiplier, needs to be tuned
to its critical value, and effectively does not appear as a coupling constant. The two
remaining bare coupling constants K_0 and Δ can be freely adjusted and depending
on their values we observe three qualitatively different behaviors of a typical configuration. The phase structure was first qualitatively described in a comprehensive
publication [2] where three phases were labeled A, B and C. The first real phase diagram obtained due to large-scale computer simulations was described in [15]. The
phase diagram, based on Monte Carlo measurements, is presented in Fig. 5.3. The
solid lines denote observed phase-transition points for configurations of size 8,0000
simplices, while the dotted lines represent an interpolation.

In the remainder of this section we describe the properties of the phases and
discuss the phase transitions.

- **Phase A.** For large values of K_0 (cf. Fig. 5.3) the universe disintegrates into un-
  correlated irregular sequences of maxima and minima with time extent of few

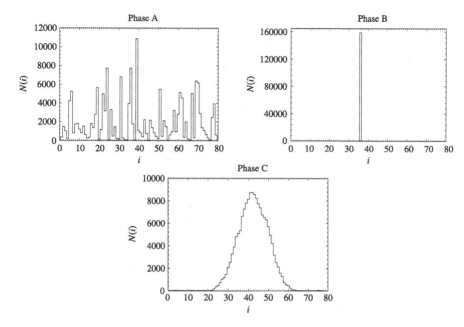

Fig. 5.4 Snapshot of a spatial volume $N(i)$ for a typical configuration of phase A, B and C. A typical configuration in phase C is bell-shaped with well-defined spatial and time extent

steps. As an example of a configuration in this phase, the spatial volume distribution $N(i)$, defined as the number of tetrahedra in a spatial slice labeled by a discrete time index i, is shown in Fig. 5.4. When looking along the time direction, we observe a number of small universes. The geometry appears to be oscillating in the time direction. They can merge and split with the passing of the Monte Carlo time. These universes are connected by *necks* not much larger than the smallest possible spatial slice. In the computer algorithm we do not allow these necks to *vanish* such that the configuration becomes disconnected. This phase is related to so-called *branched polymers phase* present in Euclidean dynamical triangulations (EDT) [9]. No spatially- nor time-extended universe, like the universe we see in reality, is observed and phase A is regarded as non-physical.

- **Phase B**. For small values of Δ nearly all simplices are localized on one spatial slice. Although we have a large three-volume collected at one spatial hypersurface of a topology of a three-sphere S^3, the corresponding slice has almost no spatial extent. The Hausdorff dimension is very high, if not infinite. In the case of infinite Hausdorff dimension the universe has neither time extent nor spatial extent, there is no geometry in a traditional sense. Phase B is also regarded as non-physical.
- **Phase C**. For larger values of Δ we observe the third, physically most interesting, phase. In this range of bare coupling constants, a typical configuration is bell-shaped and behaves like a well-defined four-dimensional manifold (cf. Fig. 5.4). The measurements of the Hausdorff dimensions confirm that at large scales the universe is genuinely four-dimensional [2]. Most results presented in this pa-

per were obtained for a point that is firmly placed in the phase C (cf. Fig. 5.3). A typical configuration has a finite time extent and spatial extent which scales as expected for a four-dimensional object. The averaged distribution of a spatial volume coincides with the distribution of Euclidean de Sitter space S^4 and thus this phase is also called the de Sitter phase.

The transitions between phases have been studied in detail in [20]. So far, there is a strong numerical evidence that the transition between phases A and C is of first order, while between phases B and C there is a second-order transition.

For the A–C phase transition, the distribution of values taken by the order parameter N_0, conjugate to K_0, reveals a two-peak structure, which corresponds to different types of geometry. The peaks become sharper with the increase of the system size, $N_4 \to \infty$. This confirms that configurations behave as if they were either in phase C or phase A and suggests that the A–C transition is of first order.

A similar two-peak distribution of the order parameter conjugate to the coupling constant Δ, namely $N_{41} - 6N_0$, is present for the B–C phase transition. But with the increasing total volume N_4 peaks become blurred and start to merge. Also, the measured value of the shift exponent $\tilde{\nu} = 2.51(3)$ [20] is far from $\tilde{\nu} = 1$ expected for a first-order transition. The above arguments strongly suggest that the B–C phase transition is of second order.

5.3 The Macroscopic de Sitter Universe

We start the quantitative description of the Universe emerging in causal dynamical triangulations by passing over local degrees of freedom of the quantum geometry, and reducing the considerations to volumes of spatial slices. The causality condition is ensured by imposing on configurations a global proper-time foliation and keeping the topology of leaves fixed. Due to the discrete structure, successive spatial slices, i.e., hypersurfaces of constant time, are labeled by a discrete time parameter i. The index i ranges from 1 to T. By construction, they are glued in the way to form a simplicial manifold with the topology of a three-sphere S^3.

5.3.1 The Spatial Volume

The spatial three-volume $N(i)$ is defined as the number of tetrahedra constituting a spatial slice $i = 1, \ldots, T$. Because each spatial tetrahedron is a base of one simplex of the type $\{1, 4\}$ and one of the type $\{4, 1\}$, the three-volumes $N(i)$ sums up to the total volume $N_{\text{tot}} \equiv \sum_{i=1}^{T} N(i) = N_{41}/2$. The spatial volume $N(i)$ is an example of the simplest observable providing information about the large-scale shape of the universe appearing in CDT path integral. An individual space-time history contributing to the partition function is not an observable, precisely in the same way as a trajectory of a particle in the quantum-mechanical path integral is not an observable either. However, it is perfectly legitimate to talk about the expectation value $\langle N(i) \rangle$ as well as about the fluctuations around the mean.

The lattice regularization present in CDT allows to adapt powerful Monte Carlo techniques to calculate expectation values, defined by Eq. (5.8). Though in two dimensions we have analytical tools, in four dimensions it is currently the only way to extract non-perturbative information about fluctuating geometries. Numerical simulations consist in generating a sequence of space-time geometries, more precisely causal triangulations $\mathcal{T}$, according to the probability distribution (5.8). Configurations are then used to calculate the average. A significant feature of the CDT approach, as shown in [11], is a dynamically emerging and physically realistic background geometry, described by the average $\langle N(i) \rangle$.

Let us focus on one particular point of the phase diagram firmly placed in phase C, and given by the following values of bare coupling constants: $K_0 = 2.2$, $\Delta = 0.6$, volume $N_{41} = 160{,}000$ and time-period $T = 80$. In this phase, the plot of $N(i)$ for an individual configuration is bell-shaped with a well-outlined *blob*. Figure 5.5 shows the volume profile $N(i)$ of a typical configuration. For the range of discrete volumes N_4 under study, the Universe *does not* extend over the entire axis, but rather is localized in a region much shorter than $T = 80$ time slices.

The Einstein–Hilbert action (5.1), and consequently the Regge action (5.4), is invariant under time translations $t \to t + \delta$. Because configurations are periodic in time, a straightforward average $\langle N(i) \rangle$ is meaningless, as it would give a uniform distribution. From Fig. 5.5 it is clear that in phase C the time translation symmetry is *spontaneously broken*. To perform a meaningful average of the spatial volume $\langle N(i) \rangle$, we thus fix the position of the center of mass of the volume distribution to be at $t = 0$. We apply this procedure to each configuration contributing to the expectation value.

The expectation value $\langle N(i) \rangle$ is measured using Monte Carlo techniques,

$$\langle N(i) \rangle \approx \frac{1}{K} \sum_{k=1}^{K} N^{(k)}(i), \tag{5.9}$$

where the brackets $\langle \ldots \rangle$ mean averaging over the whole ensemble of causal triangulations weighted with the Regge action (5.4) and the expectation value is approximated be a sum over K statistically independent Monte Carlo configurations. Figure 5.5 shows the average spatial volume $\langle N(i) \rangle$ (black thick line) measured at a point in the phase C, $K_0 = 2.2$ and $\Delta = 0.6$. The heights of the boxes visible in the plot indicate the amplitude of spatial volume fluctuations for each i given by $\sigma_i = \sqrt{\langle N(i)^2 \rangle - \langle N(i) \rangle^2}$. Results obtained by simulations show that the average geometry, in the blob and tail region, is extremely well approximated by the formula

$$\bar{N}(i) \equiv \langle N(i) \rangle = H \cdot \cos^3(i/W), \tag{5.10}$$

where W is proportional to the time extent of the Universe and H denotes its maximal spatial volume. The fit $H \cdot \cos^3(i/W)$ is also plotted in Fig. 5.5 with a dashed gray line, but it is indistinguishable from the empirical curve. The background geometry given by the solution (5.10) is consistent with the geometry of a four-sphere

Fig. 5.5 Spatial volume $N(i)$ of a randomly chosen typical configuration (*gray line*) and background geometry $\langle N(i)\rangle$ (*black line*): Monte Carlo measurements for fixed $N_{41} = 160,000$, $K_0 = 2.2$, $\Delta = 0.6$. The best fit (5.10) yields indistinguishable curves at given plot resolution. The bars height indicate the average size of quantum fluctuations

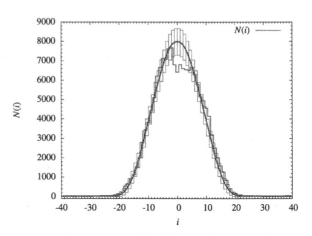

S^4 and corresponds to Euclidean de Sitter space, the maximally symmetric solution of classical Einstein equations with a positive cosmological constant [12, 13].

This is one of the most important results obtained within the CDT framework [11]. While no background was put by hand, the measurements present a direct evidence that the *background geometry* of a four-sphere emerges dynamically. Moreover, neglecting the stalk, which by construction has a non-zero volume, we spontaneously end up with the S^4 topology, although we started with $\mathcal{M} = S^1 \times S^3$.

5.3.2 The Mini-superspace Model

The shape of the three-volume $\bar{N}(i) = H \cdot \cos^3(i/W)$ emerges as a classical solution of the *mini-superspace* model. This model appears for example in quantum cosmological theories developed by Hartle and Hawking in their semi-classical evaluation of the wave function of the Universe [17]. This model assumes a spatially homogeneous and isotropic metric on a Euclidean space-time with $S^1 \times S^3$ topology,

$$ds^2 = d\tau^2 + a^2(\tau)d\Omega_3^2, \qquad (5.11)$$

where $a(\tau)$ is the *scale factor* depending on the *proper time* τ and $d\Omega_3^2$ denotes the line element on S^3. This means that all degrees of freedom except the three-volume (scale factor) are *frozen*. In CDT model we have the opposite situation, no degrees of freedom are excluded, instead we integrate out all of them but the scale factor. Nevertheless, in both cases results demonstrate high similarity. The *physical volume* of a spatial slice for a given time τ equals $v(\tau) = \int d\Omega_3\sqrt{\det g|_{S^3}} = 2\pi^2 a(\tau)^3$. The Euclidean version of the Einstein–Hilbert action (5.1) [5, 6] calculated for the metric (5.11) up to boundary terms is given by

$$S[a] = \frac{2\pi^2}{16\pi G} \int d\tau \left(-6a\dot{a}^2 - 6a + 2\Lambda a^3\right), \qquad (5.12)$$

and is called the mini-superspace action.

Although it is formally easy to perform the Wick rotation of the Einstein–Hilbert action (5.1), the corresponding Euclidean action suffers from the unboundedness of the conformal mode. This is caused by the wrong sign of the kinetic term, as is reflected in the standard mini-superspace action (5.12). Consequently, the Regge action (5.4) is also unbounded from below. Some triangulations may have very large negative values of the Regge action, but the action is always bounded from below due to the UV lattice regularization. The problem of infinities is revived when taking the continuum limit. Fortunately, in the non-perturbative approaches, like CDT, the partition function emerges as a subtle interplay of the entropic nature of triangulations and the *bare* action. The entropy factor may suppress the unbounded contributions coming from the conformal factor. There is a strong evidence [21] that, after integrating out all degrees of freedom except the scale factor, which means taking into account the non-perturbative measure, one obtains a *positive* kinetic term in (5.12). This is exactly what happens in four-dimensional causal dynamical triangulations: the *effective* action for $N(i)$ is equal to the mini-superspace action (5.12), but with an opposite sign, and is thus bounded from below. Together with a convergence of the coupling constants to their critical values, if such a point exists, the *entropic* and *action* terms should be balanced, and one hopes to obtain the proper continuum behavior.

Turning back to the spatial volume variable, the mini-superspace action (5.12) can be rewritten as

$$S[v] = -\frac{1}{24\pi G} \int d\tau \left(\frac{\dot{v}^2}{v} + \beta v^{1/3} - 3\Lambda v \right), \quad \beta = 9\left(2\pi^2\right)^{2/3}. \quad (5.13)$$

The overall sign of the action does not affect the classical solution of equations of motion. The classical trajectory, solving the Euler–Lagrange equation, is given by

$$\bar{v}(\tau) = 2\pi^2 R^3 \cos^3\left(\frac{\tau}{R}\right), \quad R = \left(\frac{\Lambda}{3}\right)^{-1/2}. \quad (5.14)$$

The physical volume $\bar{v}(\tau)$ describes the maximally symmetric space for a positive cosmological constant, namely the Euclidean *de Sitter Universe* or a geometry of a four-sphere S^4 with radius R. This result is in agreement with the relation (5.10) for $\bar{N}(i)$ found in numerical simulations. The de Sitter space *emerges dynamically* as a background geometry in the CDT model.

5.3.3 The Four-Dimensional Space-Time

The scaling properties and the measured spectral dimension of the ensemble of triangulations show that the Universe coming out in the CDT model is genuinely four-dimensional. Up to now, we have presented results for only one value of the total volume N_{tot}. Keeping the coupling constants K_0 and Δ fixed, which naïvely means that the geometry of simplices is not changed, we measure the expectation value $\bar{N}(i)$ for different total volumes N_{tot}.

Fig. 5.6 Average scaled spatial volume $\bar{n}(t)$ for a variety of total volumes N_{tot} calculated for the scaling dimension $d_H = 4$. Measured in Monte Carlo simulations for $K_0 = 2.2$ and $\Delta = 0.6$. We omit the error bars not to obscure the picture. The *dashed line* plots the fit $\bar{n}(t) = \frac{3}{4B} \cos^3(t/B)$, where $B = 0.69$

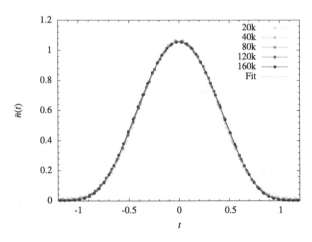

If the scaling dimension is d_H time intervals should scale as N_{tot}^{1/d_H}, which implies that the volume-independent time coordinate t scales as a function of a discrete time i as

$$t \equiv \Delta t \cdot i, \qquad \Delta t \equiv N_{\text{tot}}^{-\frac{1}{d_H}}. \tag{5.15}$$

To compare the spatial volume distributions $N(i)$ for geometries with different volumes N_{tot}, we introduce the scaled three-volume $n(t)$,

$$n(t) \equiv N_{\text{tot}}^{-1+\frac{1}{d_H}} N(i), \qquad \bar{n}(t) = \langle n(t) \rangle. \tag{5.16}$$

For very large N_{tot}, the time interval Δt is close to zero and in the continuum limit the sum over discrete time steps can be replaced by an integral,

$$\int dt \ldots \quad \leftrightarrow \quad \sum_i \Delta t \ldots . \tag{5.17}$$

The normalization condition reads $\int n(t)dt = N_{\text{tot}}^{-1} \sum_i N(i) = 1$.

Now it is possible to directly compare $n(t)$ for various total volumes and check for which value of the scaling dimension d_H the overlap is the best [2]. The estimated value of $d_H = 3.98 \pm 0.10$ minimizes the error function defined as a spread of scaled spatial volumes $n(t)$. The error of determination of d_H was estimated using the Jackknife method [18]. The expected value $d_H = 4$ is very close to the measured result, and is well within the margin of error. Figure 5.6 shows the scaled three-volumes $n(t)$ using $d_H = 4.0$ for several values of total volumes N_{tot}.

Since $\bar{n}(t)$ is normalized and is obtained by the scaling of $\bar{N}(i)$ which is given by Eq. (5.10), it is expressed by the formula

$$\bar{n}(t) = \frac{3}{4B} \cos^3\left(\frac{t}{B}\right), \tag{5.18}$$

where B depends only on the coupling constants K_0 and Δ, but not on N_{tot}. For $K_0 = 2.2$ and $\Delta = 0.6$, the measured values is $B \approx 0.69$. The curve (5.18) with adjusted B is drawn with a dashed line in Fig. 5.6, and the fit is remarkably good.

From Eqs. (5.16) and (5.18) and the scaling dimension $d_H = 4$ we obtain the following expression for the three-volume $\bar{N}(i)$:

$$\bar{N}(i) = \frac{3}{4} \frac{N_{\text{tot}}^{3/4}}{B} \cos^3\left(\frac{i}{BN_{\text{tot}}^{1/4}}\right). \tag{5.19}$$

As expected for a four-dimensional space-time, the time extent T_{univ} of the blob, measured in units of time steps, scales as $T_{\text{univ}} \sim \pi B \cdot N_{\text{tot}}^{1/4}$. The expression (5.19) specifies expression (5.10) and is only valid in the extended part of the Universe where the spatial three-volumes are larger than the minimal cut-off size.

Let us relate the discrete spatial volume $N(i)$ with the physical volume $v(\tau)$ of hypersurfaces of constant time. The classical solution $\bar{v}(\tau)$ is given by formula (5.14), while the average discrete volume $\bar{N}(i)$ is given by formula (5.19). Up to some factors they are expressed by the same function. Henceforth, we make the key assumption that the average configuration described by $\bar{N}(i)$ in fact has a geometry of a four-sphere S^4 given by $\bar{v}(\tau)$. The physical total four-volume of a four-sphere with a radius R equals

$$V_4 = \int_{-\frac{\pi}{2}R}^{\frac{\pi}{2}R} \bar{v}(\tau)d\tau = \frac{8\pi^2}{3} R^4 = C_4 a^4 N_{\text{tot}}, \tag{5.20}$$

where

$$C_4 = 2\left(\text{Vol}^{\{4,1\}} + \frac{N_{32}}{N_{41}} \text{Vol}^{\{3,2\}}\right),$$

which is interpreted as the average four-volume shared by one spatial tetrahedron. Here, $a = a_s$ is the cut-off length, i.e., the lattice constant. The continuum time t defined by (5.15) and the discrete time i are proportional to the proper time τ (cf. (5.11)),

$$\tau = \sqrt{g_{tt}}t = \sqrt{g_{tt}}\Delta t i, \quad \Delta t = N_{\text{tot}}^{-1/4}. \tag{5.21}$$

A slab between slices i and $i + 1$ has a proper-time extent $\Delta\tau$ and a four-volume

$$v(\tau)\Delta\tau = v(\tau)\sqrt{g_{tt}}\Delta t = C_4 a^4 N(i) = N_{\text{tot}}^{3/4} C_4 a^4 n(t). \tag{5.22}$$

The above equation is consistent with formula (5.20) which determines the total four-volume of the emerging de Sitter space with a radius R. The proper-time extent of the de Sitter Universe is πR, while in terms of the time t it is equal to πB, hence

$$\sqrt{g_{tt}} = \frac{\tau}{t} = \frac{R}{B}, \quad R = \left(\frac{3C_4 N_{\text{tot}}}{8\pi^2}\right)^{1/4} a. \tag{5.23}$$

Assuming such scaling relations between physical and discrete volume (cf. (5.22)), and between proper and discrete times (cf. (5.21)), we ensure that the empirically derived formulas (5.18) or (5.19) describe a Euclidean de Sitter Universe for all N_{tot}.

Another quantity revealing information about the geometry is related to the diffusion phenomena, namely the so-called spectral dimension d_S. On a d-dimensional Riemannian manifold with a metric $g_{\mu\nu}(\mathbf{x})$, let $\rho(\mathbf{x}, \mathbf{x}_0; \sigma)$ be the probability density of finding a diffusing particle at position $\mathbf{x}$ after some fictitious diffusion time σ,

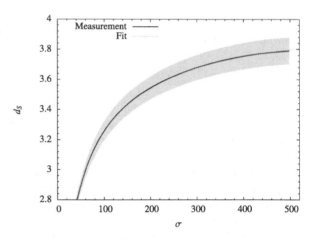

Fig. 5.7 The spectral dimension d_S of the Universe as a function of diffusion time σ, measured for $K_0 = 2.2$, $\Delta = 0.6$ and $N_4 \approx 368$k. The *thick curve* plots the average measured spectral dimension, while the *highlighted area* represents the error bars. The best fit $d_S(\sigma) = 4.02 - \frac{120}{58+\sigma}$ is drawn with a *dashed line*

with an initial position at $\sigma = 0$ fixed at $\mathbf{x}_0$. The evolution of $\rho(\mathbf{x}, \mathbf{x}_0; \sigma)$ is controlled by the diffusion equation

$$\partial_\sigma \rho(\mathbf{x}, \mathbf{x}_0; \sigma) = \triangle_g \rho(\mathbf{x}, \mathbf{x}_0; \sigma), \qquad \rho(\mathbf{x}, \mathbf{x}_0; \sigma = 0) = \frac{1}{\sqrt{\det g(\mathbf{x})}} \delta(\mathbf{x} - \mathbf{x}_0),$$
$$(5.24)$$

where $\triangle_g$ is the Laplace operator corresponding to $g_{\mu\nu}(\mathbf{x})$. The *return probability* describes the probability of finding a particle at the initial point after diffusion time σ. The *average return probability* $P(\sigma)$, supplying a global information about the geometries, is given by

$$P(\sigma) = \left\langle \frac{1}{V_4} \int d^d\mathbf{x} \sqrt{\det g(\mathbf{x})} \rho(\mathbf{x}, \mathbf{x}; \sigma) \right\rangle,$$

where $V_4 = \int d^d\mathbf{x} \sqrt{\det g(\mathbf{x})}$ is the total space-time volume and the average is also performed over the ensemble of geometries. For infinite flat manifolds the spectral dimension d_S can be extracted from the return probability due to its definition,

$$d_S \equiv -2 \frac{d \log P(\sigma)}{d \log \sigma}. \qquad (5.25)$$

For the Euclidean flat manifold $\mathscr{R}^d$, the spectral and Hausdorff dimensions are equal to the topological dimension, $d_S = d_H = d$. For the four-sphere S^4, the spectral dimension $d_S = 4$ for short diffusion times, while for very large times, because of the finite volume, the zero mode of the Laplacian will dominate and, with the above definition, d_S will tend to zero.

Definition (5.25) is particularly convenient because it is easy to perform numerical simulations which measure the return probability. In the CDT framework, the space-time geometry is regularized by piecewise flat manifolds built of four-simplices. Let us recall that after the Wick rotation space-times appearing in the model are Riemannian manifolds equipped with the positive-definite metric tensor. The diffusion process can be carried out by implementing the discretized version of the diffusion equation (5.24), $\rho(i, i_0; \sigma + 1) - \rho(i, i_0; \sigma) = \Delta \sum_{j \leftrightarrow i} (\rho(j, i_0; \sigma) - \rho(i, i_0; \sigma))$, where Δ denotes the time step and the sum is evaluated over all simplices j adjacent to i. Here variables i_0, i and j denote labels of simplices. The diffusion process is running on the dual lattice, i.e. the probability flows from a simplex to its neighbors. Since each simplex has exactly five neighbors, it is convenient to set $\Delta = 1/5$ and the diffusion equation reads $\rho(i, i_0; \sigma + 1) = \frac{1}{5} \sum_{j \leftrightarrow i} \rho(j, i_0; \sigma)$.

To evaluate $\rho(i, i_0; \sigma)$, we pick an initial four-simplex i_0 lying in the central slice i_{CV} and impose the initial condition $\rho(i, i_0; \sigma = 0) = \delta_{i i_0}$. We iterate the diffusion equation and calculate the probability density $\rho(i, i_0; \sigma)$ for consecutive diffusion steps σ [10]. Finally, we repeat the above operations for a number of random starting points i_0 ($K = 100$) and calculate the average return probability $P(\sigma) = \frac{1}{K} \sum_{i_0=1}^{K} \rho(i_0, i_0; \sigma)$. In numerical simulations the return probability $P(\sigma)$ is averaged over a number of triangulations ($\sim 1,000$) and the spectral dimension d_S is calculated from the definition (5.25). Figure 5.7 shows the spectral dimension d_S as a function of the diffusion time steps σ, in the range $40 < \sigma < 500$. For small values of σ (<30) lattice artifacts are very strong and the spectral dimension becomes irregular. Because of the finite volumes of configurations, for very large σ ($\gg 500$), the spectral dimension d_S falls down to zero. In the presented range, the measured spectral dimension d_S is very well expressed by the formula

$$d_S(\sigma) = a - \frac{b}{c + \sigma} = 4.02 - \frac{120}{58 + \sigma}, \tag{5.26}$$

where variables a, b and c were obtained from the best fit. As observed, the spectral dimension depends on a diffusion time, and thus it is *scale dependent*. Small σ, means that the diffusion process probes only the nearest vicinity of the initial point. Extrapolation of results gives the *short-distance* limit of the spectral dimension

$$d_S(\sigma \to 0) = 1.95 \pm 0.10.$$

In the *long-distance* limit the spectral dimension tends to

$$d_S(\sigma \to \infty) = 4.02 \pm 0.05.$$

The short-range value of the spectral dimension $d_S = 2$, much smaller than the scaling dimension d_H, suggests a fractal nature of geometries appearing in the path integral at short distances. At long distances $d_S = 4$, and configurations resemble a smooth manifold. Amazingly, such non-trivial scale dependence of the spectral dimension of the quantum space-time, the same infrared ($d_S = 4$) and ultraviolet ($d_S = 2$) behavior, is also present in Hořava–Lifshitz gravity [16] and in the renormalization group approach [19] (see the article by Reuter and Saueressig in this volume).

5.4 Quantum Fluctuations

As we have seen, the dynamically emerging background geometry agrees strikingly well with the solution of the mini-superspace model. By investigating properties of the semi-classical limit of the lattice approach, we will check if quantum fluctuations around the classical trajectory (5.14) are also correctly described by the effective mini-superspace action (5.13). Nevertheless, it should be clearly stated that these considerations are truly non-perturbative, and take into account both a very important influence of the *entropy factor*, which does not depend on bare coupling constants, as well as the bare action (5.4). Based on numerical data obtained by computer simulations, we construct, within the semi-classical approximation, the effective action describing discrete spatial volume $N(i)$ and compare it with the mini-superspace action (5.13). The effective action comes into existence because of a subtle interplay between the entropy and the bare action (5.4).

Let us denote the deviation of the three-volume $N(i)$ from the expectation value $\bar{N}(i)$ by

$$\eta_i = N(i) - \bar{N}(i).$$

Imitating the path integral approach to quantum mechanics, $N(i)$ describes the position at discrete time i of a non-physical particle trajectory, giving a contribution to the partition function. Likewise, η_i is a fluctuation from the classical trajectory $\bar{N}(i)$. In the semi-classical approximation, the spatial volume fluctuations η_i are described by a quadratic form $\mathbf{P}$, obtained by the quadratic expansion of the effective action around the classical trajectory:

$$S[N = \bar{N} + \eta] \approx S[\bar{N}] + \frac{1}{2} \sum_{i,j} \eta_i \, \mathbf{P}_{ij} \, \eta_j + O(\eta^3), \qquad (5.27)$$

where the sum is performed over time slices $i, j = 1, \ldots, T$.

The $\mathbf{P}$ matrix carries information about quantum fluctuations and may be extracted from numerical data. In analogy to $\langle N(i) \rangle$ (cf. (5.9)), we measure the covariance matrix $\mathbf{C}$ of volume fluctuations using Monte Carlo techniques,

$$\mathbf{C}_{ij} \equiv \langle \eta_i \eta_j \rangle \approx \frac{1}{K} \sum_{k=1}^{K} \left(N^{(k)}(i) - \bar{N}(i) \right)\left(N^{(k)}(j) - \bar{N}(j) \right). \qquad (5.28)$$

If the quadratic approximation describes properly quantum fluctuations around the average $\bar{N}$, the propagator $\mathbf{C}$ and the matrix $\mathbf{P}$ are directly related, $\mathbf{C}_{ij} = \mathbf{P}_{ij}^{-1}$.

For numerical convenience the measurements were performed only for triangulations with a fixed total volume $N_{\text{tot}} \equiv \sum_{i=1}^{T} N(i)$. This constraint imposes on the covariance matrix $\mathbf{C}$ the existence of a zero mode, with corresponding constant eigenvector $e_j^0 = 1/\sqrt{T}$, preventing the straightforward inversion of $\mathbf{C}$. In order to invert the matrix $\mathbf{C}$ we project it on the subspace orthogonal to the zero mode e^0 and then perform the inversion. Details of this procedure are described in [12].

After measuring in Monte Carlo simulations the covariance matrix $\mathbf{C}$, we get the empirical Sturm–Liouville operator $\mathbf{P}$ which can be compared with the predictions of the mini-superspace model. The empirical $\mathbf{P}$ matrix has an expected tridiagonal structure with a high accuracy. The tridiagonal form suggests that the effective action describing fluctuations of $N(i)$ is quasi-local in time. The action consists of the kinetic part, which couples volumes of successive slices providing the non-zero subdiagonal elements of $\mathbf{P}$, and the potential part, which contributes only to the diagonal.

In [11] it was shown that the effective action corresponds to a discretization of the mini-superspace action (5.13) up to an overall sign. Below we derive a discrete version of the mini-superspace action with reversed sign,

$$S[v] = \int d\tau \left(\alpha \frac{\dot{v}^2}{v} + \beta v^{1/3} - 2\Lambda v \right), \tag{5.29}$$

which later will be compared to the empirical action. We have incorporated the factor $1/(24\pi G)$ into constants α, β and Λ. The discretization procedure is not unique, but up to the order considered here, all discretizations are equivalent. We substitute the physical volume $v(\tau)$ with the discrete volume $N(i)$ which may be treated as a continuous variable inside the blob. The *stalk* region is governed by very strong lattice artifacts, and therefore is not reliably treated in the semi-classical approximation. The standard discretization of the time derivative is $\dot{v} \to N(i+1) - N(i)$, and the kinetic part is written as

$$\alpha \frac{\dot{v}^2}{v} \to g_1 \frac{(N(i+1) - N(i))^2}{N(i+1) + N(i)}.$$

Because both matrices $\mathbf{C}$ and $\mathbf{P}$ are symmetric, the discretized terms also must be symmetric in i and $i+1$. The potential part is discretized straightforwardly,

$$\beta v^{1/3} - 2\Lambda v \to g_2 N(i)^{1/3} - g_3 N(i).$$

Therefore, a discretized, dimensionless version of action (5.29) is given by

$$S[N] = \sum_i g_1 \frac{(N(i+1) - N(i))^2}{N(i+1) + N(i)} + g_2 N(i)^{1/3} - g_3 N(i). \tag{5.30}$$

Further, we show that the discrete effective action (5.30) describes not only the average $\bar{N}(i)$ (5.10), what follows from the classical trajectory of Eq. (5.29), but indeed also the measured fluctuations $\eta(i)$.

5.4.1 The Effective Action

The $\mathbf{P}$ operator can be decomposed into the *kinetic* part $\mathbf{P}^{\text{kin}}$ and the *potential* part $\mathbf{P}^{\text{pot}}$,

$$\mathbf{P} = \mathbf{P}^{\text{kin}} + \mathbf{P}^{\text{pot}}.$$

Only the kinetic part $\mathbf{P}^{\text{kin}}$ contributes to the sub-diagonal elements of the tridiagonal matrix $\mathbf{P}$. Because the square of the time derivative couples the preceding and following time steps, and because the covariance matrix is symmetric, $\mathbf{P}^{\text{kin}}$ should be

Fig. 5.8 Kinetic term: The directly measured expectation values $\bar{N}(i)$ (*black line*), compared to $\frac{g_1}{k_i}$ (*thick line*) extracted from the measured covariance matrix **C** for $K_0 = 2.2$, $\Delta = 0.6$ and various total volumes N_{tot} ranging from 20,000 to 160,000 simplices. The theoretical prediction $\frac{g_1}{k_i} = \frac{1}{2}(\bar{N}(i) + \bar{N}(i+1))$ is realized with a very high accuracy. The value of g_1 is constant for all volumes N_{tot}

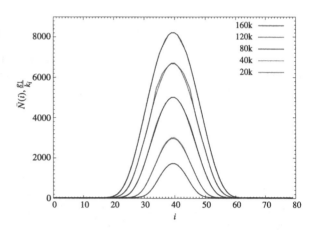

a symmetric tridiagonal matrix, such that the sum of elements in a row or a column is always zero. It can be decomposed into parts linearly dependent on k_i:

$$\mathbf{P}^{\text{kin}} = \sum_{i=1}^{T} k_i \mathbf{X}^{(i)}, \quad \mathbf{X}^{(i)}_{jk} = \delta_{ij}\,\delta_{ik} + \delta_{(i+1)j}\,\delta_{(i+1)k} - \delta_{(i+1)j}\,\delta_{ik} - \delta_{ij}\,\delta_{(i+1)k},$$

$$(5.31)$$

where $\mathbf{X}^{(i)}$ is a matrix corresponding to the discretization of the second time derivative ∂_t^2 at a time $t = i$.

Neglecting details of the zero mode removal, the potential part is diagonal,

$$\mathbf{P}^{\text{pot}} = \text{Diag}(\{u_i\}) = \sum_{i=1}^{T} u_i \mathbf{Y}^{(i)}, \quad \mathbf{Y}^{(i)}_{jk} = \delta_{ij}\delta_{ik}.$$

The decomposition of the empirical matrix **P** into a kinetic and potential part is done using the least square method. We find such values of $\{k_i\}$ and $\{u_i\}$ for which the matrix $\mathbf{P}^{\text{kin}} + \mathbf{P}^{\text{pot}}$ is as close as possible to the empirical matrix **P**, i.e., we minimize the residual sum of squares

$$\text{RSS}[\{k_i\}, \{u_i\}] \equiv \text{Tr}\big[\mathbf{P} - \big(\mathbf{P}^{\text{kin}} + \mathbf{P}^{\text{pot}}\big)\big]^2. \qquad (5.32)$$

We will omit details of the parameter fitting. Equation (5.32) is quadratic in $\{k_i\}$ and $\{u_i\}$, and the fitting boils down to calculating traces of products of matrices $\mathbf{X}^{(i)}$ and $\mathbf{Y}^{(j)}$. We will show that the fitted values of the kinetic term $\{k_i\}$, obtained by minimizing residues (5.32), are indeed in agreement with the kinetic part of the discrete mini-superspace action (5.30). The quadratic expansion (5.27) of the action (5.30) gives

$$k_i = -\mathbf{P}_{ii+1} = -\frac{\partial^2 S[N]}{\partial N(i)\partial N(i+1)}\bigg|_{N=\bar{N}} = g_1 \frac{8\bar{N}(i)\bar{N}(i+1)}{(\bar{N}(i) + \bar{N}(i+1))^3}. \qquad (5.33)$$

Fig. 5.9 The extracted potential term u_i as a function of average volume $\bar{N}(i)$. The fit $c_2\bar{v}_t^{-5/3}$ presents the behavior expected for the mini-superspace model. The visible points correspond to the blob region

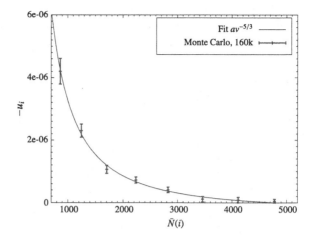

In the zeroth order approximation, $\bar{N}(i) \approx \bar{N}(i + 1)$, we expect the following behavior of the kinetic term:

$$\frac{g_1}{k_i} = \frac{(\bar{N}(i) + \bar{N}(i+1))^3}{8\bar{N}(i)\bar{N}(i+1)} \approx \frac{1}{2}\big(\bar{N}(i) + \bar{N}(i+1)\big). \tag{5.34}$$

Figure 5.8 presents the plot of g_1/k_i for the empirical values of k_i and various total volumes N_{tot}. The theoretical fit (5.34) agrees extremely well with the measured quantities. Additionally, the effective coupling constant g_1 does not depend on N_{tot} in the margin of error. For $K_0 = 2.2$, $\Delta = 0.6$, we measured $g_1 = 0.038 \pm 0.002$. The kinetic part of the quantum fluctuations is indeed described by the mini-superspace action (5.29).

Further we will directly show that values of the potential term $\{u_i\}$ extracted from the empirical inverse propagator $\mathbf{P}$ also agree with the mini-superspace model. Within this framework, we expect that

$$u_i = U''\big(\bar{N}(i)\big) = -\frac{2}{9}g_2\bar{N}(i)^{-5/3}. \tag{5.35}$$

Figure 5.9 shows the measured values of coefficients u_i extracted from the empirical matrix $\mathbf{P}^{\text{pot}}$. Because of large statistical errors, it is not an easy task to determine u_i. The physically interesting region of large volumes corresponds to relatively small values of u_i as they are expected to fall as $\bar{N}(i)^{-5/3}$. Due to the existence of the zero mode, the blob region is also affected by the huge contribution from the stalk. Moreover, in analogy with the situation in the ordinary path-integral approach to quantum mechanics, when the time step approaches zero in the continuum limit $N_{\text{tot}} \to \infty$, the potential term is sub-dominant with respect to the kinetic term for individual space-time histories in the path integral.

Nevertheless, due to a sufficiently long Monte Carlo sample, the obtained results allowed us to confirm that indeed Eq. (5.35) is in agreement with measurements. Figure 5.9 presents the measured coefficients $-u_i$ as a function of the average three-volume $\bar{N}(i)$. The error bars shown on the plot were estimated using the Jackknife

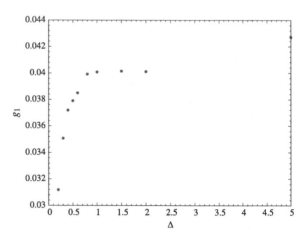

Fig. 5.10 The measured effective coupling constant g_1 as a function of bare coupling constant Δ for $K_0 = 2.2$. The B–C transition point is located at about $\Delta^{\mathrm{crit}} = 0.05$. When approaching phase B from phase C, the coupling constant g_1 diminishes and the fluctuations grow, as expected when reaching phase transition point

method. Such a form allows us to directly compare the potential coefficients with theoretical predictions $-u_i \propto \bar{N}(i)^{-5/3}$. The selected range of $\bar{N}(i)$ corresponds to the *bulk*. The best fit of the form $f(x) = a x^{-c}$ to the empirical values u_i as a function of $\bar{N}(i)$ gives $c = -1.658 \pm 0.096$. The measured exponent coefficient c is very close to the theoretical value $c = -5/3$. The fit $f(x) = a x^{-5/3}$, corresponding to the potential part of action (5.30), is presented in Fig. 5.9 with a thin line. The agreement with the data is good; the potential part of the effective action is indeed given by $U(x) = g_2 x^{1/3} - g_3 x$. Apart from obtaining the correct power $u_i \propto \bar{N}^{-5/3}(i)$, the coefficient in front of this term is also independent of N_{tot}.

5.4.2 Flow of the Gravitational Constant

The quantum fluctuations of the three-volume are very accurately described by the discrete, dimensionless effective action

$$S\big[N(i)\big] = \sum_i g_1 \frac{(N(i+1) - N(i))^2}{2N(i)} + g_2 N^{1/3}(i), \qquad (5.36)$$

where we have omitted the cosmological constant term, since during the measurements the total volume N_{tot} was fixed. This action comes out as a discretization of the mini-superspace action (5.13) with the opposite sign, which solves the problem of unboundedness. Let us note that it is justified to use the semi-classical approximation as the distribution of spatial volumes $N(i)$ in the bulk is given by Gaussian fluctuations around the mean.

Using Eqs. (5.16), (5.17) and (5.21)–(5.23), we can rewrite the above discrete action in terms of the physical volume $v(\tau)$,

$$S\big[v(\tau)\big] = \frac{g_1 g_{tt}}{2\sqrt{N_{\mathrm{tot}} C_4 a^4}} \int d\tau \left[\frac{\dot{v}^2}{v} + \tilde{g}_2 v^{1/3} \right]. \qquad (5.37)$$

It is natural to identify the coupling constant G multiplying the effective action (5.13) with Newton's gravitational constant G. Using Eq. (5.23), we get the following relations between the gravitational constant G and the effective constant g_1 [11, 12]:

$$G = \frac{2\sqrt{N_{\text{tot}}}C_4 a^4}{24\pi g_1 g_{tt}} = \frac{a^2}{g_1}\frac{\sqrt{C_4}B^2}{3\sqrt{6}}. \tag{5.38}$$

In order to keep fixed the physical constant G, and thus the amplitude of fluctuations $\sqrt{\langle(\delta v(\tau))^2\rangle} \propto g_1^{-1/2} N_{\text{tot}}^{1/4} a^2$, when taking the continuum limit $a \to 0$ one has to tune the effective coupling constant $g_1 \propto a^2$. This means that in terms of the lattice volume $N(i)$ fluctuations should diverge, and this happens when we approach a second- or higher-order transition line. Therefore it is important to determine the order of the transition. Figure 5.10 shows the measured effective coupling constant g_1 for various values of Δ. As mentioned before, the effective coupling constant g_1 does not depend on N_{tot} when the bare coupling constants are fixed, and the same is true for the classical trajectory $\bar{v}(t)$. Therefore, one also has to properly tune the bare coupling constants so that the *effective* coupling constant satisfies $g_1 a^{-2} = \text{const}$ while taking the limits $N_{\text{tot}} \to \infty$ and $a \to 0$. Indeed, when we approach the B–C transition line g_1 tends to zero.

Using relation (5.38) we can express the cut-off length a in terms of the Planck length, and thus estimate the size of the Universe generated in computer simulations. Let us recall that in natural units $G = \ell_{\text{Pl}}^2$. For the bare coupling constants $K_0 = 2.2$, $\Delta = 0.6$ we measured the quantities: $K_4^{\text{crit}} = 0.922$, $\xi = N_{32}/N_{41} = 1.30$, $\alpha = 0.5858$, $C_4 = 0.0317$, $g_1 = 0.038$, which results in $a \approx 1.9\ell_{\text{Pl}}$ and the linear size πR of the universe built from 160,000 simplices is about $20\ell_{\text{Pl}}$. The quantum de Sitter universes studied here are therefore quite small, and quantum fluctuations around their average shape are large (cf. (5.5)). Surprisingly, the semi-classical mini-superspace formulation gives an adequate description of the measured data, at least for the volume profile.

5.5 The Geometry of Spatial Slices

Let us look deeper into the geometry of spatial slices. A spatial slice is a leaf of the imposed global proper-time foliation and is labeled by a discrete time index i. Each such hypersurface is a three-dimensional triangulation built of equilateral spatial tetrahedra, more precisely, a piecewise linear manifold of topology S^3. However, it does not mean that the geometry of slices is close to the geometry of a three-dimensional sphere.

5.5.1 The Hausdorff Dimension

Let us denote the number of tetrahedra building slice i by the discrete three-volume $n_3 \equiv N(i)$. A basic observable defined on a slice is the number of tetrahedra $n(r, i_0)$

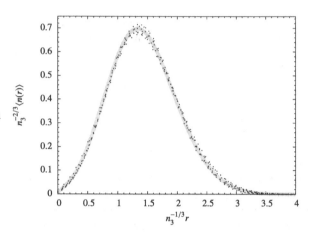

Fig. 5.11 Scaled values of the radius $n_3^{-1/d_H} r$ and shell area $n_3^{-1+1/d_H} n(r)$ for $d_H = 3$. Data points for various values of slice volume n_3 overlap. The *gray strip* plots the scaled radial volume averaged over all data points. Measurements were performed at $K_0 = 2.2$ and $\Delta = 0.6$

at a three-dimensional distance r from some initial tetrahedron i_0. At distance $r = 0$ only the initial tetrahedron is counted and $n(0, i_0) = 1$. For such definition, $n(r, i_0)$ corresponds to an area of the shell of radius r. Summing up the area over all shells gives the discrete volume of a slice n_3. Let $n(r)$ denotes the average of $n(r, i_0)$ over all n_3 initial tetrahedra i_0,

$$n(r) = \frac{1}{n_3} \sum_{i_0=1}^{n_3} n(r, i_0), \qquad n_3 = \sum_{r=0}^{r_{max}} n(r, i_0).$$

We will investigate scaling properties of $n(r)$ with respect to the slice volume n_3. The Hausdorff dimension of spatial slices may be measured by a comparison of the scaled with volume n_3 values of the radial volume $n(r)$. First, for a large number of Monte Carlo configurations, slices with the same volume n_3 (more or less few tetrahedra) are collected into groups. The average radial volume $n(r)$ within a group n_3 is denoted as $\langle n(r) \rangle_{n_3}$.

For the Hausdorff dimension d_H we expect that the radius r and the average volume $\langle n(r) \rangle_{n_3}$ scaled and normalized in the following way

$$\left(r, \langle n(r) \rangle_{n_3} \right) \to \left(n_3^{-1/d_H} r, n_3^{-1+1/d_H} \langle n(r) \rangle_{n_3} \right) \tag{5.39}$$

overlap for all n_3. We define the error of the overlap of the scaled points and find such value of d_H which minimizes the dispersion. The best fit is obtained for $d_H = 2.94 \pm 0.05$. Figure 5.11 presents the measured values of $\langle n(r) \rangle_{n_3}$ scaled according to (5.39) with $d_H = 3$ and for various values of n_3 between 1,000 and 4,000 tetrahedra.

The measured value of d_H is independent of the coupling constants K_0 and Δ, as long as we stay well inside the phase C. This results is true if we consider the ensemble average of the slice geometry, but it does not mean that individual spatial slices resemble a smooth three-dimensional geometry.

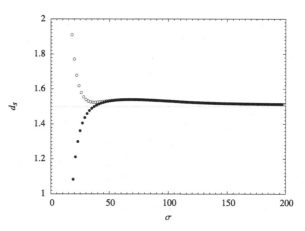

Fig. 5.12 Spectral dimension d_S of spatial slices as a function of diffusion time σ. For short diffusion times, a split for even (*empty*) and odd (*filled*) values of σ is observed arising from the discrete structure. The measured values of d_S converge to the thin line corresponding to $d_S = 1.5$

5.5.2 Spectral Dimension

We measure the spectral dimension of spatial slices in the same way as for the whole simplicial manifolds. The probability of finding a diffusing particle in tetrahedron i after a diffusion time σ and starting at tetrahedron i_0 is given by the probability density $\rho(i, i_0; \sigma)$. The discrete diffusion equation, describing the evolution of the probability density, can be written as $\rho(i, i_0; \sigma + 1) = \frac{1}{4} \sum_{j \leftrightarrow i} \rho(j, i_0; \sigma)$, where the sum is over all tetrahedra j adjacent to i. For a starting tetrahedron i_0, chosen at random, we set the initial condition $\rho(i, i_0; \sigma = 0) = \delta_{i i_0}$. By iterating the diffusion equation, we calculate the return probability $P(\sigma, i_0) \equiv \rho(i_0, i_0; \sigma)$ for successive discrete diffusion steps σ. Further, we compute the *average return probability* $P(\sigma) \equiv \langle\langle P(\sigma, i_0) \rangle_{i_0} \rangle_{MC}$ by averaging over initial points and configurations. For each configuration we consider only the central slice. The spectral dimension d_S is obtained from the return probability using the definition (5.25). For a three-sphere geometry, the spectral dimension d_S is equal 3 for short diffusion times, and d_S will tend to zero for longer times. Figure 5.12 shows the values of the spectral dimension d_S as a function of the diffusion time σ, determined by numerical simulation using the definition (5.25) for a randomly chosen typical configurations in phase C. Due to the discrete lattice structure, for small values of σ a split for even and odd diffusion times is observed. Because of the finite volumes of the spatial slices, for very large σ, d_S falls down to zero. For the intermediate region, there is a plateau of the spectral dimension at $d_S \approx 1.5$.

The significant difference between the measured Hausdorff dimension of spatial slices, $d_H \approx 3$, and the measured spectral dimension, $d_S \approx 1.5$, is an indication of the fractal nature of the slices. Indeed, this was proved in a direct way [14]. The three-dimensional spatial slices reveal a large number of *minimal necks*. A minimal neck consists of four triangles forming a tetrahedron, but where this tetrahedron does not belong to the triangulation. They provide the three-dimensional triangulation with a tree-structure (for S^3 geometry). At many random places, a branch bifurcates into two or more branches. Most probably when the size of the slice grows to infinity, we would observe the fractal structure of branched polymers. A similar structure is present in three-dimensional Euclidean dynamical triangulations, where so-called *baby Universes* separated by minimal necks are observed [22].

5.6 Conclusions

The model of causal dynamical triangulations is a non-perturbative and background-independent approach to quantum gravity. The foundations of this model are very simple. It is a mundane lattice field theory with a piecewise linear manifold serving as a regularization of general relativity. The introduction of Wick rotation allows us to use very powerful Monte Carlo techniques and calculate quantum expectation values of observables.

Based on the Monte Carlo measurements we predict the existence of three phases within the CDT model. In the physically most interesting phase, so called de Sitter phase, the time-translational symmetry is spontaneously broken and the scale factor as a function of time behaves as a bell-shaped distribution. Recent results give a strong evidence that the Universe which emerges dynamically in causal dynamical triangulations is genuinely *four-dimensional*. Its geometry corresponds to de Siter space, the maximally symmetric solution to the classical Einstein equations in the presence of a positive cosmological constant. At large scales both the Hausdorff and spectral dimensions are equal to 4. CDT presents a picture of the Universe with superimposed finite quantum fluctuations around the classical trajectory, which are well described semi-classically. The measurements of the covariance matrix allowed us to reconstruct the discrete effective action describing quantum fluctuations of the three-volume $N(i)$. This action was identified with the discretization of the mini-superspace action. In the CDT model, however, no reduction of degrees of freedom is introduced. Due to the identification, the effective coupling constant can be related to the physical gravitational constant, giving a recipe of how to obtain a meaningful continuum limit and expressing the lattice constant in terms of physical units.

The spatial slices of the imposed foliation reveal, however, a fractal structure similar to branched polymers. Although the measurements show that the Hausdorff dimension of the slices is equal to 3, the measured spectral dimension is only half of this value. Indeed, the fractality was confirmed by a direct analysis of tree structures defined in terms of so-called minimal necks.

Acknowledgements The author would like to thank Jerzy Jurkiewicz and Jan Ambjørn for introducing him into this fascinating topic and for a fruitful collaboration. The author acknowledges support by *The Danish Research Council* via the grant "Quantum Gravity and the Role of Black Holes".

References

1. J. Ambjørn, J. Jurkiewicz, R. Loll, Phys. Rev. Lett. **85**, 924 (2000). hep-th/0002050

2. J. Ambjørn, J. Jurkiewicz, R. Loll, Phys. Rev. D **72**, 064014 (2005). hep-th/0505154
3. J. Ambjørn, J. Jurkiewicz, R. Loll, Nucl. Phys. B **610**, 347 (2001). hep-th/0105267
4. J. Ambjørn, J. Jurkiewicz, R. Loll, in *Approaches to Quantum Gravity*, ed. by D. Oriti (Cambridge University Press, Cambridge, 2009), pp. 341–359. hep-th/0604212
5. S.W. Hawking, in *General Relativity: An Einstein Centenary Survey*, ed. by S.W. Hawking, W. Israel (Cambridge University Press, Cambridge, 1979), pp. 746–789
6. G.W. Gibbons, S.W. Hawking, Phys. Rev. D **15**, 2752 (1977)
7. C.W. Misner, K.S. Thorne, J.A. Wheeler, *Gravitation* (Freeman, New York, 1973)
8. R.M. Wald, *General Relativity* (Chicago University Press, Chicago, 1984)
9. J. Ambjørn, S. Varsted, Nucl. Phys. B **373**, 557 (1992)
10. J. Ambjørn, J. Jurkiewicz, R. Loll, Phys. Rev. Lett. **95**, 171301 (2005). hep-th/0505113
11. J. Ambjørn, A. Görlich, J. Jurkiewicz, R. Loll, Phys. Rev. Lett. **100**, 091304 (2008). arXiv:0712.2485
12. J. Ambjørn, A. Görlich, J. Jurkiewicz, R. Loll, Phys. Rev. D **78**, 063544 (2008). arXiv:0807.4481
13. A. Görlich, Acta Phys. Pol. B **39**, 3343 (2008)
14. J. Ambjørn, A. Görlich, J. Jurkiewicz, R. Loll, Phys. Lett. B **690**, 420 (2010). arXiv:1001.4581
15. J. Ambjørn, A. Görlich, S. Jordan, J. Jurkiewicz, R. Loll, Phys. Lett. B **690**, 413 (2010). arXiv:1002.3298
16. P. Hořava, Phys. Rev. Lett. **102**, 161301 (2009). arXiv:0902.3657
17. J.B. Hartle, S.W. Hawking, Phys. Rev. D **28**, 2960 (1983)
18. B. Efron, Ann. Stat. **7**, 1 (1979)
19. O. Lauscher, M. Reuter, J. High Energy Phys. **0510**, 050 (2005). hep-th/0508202
20. J. Ambjørn, S. Jordan, J. Jurkiewicz, R. Loll, Phys. Rev. Lett. **107**, 211303 (2011). arXiv:1108.3932
21. A. Dasgupta, R. Loll, Nucl. Phys. B **606**, 357 (2001). hep-th/0103186
22. J. Ambjørn, S. Jain, G. Thorleifsson, Phys. Lett. B **307**, 34 (1993). hep-th/9303149

Chapter 6
Massive Gravity: A Primer

E.A. Bergshoeff, M. Kovacevic, J. Rosseel, and Y. Yin

Abstract We show that the recently constructed $3D$ higher-derivative "New Massive Gravity theory" is the result of a general procedure that allows one to construct, in the free case, higher-derivative gauge theories for a wide class of "spins" in diverse dimensions. We specify the criterium that the "spin" and dimension need to satisfy in order for the construction to apply. To clarify the general procedure we present examples of higher-derivative gauge theories for the special cases of spin 1 in $D = 3$, 5 and 7 dimensions. We next apply the procedure to spin 2 in $D = 3$ dimensions and show how the New Massive Gravity and Topological Massive Gravity theories are constructed. Both theories allow interactions. We indicate how and under which conditions the $3D$ New Massive Gravity theory can be extended to $D = 4$ dimensions and the $3D$ Topological Massive Gravity theory can be extended to $D = 7$ dimensions. We discuss the issue of interactions of these two theories.

6.1 Introduction

These lectures deal with higher-derivative theories of gravity. Consider first Einstein's theory of gravity as a theory of interacting massless spin 2 particles around a Minkowski space-time background. The dynamics of this theory is described by the Einstein-Hilbert action which is second-order in the derivatives. As is well-known, Einstein's theory of gravity is perturbative non-renormalizable when expanded around a flat Minkowski spacetime. One way to try to cure this problem is

E.A. Bergshoeff (✉) · M. Kovacevic · J. Rosseel · Y. Yin
Centre for Theoretical Physics, University of Groningen, Nijenborgh 4, 9747 AG Groningen, The Netherlands
e-mail: E.A.Bergshoeff@rug.nl

M. Kovacevic
e-mail: maikovacevic@gmail.com

J. Rosseel
e-mail: J.Rosseel@rug.nl

Y. Yin
e-mail: Y.Yin@rug.nl

G. Calcagni et al. (eds.), *Quantum Gravity and Quantum Cosmology*,
Lecture Notes in Physics 863, DOI 10.1007/978-3-642-33036-0_6,
© Springer-Verlag Berlin Heidelberg 2013

by adding higher-derivative terms to the Einstein-Hilbert action in order to obtain better behaving propagators that could lead to a perturbative renormalizable theory.

Already in the seventies of the previous century a systematic investigation of the effect of adding fourth-order derivative terms to the Einstein-Hilbert action was undertaken by Stelle [1, 2]. He considered the most general such terms:

$$\mathscr{L} \sim R + a\left(R_{\mu\nu}{}^{ab}\right)^2 + b(R_{\mu\nu})^2 + cR^2. \tag{6.1}$$

Here $R_{\mu\nu}{}^{ab}$, $R_{\mu\nu}$, R are the Riemann tensor, Ricci tensor, Ricci scalar, respectively, and a, b and c are generic coefficients with dimension of one over mass squared. The outcome of his studies was that for generic coefficients the theory is renormalizable[1] but not unitary. It is easy to understand why this is the case. In the above Lagrangian the fourth order derivative terms act like the kinetic terms and the Einstein-Hilbert term as the mass term. Since the kinetic terms are fourth-order in derivatives they can generically be written as the product of two second-order operators. It turns out that one operator corresponds to a massless graviton and the other one to a *massive* graviton. Unfortunately, it turns out that the signs of the two kinetic terms are opposite and that is why ghosts cannot be avoided.

The occurrence of a massive and massless graviton with opposite signs is a generic feature of any dimension. For each dimension this would imply a break-down of unitarity except for three dimensions since in three dimensions there is no massless graviton! This implies that one is only left with the massive graviton only whose kinetic term can always be given the correct sign by adjusting the over-all sign of the Lagrangian. This is the reason that unitary higher-derivative theories of gravity do exist in three dimensions. There is one more special situation that is less obvious. It turns out that when expanding around an AdS vacuum solution instead of a Minkowski space-time the coefficient in front of the linearized Einstein-Hilbert term gets shifted with a term involving the cosmological term Λ. The value of Λ can be chosen such that the coefficient in front of this term vanishes which has the effect that there is no massive graviton! This special so-called "critical" point in parameter space leads to the so-called "critical" gravity theories. Note that these critical gravity theories are not limited to three dimensions. They will be shortly discussed later in these lectures.

It turns out that in three dimensions there are not one but two unitary higher-derivate gravity theories. They are called Topological Massive Gravity (TMG) [3] and New Massive Gravity (NMG) [4, 5]. An important difference between the two theories is that only one of them (NMG) is parity-invariant. In these lectures we will discuss the general procedure for constructing these TMG and NMG theories. This also shows the way of how to extend these constructions, at least at the linearized level, to higher than three dimensions.

---

[1]This is not the case for special choices of the coefficients. In particular, scalar gravity, with $a = b = 0$ and Weyl gravity, in which case a, b and c are chosen such that the Weyl tensor squared combination is obtained, are *not* perturbative renormalizable.

The organization of these lectures is as follows. In Sect. 6.2 we will discuss the general procedure of constructing higher-derivative gauge theories mentioned above for general dimensions and general spin. We will do this on hand of Young tableaux thereby avoiding too many explicit (and complicated!) formulae. In Sect. 6.3 we will elucidate this procedure by working out several examples corresponding to "spin 1" fields. By this we mean fields that carry an index structure corresponding to a Young tableaux with one column. Subsequently, in Sect. 6.4 we will discuss the "spin 2" case, i.e. we will discuss fields whose symmetry structure correspond to Young tableaux with two columns. This will include the construction of the $3D$ NMG and $3D$ TMG theories and a discussion of the higher-dimensional generalization (at the linearized level) of these theories. This will lead to the construction of a new $4D$ NMG and $7D$ TMG theory which will be briefly discussed. In the conclusions we will address a few open issues. We have included an Appendix which contains the four exercises that were mentioned during the lectures together with their answers.

6.2 General Spin

In this section we will explain the general procedure of how to construct a higher-derivative gauge theory for a massive field in a pictorial way using Young tableaux. The precise formulae, corresponding to specific examples, will be presented in the following sections. First, we will explain in Sect. 6.2.1 how to "boost up the derivatives" of a given massive theory. Next, in Sect. 6.2.2 we will explain how to "take the square root" of a massive theory. The techniques of the first subsection may then be applied to boost up the derivatives of this "square root" theory.

6.2.1 "Boosting up the Derivatives"

Following [6, 7],[2] the starting point is a field S in D dimensions with indices corresponding to a $GL(D, \mathbb{R})$ Young tableau with s columns. In order to elucidate the general procedure, we consider as an example a $4D$ field with indices corresponding to the following Young tableau with $s = 2$ columns:

$$S \sim \boxed{}\!\boxed{} \tag{6.2}$$

For simplicity, we will restrict in the discussion below to the cases $s = 1$ and $s = 2$ only. Most of the discussion, however, is valid for any s. In order that the field S

---

[2]For a recent discussion, see also [8].

describes a massless spin[3] corresponding to the same Young tableau but with the indices now referring to the $SO(D-2)$ little group[4] the field S should transform under a set of gauge transformations whose parameters λ correspond to $GL(D, \mathbb{R})$ Young tableaux that are obtained from the original tableau by deleting one box in all possible ways such as to obtain an allowed Young tableau. For our example (6.2) given above this leads to gauge parameters λ_1 and λ_2 corresponding to the following two $GL(D, \mathbb{R})$ Young tableaux

$$\lambda_1 \sim \boxed{} \qquad \lambda_2 \sim \boxed{} \tag{6.3}$$

This corresponds to a generic 2-tensor gauge parameter $\lambda = \lambda_1 + \lambda_2$. The transformation rule of the gauge field S is obtained by hitting the parameters $\lambda_{1,2}$ with a derivative and projecting to the original Young tableau:

$$\delta \; \boxed{} = \boxed{}^{\partial} + \boxed{}_{\partial} \tag{6.4}$$

or, shortly, $\delta S = \partial \lambda_1 + \partial \lambda_2$.

For a Young tableau with s columns a gauge-invariant curvature is obtained by adding one box, representing a derivative, to each column. This leads to a curvature with s derivatives. Following the $4D$ spin 2 case we will call this curvature the "generalized" Riemann tensor $R(S)$ or, shortly, the Riemann tensor. For our example (6.2) we obtain

$$R(S) \sim \boxed{\begin{array}{c} \, \partial \\ \partial \end{array}} \tag{6.5}$$

That this Riemann tensor is gauge-invariant can be seen from the fact that the substitution of the transformation rule (6.4) into the expression (6.5) always leads to a column with two derivatives and hence a vanishing result since two derivatives commute [8].

We now construct out of the Riemann tensor $R(S)$ another tensor $G(S)$ by taking the dual of each column. Due to the Bianchi identities of the Riemann tensor this new tensor is divergence-free in *each* of its indices. We now assume that the field S and the tensor $G(S)$ have indices corresponding to the *same* Young tableau. For the example given in Eq. (6.2) this assumption is valid. Assuming this property we can identify $G(S)$ with the "generalized Einstein" tensor for S and write down the following equations of motion for S:

$$G(S) = 0. \tag{6.6}$$

---

[3]In $3D$ there is no concept of massless spin. In $D = 3, 4$ a Young tableau with s columns always describes (massless or massive) degrees of freedom of spin s or less. For $D > 4$ the specification of spin requires more than one number. For ease of notation we will call in these lectures any field with indices corresponding to a $GL(D, \mathbb{R})$ Young tableau with s columns a "spin-s" field.

[4]To obtain an irreducible $SO(D-2)$ representation from the field S one should first require that all indices only take values in the $(D-2)$ transverse directions and, next, that all traces in any of these transverse directions vanish.

Table 6.1 This table lists, for $s = 1, 2$, all the $GL(D, \mathbb{R})$ representations of S in $3 \leq D \leq 7$ dimensions for which the massless representation describes zero physical degrees of freedom. The star indicates that the equation of motion of the corresponding field S cannot be integrated to a Lagrangian. The $s = 2$ Young tableaux with a † indicate the family of fields S that are all dual to a symmetric tensor

	$D = 3$	$D = 4$	$D = 5$	$D = 6$	$D = 7$
$s = 1$	☐		⬚*		⬚
$s = 2$	☐☐†	⬚†	⬚ ⬚†	⬚ ⬚†	⬚ ⬚†

Restricting to $s = 1, 2$, we find that for a single-column $s = 1$ Young tableau with p boxes (p odd) and for any two-column $s = 2$ Young tableau these equations of motion can be integrated to the following Lagrangian for S:[5]

$$\mathcal{L} \sim SG(S). \tag{6.7}$$

Making use of the property that the Einstein tensor $G(S)$ is divergence-free in each of its indices one can show that this Lagrangian is invariant under the gauge transformations (6.4). The corresponding Euler-Lagrange equations imply the equations of motion (6.6). To derive these equations we use the fact that the Einstein tensor $G(S)$ defines a rank s self-adjoint differential operator. The special thing about the cases described by the Lagrangian (6.7) is that the vanishing of the Einstein tensor $G(S)$ implies the vanishing of the Riemann tensor $R(S)$ since, by construction, the two are dual to each other. Since the Riemann tensor is zero, the original field S is a pure gauge and, therefore, does not describe any massless physical degrees of freedom. The fact that there are no non-trivial solutions S of the equation $G(S) = 0$ is the crucial property that underlies the construction of the higher-derivative massive gauge theories we are going to describe below.

For a single-column $s = 1$ Young tableau with p boxes the fact that S and $G(S)$ correspond to the same Young tableau implies that the following relation between p and D must hold:

$$p = \frac{1}{2}(D - 1). \tag{6.8}$$

Similarly, for a Young tableau with $s = 2$ columns, of height p and q, we obtain the condition

$$p + q = D - 1, \quad p, q \neq 0. \tag{6.9}$$

Consider now a field S corresponding to a given $GL(D, \mathbb{R})$ Young tableaux. Following [6, 9] we may write down the massive "generalized" Fierz-Pauli (FP) equa-

---

[5]For $s = 1$ and p even this Lagrangian would be a total derivative.

tions for this field as follows:

$$\left(\Box - m^2\right)S = 0, \qquad S^{\text{tr}} = 0, \qquad \partial \cdot S = 0. \tag{6.10}$$

Here S^{tr} indicates the trace of any of the two indices carried by S while $\partial \cdot S$ denotes the divergence taken with respect to any of the indices of S. The effect of the algebraic and differential subsidiary conditions given in Eq. (6.10) is that the massive physical degrees of freedom described by S transform according to a $SO(D-1)$ Young tableau that is equal to the original $GL(D, \mathbb{R})$ Young tableau that corresponds to S. We now assume that the massless representation corresponding to S describes zero degrees of freedom. This requires imposing the restrictions (6.8) and (6.9), for $s = 1$ and $s = 2$, respectively. For $3 \leq D \leq 7$ this leads to the cases listed in Table 6.1. Note that for $s = 2$ we obtain in each dimension a mixed-symmetry tensor that is the massive dual of a symmetric tensor [10]. This family of fields is indicated with a dagger in Table 6.1. They play a special role in the construction of "New Massive Gravity" theories beyond $3D$, see Sect. 6.4.3 [11].

Assuming from now on that we restrict to the cases listed in Table 6.1 we know that the Einstein tensor $G(S)$ is in the same representation as S. We may now exploit this fact and solve the divergence-free condition $\partial \cdot S = 0$ by making the following replacement in the massive equations of motion (6.10):

$$S = G(T), \tag{6.11}$$

for some other field T that is in the same $GL(D, \mathbb{R})$ representation as S. Note that after the replacement (6.11) one ends up with a gauge-theory for T although the starting point (6.10) is not a gauge theory. The important thing is that the equation $G(T) = 0$ does not have any non-trivial solution which is not a pure gauge. Therefore, the replacement (6.11) represents *all* solutions of the equation $\partial \cdot S = 0$. This implies that the degrees of freedom remain the same independent of whether they are described in terms of S or T. The substitution (6.11) therefore leads us to an equivalent higher-derivative gauge theory for the massive field T with the following equations of motion:

$$\left(\Box - m^2\right)G(T) = 0, \qquad G(T)^{\text{tr}} = 0. \tag{6.12}$$

For $s = 2$ the above procedure was first applied to the case of a symmetric tensor in $3D$ in which case it leads to the (linearized) equations of motion of NMG [4, 5].

For Young tableaux with $s = 1$ or $s = 2$ columns one can write down actions corresponding to the equations of motion (6.10) and the boosted up equations of motion (6.12).[6] However, it is not guaranteed that after boosting up the derivatives the action will not contain ghosts. We consider first the $s = 1$ case. It turns out that for a $(2k - 1)$-form T in $D = 4k - 1$ dimensions ghosts will occur. The reason for this is that in these dimensions the "helicities" described by the $(2k-1)$-form T split

---

[6]Starting from $s = 3$ one needs to introduce an extra set of auxiliary fields to write down such actions.

into two groups which are not in the same induced representation of the Poincaré group. They can only be mapped to each other by a parity transformation. Since the replacement (6.11) breaks parity in these cases one does end up with a relative minus sign between the kinetic terms of these two groups of helicities. Therefore, one cannot adapt the overall sign of the action such as to avoid ghosts. On the other hand, for a $2k$-form in $D = 4k + 1$ dimensions the equations of motion cannot be integrated to an action and the issue does not arise. It turns out that for $s = 2$ the issue of ghosts does not arise since the replacement (6.11) never breaks parity for $s = 2$. It has been conjectured that the same is true for any even s [12].

6.2.2 "Taking the Square Root"

The feature described at the end of the previous subsection, namely that the helicities described by a field S, for given s, split into two groups which are only connected by a parity transformation, manifests itself in a factorization of the Klein-Gordon operator acting on that field. To be explicit, for $D = 4k - 1$ one can show that the Klein-Gordon operator $\Box - m^2$, when acting on a field S corresponding to a Young tableaux with s columns of height $2k - 1$ each, that satisfies the massive FP equation (6.10), can be factorized in terms of two first-order matrix operators $\mathscr{D}(\pm m)_{\mu_1 \cdots \mu_{2k-1}}{}^{\nu_1 \cdots \nu_{2k-1}}$ as follows:[7]

$$\mathscr{D}(m)\mathscr{D}(-m)S = 0, \qquad S^{\mathrm{tr}} = 0, \qquad \partial \cdot S = 0, \tag{6.13}$$

where the full index structure of the operator $\mathscr{D}(m)$ is given by

$$\mathscr{D}(m)_{\mu_1 \cdots \mu_{2k-1}}{}^{\nu_1 \cdots \nu_{2k-1}} = \frac{1}{(2k-1)!} \varepsilon_{\mu_1 \cdots \mu_{2k-1}}{}^{\alpha \nu_1 \cdots \nu_{2k-1}} \partial_\alpha + m \delta_{\mu_1 \cdots \mu_{2k-1}}^{\nu_1 \cdots \nu_{2k-1}}. \tag{6.14}$$

It is understood that this operator acts on the first column of the Young tableaux corresponding to S. It is an on-shell projector:

$$\mathscr{D}^2(m)S = \mathscr{D}(m)S \quad \text{if } S \text{ satisfies (6.13).} \tag{6.15}$$

One can show that the symmetry properties of $\mathscr{D}(m)\mathscr{D}(-m)S$ are the same as that of S itself as a consequence of the algebraic and differential subsidiary conditions.

One could try to write down a similar factorization in $D = 4k + 1$ dimensions for a Klein-Gordon operator when acting on a Young tableau with s columns of height $2k$ each. However, in this case one finds that the Klein-Gordon operator with the "wrong" sign of the mass term factorizes:

$$(\Box + m^2)S = -\mathscr{D}(m)\mathscr{D}(-m)S = 0, \qquad S^{\mathrm{tr}} = 0, \qquad \partial \cdot S = 0. \tag{6.16}$$

---

[7] We do not indicate indices. In later sections we will give the precise form of the equations in specific examples, including the indices.

The factorization (6.13) of the Klein-Gordon operator $\Box - m^2$ in $D = 4k - 1$ dimensions shows that one can take the "square root" of the generalized FP equations (6.10) and describe the dynamics of only half of the degrees of freedom by the first-order differential equations

$$\mathscr{D}(m)S = 0. \tag{6.17}$$

Note that this equation is not in the same representation as that of S. One can show that it implies the algebraic conditions $S^{\mathrm{tr}} = 0$ and the differential subsidiary conditions $\partial \cdot S = 0$. The other half of the degrees of freedom are described by a similar set of equations but with m replaced by $-m$. Under parity the two equations are mapped into each other. For $s = 1$ these equations reduce to the massive self-duality equations [13, 14]

$$R(S) = \pm m^{\star} S. \tag{6.18}$$

Such massive self-duality equations occur for instance in seven-dimensional gauged supergravity theories where S is a 3-form and m plays the role of the gauge coupling constant [13].

One can play the same trick of "boosting up the derivatives" not only on the generalized FP equations (6.10) but also, in $D = 4k - 1$ dimensions, on the "square root" of these equations, see Eq. (6.17). One thus arrives at the following higher-order derivative equations describing the same degrees of freedom:

$$\mathscr{D}(m)G(T) = 0. \tag{6.19}$$

The integration of these equations of motion to an action in this case does not lead to ghosts since the degrees of freedom are always in the same irreducible induced representation of the Poincaré group. In $D = 3$ dimensions this leads to Topological Massive Electrodynamics (TME) for $s = 1$ [15, 16] and Topological Massive Gravity (TMG) for $s = 2$ [3]. The analogue of Eqs. (6.19) does not exist in $D = 4k + 1$ dimensions since the integration of these equations would lead to a Klein-Gordon equation with the "wrong" sign in front of the mass term.

This ends our discussion of the general procedure of how to obtain out of a generalized massive FP theory for a massive field S, or its "square root", a massive higher-derivative gauge theory for a field T without ghosts. In the next sections we will further explain the general expressions introduced in this section at the hand of the one-column Young tableaux, i.e. $s = 1$.

6.3 Spin 1

In this section we consider the general case of a field S in D dimensions with indices corresponding to a one-column $s = 1$ Young tableau. As explained in footnote 3 we will generically denote this set of fields as "spin-1" fields. In these cases we are dealing with a p-form gauge field $S_{\mu_1 \cdots \mu_p}(x)$ with gauge transformation

$$\delta S_{\mu_1 \cdots \mu_p}(x) = p \partial_{[\mu_1} \lambda_{\mu_2 \cdots \mu_p]}(x). \tag{6.20}$$

The gauge-invariant curvature or "Riemann tensor" of S is given by the curl of this gauge field:

$$R_{\mu_1\cdots\mu_{p+1}}(S) = (p+1)\partial_{[\mu_1}S_{\mu_2\cdots\mu_{p+1}]}. \tag{6.21}$$

In the following we discuss the cases $p=1$, $p=2$ and $p=3$ in more detail.

$p=1$ The simplest case that satisfies the condition (6.8) is a vector ($p=1$) in $D=3$ dimensions. In that case the curvature or "Riemann tensor" $R(S)$ and the "Einstein tensor" $G(S)$ are given by

$$R_{\mu\nu}(S) = 2\partial_{[\mu}S_{\nu]}, \qquad G_\mu(S) = \frac{1}{2}\varepsilon_\mu{}^{\nu\rho}R_{\nu\rho}(S). \tag{6.22}$$

The massless Lagrangian (6.7) is now given by

$$\mathscr{L} = \frac{1}{2}\varepsilon^{\mu\nu\rho}S_\mu R_{\nu\rho}(S), \tag{6.23}$$

which indeed does not describe any massless spin 1 degree of freedom.

We next consider the massive Proca equation for a $3D$ massive vector field S_μ:

$$\left(\Box - m^2\right)S_\mu = 0, \qquad \partial^\mu S_\mu = 0. \tag{6.24}$$

These equations are derivable from the Proca Lagrangian

$$\mathscr{L} = \frac{1}{2}G^\mu(S)G_\mu(S) - \frac{1}{2}m^2 S^\mu S_\mu. \tag{6.25}$$

This Lagrangian describes the unitary propagation of two states, one with helicity $+1$ and one with helicity -1, see Exercise 1.[8] The differential subsidiary condition is solved by making the substitution:

$$S_\mu = G_\mu(T) \tag{6.26}$$

in terms of another vector field T_μ. Note that T_μ is a gauge field with gauge transformations $\delta T_\mu = \partial_\mu\lambda$. The substitution (6.26) leads to the following higher-derivative so-called "extended Proca" equation for T_μ:

$$\left(\Box - m^2\right)G_\mu(T) = 0, \tag{6.27}$$

which can be integrated to the following Lagrangian containing the "extended Chern-Simons" term introduced in [17]:

$$\mathscr{L} = -\frac{1}{2}T^\mu G_\mu(T) + \frac{1}{2m^2}\varepsilon^{\mu\nu\rho}G_\mu(T)\partial_\nu G_\rho(T). \tag{6.28}$$

---

[8]The exercises, together with their solutions, are given in the Appendix.

Table 6.2 This table lists all the $s = 1$ cases, with $3 \leq D \leq 7$, where the "boosting up the derivatives" trick works without encountering ghosts. This leads to the $3D$ and $7D$ "Topological Massive Electrodynamics" (TME) theories indicated in the table. The $5D$ "Extended Proca" (EP) theory, indicated by a star in the table, is special in the sense that the equation of motion of this theory cannot be integrated to a Lagrangian

	$D = 3$	$D = 4$	$D = 5$	$D = 6$	$D = 7$
EP			▯*		
TME	□				▤

A canonical analysis shows that this higher-derivative gauge theory contains ghosts [12, 17]. For a proof of this statement, see Exercise 2.

To avoid ghosts one should first take the "square root" and consider the massive self-duality equations

$$R_{\mu\nu}(S) = m\varepsilon_{\mu\nu}{}^{\rho}S_{\rho}. \tag{6.29}$$

Boosting up the derivatives and integrating the equations of motion leads to the Lagrangian of $3D$ TME [15, 16], see Table 6.2

$$\mathscr{L} = -\frac{1}{4m}R^{\mu\nu}(T)R_{\mu\nu}(T) + \frac{1}{2}\varepsilon^{\mu\nu\rho}T_{\mu}\partial_{\nu}T_{\rho}. \tag{6.30}$$

p = 2 We now move on and consider the next simplest case of a 2-form ($p = 2$) in $5D$. In this case we are dealing with gauge fields S, gauge parameters λ and Riemann tensors $R(S)$ corresponding to the following Young tableaux

$$S \sim \boxed{} \qquad \lambda \sim \square \qquad R(S) \sim \boxed{} \tag{6.31}$$

These expressions correspond to the following formulae:

$$\delta S_{\mu\nu} = 2\partial_{[\mu}\lambda_{\nu]}, \qquad R_{\mu\nu\rho}(S) = 3\partial_{[\mu}S_{\nu\rho]}, \tag{6.32}$$

while the Einstein tensor $G_{\mu\nu}(S)$ is given by

$$G_{\mu\nu}(S) = \frac{1}{3}\varepsilon_{\mu\nu}{}^{\rho\sigma\tau}R_{\rho\sigma\tau}(S). \tag{6.33}$$

In this case the equation $G_{\mu\nu}(S) = 0$ cannot be integrated to a Lagrangian since the candidate kinetic term $S^{\mu\nu}G_{\mu\nu}(S)$ is a total derivative, see Table 6.2. This is similar to the self-dual 2-form in IIB string theory whose dynamics can be described by an equation of motion without having a Lagrangian.

We next consider the equations of motion for a massive $5D$ two-form $S_{\mu\nu}$:

$$(\Box - m^2)S_{\mu\nu} = 0, \qquad \partial^{\mu}S_{\mu\nu} = 0. \tag{6.34}$$

These equations are derivable from the following Lagrangian:

$$\mathcal{L} = \frac{1}{8} G^{\mu\nu}(S) G_{\mu\nu}(S) + \frac{1}{2} m^2 S^{\mu\nu} S_{\mu\nu}. \tag{6.35}$$

The differential subsidiary condition given in (6.34) is solved by making the following substitution:

$$S_{\mu\nu} = G_{\mu\nu}(T) \tag{6.36}$$

in terms of another 2-form field $T_{\mu\nu}$. Note that $T_{\mu\nu}$ is a gauge field with gauge transformations $\delta T_{\mu\nu} = 2\partial_{[\mu}\lambda_{\nu]}$. The substitution (6.36) leads to the following higher-derivative equations of motion for T:

$$(\Box - m^2) G_{\mu\nu}(T) = 0. \tag{6.37}$$

Again, these equations cannot be integrated. Trying a Lagrangian of the form $\mathcal{L} \sim \alpha T^{\mu\nu} G_{\mu\nu}(T) + \beta \varepsilon^{\mu\nu\rho\sigma\tau} G_{\mu\nu}(T)\partial_\rho G_{\sigma\tau}(T)$ one finds that both terms are total derivatives. The dynamics of this case can only be described by a set of equations of motion without having a Lagrangian. Taking the "square root" is not an option in this case since the integrability conditions of the massive self-duality equations would lead to a Klein-Gordon equation with the wrong sign in front of the mass term.

p = 3   Finally, we consider a 3-form ($p = 3$) in $D = 7$ dimensions. We are now dealing with gauge fields S, gauge parameters λ and Riemann tensors $R(S)$ given by the following Young tableaux:

$$S \sim \boxed{} \qquad \lambda \sim \boxed{} \qquad R(S) \sim \boxed{} \tag{6.38}$$

These expressions correspond to the following formulae:

$$\delta S_{\mu\nu\rho} = 3\partial_{[\mu}\lambda_{\nu\rho]}, \qquad R_{\mu\nu\rho\sigma}(S) = 4\partial_{[\mu}S_{\nu\rho\sigma]}, \tag{6.39}$$

while the Einstein tensor $G_{\mu\nu\rho}(S)$ is given by

$$G_{\mu\nu\rho}(S) = \frac{1}{4}\varepsilon_{\mu\nu\rho}{}^{\alpha\beta\gamma\delta} R_{\alpha\beta\gamma\delta}(S). \tag{6.40}$$

This leads to the following massless Lagrangian

$$\mathcal{L} = S^{\mu\nu\rho}(S) G_{\mu\nu\rho}(S), \tag{6.41}$$

which does not describe any massless degrees of freedom.

We next consider the massive Proca equation for a 7D massive 3-form $S_{\mu\nu\rho}$:

$$(\Box - m^2) S_{\mu\nu\rho} = 0, \qquad \partial^\mu S_{\mu\nu\rho} = 0. \tag{6.42}$$

These equations are derivable from the Lagrangian

$$\mathcal{L} = G^{\mu\nu\rho}(S)G_{\mu\nu\rho}(S) + \frac{1}{2}m^2 S^{\mu\nu\rho} S_{\mu\nu\rho}.$$ (6.43)

The differential subsidiary condition is solved by making the substitution:

$$S_{\mu\nu\rho} = G_{\mu\nu\rho}(T)$$ (6.44)

in terms of another 3-form field $T_{\mu\nu\rho}$. This substitution leads to the following higher-derivative equations for $T_{\mu\nu\rho}$:

$$(\Box - m^2)G_{\mu\nu\rho}(T) = 0,$$ (6.45)

which can be integrated to the following Lagrangian

$$\mathcal{L} = \frac{1}{2}T^{\mu\nu\rho}(\Box - m^2)G_{\mu\nu\rho}(T).$$ (6.46)

To see whether the Lagrangian (6.46) describes ghosts or not we perform a canonical analysis. We first fix all gauge degrees of freedom by imposing the following gauge-fixing conditions on the 3-form T and the 2-form gauge parameters λ:

$$\partial^i T_{i\mu\nu} = 0, \qquad \partial^i \lambda_{i\mu} = 0, \quad i = 1,\ldots,6.$$ (6.47)

Using these conditions it follows that $\delta(\partial^i T_{i\mu\nu}) = \nabla^2 \lambda_{\mu\nu}$, which shows that indeed all gauge degrees of freedom in T are fixed.

Taking the gauge-fixing conditions (6.47) into account, we decompose T as follows:

$$T_{0ij} = T_{ij}, \qquad T_{ijk} = \varepsilon_{ijk}{}^{lmn} \partial_l U_{mn},$$ (6.48)

where $T_{ij} = -T_{ji}$, $U_{ij} = -U_{ji}$, $\partial^i T_{ij} = 0$ and $\partial^i U_{ij} = 0$. Therefore, T_{ij} and U_{ij} each describe 10 components.[9]

Using the decomposition (6.48) and dropping all terms with a spatial divergence of T or U, the Lagrangian (6.46) can be rewritten as follows:

$$\mathcal{L} = 36T^{ij}(\Box - m^2)\nabla^2 U_{ij}.$$

The off-diagonal nature of this expression shows that this Lagrangian describes 20 massive degrees of freedom but that half of them are ghosts.

To avoid ghosts one should first take the "square root" and consider the massive self-duality equations

$$R_{\mu\nu\rho\sigma}(S) = \frac{1}{3!}m\varepsilon_{\mu\nu\rho\sigma}{}^{\alpha\beta\gamma} S_{\alpha\beta\gamma}.$$ (6.49)

---

[9]It is always understood that T_{ij} and U_{ij} are spatially divergenceless. This means that when we apply the variational principle, we should not vary the "divergenceful degrees of freedom".

Table 6.3 This table lists all the $s = 2$ cases where the "boosting up the derivatives" trick works without introducing ghosts. This leads to the different NMG and TMG theories indicated in the table for $3 \leq D \leq 7$. The cases with the sub-indices 1–3 are discussed in Sects. 6.4.1–6.4.3

	$D = 3$	$D = 4$	$D = 5$	$D = 6$	$D = 7$
NMG	$\square_1$	$\square_3$			
TMG	$\square_2$				$\square_3$

Boosting up the derivatives and integrating the equations of motion leads to the $7D$ higher-derivative TME Lagrangian, see Table 6.2

$$\mathscr{L} = -\frac{3}{4m} R^{\mu\nu\rho\sigma}(T) R_{\mu\nu\rho\sigma}(T) + \frac{1}{2} \varepsilon^{\mu\nu\rho\sigma\alpha\beta\gamma} T_{\mu\nu\rho} \partial_\sigma T_{\alpha\beta\gamma}. \tag{6.50}$$

This finishes our discussion of the one-column Young tableaux.

6.4 Spin 2

We now consider fields corresponding to two-column Young tableaux, i.e. $s = 2$. For $3 \leq D \leq 7$ the cases where the "boosting up the derivatives" procedure does not lead to ghosts are indicated in Table 6.3. In the first subsection we will discuss the $3D$ NMG theory [4, 5]. In the next subsection we will review the $3D$ TMG theory [3]. In Sect. 6.4.3 we will briefly discuss the extensions of the $3D$ NMG and TMG theories to higher dimensions. To keep in line with notational conventions we will denote the two-column fields with the letter h instead of S since in specific cases h can be viewed as the linearization of a metric tensor g.

6.4.1 3D New Massive Gravity

It is well-known that the pure Einstein-Hilbert term in three dimensions does not describe any physical degrees of freedom: there are no gravitational waves in three dimensions. For a proof of this, see Exercise 3. This is consistent with our analysis in Sect. 6.2 where we concluded that setting the Einstein tensor corresponding to a $3D$ symmetric tensor to zero implies that there are only gauge degrees of freedom left. In this section we will show that adding a specific combination of higher-derivative terms quadratic in the Riemann tensor has the effect that *massive* gravitons, with helicities $+2$ and -2, start propagating unitarily. The corresponding model is called NMG [4, 5]. The mass parameter is related to the dimension-full parameter in front of the higher-derivative terms. Effectively, the higher-derivative term acts as the kinetic term and the original Einstein-Hilbert term behaves like a mass term.

It is surprising that NMG, given the fact that it contains higher derivatives, does not contain ghosts. The same is not true for similar higher-derivative models in four spacetime dimensions [1, 2]. In general our method of "boosting up the derivatives" does not guarantee that this is the case. However, since in this case the theory is parity-preserving, there are no ghosts to be expected. Below we will give a separate proof that integrating the NMG equations of motion leads to a Lagrangian without ghosts. But first we will describe how NMG is obtained by the boosting up procedure.

Our starting point are the Fierz-Pauli (FP) equations for a symmetric tensor $\tilde{h}_{\mu\nu}$ in D dimensions:

$$(\Box - m^2)\tilde{h}_{\mu\nu} = 0, \qquad \eta^{\mu\nu}\tilde{h}_{\mu\nu} = 0, \qquad \partial^\mu \tilde{h}_{\mu\nu} = 0. \qquad (6.51)$$

The last two of these FP equations are algebraic and differential subsidiary conditions that have to the imposed in order to obtain the correct counting of degrees of freedom. This counting is as follows:

$$\frac{1}{2}D(D+1) - 1 - D = \begin{cases} 5 & \text{for } 4D, \\ 2 & \text{for } 3D. \end{cases} \qquad (6.52)$$

In $3D$, the Lagrangian that gives these FP equations is given by

$$\mathscr{L}_{\text{FP}} = \frac{1}{2}\tilde{h}^{\mu\nu} G^{\text{lin}}_{\mu\nu}(\tilde{h}) - \frac{1}{2}m^2(\tilde{h}^{\mu\nu}\tilde{h}_{\mu\nu} - \tilde{h}^2), \qquad (6.53)$$

where we denote $\tilde{h} \equiv \eta^{\mu\nu}\tilde{h}_{\mu\nu}$. The $3D$ linearized Einstein tensor $G^{\text{lin}}_{\mu\nu}(\tilde{h})$ for any symmetric tensor $\tilde{h}_{\mu\nu}$ is defined as

$$G^{\text{lin}}_{\mu\nu}(\tilde{h}) \equiv \varepsilon_\mu{}^{\alpha\beta}\varepsilon_\nu{}^{\gamma\delta}\partial_\alpha\partial_\gamma \tilde{h}_{\beta\delta}. \qquad (6.54)$$

We note that the trace $\tilde{h}$ plays the role of an *auxiliary field*: it is needed to write down a Lagrangian but it does not describe a physical degree of freedom. Such auxiliary fields become more and more abundant when one considers fields with spin higher than two. It is instructive to see what goes wrong if one actually tries to write down a FP Lagrangian in terms of a symmetric and traceless tensor $H_{\mu\nu}$ alone. The Klein-Gordon equation and the differential subsidiary condition for $H_{\mu\nu}$ would read:

$$(\Box - m^2)H_{\mu\nu} = 0, \qquad \partial^\mu H_{\mu\nu} = 0. \qquad (6.55)$$

In analogy with the spin-1 case one could try to combine the above equations into the following single equation of motion:

$$\partial^\rho(\partial_\rho H_{\mu\nu} - \partial_\mu H_{\rho\nu}) - m^2 H_{\mu\nu} = 0. \qquad (6.56)$$

The nice thing about this equation is that it implies the differential subsidiary condition. However, unlike the spin 1 case, this equation can never serve as the equation of motion for $H_{\mu\nu}$ since, unlike $H_{\mu\nu}$ itself, it is not symmetric in the free indices μ and ν. One could next try to write down the most general symmetric and traceless equation but it turns out that that does not work. The problem is that, in order to derive the differential subsidiary condition $\partial^\mu H_{\mu\nu} = 0$ one needs to make use of the constraint $\partial^\rho \partial^\sigma H_{\rho\sigma} = 0$ first. In order to impose this constraint we must extend the field content and introduce an additional auxiliary scalar H. Making the most general Ansatz in terms of $H_{\mu\nu}$ and H one can indeed arrange things such that the equations of motion imply both the constraint $\partial^\rho \partial^\sigma H_{\rho\sigma} = 0$ as well as $H = 0$. Hence, the Lagrange multiplier H does not introduce a new physical degree of freedom. Having derived the Lagrangian in terms of $H_{\mu\nu}$ and H one can go back to the $\tilde{h}$-basis via the transformation

$$\tilde{h}_{\mu\nu} \equiv H_{\mu\nu} + \frac{1}{3}\eta_{\mu\nu} H \tag{6.57}$$

and recover the FP Lagrangian (6.53).

We now apply our "boosting up the derivatives" procedure as explained in Sect. 6.2.1. We take the standard FP equations in terms of a symmetric tensor $\tilde{h}_{\mu\nu}$. We next solve the differential subsidiary condition $\partial^\mu \tilde{h}_{\mu\nu} = 0$ by expressing $\tilde{h}_{\mu\nu}$ in terms of the Einstein tensor of another symmetric field $h_{\mu\nu}$:

$$\tilde{h}_{\mu\nu} = G^{\text{lin}}_{\mu\nu}(h), \tag{6.58}$$

with the linearized Einstein tensor $G^{\text{lin}}_{\mu\nu}(h)$ defined in (6.54). Substituting this solution of the constraint into the original FP equations (6.51) we obtain the following equivalent higher-order equations of motion:

$$(\Box - m^2)G^{\text{lin}}_{\mu\nu}(h) = 0, \qquad R^{\text{lin}}(h) = 0, \tag{6.59}$$

where $R^{\text{lin}}(h)$ is the linearized Ricci scalar, i.e. the trace of the linearized Ricci tensor

$$R^{\text{lin}}_{\mu\nu}(h) = \Box h_{\mu\nu} - 2\partial_{(\mu}\partial^\rho h_{\nu)\rho} + \partial_\mu \partial_\nu h. \tag{6.60}$$

The linearized Einstein tensor can be written as $G^{\text{lin}}_{\mu\nu}(h) = R^{\text{lin}}_{\mu\nu}(h) - \frac{1}{2}g_{\mu\nu}R^{\text{lin}}$, so $R^{\text{lin}}(h) = 0$ is equivalent to $\eta^{\mu\nu}G^{\text{lin}}_{\mu\nu}(h) = 0$. The equations of motion (6.59) can be integrated to a Lagrangian. At this point there are two surprises. First of all, as we will show below, this Lagrangian does not contain ghosts. Secondly, it turns out that the Lagrangian can be extended to a more general non-linear one with interactions. More precisely, the quadratic (in $h_{\mu\nu}$) Lagrangian corresponding to (6.59) can be viewed as the linearization of a non-linear quadratic curvature Lagrangian where the metric $g_{\mu\nu}$ is expanded around a flat Minkowski spacetime metric $\eta_{\mu\nu}$ as follows:

$$g_{\mu\nu} = \eta_{\mu\nu} + h_{\mu\nu}. \tag{6.61}$$

Upon making this substitution into this quadratic curvature Lagrangian and retaining only the terms quadratic in $h_{\mu\nu}$ one obtains the quadratic Lagrangian that yields the equations of motion (6.59). It turns out that the quadratic curvature Lagrangian in question is the NMG Lagrangian given by:

$$\mathcal{L} = \sqrt{-g}\left[-R - \frac{1}{2m^2}\left(R^{\mu\nu}R_{\mu\nu} - \frac{3}{8}R^2 \right) \right].$$ (6.62)

A noteworthy feature of this NMG Lagrangian is that the Einstein Hilbert term has the so-called "wrong" sign in the sense that it is not the sign it should have in four spacetime dimensions. Note that this is possible due to the fact that the Einstein-Hilbert term plays the role of a mass term and not of a kinetic term.

Before we prove that the NMG Lagrangian (6.62) describes unitarily the helicity states $+2$ and -2 it is convenient to introduce the following generalization of this Lagrangian:

$$\mathcal{L} = \sqrt{-g}\left[\sigma R + 4\lambda m^2 - \frac{1}{2m^2}\left(R^{\mu\nu}R_{\mu\nu} - \frac{3}{8}R^2 \right) \right].$$ (6.63)

We have introduced here two new parameters: a sign parameter $\sigma = \pm$ and a cosmological parameter λ. The Lagrangian (6.63) is sometimes referred to as "Cosmological New Massive Gravity" (CNMG). Note that the cosmological parameter λ we have introduced is not necessarily equal to the cosmological constant Λ characterizing a maximally symmetric background. This is typical for higher-derivative theories. Substituting the Ansatz

$$G_{\mu\nu} = 2\Lambda g_{\mu\nu}$$ (6.64)

into the NMG equations of motion leads to the following quadratic relationship between λ and Λ:

$$4m^4\lambda = \Lambda(\Lambda + 4m^2\sigma).$$ (6.65)

To analyze the modes propagated by the CNMG Lagrangian (6.63) it is convenient to first lower the number of derivatives by introducing a second auxiliary symmetric tensor field $f_{\mu\nu}$. In terms of $g_{\mu\nu}$ and $f_{\mu\nu}$ one can write down the following equivalent Lagrangian:

$$\mathcal{L} = \sqrt{-g}\left[\sigma R + 4\lambda m^2 + f^{\mu\nu}G_{\mu\nu} + \frac{1}{2}m^2\left(f^{\mu\nu}f_{\mu\nu} - f^2 \right) \right].$$ (6.66)

The equation of motion of $f_{\mu\nu}$ may be used to solve for $f_{\mu\nu}$ in terms of $G_{\mu\nu}(g)$. Substituting this solution back into the Lagrangian (6.66) one obtains the CNMG Lagrangian (6.63).

We now consider the linearization of (6.66) around a maximally symmetric background with metric $\bar{g}_{\mu\nu}$ with cosmological constant Λ. We first expand the metric $g_{\mu\nu}$ around this background as follows:

$$g_{\mu\nu} = \bar{g}_{\mu\nu} + h_{\mu\nu}.$$ (6.67)

It turns out to be convenient to expand the auxiliary field $f_{\mu\nu}$ as

$$f_{\mu\nu} = \frac{1}{m^2}\{\Lambda[\bar{g}_{\mu\nu} + h_{\mu\nu}] - k_{\mu\nu}\},\tag{6.68}$$

where $k_{\mu\nu}$ is an independent symmetric tensor fluctuation field. Substituting the expansions (6.67) and (6.68) into the CNMG Lagrangian (6.63) one obtains (the details can be found in [4, 5]) the following quadratic Lagrangian in terms of the fluctuations $h_{\mu\nu}$ and $k_{\mu\nu}$:

$$\mathscr{L}_{\text{quadr}} \sim -\frac{1}{2}\bar{\sigma}h^{\mu\nu}\mathscr{G}^{\text{lin}}_{\mu\nu}(h) - \frac{1}{m^2}k^{\mu\nu}\mathscr{G}^{\text{lin}}_{\mu\nu}(h) + \frac{1}{2m^2}\left(k^{\mu\nu}k_{\mu\nu} - k^2\right).\tag{6.69}$$

Here

$$\bar{\sigma} = \sigma - \frac{\Lambda}{2m^2}\tag{6.70}$$

is a shifted σ parameter and $\mathscr{G}^{\text{lin}}_{\mu\nu}(h)$ is the linearized Einstein tensor in the presence of a cosmological constant:

$$\mathscr{G}^{\text{lin}}_{\mu\nu}(h) = \mathscr{R}^{\text{lin}}_{\mu\nu}(h) - \frac{1}{2}\bar{g}_{\mu\nu}\bar{g}^{\rho\sigma}\mathscr{R}^{\text{lin}}_{\rho\sigma}(h) + 4\Lambda h_{\mu\nu} - 2\Lambda\bar{g}_{\mu\nu}h.\tag{6.71}$$

The linearized Ricci tensor $\mathscr{R}^{\text{lin}}_{\mu\nu}$ is given by

$$\mathscr{R}^{\text{lin}}_{\mu\nu}(h) = \Box h_{\mu\nu} - \nabla^\rho\nabla_\mu h_{\rho\nu} - \nabla^\rho\nabla_\nu h_{\rho\mu} + \nabla_\mu\nabla_\nu h.\tag{6.72}$$

Some general properties of the generalized Einstein tensor (6.71) are given in Exercise 4.

One can show that, after an appropriate diagonalization, the Lagrangian (6.69) can be written as the sum of a massless spin 2 Lagrangian and a *massive* spin 2 Lagrangian, with mass

$$M^2 = -m^2\bar{\sigma}.\tag{6.73}$$

This is related to the fact that the kinetic operator, which is of fourth-order in the derivatives, can be written as the product of two second-order derivative operators. One of these operators describes a massless graviton while the other factor describes a massive graviton. In general the Lagrangian (6.69) contains a ghost because the signs of the kinetic terms of the massless and massive graviton turn out to be of opposite sign. There are now two special situations where this does not cause any problem:

$D = 3$ In this case there is no massless graviton but only a massive graviton. This implies that one can always adapt the overall sign of the Lagrangian such that the kinetic term of the massive graviton has the correct sign. This case leads to the $3D$ NMG theory of [4, 5].

Fig. 6.1 This figure indicates the unitary bulk region (*the boldface line*) for the choice $\sigma = -1$. The boundaries of this unitary region occur for $\lambda = -1$ and $\lambda = 3$ and are discussed in the text

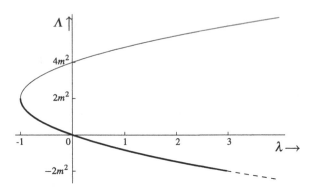

$\Lambda = 2m^2\sigma$ For this special value of the cosmological constant the coefficient $\bar{\sigma}$ in front of the linearized Einstein-Hilbert term vanishes and the massive graviton becomes massless. This special point in the parameter space is more subtle in the sense that it is a degenerate point in the spectrum where one mode, the massive graviton, gets replaced by another, so-called logarithmic mode. This leads to a $3D$ so-called "critical gravity" theory. The interesting thing about this critical point is that it allows a natural generalization to $D > 3$ dimensions [18, 19].

Due to the fact that the sign parameter σ gets shifted to a $\bar{\sigma}$ in a cosmological background one can have unitary bulk models for both signs of σ depending on the value of Λ. There are now several situations. As an example we have given the unitary bulk region in Fig. 6.1 for the choice of $\sigma = -1$. Note that for each choice of the cosmological parameter λ there may be two distinct values of the cosmological constant Λ. The boundary points $\lambda = -1$ and $\lambda = 3$ are special. For $\lambda = -1$ there is an enhanced gauge symmetry leading to a so-called "partial massless" graviton [4, 5] while the $\lambda = 3$ case corresponds to the critical gravity case discussed above.

6.4.2 3D Topological Massive Gravity

In this section we consider the "square root", as described in Sect. 6.2.2, of the $3D$ massive FP equation and show how the procedure of "boosting up the derivatives", as described in Sect. 6.2.1, leads to $3D$ TMG. Our starting point is the massive spin 2 FP equations (6.51). Following the general procedure as described in Sect. 6.2.2 we write the $3D$ Klein-Gordon operator as the product of two first-order matrix operators

$$\left[\mathcal{O}(\pm m)\right]_\mu{}^\rho = \varepsilon_\mu{}^{\tau\rho}\partial_\tau \pm m\delta_\mu^\rho. \tag{6.74}$$

Using such first-order operators, the Klein-Gordon operator acting on a symmetric, traceless and divergenceless rank-2 tensor factorizes as follows:

$$(\Box - m^2)\tilde{h}_{\mu\nu} = \left[\mathcal{O}(m)\right]_\mu{}^\sigma\left[\mathcal{O}(-m)\right]_\sigma{}^\rho\tilde{h}_{\rho\nu}. \tag{6.75}$$

To show this factorization one must use that $\tilde{h}_{\mu\nu}$ satisfies the algebraic and differential subsidiary conditions of the FP equations.

We now take only one of the two first-order operators and consider the $\sqrt{\text{FP}}$ equation $[\mathscr{O}(-m)]_\mu{}^\rho \tilde{h}_{\rho\nu} = 0$:

$$m\tilde{h}_{\mu\nu} = \varepsilon_\mu{}^{\rho\sigma} \partial_\rho \tilde{h}_{\sigma\nu}. \tag{6.76}$$

One can easily prove from this equation that the symmetric tensor $\tilde{h}_{\mu\nu}$ satisfies the FP subsidiary conditions of tracelessness and divergencefreeness. This equation can be integrated to the following first-order action [20]

$$S = \frac{1}{2} \int d^3x \left\{ \varepsilon^{\mu\nu\rho} \tilde{h}_\mu{}^\sigma \partial_\nu \tilde{h}_{\rho\sigma} - m\left(\tilde{h}^{\nu\mu} \tilde{h}_{\mu\nu} - \tilde{h}^2 \right) \right\}, \tag{6.77}$$

which contains a *non-symmetric* tensor $\tilde{h}_{\mu\nu} \neq \tilde{h}_{\nu\mu}$. The tensor $\tilde{h}_{\mu\nu}$ can be proven to be symmetric after applying the variational principle and then manipulating its equations of motion, but being a fundamental field in the action, it's not symmetric. Its anti-symmetric part behaves like the kind of auxiliary fields we discussed in the case of NMG, see Sect. 6.4.1.

We now apply the "boosting up" procedure and consider the $\sqrt{\text{FP}}$ equations (6.76) in terms of a symmetric tensor $\tilde{h}_{\mu\nu}$. We next solve for the divergence-less condition by expressing the tensor $\tilde{h}_{\mu\nu}$ in terms of a linearized second-order Einstein operator acting on another symmetric tensor $h_{\mu\nu}$:

$$\tilde{h}_{\mu\nu} = G^{\text{lin}}_{\mu\nu}(h). \tag{6.78}$$

Substituting this solution of the differential subsidiary condition into the original $\sqrt{\text{FP}}$ equations (6.76) one obtains the following equivalent set of higher-order equations of motion:

$$m G^{\text{lin}}_{\mu\nu}(h) = \varepsilon_\mu{}^{\rho\sigma} \partial_\rho G^{\text{lin}}_{\sigma\nu}(h). \tag{6.79}$$

These equations can be integrated to a Lagrangian that can be viewed as the linearization of the Lagrangian of TMG [3] around a Minkowski spacetime. Writing $g_{\mu\nu} = \eta_{\mu\nu} + h_{\mu\nu}$ the TMG Lagrangian in terms of $g_{\mu\nu}$ is given by [3]

$$\mathscr{L} = -\sqrt{-g}R + \frac{1}{m}\mathscr{L}_{\text{LCS}}, \tag{6.80}$$

where the last term represents a Lorentz Chern Simons term:

$$\mathscr{L}_{\text{LCS}} = -\varepsilon^{\mu\nu\rho} \left[\Gamma^\alpha_{\mu\beta} \partial_\nu \Gamma^\beta_{\rho\alpha} + \frac{2}{3} \Gamma^\alpha_{\mu\gamma} \Gamma^\gamma_{\nu\beta} \Gamma^\beta_{\rho\alpha} \right]. \tag{6.81}$$

Here Γ is the usual Levi-Civita connection for the spacetime metric g:

$$\Gamma^\rho_{\mu\nu} = \frac{1}{2} g^{\rho\sigma} (\partial_\mu g_{\nu\sigma} + \partial_\nu g_{\mu\sigma} - \partial_\sigma g_{\mu\nu}). \tag{6.82}$$

The Riemann curvature tensor is determined, in $3D$, by the Ricci tensor, which is

$$R_{\mu\nu} \equiv R_{\rho\mu}{}^{\rho}{}_{\nu} = -2\left(\partial_{\rho}\Gamma^{\rho}_{\mu\nu} - \partial_{\mu}\Gamma^{\rho}_{\rho\nu} + \Gamma^{\rho}_{\rho\lambda}\Gamma^{\lambda}_{\mu\nu} - \Gamma^{\rho}_{\mu\lambda}\Gamma^{\lambda}_{\rho\nu}\right). \tag{6.83}$$

Note that, like in the NMG case, the Einstein-Hilbert term in the TMG Lagrangian has the "wrong" sign.

6.4.3 Extensions

It turns out that, using our general procedure described in Sect. 6.2, both the $3D$ NMG as well as the $3D$ TMG theories constructed in the previous two subsections allow, at least at the linearized level, a natural extension to $D > 3$ dimensions. The case of NMG has recently been discussed in [11]. Since this result was published only after the Naxos lectures we will be rather brief here. The basic idea, needed to extend NMG beyond three dimensions, is to use exotic representations to describe the massive spin 2 states. Only in three dimensions the usual symmetric tensor description suffices. In $D > 3$ dimensions one should use, instead, a massive dual representation. These are the mixed-symmetry representations indicated by a dagger in Table 6.1. A common feature of all these representations is that the corresponding Einstein tensor does not describe any massless degrees of freedom. Starting from the generalized FP equations of these fields one can therefore use our "boosting up the derivatives" procedure and construct an equivalent higher-order in derivatives Lagrangian that describes, unitarily, the same massive degrees of freedom as the original massive spin 2 FP equation. The $4D$ NMG Lagrangian makes use of the following exotic representation

$$h \sim \begin{array}{c}\boxed{}\boxed{}\\\boxed{}\end{array} \quad 4D \text{ NMG} \tag{6.84}$$

For the actual construction of the $4D$ NMG Lagrangian and for more details we refer to [11].

The $3D$ TMG theory can also be extended, at the linearized level, to $D > 3$ dimensions but it requires the use of different exotic representations of the massive spin 2 states. Instead of using the massive dual of the symmetric tensor representation one should use a *self-dual* representation. Only in three dimensions these two representations coincide and that is why both the $3D$ NMG and the $3D$ TMG theories can be formulated in terms of the symmetric tensor representation. Such massive self-dual representations exist in odd dimensions only and only in $D = 4k - 1$ dimensions, with k integer, do the integrability conditions that follow from the corresponding self-duality equations yield the desired Klein-Gordon operators with the correct sign in front of the m^2 term. The first dimensions beyond $3D$ where this occurs is the $7D$ case. In that case the h field corresponds to the following self-dual representation:

$$h \sim \begin{array}{c}\boxed{}\boxed{}\\\boxed{}\boxed{}\\\boxed{}\boxed{}\end{array} \quad 7D \text{ TMG} \tag{6.85}$$

The details of the construction of the corresponding (linearized) $7D$ TMG theory will be discussed elsewhere [21].

6.5 Conclusions

In these lecture we have indicated the general procedure that can be applied to construct higher-derivative theories of gravity. The method is based on the assumption that the field involved does not describe any degrees of freedom as a massless representation. This requires the property that setting its Einstein tensor to zero implies that the field in question is a pure gauge. We derived the general criterium that needs to be satisfied in order for this to be true. We exemplified our procedure by first working out several cases involving "spin" 1 fields. Next, we applied the procedure to the "spin 2" case. We first reviewed the constructions of the $3D$ NMG and the $3D$ TMG theories and, subsequently, showed how the procedure can also be applied to construct higher-dimensional generalizations of these theories, at least at the linearized level. The lowest-dimensional examples beyond $3D$ that we discussed were the $4D$ NMG and the $7D$ TMG theories.

It is an open question whether interactions can be introduced for the $D > 3$ NMG and TMG theories. An example of a $4D$ non-linear theory that makes use of an exotic representation is the Eddington-Schrödinger theory which is equivalent to general relativity with a cosmological constant, see [11]. This is some encouragement that it might be possible to introduce interactions for the case of $4D$ NMG. It might necessitate that we need to consider an AdS background instead of a flat Minkowski spacetime. It would be interesting to see whether the $7D$ TMG theory can be reformulated as a Chern-Simons theory like the $3D$ case. This could facilitate the introduction of interactions in that case. Clearly, at the time of writing these lectures the issue of interactions is unresolved and requires a further investigation.

Acknowledgements We would like to thank our collaborators Olaf Hohm and Paul Townsend, with whom we wrote most of the papers upon which these lectures are based, for the many fruitful interactions we had. These proceedings are based on two lectures given by one of us (E.B.). He wishes to thank the organizers of the 6th Aegean Summer School for their kind invitation to deliver these lectures and for the stimulating atmosphere provided by the School.

Appendix: Exercises

During the lectures several exercises were given. They are repeated here together with their solutions.

Exercise 1 Show that the kinetic terms of the two degrees of freedom described by the $3D$ Lagrangian (6.25) have the same sign. Hint: Use the following decomposition:

$$S_0 = \frac{1}{\sqrt{-\nabla^2}}(\phi_0 + \dot{\lambda}), \qquad S_i = \frac{1}{\sqrt{-\nabla^2}}\left(\varepsilon^{ij}\partial_j\phi_1 + \partial_i\lambda\right), \quad i = 1, 2. \qquad (6.86)$$

Solution The $3D$ Lagrangian is given by:

$$\mathscr{L} = \frac{1}{2} G^\mu(S) G_\mu(S) - \frac{1}{2} m^2 S^\mu S_\mu, \tag{6.87}$$

where $G_\mu(S) = \frac{1}{2} \varepsilon_\mu{}^{\nu\rho} R_{\nu\rho}(S)$ and $R_{\mu\nu}(S) = 2\partial_{[\mu} S_{\nu]}$. Now use the decomposition (6.86) to calculate both terms in the Lagrangian (6.87). For the mass term we obtain

$$S^\mu S_\mu = S^0 S_0 + S^i S_i = -S_0 S_0 + S_i S_i \tag{6.88}$$

$$= (\phi_0 + \dot\lambda) \frac{1}{\nabla^2} (\phi_0 + \dot\lambda) - (\hat\partial_i \phi_1 + \partial_i \lambda) \frac{1}{\nabla^2} (\hat\partial_i \phi_1 + \partial_i \lambda), \tag{6.89}$$

where $\hat\partial_i \equiv \varepsilon^{ij} \partial_j$. This can be rewritten as follows:

$$S^\mu S_\mu = \phi_0 \frac{1}{\nabla^2} \phi_0 + \dot\lambda \frac{1}{\nabla^2} \dot\lambda + 2\phi_0 \frac{1}{\nabla^2} \dot\lambda + \phi_1^2 + \lambda^2. \tag{6.90}$$

Similarly, the first term in (6.87) reduces to:

$$G^\mu(S) G_\mu(S) = -\frac{1}{2} R_{\mu\nu} R^{\mu\nu} = -2R_{0i} R_{0i} + R_{ij} R_{ij} = -2(\phi_1 \Box \phi_1 + \phi_0^2). \tag{6.91}$$

Substituting the expressions

$$R_{0i} = \frac{1}{\sqrt{-\nabla^2}} (\hat\partial_i \dot\phi_1 - \partial_i \phi_0), \qquad R_{ij} = \frac{1}{\sqrt{-\nabla^2}} (\partial_i \hat\partial_j \phi_1 - \partial_j \hat\partial_i \phi_1) \tag{6.92}$$

we obtain

$$G^\mu(S) G_\mu(S) = -2(\phi_1 \Box \phi_1 + \phi_0^2). \tag{6.93}$$

Putting everything together, the Lagrangian becomes:

$$\mathscr{L} = \frac{1}{2} \left(\phi_0^2 - m^2 \phi_0 \frac{1}{\nabla^2} \phi_0 - 2m^2 \phi_0 \frac{1}{\nabla^2} \dot\lambda - m^2 \lambda^2 - m^2 \dot\lambda \frac{1}{\nabla^2} \dot\lambda + \phi_1 \Box \phi_1 - m^2 \phi_1^2 \right). \tag{6.94}$$

From this Lagrangian we obtain the EOM for the field ϕ_0:

$$\phi_0 - \frac{m^2}{\nabla^2} \phi_0 - \frac{m^2}{\nabla^2} \dot\lambda = 0 \quad \text{or} \quad \phi_0 = \frac{m^2}{\nabla^2 - m^2} \dot\lambda. \tag{6.95}$$

Substituting this back into the Lagrangian and using that

$$\phi_0^2 = -\dot\lambda^2 + 2\dot\lambda \frac{1}{1 - \frac{\nabla^2}{m^2}} \dot\lambda = -\dot\lambda^2 - 2\dot\lambda \phi_0, \tag{6.96}$$

$$\frac{1}{\nabla^2} \phi_0 = -\frac{1}{\nabla^2} \dot\lambda + \frac{1}{m^2} \phi_0 \tag{6.97}$$

we obtain the following expression for the Lagrangian:

$$\mathscr{L} = \frac{1}{2}\left(\phi_1 \Box \phi_1 - m^2\phi_1^2 - \lambda\frac{m^2}{\nabla^2 - m^2}\lambda - m^2\lambda^2\right).$$

(6.98)

We next redefine λ in terms of a $\tilde{\lambda}$

$$\lambda = \sqrt{-\nabla^2 + m^2}\,\tilde{\lambda}$$

(6.99)

to obtain the simpler expression

$$\mathscr{L} = \frac{1}{2}\left(\phi_1\Box\phi_1 - m^2\phi_1^2 - m^2\tilde{\lambda}\partial_0\partial_0\tilde{\lambda} + m^2\tilde{\lambda}\nabla^2\tilde{\lambda} - m^4\tilde{\lambda}^2\right),$$

(6.100)

which reduces to

$$\mathscr{L} = \frac{1}{2}\phi_1(\Box - m^2)\phi_1 + \frac{1}{2}\lambda_1(\Box - m^2)\lambda_1$$

(6.101)

in terms of $\lambda_1 = m^2\tilde{\lambda}$.

We deduce that the Lagrangian (6.101) describes two degrees of freedom, where both of them have the same sign in front of the kinetic terms, and therefore, there are no ghosts.

Exercise 2 We have seen that "boosting up the derivatives" in the $3D$ Lagrangian (6.25) leads to the Lagrangian (6.28). Show that this Lagrangian describes two propagating degrees of freedom, one of which is a ghost. Hint: Work in the transverse gauge $\partial_i T_i = 0$ and use the following decomposition:

$$T_0 = \frac{1}{\sqrt{-\nabla^2}}\phi_0, \qquad T_i = \frac{1}{\sqrt{-\nabla^2}}\varepsilon^{ij}\partial_j\phi_1.$$

(6.102)

Solution We consider the Lagrangian:

$$\mathscr{L} = -\frac{1}{2}T^\mu G_\mu(T) + \frac{1}{2m^2}\varepsilon^{\mu\nu\rho}G_\mu(T)\partial_\nu G_\rho(T),$$

(6.103)

where

$$G_\mu(T) = \frac{1}{2}\varepsilon_\mu{}^{\nu\rho}R_{\nu\rho}(T), \qquad R_{\mu\nu}(T) = 2\partial_{[\mu}T_{\nu]}.$$

(6.104)

Using the decomposition (6.102) the first term in the Lagrangian becomes:

$$T^\mu G_\mu(T) = -\varepsilon_0{}^{jk}T_0\partial_j T_k + \varepsilon^{oij}T_j\partial_0 T_i + \varepsilon^{ij0}T_i\partial_j T_0 + \varepsilon^{ijk}T_i\partial_j T_k$$

$$= -\left(\frac{1}{\sqrt{-\nabla^2}}\phi_0\right)\hat{\partial}^k\left(\frac{1}{\sqrt{-\nabla^2}}\hat{\partial}_k\phi_1\right) + \left(\frac{1}{\sqrt{-\nabla^2}}\hat{\partial}_i\phi_1\right)\left(\frac{1}{\sqrt{-\nabla^2}}\hat{\partial}^i\phi_0\right)$$

$$= 2\phi_0\phi_1.$$

(6.105)

Similarly, the second term in the Lagrangian (6.103) becomes

$$\varepsilon^{\mu\nu\rho} G_\mu(T) \partial_\nu G_\rho(T) = \varepsilon_0{}^{ij} T_j \Box \partial_i T_0 - \varepsilon_i{}^{0j} T_j \partial_0 T^i - \varepsilon_i{}^{j0} T_0 \Box \partial_j T^i - \varepsilon_i{}^{jk} T_k \Box \partial_j T^i$$

$$= 2\varepsilon_0{}^{ij} T_j \Box \partial_i T_0 = 2T_j \Box \hat{\partial}^j T_0$$

$$= 2\left(\frac{1}{\sqrt{-\nabla^2}} \partial^j \phi_1\right) \Box \hat{\partial}^j \left(\frac{1}{\sqrt{-\nabla^2}} \phi_0\right) = 2\phi_1 \Box \phi_0. \qquad (6.106)$$

Substituting the calculated terms into the original Lagrangian we obtain:

$$\mathscr{L} = \frac{1}{m^2} \phi_1 (\Box - m^2) \phi_0. \qquad (6.107)$$

Writing

$$\phi_0 = \eta - \psi, \qquad \phi_1 = \eta + \psi \qquad (6.108)$$

the Lagrangian reads

$$\mathscr{L} = \frac{1}{m^2} \left[\eta (\Box - m^2) \eta - \psi (\Box - m^2) \psi \right]. \qquad (6.109)$$

The relative minus sign between the two kinetic terms shows that there are two propagating degrees of freedom, one of which is a ghost.

Exercise 3 Consider the $3D$ linearized Einstein-Hilbert Lagrangian when linearized around a Minkowski spacetime:

$$\mathscr{L} = \frac{1}{2} S^{\mu\nu} G_{\mu\nu}(S), \qquad (6.110)$$

with the Einstein tensor $G_{\mu\nu}(S)$ defined by

$$G_{\mu\nu}(S) = \varepsilon_\mu{}^{\alpha\beta} \varepsilon_\nu{}^{\gamma\delta} \partial_\alpha \partial_\gamma S_{\beta\delta}. \qquad (6.111)$$

Show that this Lagrangian does not describe any physical degrees of freedom. Hint: use the following decomposition:

$$S_{00} = -\frac{1}{\nabla^2} \phi_0, \qquad S_{0i} = -\frac{1}{\nabla^2} \hat{\partial}_i \phi_1, \qquad S_{ij} = -\frac{1}{\nabla^2} \hat{\partial}_i \hat{\partial}_j \phi_2, \qquad (6.112)$$

with $\hat{\partial}_i \equiv \varepsilon^{ij} \partial_j$.

Solution We first rewrite the Lagrangian (6.110) as follows

$$\mathscr{L} = \frac{1}{2} \left(S^{00} G_{00} + 2S^{0i} G_{0i} + S^{ij} G_{ij} \right) = \frac{1}{2} \left(S_{00} G_{00} - 2S_{0i} G_{0i} + S_{ij} G_{ij} \right). \qquad (6.113)$$

Using the definition (6.111) of the Einstein tensor and the decomposition (6.112) of $S_{\mu\nu}$ we obtain:

$$G_{00}(S) = \hat{\partial}^i \hat{\partial}^j S_{ij} = -\nabla^2 \phi_2,$$

$$G_{0i}(S) = -\partial_i \dot{\phi}_2 - \hat{\partial}_i \phi_1,$$

$$(6.114)$$

$$G_{ij}(S) = -\frac{1}{\nabla^2}(\hat{\partial}_i \hat{\partial}_j \ddot{\phi}_0 + \hat{\partial}_i \partial_j \dot{\phi}_1 + \partial_i \hat{\partial}_j \dot{\phi}_1 + \partial_i \partial_j \ddot{\phi}_2)$$

and hence

$$S_{00} G_{00}(S) = \left(-\frac{1}{\nabla^2}\phi_0\right)\left(-\nabla^2 \phi_2\right) = \phi_0 \phi_2,$$

$$S_{0i} G_{0i}(S) = \left(-\phi_1 \frac{1}{\nabla^2}\hat{\partial}_i \partial_i \dot{\phi}_2 - \phi_1 \frac{1}{\nabla^2}\hat{\partial}_i \hat{\partial}_i \phi_1\right) = -\phi_1^2,$$

$$(6.115)$$

$$S_{ij} G_{ij}(S) = \phi_2 \frac{1}{\nabla^2}\frac{1}{\nabla^2}\nabla^2 \nabla^2 \phi_0 = \phi_0 \phi_2.$$

Using all these expressions the Lagrangian as given in Eq. (6.113) reduces to

$$\mathscr{L} = \frac{1}{2}\left(2\phi_0\phi_2 + 2\phi_1^2\right) = \phi_0\phi_2 + \phi_1^2.$$

$$(6.116)$$

This Lagrangian does not describe any propagating degrees of freedom.

Exercise 4 Show that the linearized generalized Einstein tensor $\mathscr{G}_{\mu\nu}^{\text{lin}}(h)$ defined in (6.71) satisfies the Bianchi identities

$$\nabla^\mu \mathscr{G}_{\mu\nu}^{\text{lin}}(h) = 0.$$

$$(6.117)$$

Show that the tensor $\mathscr{G}_{\mu\nu}^{\text{lin}}(h)$ is invariant under the linear diffeomorphisms

$$\delta h_{\mu\nu} = \nabla_\mu \varepsilon_\nu + \nabla_\nu \varepsilon_\mu.$$

$$(6.118)$$

Hint: Use that

$$[\nabla_\mu, \nabla_\nu]V_\rho = \Lambda(\bar{g}_{\mu\rho}V_\nu - \bar{g}_{\nu\rho}V_\mu).$$

$$(6.119)$$

Solution Taking the divergence ∇^μ of the generalized Einstein tensor gives:

$$\nabla^\mu \mathscr{G}_{\mu\nu}^{\text{lin}}(h) = \nabla^\mu \mathscr{R}_{\mu\nu}^{\text{lin}} - \frac{1}{2}\nabla_\nu \bar{g}^{\rho\sigma}\mathscr{R}_{\rho\sigma}^{\text{lin}} + 4\Lambda\nabla^\mu h_{\mu\nu} - 2\Lambda\nabla_\nu h$$

$$= \nabla^\mu \Box h_{\mu\nu} - \nabla^\mu \nabla^\rho \nabla_\mu h_{\rho\nu} - \nabla^\mu \nabla^\rho \nabla_\nu h_{\rho\mu} + \Box \nabla_\nu h$$

$$- \nabla_\nu\left(\Box h - \nabla^\alpha \nabla^\sigma h_{\alpha\sigma}\right) + 4\Lambda\nabla^\mu h_{\mu\nu} - 2\Lambda\nabla_\nu h.$$

$$(6.120)$$

Using the property $[\nabla_\mu, \nabla_\nu]V_\rho = \Lambda(\bar{g}_{\mu\rho}V_\nu - \bar{g}_{\nu\rho}V_\mu)$ together with

$$[\nabla_\mu, \nabla_\nu]V_{\rho\sigma} = 2\Lambda(\bar{g}_{\rho[\mu}V_{\nu]\sigma} + \bar{g}_{\sigma[\mu}V_{\nu]\rho}),$$

$$(6.121)$$

the previous relation reduces to:

$$
\begin{aligned}
\nabla^\mu \mathscr{G}^{\text{lin}}_{\mu\nu}(h) &= -\Lambda \nabla^\mu h_{\mu\nu} + \Lambda \nabla_\nu h - \bar{g}^{\mu\alpha} \bar{g}^{\rho\beta} [\nabla_\alpha, \nabla_\beta] \nabla_\mu h_{\rho\nu} \\
&= -\Lambda \nabla^\mu h_{\mu\nu} + \Lambda \nabla_\nu h - \Lambda \big(3\nabla^\rho h_{\rho\nu} - \nabla^\rho h_{\rho\nu} + \nabla_\nu h \big) \\
&\quad - \Lambda \big(-\nabla^\alpha h_{\nu\alpha} + \nabla^\rho h_{\rho\nu} - 3\nabla^\alpha h_{\alpha\nu} \big) \\
&= -\Lambda \nabla^\mu h_{\mu\nu} + \Lambda \nabla^\alpha h_{\alpha\nu} + \Lambda \nabla_\nu h - \Lambda \nabla_\nu h \\
&= 0,
\end{aligned}
\tag{6.122}
$$

where we used:

$$
[\nabla_\alpha, \nabla_\beta] \nabla_\mu h_{\rho\nu} = 2\Lambda (\bar{g}_{\mu[\alpha} \nabla_{\beta]} h_{\rho\nu} + \bar{g}_{\nu[\alpha} \nabla_{|\mu|} h_{\rho|\beta]} + \bar{g}_{\rho[\alpha} \nabla_{|\mu|} h_{\beta]\nu}).
\tag{6.123}
$$

We now calculate the variation of the generalized Einstein tensor $\mathscr{G}^{\text{lin}}_{\mu\nu}(h)$ under the linearized diffeomorphisms (6.118). We first calculate $\delta \mathscr{R}^{\text{lin}}_{\mu\nu}$:

$$
\begin{aligned}
\delta \mathscr{R}^{\text{lin}}_{\mu\nu} &= \Box \delta h_{\mu\nu} - \nabla^\rho \nabla_\mu \delta h_{\rho\nu} - \nabla^\rho \nabla_\nu \delta h_{\rho\mu} + \nabla_\mu \nabla_\nu \bar{g}^{\alpha\beta} \delta h_{\alpha\beta} \\
&= -4\Lambda \nabla_\mu \varepsilon_\nu - 4\Lambda \nabla_\nu \varepsilon_\mu,
\end{aligned}
\tag{6.124}
$$

where we used (6.118) and the following relation:

$$
[\nabla^\rho, \nabla_\mu] \nabla_\nu \varepsilon_\rho = 3\Lambda \nabla_\mu \varepsilon_\nu - \Lambda \bar{g}_{\mu\nu} \nabla^\rho \varepsilon_\rho.
\tag{6.125}
$$

It then follows that:

$$
\begin{aligned}
\delta \mathscr{G}^{\text{lin}}_{\mu\nu}(h) &= -4\Lambda (\nabla_\mu \varepsilon_\nu + \nabla_\nu \varepsilon_\mu) + \Lambda \bar{g}_{\mu\nu} \bar{g}^{\rho\sigma} (2\nabla_\rho \varepsilon_\sigma + 2\nabla_\sigma \varepsilon_\rho) \\
&\quad + 4\Lambda (\nabla_\mu \varepsilon_\nu - 2\nabla_\nu \varepsilon_\mu) - 2\Lambda \bar{g}_{\mu\nu} \bar{g}^{\rho\sigma} (\nabla_\rho \varepsilon_\sigma + \nabla_\sigma \varepsilon_\rho) \\
&= -4\Lambda \bar{g}_{\mu\nu} \nabla^\rho \varepsilon_\rho + 4\Lambda \bar{g}_{\mu\nu} \nabla^\rho \varepsilon_\rho = 0,
\end{aligned}
\tag{6.126}
$$

which is what we wanted to proof.

References

1. K.S. Stelle, Renormalization of higher derivative quantum gravity. Phys. Rev. D **16**, 953 (1977)
2. K.S. Stelle, Classical gravity with higher derivatives. Gen. Relativ. Gravit. **9**, 353 (1978)
3. S. Deser, R. Jackiw, S. Templeton, Topologically massive gauge theories. Ann. Phys. **140**, 372 (1982) [Erratum-ibid. **185**, 406 (1988), Ann. Phys. **281**, 409 (2000)]
4. E.A. Bergshoeff, O. Hohm, P.K. Townsend, Massive gravity in three dimensions. Phys. Rev. Lett. **102**, 201301 (2009). arXiv:0901.1766 [hep-th]
5. E.A. Bergshoeff, O. Hohm, P.K. Townsend, More on massive $3D$ gravity. Phys. Rev. D **79**, 124042 (2009). arXiv:0905.1259 [hep-th]
6. T. Curtright, Generalized gauge fields. Phys. Lett. B **165**, 304 (1985)

7. J.M.F. Labastida, Massless particles in arbitrary representations of the Lorentz group. Nucl. Phys. B **322**, 185 (1989)
8. M. Henneaux, A. Kleinschmidt, H. Nicolai, Higher spin gauge fields and extended Kac-Moody symmetries. arXiv:1110.4460 [hep-th]
9. T.L. Curtright, P.G.O. Freund, Massive dual fields. Nucl. Phys. B **172**, 413 (1980)
10. B. Gonzalez, A. Khoudeir, R. Montemayor, L.F. Urrutia, Duality for massive spin two theories in arbitrary dimensions. J. High Energy Phys. **0809**, 058 (2008). arXiv:0806.3200 [hep-th]
11. E.A. Bergshoeff, J.J. Fernandez-Melgarejo, J. Rosseel, P.K. Townsend, On 'new massive' 4D gravity. arXiv:1202.1501 [hep-th]
12. E.A. Bergshoeff, O. Hohm, P.K. Townsend, On higher derivatives in 3D gravity and higher spin gauge theories. Ann. Phys. **325**, 1118 (2010). arXiv:0911.3061 [hep-th]
13. P.K. Townsend, K. Pilch, P. van Nieuwenhuizen, Selfduality in odd dimensions. Phys. Lett. B **136**, 38 (1984). Addendum-ibid. B **137**, 443 (1984)
14. S. Deser, R. Jackiw, 'Selfduality' of topologically massive gauge theories. Phys. Lett. B **139**, 371 (1984)
15. J.F. Schonfeld, A mass term for three-dimensional gauge fields. Nucl. Phys. B **185**, 157 (1981)
16. S. Deser, R. Jackiw, S. Templeton, Three-dimensional massive gauge theories. Phys. Rev. Lett. **48**, 975 (1982)
17. S. Deser, R. Jackiw, Higher derivative Chern-Simons extensions. Phys. Lett. B **451**, 73 (1999). hep-th/9901125
18. H. Lu, C.N. Pope, Critical gravity in four dimensions. Phys. Rev. Lett. **106**, 181302 (2011). arXiv:1101.1971 [hep-th]
19. S. Deser, H. Liu, H. Lu, C.N. Pope, T.C. Şişman, B. Tekin, Critical points of D-dimensional extended gravities. Phys. Rev. D **83**, 061502 (2011). arXiv:1101.4009 [hep-th]
20. C. Aragone, A. Khoudeir, Selfdual massive gravity. Phys. Lett. B **173**, 141 (1986)
21. E.A. Bergshoeff, M. Kovacevic, J. Rosseel, Y. Yin, On topologically massive spin-2 gauge theories beyond three dimensions. J. High Energy Phys. **1210**, 055 (2012). arXiv:1207.0192 [hep-th]

Part II
Quantum Cosmology

Chapter 7
Loop Quantum Cosmology, Space-Time Structure, and Falsifiability

Martin Bojowald

Abstract Loop quantum cosmology attempts to understand the full dynamics of loop quantum gravity by realizing crucial effects in simpler, usually symmetric settings. Several subtleties arise especially when cosmological questions are to be addressed, related to possible mini-superspace artefacts, consistent cosmological perturbation theory, and quantum space-time structure. Recent work on inhomogeneous perturbations has highlighted some of the dangers of an over-reliance on simple models, sometimes not just reduced by symmetry but also in the possible forms of matter or quantum corrections. Only a consistent treatment of inhomogeneity, taking into account the full gauge structure related to general covariance, can show what happens at high densities in quantum gravity. The relevant methods and results (especially effective equations, potential observational signatures, singularity resolution and signature change) are surveyed in here.

7.1 Introduction

In order to understand quantum gravity and its implications for cosmology, the strategy of loop quantum cosmology [1–3] is to start with mini-superspace models of high degrees of symmetry. In this way, one can probe quantum-geometry effects suggested by the full theory of loop quantum gravity [4–6]. In a second set of (usually many and long) steps, one should then test what effects are reliable within the full setting, how they may change qualitatively or quantitatively as more degrees of freedom are taken into account, and eventually what the cosmological implications of the full theory itself are.

There is justified skepticism: Classically, symmetric solutions are exact, if special, realizations of space-time according to general relativity. With any kind of quantization, however, we necessarily violate uncertainty relations when we say that one field, such as the spatial metric, and its canonically conjugate one, for instance extrinsic curvature, have exactly vanishing non-symmetric contributions. For

M. Bojowald (✉)
The Pennsylvania State University, 104 Davey Lab, University Park, PA 16802, USA
e-mail: bojowald@gravity.psu.edu

G. Calcagni et al. (eds.), *Quantum Gravity and Quantum Cosmology*,
Lecture Notes in Physics 863, DOI 10.1007/978-3-642-33036-0_7,
© Springer-Verlag Berlin Heidelberg 2013

all symmetric solutions, at least some components of a canonical pair must both be restricted to vanish, or else the solution would evolve away from a symmetric metric. In quantum gravity, symmetry reduction can be considered as an approximation assuming that violations of the uncertainty relation do not matter much, but the question of what kind of approximation exactly one is dealing with remains open.

It has been difficult to quantify what one is missing in a symmetry-reduced context, mainly owing to a lack of knowledge of and contact with a possible full theory of quantum gravity. The main reason of doubt, for a long time, has been essentially a classical one: If one cannot impose exact symmetries, unstable dynamics may quickly move non-symmetric solutions away from those of symmetric models, and the latter's implications would be too special and fine-tuned [7]. In other words, classical instability can enlarge violations of the uncertainty relation committed by mini-superspace models.

The situation has been improved by detailed developments in the full theory of loop quantum gravity and its method of symmetry reduction at the quantum level [8]. Although the full formulation as well as the connection with reduced models remains tentative at a dynamical level and is still being developed, several general aspects of quantum space-time structure have been uncovered. By now, we have arrived at several statements telling us what can and what cannot be trusted about mini-superspace models. The fact that mini-superspace models are always preliminary should not be forgotten, especially when it comes to high-density and presumably strongly quantum regimes such as the big bang. At present, no theory, including loop quantum cosmology, is able to tell us what the actual quantum nature of the big bang might be. Nevertheless, during more than a decade of following the strategy of loop quantum cosmology, we have found several promising indications of new effects that not only make the classical picture more consistent (removing singularities) but also have the potential of being tested observationally.

At a mathematical level, one quickly finds that loop quantum cosmology, as motivated by full loop quantum gravity, is based on a quantum representation of states and basic operators that is inequivalent [1, 9, 10] to what is used in traditional Wheeler–DeWitt quantizations [11, 12]. Based on new expressions of basic operators [1, 9, 13], also the dynamics shows characteristic new features and effects [14, 15]. However, more recently it has become clear that isotropic models on their own cannot be reliable for cosmological predictions, for several reasons. They do not capture all crucial aspects of spatial discreteness as implied by the full theory of loop quantum gravity. Moreover, many aspects of the dynamics (at a quantum, effective, or semi-classical level) become so trivial in isotropic models that they simply hide crucial full effects; oftentimes, the reduced nature especially of the dynamics of isotropic models lulls us into believing in an oversimplified world.

It turned out that the isotropic lullaby is even more potent than symmetries would suggest: Some popular models, with suitable matter contributions as discussed in detail in here, are harmonic. Implicitly, they eliminate many generic quantum features just as the harmonic oscillator does compared with more general quantum systems. In such models and closely related ones, the cosmological dynamics appears much more regular than it is generally. One of the problems shown by recent work on loop quantum cosmology is that a detailed analysis of specific models, as one was naturally driven to when mathematical questions of physical Hilbert spaces were being addressed, turned out to be at odds with the generality required for a reliable picture of quantum space-time.

Especially investigations in [16, 17], which applied methods of constructing physical Hilbert spaces as laid out in [18], had to restrict models even beyond the symmetry assumptions traditionally used in (loop) quantum cosmology. In [16] and the rather large set of articles that followed, there are three levels of reduction:

1. symmetry reduction to isotropy or homogeneity, but also
2. a specific form of matter given by a free, massless scalar and
3. in many cases, a strict focus not only on one type of corrections (holonomy corrections) but also a specific form of them chosen *ad hoc*.

The restriction on matter (2nd level) was necessary because a deparameterization approach was chosen to construct physical Hilbert spaces, using evolution by one of the phase-space degrees of freedom instead of coordinate time. A free, massless scalar φ has a constant momentum which never becomes zero; the scalar has a strictly monotonic relationship with coordinate time and can directly be substituted in equations of motion. The focus on one type of corrections and one specific realization of them (3rd level) was partially based on an incomplete understanding of other corrections at that time, and partially on being misled by mini-superspace artefacts. Initially, the results, a smooth bounce replacing the classical singularity with wavefunctions evolving nearly semi-classically without much apparent influence of strong quantum behavior, seemed so promising that questions of how reliable all the choices are were put on the back burner; or when necessary questions of robustness were addressed, they were done in such a restricted context that they could be considered only as *pro forma*.

This approach has led to some valuable results regarding mathematical questions, for instance about self-adjointness properties of Hamiltonians [19, 20], possible ways to incorporate inhomogeneity in Gowdy models [21, 22], relationships with path-integral quantization [23–26], and also for numerical techniques as applied in several examples to shed light on wavefunction evolution [27–29]. Unfortunately, however, although mathematics and physics often go hand in hand, the specializations required for detailed mathematical results in isotropic loop quantum cosmology turned out to be diametrically opposed to what is needed for reliable physics. General methods of effective descriptions, which also allow one to address interesting mathematical and computational questions, have highlighted many shortcomings, in particular:

- The matter choice of a free massless scalar was serendipitous in that it not only trivializes the problem of time, it also eliminates quantum back-reaction and makes the initial model analyzed in [16] harmonic. Although this property was not realized initially but a little later, in [30] which was partially motivated by numerical evolutions of states shown in [16], it is important for understanding aspects of bounces properly. In these models, the Friedmann equation

$$\left(\frac{\dot{a}}{a}\right)^2 = \frac{4\pi G}{3}\frac{p_\varphi^2}{a^6} = \frac{8\pi G}{3}\rho_{\text{free}} \tag{7.1}$$

is written canonically as a constraint $12\pi G(p_V V)^2 - p_\varphi^2 = 0$, with $V = a^3$ (the volume of any region with coordinate volume one) and momentum $p_V = \mathcal{H}/4\pi G$ proportional to the Hubble parameter $\mathcal{H} = \dot{a}/a$. If φ is used as internal time for deparameterization, evolution is generated by $p_\varphi(V, p_V) = \sqrt{12\pi G}|p_V V|$ as the Hamiltonian. Solving the constraint for p_φ in terms of V and p_V yields an expression quadratic in canonical variables, and quantum dynamics following the classical trajectories no matter what states we use (just as it is realized for the harmonic oscillator in quantum mechanics). The harmonic nature is preserved, although not obviously so, if one includes modifications motivated by quantum geometry in loop quantum gravity [30, 31]: holonomy corrections are implemented by replacing p_V with $\sin(Lp_V)/L$ with some discreteness parameter L, possibly of Planckian scale.

- Holonomy corrections, realized by the replacement of p_V by $\sin(Lp_V)/L$ or in resummed form eliminating p_V via equations of motion by a modified Friedmann equation [32, 33]

$$\left(\frac{\dot{a}}{a}\right)^2 = \frac{8\pi G}{3}\rho_{\text{free}}\left(1 - \frac{\rho_{\text{free}}}{\rho_{\text{QG}}}\right) \tag{7.2}$$

with the free scalar density ρ_{free} and a quantum-gravity scale $\rho_{\text{QG}} = 3/8\pi GL^2$ in terms of L, are not the main or dominant source of effects in loop quantum gravity. They depend on the curvature scale $p_V \propto \mathcal{H}$, just as higher-curvature corrections do which are independent and have been ignored in work related to [16] either by explicit choices or by considering only special classes of states. Such corrections do not appear in harmonic models as realized by the free massless scalar in a spatially flat isotropic background with vanishing cosmological constant, but they do show up when any one of these assumptions is violated. With self-interacting matter, it is not clear what bounce behavior is realized, or if there even is a bounce [34, 35].

- While deparameterization in the presence of a global clock variable can always be used to describe phase-space trajectories, a proper discussion of evolution in quantum gravity requires good knowledge of quantum space-time structure. Singularity avoidance in loop quantum cosmology [36], including bounce solutions in restricted contexts, has the advantage of being independent of modified matter sources violating energy conditions, as often used in other bounce models [37]. Rather, geometry is modified, for instance by replacing p_V with $\sin(Lp_V)/L$, and therefore the Raychaudhuri equation for effective isotropic geometries takes a different form compared to the classical one, sometimes allowing minima of the scale factor. Quantum geometry, however, is a double-edged sword. It modifies space-time structure in ways that cannot be seen when only homogeneous configurations are considered. Also wave equations for inhomogeneous modes should be expected to change, and here finding the correct modifications is much more difficult because there are strong consistency requirements from general covariance or anomaly freedom. When all this is taken into account, quantum gravity significantly modifies the propagation of modes, and might even induce signature change implying that no time (coordinate or internal) exists in certain regimes. This consequence turned out to be realized in loop quantum cosmology, right in the putative bounce phase of isotropic models [38, 39]. The universe does not evolve through a bounce at high density; it rather emerges from Euclidean space.

Instead of using simple isotropic models as a stand-alone setting for cosmological model-building, their modern application is as a jumpboard, as a well-understood basis for cosmological perturbation theory. In this context, a combination of results using numerical methods following [16] and analytic approximations in the effective context, following [30], has contributed to a detailed understanding of background models originally proposed in loop quantum cosmology. The existing solvable and harmonic models then play a role similar to the harmonic oscillator as a starting point for quantum field theory. As in this more familiar case, interactions significantly change effects seen in free and harmonic models.

At the present stage, perturbation theory has not been fully worked out yet, but several characteristic features have been found. Experience with loop quantum cosmology shows that one should always keep an open eye on the unexpected, rather than merely developing and defending one's favorite scenario. Low-energy implications of loop quantum cosmology, for instance during inflation, are rather well-understood (and promising), but the nature of strong quantum regimes such as the big bang remains wide open even regarding qualitative scenarios.

As one recent highlight that illustrates the necessity of always keeping up a shield of caution, it was seen that some models of loop quantum cosmology eliminate not only the singularity at high density, but even time. Lorentzian space-time has a well-defined, non-singular beginning at which one may pose initial values. Although the isotropic models used also have a collapse phase, it is not deterministically connected to expansion through a dynamical bounce. The correct interpretation can be seen only when inhomogeneity is added to the model in a consistent way, a procedure in which one has to face the most daunting problems of full quantum gravity even if inhomogeneity is only perturbative. These issues and related aspects of quantum space-time structure will be the main focus of this contribution. For other up-to-date reviews, we refer to [2, 3, 40, 41].

7.2 Canonical Gravity

Loop quantum gravity follows a canonical quantization strategy, although its results are sometimes combined with space-time summation and path-integral techniques in order to shed additional light on dynamical aspects [4, 42, 43]. At its present stage, the theory does not fully support the hope that canonical quantization could lead to a fundamental theory of quantum gravity: It is beset by ambiguities in formulating quantum dynamics.

To test the theory further, beyond the question of how its dynamics may be constructed, we need to investigate its low-energy behavior for instance in a cosmological background. In such a situation, loop quantum gravity suggests several new candidates for quantum-geometry corrections which, even if they cannot fully be specified in the presence of ambiguities, make the theory and its general approach promising for physics. But before such effects can be evaluated, the key issue of consistency must be addressed: We must ensure that the theory, even at the quantum level, obeys some notion of general covariance.

General covariance corresponds to gauge freedom of general relativity. If the theory is quantized, the same amount of gauge freedom must still be present, or else the theory is anomalous and does not allow the right number and form of solutions. At the quantum level, gauge transformations may not be exactly of the classical form and there can be quantum effects even in them, not just in the dynamical equations. But for every classical gauge transformation, there must be a quantum analog in order to ensure that spurious degrees of freedom are still removed after the theory is solved. The correct treatment of gauge and covariance is also necessary in cosmology for self-consistent perturbation equations, but it is not obvious in a canonical setting in which one starts with different implementations of space and time derivatives.

7.2.1 Cosmic Subtleties

The classical scalar cosmological perturbation equations are

$$\partial_c(\dot{\psi} + \mathscr{H}\phi) = 4\pi G \dot{\bar{\varphi}} \partial_c \delta\varphi, \tag{7.3}$$

$$\Delta\phi - 3\mathscr{H}(\dot{\psi} + \mathscr{H}\phi) = 4\pi G\left[\dot{\bar{\varphi}}\delta\dot{\varphi} - \dot{\bar{\varphi}}^2\phi + a^2 V'(\bar{\varphi})\delta\varphi\right], \tag{7.4}$$

$$\ddot{\psi} + \mathscr{H}(2\dot{\psi} + \dot{\phi}) + \left(2\dot{\mathscr{H}} + \mathscr{H}^2\right)\phi = 4\pi G\left[\dot{\bar{\varphi}}\delta\dot{\varphi} - a^2 V'(\bar{\varphi})\delta\varphi\right], \tag{7.5}$$

$$\delta\ddot{\varphi} + 2\mathscr{H}\delta\dot{\varphi} - \Delta\delta\varphi + a^2 V''(\bar{\varphi})\delta\varphi + 2a^2 V'(\bar{\varphi})\phi - \dot{\bar{\varphi}}(\dot{\phi} + 3\dot{\psi}) = 0, \tag{7.6}$$

together with $\phi = \psi$ (in the absence of anisotropic stress) for the two metric perturbations ϕ of the time-time component and ψ of the space-space components, assuming a scalar field $\varphi = \bar{\varphi} + \delta\varphi$ as the matter source. We have four equations for two functions ψ and $\delta\varphi$ if we set $\phi = \psi$, identified by the number of time derivatives they contain as two constraints and two evolution equations. (The equation $\phi = \psi$ follows from the off-diagonal spatial part of Einstein's equation and is therefore part of another evolution equation, not a constraint.)

Classically the system of equations is consistent: the constraints are automatically preserved by evolution (they have to be imposed just for initial values) and evolution is gauge invariant and insensitive to coordinate changes. Also for the second aspect, the presence of preserved constraints is important, for they function as the canonical generators of gauge transformations. If they were not preserved by evolution, it would be impossible to express the equations by gauge-invariant variables without coupling to gauge artefacts.

But what happens if we try to bring in quantum corrections? It is easy to modify the background equations (in isotropy, the Friedmann equation and the Klein–Gordon equation for a scalar field) by all kinds of effects, for instance those in (7.2). There is a gauge transformation even in isotropic models, corresponding to time reparameterizations. Indeed, the Friedmann equation contains only first-order time derivatives and plays the role of a constraint. One can take another time derivative of the equation and eliminate second-order time derivatives of the scalar so as to obtain the Raychaudhuri equation, an equation of motion that automatically preserves the constraint presented by the Friedmann equation and respects the gauge transformation.

Evolution is consistent if it is derived from a consistent set of constraints, as in the example of the Raychaudhuri equation. If we bring in perturbative inhomogeneity, however, the set of constraints and corresponding gauge transformations becomes much larger than just a single one. Not only must the evolution equations respect the constraints and their gauge transformations, the constraints must also respect one another's gauge transformations. If the constraints are consistent in this sense (or first class), the evolution equations they generate are guaranteed to be consistent. But in contrast to isotropic models, ensuring that the constraints themselves are consistent with one another is a highly non-trivial task. Quantum corrections then cannot be inserted at will, for instance by correcting just the background equations for $\mathcal{H}$ and $\bar{\varphi}$ in (7.3)–(7.6) and using the classical form of perturbation equations. Ensuring consistency and covariance becomes a highly restrictive condition, required for cosmological perturbation theory or any analysis of inhomogeneous situations to proceed.

There can be no shortcuts: Sometimes one uses gauge fixing in order to eliminate the gauge freedom before analyzing the equations. This procedure is valid classically because we already know the gauge system and its consistency. But if we use gauge fixing before we quantize the theory or insert quantum corrections in the classical equations, there is no way to tell whether the results are consistent. We use the classical gauge structure in order to fix the gauge, but then modify equations and constraints by quantum corrections. The new gauge transformations generated by modified constraints may or may not be consistent with the gauge structure used to fix the gauge. There is no way of telling after the gauge has been fixed; one may be lucky, but in many cases one is not. Many examples are known already in the context of cosmological perturbation theory in which gauge-fixed models do not agree with consistent treatments, sometimes crucially so. (Several examples will be provided below.)

A milder form of fixing the structure related to covariance before quantization is deparameterization, a process in which one chooses one of the degrees of freedom as a measure for change instead of coordinate time. Equations can then be formulated in partially gauge-invariant terms, but there is still a remnant of the anomaly problem. Equations will be invariant under changes of coordinates, but not under changes of frame once the time degree of freedom has been chosen. If one can show that physical results do not depend on the choice of internal time, one can confidently use the theory. But showing that is usually a highly non-trivial procedure; at the quantum level, it is not even clear in all cases how to transform from one internal time to another one. So far, semi-classical regimes can be treated in this manner using effective techniques [44–46], but the question of insensitivity to the choice of internal time is rarely addressed in deparameterized models. Also deparameterization does not avoid a detailed discussion of the anomaly problem and of the insensitivity of physics to mathematical choices.

The form of gauge transformations, the way they arise in general relativity, and the question of how they can be dealt with in quantum theory are thus important issues for any physical evaluation of quantum gravity, including questions of cosmology. We will now focus on the structure of space-time in order to highlight the relevant features.

7.2.2 Deformations of Space

Before entering general covariance, it is instructive to have a look at the space-time structure of special relativity. The key notion to relate space and time is a Lorentz boost of velocity v, mapping time t to another function

$$ct' = \frac{ct - vx/c}{\sqrt{1 - v^2/c^2}} \tag{7.7}$$

of the original coordinates. In geometrical terms, we may interpret the algebraic relation between ct and ct' as a linear deformation of a spatial slice $ct = \text{const}$ to $ct' = \text{const}$, so that $ct = \text{const} + vx/c$ on a spatial slice after the deformation; see Fig. 7.1. Completing this translation from algebraic transformations to geometrical operators for all Poincaré transformations, one can confirm that all possible forms of spatial deformations agree with generators P_μ and $M_{\mu\nu}$ of the Poincaré algebra

$$[P_\mu, P_\nu] = 0, \tag{7.8}$$

$$[M_{\mu\nu}, P_\rho] = \eta_{\mu\rho} P_\nu - \eta_{\nu\rho} P_\mu, \tag{7.9}$$

$$[M_{\mu\nu}, M_{\rho\sigma}] = \eta_{\mu\rho} M_{\nu\sigma} - \eta_{\mu\sigma} M_{\nu\rho} - \eta_{\nu\rho} M_{\mu\sigma} + \eta_{\nu\sigma} M_{\mu\rho}. \tag{7.10}$$

With the geometrical reinterpretation, one can extend the setting easily to general covariance: we use non-linear deformations of spatial slices instead of just linear ones as illustrated in Fig. 7.2. We then have a much larger class of generators, $D[N^a]$ for tangential deformations along spatial vector fields $N^a(x)$ and $H[N]$ for normal deformations by functions $N(x)$. Geometry of hypersurface deformations then provides the algebra

$$[D[N^a], D[M^a]] = D[\mathscr{L}_{M^a} N^a], \tag{7.11}$$

$$[H[N], D[M^a]] = H[\mathscr{L}_{M^a} N], \tag{7.12}$$

$$[H[N_1], H[N_2]] = D[q^{ab}(N_1 \partial_b N_2 - N_2 \partial_b N_1)], \tag{7.13}$$

Fig. 7.1 A Lorentz boost changes t to a new time coordinate t'. Geometrically, the transformation can be interpreted as a linear deformation of a spatial slice $ct = $ const along its normal n^a by a linear amount $N(x) = $ const $+ vx/c$

Fig. 7.2 Non-linear normal deformations of spatial hypersurfaces with lapse functions N_1 and N_2 always commute to a spatial deformation with shift vector $N^a = q^{ab}(N_1 \partial_b N_2 - N_2 \partial_b N_2)$

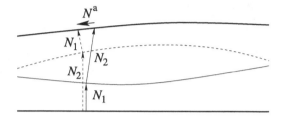

using Lie derivatives and the induced metric q_{ab} on spatial slices. (See, e.g., [39] for a recent explicit demonstration.) The metric appears because one refers to the normal vector in order to identify $H[N]$ as normal deformations.

In order to confirm the hypersurface-deformation algebra as a generalization of the Poincaré algebra, we can evaluate it for linear functions in some Cartesian coordinate patch (Fig. 7.3). With linear functions $N = P_0 + x^a \tilde{M}_{a0}$, $N_a = P_a + x^b \tilde{M}_{ba}$ in a Minkowski background, we obtain the correct Poincaré algebra from the hypersurface-deformation algebra.

7.2.3 Gauge Theory

The hypersurface-deformation algebra generalizes the Poincaré algebra, and necessarily makes it local with generators depending on the position. Any theory invariant under hypersurface deformations must be a gauge theory. We may choose different fields for the theory to be based on, but there is one distinguished candidate: The induced metric q_{ab} changes under deformations of spatial slices, and appears in structure functions of (7.13). When looking for a theory invariant under hypersurface deformations, it is then natural to take q_{ab} as a canonical field, combined with a momentum field π^{ab}. Hypersurface deformations will no longer be abstract generators of an algebra, but appear as constraint functions $D[N^a] = 0$ and $H[N] = 0$ on phase space, that is as functionals of q_{ab} and π^{ab}, such that

$$\{D[N^a], D[M^a]\} = D[\mathscr{L}_{M^a} N^a], \tag{7.14}$$

$$\{H[N], D[M^a]\} = H[\mathscr{L}_{M^a} N], \tag{7.15}$$

$$\{H[N_1], H[N_2]\} = D[q^{ab}(N_1 \partial_b N_2 - N_2 \partial_b N_1)], \tag{7.16}$$

Fig. 7.3 Two linear normal deformations with lapse functions $N_1(x) = vx/c$ and $N_2(x) = c\Delta t - vx/c$ commute to a spatial translation with shift $N^a = -v^a\Delta t$: Boosting to velocity v, waiting some time Δt and then boosting back to velocity zero makes all objects appear to move a distance $v\Delta t$

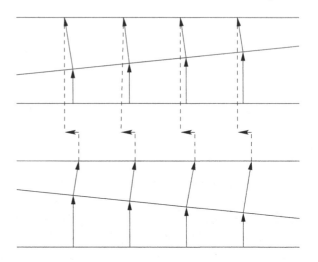

under Poisson brackets. Gauge transformations of phase-space functions then have the general form $\delta f = \{f, H[N] + D[N^a]\}$, interpreted as the generators of space-time diffeomorphisms.

Local hypersurface-deformation invariance is equivalent to general covariance: The requirement of (7.14)–(7.16) as a gauge algebra is the canonical analog of having a space-time scalar action. (As argued by Dirac [47], the canonical notion is perhaps even more fundamental.) The hypersurface-deformation algebra is a very strong kind of symmetry: it determines the dynamical structure of general relativity. As shown in [48, 49], a gauge theory with local generators of hypersurface deformations and second-order equations of motion for the spatial metric q_{ab} must be of Einstein–Hilbert form. To see this, one performs a Legendre transformation from (π^{ab}, H) to (K_{ab}, L) with extrinsic curvature $K_{ab} = N^{-1}\delta H[N]/\delta\pi^{ab}$ and the Lagrangian density $L = \pi^{ab}K_{ab} - H$, uses the diffeomorphism constraint $D^a = -2\nabla_b\pi^{ab}$ and writes (7.16) in unsmeared form as

$$\frac{\delta L(x)}{\delta q_{ab}(x')}K_{ab}(x')\delta(x, x') - (x \leftrightarrow x')$$

$$= -2\frac{\delta L}{\delta K_{ab}}(\nabla_b\beta\nabla_a\delta(x, x') + \beta\nabla_a\nabla_b\delta(x, x')) - (x \leftrightarrow x').$$

This differential equation for L can be solved in an expansion by powers of K_{ab}. Only the values of Newton's and the cosmological constant remain free.

With higher-derivative corrections there are new degrees of freedom, with some of the time derivatives of q_{ab} canonically independent of π_{ab}. Functional derivatives and terms in the above equation will have to be changed accordingly, as well as the expansion of L by powers of all the quantities related to time derivatives. The quantum origin of new degrees of freedom that may give rise to higher-derivative corrections will be explained in Sect. 7.4. If the condition on the order of derivatives is dropped, there are many more action principles of higher-curvature form. Including them, all classical actions of gravity have the same gauge algebra (7.14)–(7.16) unless they explicitly break covariance.

7.2.4 Quantum Corrections

With these results, there are three options for corrections from quantum gravity.

7.2.4.1 Break Covariance

One may decide to dispense with gauge generators and covariance altogether. However, since covariance is implemented by gauge transformations, the theory is anomalous if the gauge is broken. For instance, we would have inconsistent cosmological perturbation equations: constraints $D[N^a] = 0$ and $H[N] = 0$ still appear but are not guaranteed to be preserved by evolution equations. Moreover, it will remain unknown how to correctly combine degrees of freedom to physical ones.

Inconsistency can formally be avoided by choosing the gauge or frame before quantization (usually referred to as gauge fixing in the former case, and deparameterization in the latter). Often, a hidden frame dependence results even if covariance is not broken explicitly, for instance from deparameterization using a classical internal time φ so as to write the Hamiltonian constraint as $C = p_\varphi + H(q, p)$. It is then not guaranteed that physical results after quantization are insensitive to the choice of internal time φ; if they are not, the theory is anomalous.

With broken covariance, the quantum "corrected" theory is not consistent unless there is a classically distinguished frame.

7.2.4.2 Preserve Hypersurface-Deformation Algebra but Allow Equations of Motion to Be of Higher Than Second Order

The action principle follows uniquely from the hypersurface-deformation algebra only if equations of motion are restricted to be of second order in derivatives. Without this assumption, there is an infinite set of covariant action principles given by higher-curvature theories.

Indeed, quantum corrections are most often thought of as terms that arise from a low-energy effective action for the propagation of gravitons [50, 51]. The classical form of covariance and the hypersurface-deformation algebra are then preserved, but as always with effective actions one expects higher time derivatives. Combining these two properties, we necessarily arrive at higher-curvature effective actions.

These theories are perfectly consistent from a gauge perspective. However, possible quantum corrections in nearly isotropic cosmology are of a tiny size, given by ratios of the quantum-gravity scale (such as the Planck length) to the Hubble distance. (See e.g. [52].)

This leaves us with only one interesting option in the cosmological context.

7.2.4.3 Non-trivial Consistent Deformation of the Hypersurface-Deformation (and Poincaré) Algebra

If we consistently deform the hypersurface-deformation algebra, allowing quantum corrections but respecting the number of gauge generators, we obtain a setting more general even than the large class of higher-curvature effective actions. With a modified notion of hypersurface deformations and covariance, we have quantum corrections in the space-time structure, not just in the dynamics. Potentially, we may obtain new, not extremely suppressed corrections.

The case of consistently deformed hypersurface-deformation algebras is not often found, but it is realized in loop quantum gravity: In all models consistently analyzed so far, starting with [53], two normal deformations obey the Poisson bracket

$$\{H_{(\beta)}[N_1], H_{(\beta)}[N_2]\} = D[\beta q^{ab}(N_1\partial_b N_2 - N_2\partial_b N_1)] \qquad (7.17)$$

with a correction function β depending on the spatial metric or extrinsic curvature. In order to see how such an algebra arises, we need to look at more details of loop quantum gravity.

7.3 Loop Quantum Gravity

We start with the classical theory of general relativity in canonical variables A_a^i and E_j^b with Poisson brackets

$$\{A_a^i(x), E_j^b(y)\} = 8\pi\gamma G\delta_a^b\delta_j^i\delta(x, y), \qquad (7.18)$$

using the Barbero–Immirzi parameter γ [54, 55]. The parameter appears in the Ashtekar–Barbero connection $A_a^i = \Gamma_a^i + \gamma K_a^i$, split into the spin connection Γ_a^i and extrinsic curvature K_a^i [55, 56]. The spin connection is by definition compatible with the densitized triad E_i^a: $\mathcal{D}_a^{(\Gamma)}E_i^a = 0$.

For a field theory, we cannot directly quantize the field values to well-defined operators. Rather, we smear fields by spatial integrations so as to remove delta functions in their Poisson brackets. Smeared fields, or mode functions on a suitable background, become well-defined operators. When the metric is one of the fields to be quantized, here related to the densitized triad E_j^b with $q^{ab} \det q_{cd} = E_i^a E_i^b$, we cannot do arbitrary integrations because there is no other metric to define integration measures. Loop quantum gravity has solved this problem [57] by using holonomies and fluxes

$$h_e(A) = \mathcal{P}\exp\left(\int_e A_a^i \tau_i t^a d\lambda\right), \qquad F_S^{(f)}(E) = \int_S n_a E_i^a f^i d^2y, \qquad (7.19)$$

with integrations over curves e of tangent vector t^a and surfaces S of co-normal n_a in space (and $su(2)$-generators $\tau_j = -i\sigma_j/2$ in terms of Pauli matrices together with $su(2)$-valued smearing functions f^i on surfaces). The smeared objects are quantized by using a representation of the holonomy-flux algebra obtained for their Poisson brackets.

In this way, the theory provides operators $\hat{h}_e$ for holonomies, but none directly for A_a^i. Moreover, flux operators $\hat{F}_S$ (understood as elementary areas) have discrete spectra containing zero [58, 59]. First, one can use the (eigen)values $F_n \sim 4\pi\gamma G\hbar n$ of flux, with integers n, to set the quantum-gravity scale, with states forming (possibly irregular) lattices in space each of whose Plaquette areas has some value F_n. The integer n appears as a quantum number, determining the excitation level of geometry. Depending on the state, the typical or average scale $\langle \hat{F} \rangle$ may differ from the Planck scale $\ell_P^2 = G\hbar$ if quantum geometry is excited to $n > 1$.

Secondly, also depending on this scale, the form of basic operators available tells us what corrections of the dynamics we can expect. For quantum dynamics, we write classical expressions of the constraints and Hamiltonians in terms of holonomies and fluxes. Classically, constraints and Hamiltonians are polynomial in A_a^i, but holonomies are not. This mismatch of classical and quantum properties requires regularizations or modifications of the classical theory by higher-order corrections, motivated by background-independent quantum geometry.

Initially, however, the resulting dynamics is far from being unique. There are many different ways to modify the classical Hamiltonian constraint so as to express it in terms of holonomies rather than polynomials in connection components. There are also factor-ordering choices, as always in canonical quantization. All the choices involved constitute a large set of quantization ambiguities which may spell great danger for the predictability of the theory.

On the other hand, strong consistency conditions are expected from the algebra of constraint operators, which generically does not remain anomaly-free after the classical constraints in (7.16) are modified. One may therefore hope that successfully implemented consistency conditions significantly reduce the amount of ambiguities—if one can find any consistent modification at all. At the present stage, the issue remains insufficiently evaluated (but see [60] for an example, and recent progress in [61]). Consistency conditions may be too strong to allow any modification of the classical theory, they may be too weak to reduce the level of ambiguity significantly, or ideally they may just be strong enough to leave one unique theory, or at least a highly constrained one.

Without a verdict on which outcome is realized, the arbitrariness of quantization ambiguities so far is countered, depending on one's viewpoint, by (i) the arbitrariness of *ad hoc* choices or (ii) the generality of phenomenological parameterizations. Even in some isotropic cosmological models, in which the number of ambiguities is smaller, articles suggesting to use *ad hoc* choices (usually implicitly) are plentiful. For instance, a large part of the recent literature is based on several crucial but *ad hoc* assumptions about the so-called area gap [16], corresponding to the step-size in the flux spectrum F_n for a changing lattice. It enters isotropic equations via the length scale L in ρ_{QG} of (7.2), and affects quantum dynamics. General phenomenological parameterizations, on the other hand, try to capture all expected ambiguities, for instance by considering lattice refinement [62, 63]. The resulting freedom in predictions might, *a priori*, be expected to be too large to obtain any interesting results, but several examples have been found in which loop quantum cosmology is predictive even in the presence of many fundamental ambiguities. While the theory may not qualify as a fundamental one, it is phenomenologically falsifiable.

7.3.1 Corrections from Loop Quantum Gravity

With quantization and other modifications, we should expect corrections of different types in the dynamics of loop quantum gravity:

- Using holonomies for connection components implies that there are higher-order corrections of the connection when holonomies are expanded to regain the classical terms at leading order. In cosmology, such corrections are sensitive to the energy density, relative to a length scale used in holonomies. For instance, to arrive at (7.2) we used $\sin^2(Lp_V)/L^2 \sim p_V^2 - \frac{2}{3}L^2 p_V^4 + \cdots$ with $p_V \propto \mathcal{H}$. The sine is a combination of holonomies evaluated in isotropic connections along straight lines of length L/a in coordinates, or L measured in Friedmann geometry. This length scale should be taken as the typical discreteness scale $L = \sqrt{|\langle \hat{F} \rangle|}$ of the underlying theory, usually assumed of Planckian order. However, it is not completely fixed by the theory itself because its size depends on the state and its excitation level, in addition to specific choices for Hamiltonians. The discreteness scale L is one of the most important parameters for phenomenological modelling (or the main target of *ad hoc* choices, such as the area gap).
- Another type of corrections related to the energy density is more familiar: Quantum back-reaction of moments of the state on its expectation values provides higher-derivative terms [64, 65]. It constitutes the canonical analog of the low-energy effective action, resulting in higher-curvature corrections. These corrections are different from holonomy corrections because they include higher time derivatives, but in nearly isotropic cosmology they are expected to be of the same, usually tiny order.
- Finally, there is another type of quantum-geometry corrections, called inverse-triad corrections, that results from quantizing

$$\left\{ A_a^i, \int \sqrt{|\det E|} d^3 x \right\} = 2\pi \gamma G \varepsilon^{ijk} \varepsilon_{abc} \frac{E_j^b E_k^c}{\sqrt{|\det E|}}. \tag{7.20}$$

Inverses of the densitized triad as on the right-hand side are needed for the Hamiltonian constraint of gravity as well as matter Hamiltonians, but there is no direct inverse of flux operators as the E-quantization: flux operators have discrete spectra containing zero. The left-hand side of the above equation, on the other hand, is directly quantizable. It does not require an inverse of the densitized triad, the connection can be expressed in terms of holonomies, and the Poisson bracket be turned into a commutator divided by $i\hbar$ [66].

Inverse-triad corrections, via holonomies for A_a^i and elementary flux operators used to quantize the volume $\int \sqrt{|\det E|} d^3 x$ depend, like holonomy corrections, on the underlying discreteness scale L of a state. But they are not directly related to curvature, the Hubble parameter $\mathcal{H}$, or the energy density, and therefore can more easily be separated from the other two types. They could also be larger in sub-Planckian curvature regimes. It is therefore of significant interest to derive their phenomenological implications. This task, however, is not easy because inverse-triad corrections are not implementable completely in mini-superspace models; one must include at least perturbative inhomogeneity. (Holonomy corrections can more easily be implemented in homogeneous models if L is simply related to the Planck length, a choice which would make inverse-triad corrections unacceptably large [67]. Avoiding the inconsistency of these two types of corrections with full reference to the lattice scale and possible dynamical changes also requires an inhomogeneous viewpoint.)

7.3.2 Construction of Inverse-Triad Corrections

Let us look at a lattice state with $U(1)$-holonomies $\hat{h}_e$ and fluxes $\hat{F}_e$. (Fluxes are normally associated with surfaces instead of curves e. However, for a regular lattice one can pick a dual surface to each edge, and then assign flux operators to edges of the lattice.) With a real parameter $0 < r < 1$, we quantize the inverse-triad formula (7.20) to

$$\left(\widehat{|F|^{r-1} \mathrm{sgn}\, F}\right)_e = \frac{\hat{h}_e^\dagger |\hat{F}_e|^r \hat{h}_e - \hat{h}_e^\dagger |\hat{F}_e|^r \hat{h}_e}{8\pi G r \gamma \ell_{\mathrm{Pl}}^2} =: \hat{I}_e,$$

the numerator representing the classical Poisson bracket of holonomy and flux as a commutator. Compared with (7.20), the formula here is more general by allowing a range of powers $0 < r < 1$ for quantization ambiguities. To evaluate these operators, we use the relations $[\hat{h}_e, \hat{F}_e] = -4\pi \gamma \ell_{\mathrm{Pl}}^2 \hat{h}_e$ and $\hat{h}_e \hat{h}_e^\dagger = 1$ of the holonomy-flux algebra together with the unitarity of $U(1)$-holonomies, such that

$$\hat{h}_e^\dagger |\hat{F}_e|^r \hat{h}_e = \left| \hat{F}_e + 4\pi \gamma \ell_{\mathrm{Pl}}^2 \right|^r, \qquad \hat{h}_e |\hat{F}_e|^r \hat{h}_e^\dagger = \left| \hat{F}_e - 4\pi \gamma \ell_{\mathrm{Pl}}^2 \right|^r.$$

Effective Hamiltonians and constraints, obtained from expectation values of Hamiltonian operators, contain an expectation value

$$\langle \hat{I}_e \rangle = \frac{\left| \langle \hat{F}_e \rangle + 4\pi \gamma \ell_{\mathrm{Pl}}^2 \right|^r - \left| \langle \hat{F}_e \rangle - 4\pi \gamma \ell_{\mathrm{Pl}}^2 \right|^r}{8\pi G r \gamma \ell_{\mathrm{Pl}}^2} + \text{moment terms} \qquad (7.21)$$

instead of the classical inverse, with 'moment terms' indicating the presence of additional fluctuations and moments that would contribute to quantum back-reaction (see Sect. 7.4) but leave the main form of inverse-triad corrections with their characteristic dependence on $\langle \hat{F} \rangle$ unchanged. This derivation shows that inverse-triad corrections depend crucially on the elementary (edge-wise) flux values in a given state, relative to the Planckian flux $4\pi \gamma \ell_{\mathrm{Pl}}^2$ [68].

For further analysis of effective constraints including inverse-triad corrections, we refer to the quantum-gravity scale $|\langle \hat{F} \rangle| = L^2$ and define the correction function

$$\alpha(L) := \frac{|\langle \hat{I} \rangle|}{I_{\mathrm{class}}} = \frac{\left| L^2 + 4\pi \gamma \ell_{\mathrm{Pl}}^2 \right|^r - \left| L^2 - 4\pi \gamma \ell_{\mathrm{Pl}}^2 \right|^r}{8\pi \gamma r \ell_{\mathrm{Pl}}^2} L^{2(1-r)}. \qquad (7.22)$$

One can easily evaluate these functions for different choices of r to see key properties. They approach the classical value 1 for large fluxes, and show significant corrections for small fluxes. Somewhat counter-intuitively, inverse-triad corrections are large if the discreteness scale L is nearly Planckian. One can understand this consequence by considering inverse-triad corrections as a natural cut-off, implied by the background-independent quantization used, of diverging inverses of E_i^a near classical singularities. By referring to the discreteness scale L, they put constraints also on holonomy corrections, which depend on the same value but are usually small for Planckian L. The interplay of different corrections with disjoint regimes of dominance is important for restrictive phenomenology.

Intuitively, inverse-triad quantization eliminates classical divergences at degenerate E_i^a. Accordingly, the dynamics is most sensitive to these corrections when fluxes are small, near classical singularities, but there are still effects at larger fluxes and in more semi-classical regimes. The relation to the quantum-gravity scale L means that the size of corrections cannot easily be estimated, unlike other corrections related to the classical density. We must know properties of the underlying quantum-gravity state to find a possible value for L, which in nearly classical regimes must be larger than the Planck length for inverse-triad corrections to be sufficiently small. The same is true for holonomy corrections for which L must be known as well. For holonomy corrections treated in isolation, L is often (ad hoc) assumed to be Planckian, but this value is not consistent with small inverse-triad corrections [67]. With observations or other input to bound the size of corrections, properties such as the discreteness scale L of an underlying quantum-gravity state can be discerned.

The interrelation of different corrections implies one crucial feature: the theory imposes two-sided bounds on its parameters. Inverse-triad corrections are large for small L, and holonomy as well as discretization effects become large for large L. There is only a finite range for allowed values, not just an upper bound which could always be evaded. The theory is falsifiable.

7.3.3 Anomaly-Freedom

Before using any corrections for cosmological phenomenology, we have to show that they are dynamically consistent, in that they do not break the gauge algebra. To that end, we first parameterize the more promising ones, inverse-triad corrections, assuming that they are small, as $\alpha = 1 + \delta(a)$ with $\delta(a) \ll 1$, written for a nearly isotropic universe with scale factor a. With a power-law parameterization in suitable phases of cosmological evolution, just two parameters δ_0 and σ then determine the effects of

$$\delta(a) \sim \delta_0 a^\sigma. \tag{7.23}$$

Crucially for their applicability, inverse-triad corrections provide a non-trivial consistent deformation of the hypersurface-deformation algebra, and accordingly a new form of quantum space-time structure. This property allows us to develop gauge-invariant and consistent cosmological perturbation equations in full detail. As shown in [53], the Hamiltonian constraint

$$H_{(\alpha^2)}[N] = \frac{1}{16\pi\gamma G} \int_\Sigma d^3x \, N\alpha \left(\varepsilon_{ijk} F_{ab}^i \frac{E_j^a E_k^b}{\sqrt{|\det E|}} + C_L \right)$$
$$+ H_{\mathrm{matter}}^{(\alpha^2)}[N] + \text{'counterterms'},$$

corrected by inverse-triad corrections α and associated counterterms to close the algebra, obeys the deformed constraint algebra

$$\{ D[N^a], D[M^a] \} = D[\mathscr{L}_{M^a} N^a],$$
$$\{ H_{(\alpha^2)}[N], D[M^a] \} = H_{(\alpha^2)}[\mathscr{L}_{M^a} N], \qquad (7.24)$$
$$\{ H_{(\alpha^2)}[N], H_{(\alpha^2)}[M] \} = D[\alpha^2 q^{ab}(N\partial_b M - M\partial_b N)],$$

of the form (7.17) with $\beta = \alpha^2$, to second order in inhomogeneity in the constraints (and therefore first order in equations of motion).

The existence of a consistent deformation implies that no gauge transformation has been broken. The same number of gauge degrees of freedom and gauge transformations as in the classical theory is realized, but with the correction by α^2, they cannot correspond to ordinary hypersurface deformations or space-time diffeomorphisms. In particular, there is no effective line element $ds_{\mathrm{eff}}^2 = \tilde{q}_{ab} dx^a dx^b$ because coordinate differentials dx^a would still transform by classical coordinate changes, not by the deformed gauge transformations changing $\tilde{q}_{ab}$. Dynamically, the theory cannot be purely of higher-curvature type, and it becomes more difficult (but not impossible [39]) to derive effective actions. Instead of using an action principle, one can work with equations of motion for observables invariant under gauge transformations generated by corrected constraints such as $H_{(\alpha^2)}[N]$.

After quantum-geometry corrections have been implemented consistently, the gauge of the new constrained system may be fixed for further analysis. Gauge fixing may therefore be used consistently, but this procedure is *not* equivalent to fixing the gauge before quantization. (Classical gauge fixing assumes the classical gauge structure, which is however deformed.) For instance, longitudinal gauge leads to $\phi = \psi$ in the classical line element

$$ds^2 = -(1 - \phi)dt^2 + a^2(1 + \psi)(dx^2 + dy^2 + dz^2),$$

which is often assumed when the gauge is fixed before quantization. But $\phi = (1 + h)\psi$ with $h \neq 0$ follows for inverse-triad corrections [53, 69], making results with classical gauge fixing before quantization inconsistent. (This consequence of inverse-triad corrections may be interpreted as an effective anisotropic stress, which is missed by classically gauge-fixed treatments.)

The modified constraints generate equations of motion, automatically being gauge invariant under local changes of hypersurfaces and preserving the constraints for a consistently deformed system. As in the classical case, one can condense the set of all equations to a single equation for the gauge-invariant scalar mode u, correcting the classical Mukhanov equation [70]:

$$-\ddot{u} + s^2 \Delta u + \frac{\ddot{z}}{z} u = 0, \qquad (7.25)$$

$$-\ddot{w} + \alpha^2 \Delta w + \frac{\ddot{\tilde{a}}}{\tilde{a}} w = 0, \tag{7.26}$$

combined with the second equation for tensor modes w [71, 72]. Here, we have corrections $s \neq \alpha = 1 + \delta$, as well as corrected $\tilde{z}(a)$, $\tilde{a}(a)$.

7.3.4 Falsifiability

The new Mukhanov equations imply different corrections for scalar and tensor modes, and therefore corrections to the tensor-to-scalar ratio. Following the derivation, one can see that counterterms are responsible for the difference. If they were zero, as always assumed in gauge-fixed treatments that do not take into account the constraint algebra, another key form of quantum corrections would be missed. We can also see from the corrections that the scalar and tensor propagation speeds differ from the classical speed of light. However, general covariance is not broken but deformed, according to the consistent constraint algebra. Physically, while gravitational waves travel at speeds different from the classical speed of light, electromagnetic waves are affected by the same type of corrections from inverse-triad quantization in the Maxwell Hamiltonian. With a consistently deformed constraint algebra, they move at the same corrected speed as gravitational waves [71]. The speed of all massless modes therefore has the same value, and causality is preserved [39].

Finally, for phenomenology we note that inverse-triad corrections according to (7.22) are sensitive to L^2/ℓ_P^2, not directly to the energy density. They can therefore be sizeable during inflation with its significantly sub-Planckian densities, more so than holonomy corrections which are of the size $L^2/\ell_{\mathcal{H}}^2$. (If $L \sim \ell_P$, inverse-triad corrections are huge even at weak curvature. Therefore, the discreteness scale L of a quantum-gravity state must be sufficiently larger than the Planck length, in contrast to what is often assumed in *ad hoc* constructions.) A theoretical estimate, based on the interplay of inverse-triad corrections with holonomy corrections, provides $\delta = \alpha - 1 > 10^{-8}$ [68]. With observational data, one complements the theoretical lower bound with an upper bound: $10^{-8} < \delta = \alpha - 1 < 10^{-4}$ [73]. There are still several orders of magnitude allowed for the range of inverse-triad corrections, but given the adverse circumstances of quantum gravity regarding observations, we have a reasonably small number of orders of magnitude. Additional effects have been analyzed in [74–82], although investigations have not always been restricted to regimes of small $\alpha - 1$ where perturbations are under control. With inverse-triad corrections, the situation regarding observational tests is much more promising than usually assumed in quantum gravity, when only curvature-related corrections are considered.

We are still not certain whether loop quantum gravity can be fundamental, and the deep quantum regime (for instance around the big bang) remains too ambiguous for reliable physics. But still, the theory can be tested thanks to inverse-triad corrections, an unexpected but unavoidable quantum-geometry effect sensitive to the microscopic quantum-gravity scale in relation to the Planck length, not to the energy density. A complete and consistent implementation of these corrections requires an inhomogeneous model: in a mini-superspace setting, the only available choices $L \sim \ell_P$ or (worse) a macroscopic scale are not consistent. A Planckian value makes inverse-triad corrections too large even in semi-classical regimes; a macroscopic value makes them negligible but does not agree with the derivation leading to (7.22). (Sometimes, inverse-triad corrections in non-compact spaces, but not in compact ones, are claimed to depend on spurious averaging volumes for homogeneity, but this happens only if the choice of scale refers to it. Moreover, inverse-triad corrections refer to elementary fluxes and cannot depend on global properties such as the spatial topology. If this were the case, the theory would be unpredictive in realistic cosmology.) At the inhomogeneous level, on the other hand, we can (and must) refer to the discreteness scale of an underlying state as the main quantum-gravity parameter. Here, a non-trivial consistent deformation of the hypersurface-deformation algebra provides conceptual consistency as well as viable phenomenology thanks to a small number of parameters in the resulting phenomenological description using (7.23).

There are additional examples of consistent deformations in the literature on loop quantum gravity. Inverse-triad corrections have been implemented consistently for linear cosmological perturbations and in spherically symmetric models [83, 84] without requiring linearization. There are also operator calculations in $(2 + 1)$-dimensional models [61].

7.3.5 Anomaly-Free Holonomy Corrections

Holonomy corrections have been consistently implemented in spherically symmetric models [84], in $2 + 1$ dimensions [85], and for cosmological perturbations [38]. In the latter case, Mukhanov-type equations are available [86]. While [85] uses calculations at an operator level, including non-local holonomies integrated along curves, spherically symmetric [84] and cosmological space-times [38] so far require a local approximation of holonomy corrections. It remains unclear if the non-locality of holonomy corrections can consistently be implemented as well, and whether the local approximations are phenomenologically significant. Local correction functions are of the form $\sin(Lp_V)/L$ for p_V as in isotropic models, except that $p_V(x)$ may be allowed to depend on the spatial position. In general, however, holonomy corrections are non-local, with integrated $\sin\{\int_e a p_V[e(\lambda)]d\lambda\}$ as matrix elements. One may use a derivative expansion for a local approximation to give rise to contributions from $\sin(Lp_V)$, but there are also new terms $L^2 \partial_a p_V$ which in strong-curvature regimes, where holonomy corrections are important, cannot be expected to be much smaller than $L^2 p_V^2$ or even higher-order terms in an expansion of the local modification.

Nevertheless, consistent deformations for local correction functions show interesting consequences for space-time structure. The general form of (7.17) for holonomy corrections has a correction function $\beta(p_V) = \cos(2Lp_V)$, written for the case of perturbations around isotropic models with curvature parameter p_V related to the Hubble parameter [38, 84]. An interesting difference with inverse-triad corrections is that β can now be negative, realized in high-density regimes where the replacement $\sin(Lp_V)/L$ of p_V in the modified Friedmann equation reaches its maximum. A negative β in the deformed constraint algebra means (7.17) that we are now dealing with Euclidean space rather than space-time [39]. Also the corresponding Mukhanov-type equations, analogous to (7.25), show this clearly, with the analog of s^2 or α^2 negative; an elliptic differential operator appears in the Mukhanov equation, not a hyperbolic one as in a wave equation.

As this example shows, there can be extremely strong and unexpected modifications of space-time structure, which remain invisible at the level of homogeneous background models or even with inhomogeneity if gauge-fixing or deparameterization is used. The form of signature change observed here means that the strong-curvature regime of loop quantum cosmology cannot directly be related to phenomenology: there is no evolution in Euclidean space, and no related notions of bounces or structure formation.

In all these cases, it is a general theory of effective equations and constraints that provides closer contact between quantum theory and phenomenology, formally analyzing quantum back-reaction of moments of states. In this way, higher-time derivatives and higher-curvature terms can be realized canonically. So far, calculations for consistent constraint algebras have remained at the level of expectation values because quantum geometry already provides interesting corrections as in (7.22). We now enter a discussion of details relevant for fluctuations and moments at higher orders.

7.4 Effective Theories

Effective descriptions always provide useful procedures to extract physical information from quantum theories at low energies, taking into account state dynamics in more intuitive terms. They allow one to derive corrections to classical equations and observables instead of working with wavefunctions. For instance, for low-energy effective actions we may start with the path integral

$$Z[J] = \int \mathscr{D}q \exp\left[-i\hbar^{-1}\left(S_{\text{class}}[q] + \int d^4x q J\right)\right], \qquad (7.27)$$

introduce $W[J] = i\hbar \log Z[J]$,

$$Q(t) = \frac{\delta W[J]}{\delta J}\bigg|_{J=0}, \qquad (7.28)$$

and define the low-energy effective action as the Legendre transform

$$\Gamma_{\text{eff}}[q] = W[J(q)] - \int d^4x q J. \qquad (7.29)$$

In perturbative quantum gravity on Minkowski space-time, we obtain higher-curvature effective actions in this way.

One can already see typical properties for systems with finitely many degrees of freedom as in quantum mechanics. For an anharmonic oscillator, one starts with the classical Hamiltonian $H = p^2/2m + m\omega^2 q^2/2 + U(q)$ and obtains

$$\Gamma_{\text{eff}}[Q(t)] = \int dt \left(\frac{1}{2} m \left\{ 1 + \frac{\hbar \omega U'''(Q)^2}{32(m\omega^2)^3[1 + m^{-1}\omega^{-2}U''(Q)]^{5/2}} \right\} \dot{Q}^2 \right.$$
$$\left. - \frac{1}{2} m\omega^2 Q^2 - U(Q) - \frac{\hbar \omega}{2} \left[1 + \frac{U''(Q)}{m\omega^2} \right]^{1/2} \right), \qquad (7.30)$$

to first order in $\hbar$ and in a derivative expansion [87]. Although this effective action looks classical to zeroth order in $\hbar$, on closer inspection this is not really the case: $Q(t)$ in (7.28) is

$$Q(t) = \frac{\int \mathcal{D}q \, q(t) \exp(-i\hbar^{-1} S_{\text{class}}[q])}{\int \mathcal{D}q \exp(-i\hbar^{-1} S_{\text{class}}[q])} = \langle \text{fin}|\hat{q}|\text{in}\rangle, \qquad (7.31)$$

an off-diagonal matrix element as the transition amplitude from an initial to a final state. The variable Q for which the low-energy effective action is written, therefore, is not guaranteed to be real-valued and the classical limit becomes problematic despite appearance. (One sometimes interprets complex variables in effective actions as a sign of particle creation. However, in quantizations of single-particle systems and especially in quantum-cosmological models this notion is somewhat obscure.)

A detailed look at the procedure reveals the following properties and problems: The low-energy effective action assumes an expansion around the ground state of the harmonic oscillator (or around the vacuum state of free field theory). Is there a ground state of quantum gravity, and what could it look like? The classical contribution from gravity to the Hamiltonian constraint is unbounded from below, suggesting that there may be no ground state. More general states are possible by adapting boundary conditions or the measure of path integrations. But the integrations will no longer be Gaussian in general, and thus be more complicated. Also, such general effective actions are no longer as unique-looking as the low-energy one: Their expansion coefficients depend on the state. In quantum field theory, an effective action may not be manifestly Poincaré covariant if the non-vacuum state is not Poincaré invariant (Minkowskian). In a Poincaré covariant quantum theory, the state and not just effective fields would change by a transformation, mapping one effective action to a new one. A single effective action on its own is Poincaré covariant only if it is based on a Poincaré-invariant vacuum state to expand around, a situation too restrictive for quantum gravity. Finally, in generally covariant theories we must deal with totally constrained systems. It is not clear how to deal with this situation for the low-energy effective action.

All these properties show disadvantages when it comes to (canonical) quantum gravity, for which path-integral techniques would have to be adapted, anyway.

7.4.1 Effective Canonical Dynamics

For loop quantum gravity, a canonical reformulation of effective-action techniques, combined with a generalization to a more general class of states and to constrained systems, is essential. Although especially field-theory aspects still have to be worked out, much of the procedure is available and resolves the disadvantages seen with the usual low-energy effective action. Even disregarding aspects of gravity, the canonical method has several advantages. It is more intuitive, for instance in that its effective equations of motion are written for the real expectation values $\langle \hat{q} \rangle(t)$ and $\langle \hat{p} \rangle(t)$ instead of the potentially complex $Q(t)$ in (7.28). It is also amenable to numerical techniques for deriving and solving its equations [89], and it provides properties of dynamical quantum states, not just corrections to the classical equations.

The basic idea is as follows: We aim to solve equations of motion

$$\frac{d\langle \hat{q} \rangle}{dt} = \frac{\langle [\hat{q}, \hat{H}] \rangle}{i\hbar}, \qquad \frac{d\langle \hat{p} \rangle}{dt} = \frac{\langle [\hat{p}, \hat{H}] \rangle}{i\hbar} \tag{7.32}$$

for expectation values without taking a detour of wavefunctions. The coupled dynamics of expectation values and, in general, moments of a state can be reformulated systematically as quantum-corrected equations of motion for $\langle \hat{q} \rangle(t)$ and $\langle \hat{p} \rangle(t)$ of the classical type.

For the example of the harmonic oscillator, $\hat{H} = \frac{1}{2m}\hat{p}^2 + \frac{1}{2}m\omega^2\hat{q}^2$, the equations, for which we do not expect any quantum corrections, are straightforward to solve:

$$\frac{d}{dt}\langle \hat{q} \rangle = \frac{1}{i\hbar}\langle [\hat{q}, \hat{H}] \rangle = \frac{1}{m}\langle \hat{p} \rangle, \qquad \frac{d}{dt}\langle \hat{p} \rangle = \frac{1}{i\hbar}\langle [\hat{p}, \hat{H}] \rangle = -m\omega^2\langle \hat{q} \rangle \tag{7.33}$$

indeed has solutions $\langle \hat{q} \rangle(t)$ and $\langle \hat{p} \rangle(t)$ of the classical form, free of quantum corrections. The harmonic oscillator is easily solvable thanks to a linear dynamical algebra $[\hat{q}, \hat{p}] = i\hbar$, $[\hat{q}, \hat{H}] = i\hbar m^{-1}\hat{p}$, $[\hat{p}, \hat{H}] = -i\hbar m\omega^2\hat{q}$. Therefore, only expectation values appear on the right-hand side of (7.33), no fluctuations or higher moments.

Quantum states are not uniquely determined by expectation values of basic operators, but we can derive equations for fluctuations $(\Delta O)^2 = \langle \hat{O}^2 \rangle - \langle \hat{O} \rangle^2$ and the covariance $C_{qp} = \frac{1}{2}\langle \hat{q}\hat{p} + \hat{p}\hat{q} \rangle - \langle \hat{q} \rangle\langle \hat{p} \rangle$:

$$\frac{d}{dt}(\Delta q)^2 = \frac{\langle [\hat{q}^2, \hat{H}] \rangle}{i\hbar} - 2\langle \hat{q} \rangle\frac{d\langle \hat{q} \rangle}{dt} = \frac{2}{m}C_{qp}, \tag{7.34}$$

$$\frac{d}{dt}C_{qp} = -m\omega^2(\Delta q)^2 + \frac{1}{m}(\Delta p)^2, \tag{7.35}$$

$$\frac{d}{dt}(\Delta p)^2 = -2m\omega^2 C_{qp}, \tag{7.36}$$

and also subject them to the (generalized) uncertainty relation

$$(\Delta q)^2(\Delta p)^2 - C_{qp}^2 \geq \frac{\hbar^2}{4}.$$

Fig. 7.4 Coherent states of the harmonic oscillator, shown by their moments: the central line $\langle\hat{q}\rangle(t)$ oscillates classically; the outer lines $\langle\hat{q}\rangle(t) \pm \Delta q(t)$ show the spread. *Solid lines: Δq* constant, unsqueezed state. *Dashed lines:* Squeezed state with covariance $C_{qp} \neq 0$

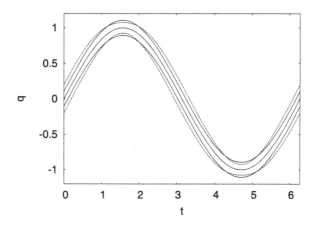

Looking for stationary states, these equations allow non-spreading solutions with $C_{qp} = 0$ and

$$\Delta p = m\omega\Delta q. \tag{7.37}$$

They satisfy the uncertainty relation if $\Delta q \geq \sqrt{\hbar/2m\omega}$. At saturation, we obtain the correct value for the ground state. More general coherent and squeezed states with oscillating spread are obtained if $C_{qp} \neq 0$, illustrated in Fig. 7.4.

The equations and calculations are much more involved in general, for instance for an anharmonic oscillator with Hamiltonian

$$\hat{H} = \frac{1}{2m}\hat{p}^2 + V(\hat{q}) = \frac{1}{2m}\hat{p}^2 + \frac{1}{2}m\omega^2\hat{q}^2 + \frac{1}{3}\lambda\hat{q}^3.$$

Now, equations of motion

$$\frac{d}{dt}\langle\hat{q}\rangle = \frac{1}{m}\langle\hat{p}\rangle, \qquad \frac{d}{dt}\langle\hat{p}\rangle = -m\omega^2\langle\hat{q}\rangle - \lambda\langle\hat{q}\rangle^2 - \lambda(\Delta q)^2 = -V'(\langle\hat{q}\rangle) - \lambda(\Delta q)^2$$

are not only non-linear, they also couple expectation values to the position fluctuation. The position fluctuation itself is dynamical and couples to the covariance, while the covariance in

$$\frac{d}{dt}(\Delta q)^2 = \frac{2}{m}C_{qp},$$

$$\frac{d}{dt}C_{qp} = \frac{1}{m}C_{qp} + m\omega^2(\Delta q)^2 + 6\lambda\langle\hat{q}\rangle(\Delta q)^2 + 3\lambda G^{0,3}$$

couples to a higher moment $G^{0,3}$ (the skewness). Continuing the process of calculating evolution equations for all moments involved, we end up with an infinite number of coupled moments [64, 65].

7.4.2 Moment Dynamics

For a systematic discussion, a useful set of coordinates is given by expectation values of basic operators, which we now denote as $q = \langle \hat{q} \rangle$ and $p = \langle \hat{p} \rangle$ for notational simplicity, together with the moments

$$\Delta(q^a p^b) := \left\langle (\hat{q} - \langle \hat{q} \rangle)^a (\hat{p} - \langle \hat{p} \rangle)^b \right\rangle_{\text{Weyl}} \tag{7.38}$$

defined with totally symmetric operator orderings, for all $a + b \geq 2$ [64] (with the notation of [88]). For semi-classical states, the moments can be arranged by orders of magnitude according to the ℏierarchy $\Delta(q^a p^b) \sim \hbar^{(a+b)/2}$, which can easily be confirmed for Gaussian states. These orders are the main ingredient to specify semi-classical regimes for effective equations, allowing for states much more general than Gaussians: A Gaussian form of the wavefunction would fix all infinitely many moments in terms of just one parameter (the variance Δq) for an unsqueezed state, or at most two if the state is squeezed (the variance Δq and the covariance $\Delta(qp)$). Moreover, the general semi-classicality condition for moments includes mixed states as well as pure ones, an especially important fact in cosmological models which are supposed to be obtained from the full theory by some kind of averaging procedure.

These variables define a phase space, with Poisson brackets following from commutators. We have

$$\{q, p\} = 1, \qquad \left\{ q, \Delta(q^a p^b) \right\} = 0 = \left\{ p, \Delta(q^a p^b) \right\},$$

and $\{\Delta(q^a p^b), \Delta(q^c p^d)\} = \cdots$ a more lengthy formula [89]. If we use the semi-classical ℏierarchy, we can determine and solve equations order by order in $\hbar$, making use of only finitely many variables and equations at each step. Restricting the Poisson brackets to moments up to a finite order in $\hbar$ in general leads to degenerate Poisson manifolds which are not symplectic. (For instance, to order $\hbar$ we include the two expectation values and three second-order moments Δq^2, $\Delta(qp)$ and Δp^2. With an odd number of dimensions, the Poisson manifold cannot be symplectic and there is no symplectic form.) Therefore, effective equations, which necessarily rely on a truncation of the infinitely many quantum degrees of freedom, cannot be derived by symplectic methods. The usual canonical tools are available for degenerate Poisson manifolds as well, including constraint classifications and analysis; see, e.g., [90, 91].

The dynamics of the moments is governed by the quantum Hamiltonian

$$H_Q\big(q, p, \Delta(\cdots)\big) = \langle H(\hat{q}, \hat{p}) \rangle = \langle H\big(q + (\hat{q} - q), p + (\hat{p} - p)\big) \rangle \tag{7.39}$$

$$= H(q, p) + \sum_{a+b \geq 2} \frac{1}{a!b!} \frac{\partial^{a+b} H(q, p)}{\partial q^a \partial p^b} \Delta(q^a p^b), \tag{7.40}$$

with coupling terms of expectation values and moments. With a general classical Hamiltonian $H = p^2/2m + m\omega^2 q^2/2 + U(q)$, introducing dimensionless variables $\tilde{\Delta}(q^a p^b) = \hbar^{-(a+b)/2} (m\omega)^{(a-b)/2} \Delta(q^a p^b)$ for convenience, the quantum Hamiltonian becomes

$$H_Q = \frac{1}{2m}p^2 + \frac{1}{2}m\omega^2 q^2 + U(q) + \frac{\hbar\omega}{2}\left(\tilde{\Delta}q^2 + \tilde{\Delta}p^2\right) \tag{7.41}$$

$$+ \sum_{a>2} \frac{1}{a!}\left(\frac{\hbar}{m\omega}\right)^{a/2} U^{(a)}(q)\tilde{\Delta}(q^a). \tag{7.42}$$

We readily identify terms of the zero-point energy $\frac{1}{2}\hbar\omega(\tilde{\Delta}q^2 + \tilde{\Delta}p^2)$: with $\Delta q = \sqrt{\hbar/2m\omega}$ and $\Delta p = m\omega\Delta q = \sqrt{m\omega\hbar/2}$ as derived for saturating stationary states in (7.37), we have

$$\tilde{\Delta}q^2 = \frac{m\omega\Delta q^2}{\hbar} = \frac{1}{2} \quad \text{and} \quad \tilde{\Delta}p^2 = \frac{\Delta p^2}{m\omega\hbar} = \frac{1}{2},$$

such that the zero-point energy in (7.41) is $E_0 = \hbar\omega/2$. In the sum (7.42), we have coupling terms $U^{(a)}(q)\tilde{\Delta}(q^a)$ of expectation values and moments. The quantum Hamiltonian H_Q generates Hamiltonian equations of motion $\dot{f} = \{f, H_Q\}$,

$$\dot{q} = \frac{p}{m}, \tag{7.43}$$

$$\dot{p} = -m\omega^2 q - U'(q) - \sum_a \frac{1}{a!}\left(\frac{\hbar}{m\omega}\right)^{a/2} U^{(a+1)}(q)\tilde{\Delta}(q^a), \tag{7.44}$$

$$\begin{aligned}
\dot{\tilde{\Delta}}(q^a p^b) &= -b\omega\tilde{\Delta}(q^{a+1}p^{b-1}) + a\omega\tilde{\Delta}(q^{a-1}p^{b+1}) - b\frac{U''(q)}{m\omega}\tilde{\Delta}(q^{a+1}p^{b-1}) \\
&\quad + \frac{\sqrt{\hbar}bU'''(q)}{2(m\omega)^{3/2}}\tilde{\Delta}(q^a p^{b-1})\tilde{\Delta}q^2 + \frac{\hbar bU''''(q)}{3!(m\omega)^2}\tilde{\Delta}(q^a p^{b-1})\tilde{\Delta}(q^3) \\
&\quad - \frac{b}{2}\left[\frac{\sqrt{\hbar}U'''(q)}{(m\omega)^{3/2}}\tilde{\Delta}(q^{a+2}p^{b-1}) + \frac{\hbar U''''(q)}{3(m\omega)^2}\tilde{\Delta}(q^{a+3}p^{b-1})\right] + \cdots,
\end{aligned} \tag{7.45}$$

as infinitely many coupled equations for infinitely many variables.

While it is challenging to solve this whole system of equations exactly, the arrangement by moments makes it amenable to semi-classical approximations in direct terms. To first order in $\hbar$, only second-order moments and their dynamics contribute to (7.44), and only the top line (7.45) is needed for the moments. Combined with an adiabatic approximation that treats the evolution of moments as slow compared to that of expectation values (as a systematic way of implementing the derivative expansion of effective actions), we obtain to second order the equation of motion

$$\begin{aligned}
&m\left\{1 + \frac{\hbar\omega U'''(q)^2}{32(m\omega^2)^3[1 + m^{-1}\omega^{-2}U''(q)]^{5/2}}\right\}\ddot{q} \\
&\quad + \frac{\hbar\dot{q}^2\{4m\omega^2 U'''(q)U''''(q)[1 + m^{-1}\omega^{-2}U''(q)] - 5U'''(q)^3\}}{128m^3\omega^7[1 + m^{-1}\omega^{-2}U''(q)]^{7/2}} \\
&\quad + m\omega^2 q + U'(q) + \frac{\hbar U'''(q)}{4m\omega[1 + m^{-1}\omega^{-2}U''(q)]^{1/2}} = 0,
\end{aligned}$$

if at zeroth adiabatic order we assume values corresponding to the moments of the harmonic-oscillator ground state [64]. This equation can be seen to result from the low-energy effective action

$$
\Gamma_{\text{eff}}[q(t)] = \int dt \left(\frac{1}{2} m \left\{ 1 + \frac{\hbar \omega U'''(q)^2}{32(m\omega^2)^3[1 + m^{-1}\omega^{-2}U''(q)]^{5/2}} \right\} \dot{q}^2 \right.
$$
$$
\left. - \frac{1}{2} m\omega^2 q^2 - U(q) - \frac{\hbar \omega}{2} \left[1 + \frac{U''(q)}{m\omega^2} \right]^{1/2} \right),   \qquad (7.46)
$$

which is formally identical to (7.30) but has the advantage of referring to real expectation values $q(t) = \langle \hat{q} \rangle(t)$ instead of complex matrix elements $Q(t)$ defined in (7.28). Properties of the state used for $\langle \hat{q} \rangle$ follow from the moments solving (7.45). See [40] for examples.

Canonical techniques based on quantum back-reaction are thereby shown to reproduce better known effective-action methods based on path integrals. At the same time, the methods are generalized because there are no restrictions on states or requirements on the existence of ground states. For instance, we may use other zeroth-order (adiabatic) values when we solve (7.45), such as ones for squeezed states as in an example provided in [64], or directly solve coupled equations for expectation values and moments without an adiabatic expansion. In the latter case, one refers to a state by initial values of moments, avoiding assumptions on how they evolve. By solving the differential equations, as used in quantum cosmology for instance in [89, 92–94], one derives detailed dynamical-state properties and not just expectation values. The procedure has been applied for high orders of the moments in [89]; see Fig. 7.5 for an illustration.

7.4.3 Effective Constraints

Also constrained systems can be dealt with systematically, providing access to observables in physical Hilbert spaces [88, 95]. In analogy with quantum Hamiltonians (7.39), we define a quantum constraint $C_Q = \langle \hat{C} \rangle = C_{\text{class}}(\langle \hat{q} \rangle, \langle \hat{p} \rangle) + \cdots$ for every constraint operator $\hat{C}$ of the quantum system. The quantum constraint must vanish in physical states, and the expansion by moments shows quantum corrections to the classical constraint surface. However, a single constraint on the quantum phase space of expectation values and moments removes only two degrees of freedom, or one canonical pair by constraining and factoring out the gauge, but not the corresponding moments. We must introduce additional constraints, in fact infinitely many ones just as there are infinitely many moments. In a physical state, all

$$
C_{f(q,p)} := \langle f(\hat{q}, \hat{p})\hat{C} \rangle
$$

with arbitrary phase-space functions $f(q, p)$ must vanish and in general are independent of C_Q on the quantum phase space. Every classical gauge transformation implies infinitely many quantum gauge transformations. Figure 7.6 illustrates the

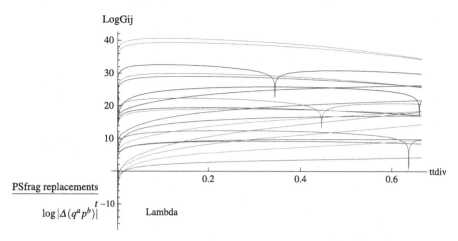

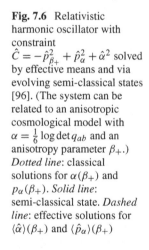

Fig. 7.5 Example for the evolution of moments up to order $a + b = 10$, with a and b odd [89]. At time $t = 0$, the initial state was chosen as an unsqueezed Gaussian, with vanishing moments of odd a and b. These moments quickly increase, showing that the state rapidly departs from its initial simple form. The different magnitudes of the moments also indicate that a new moment hierarchy with subsets of similar sizes is dynamically attained. The dynamics corresponds to the large-volume behavior of an isotropic model with a positive cosmological constant

Fig. 7.6 Relativistic harmonic oscillator with constraint $\hat{C} = -\hat{p}_{\beta_+}^2 + \hat{p}_\alpha^2 + \hat{\alpha}^2$ solved by effective means and via evolving semi-classical states [96]. (The system can be related to an anisotropic cosmological model with $\alpha = \frac{1}{6} \log \det q_{ab}$ and an anisotropy parameter β_+.) *Dotted line*: classical solutions for $\alpha(\beta_+)$ and $p_\alpha(\beta_+)$. *Solid line*: semi-classical state. *Dashed line*: effective solutions for $\langle \hat{\alpha} \rangle (\beta_+)$ and $\langle \hat{p}_\alpha \rangle (\beta_+)$

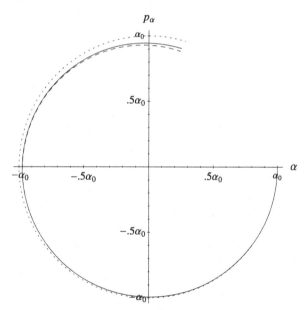

validity of the methods, which can also be confirmed impressively by comparing the evolution of moments in Fig. 7.7 [96].

Effective-constraints methods (or deparameterized effective equations) can also be applied to non-canonical holonomy-type variables in loop quantum cosmology.

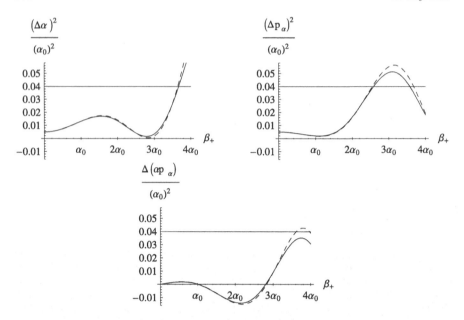

Fig. 7.7 Moments for the relativistic harmonic oscillator as in Fig. 7.6 [96]. *Solid line*: semi-classical state initially Gaussian. *Dashed line*: effective solution with the same initial values. *Horizontal lines* indicate values of moments no longer considered small

The modified Friedmann equation (7.2) is then realized at zeroth order in $\hbar$: it reflects merely modifications from quantum geometry but ignores possible corrections for quantum dynamics. In analogy with the system in Fig. 7.6, solutions to (7.2) correspond to the dotted, not the dashed line. The modified Friedmann equation therefore does not capture deviations of quantum-state evolution from the classical trajectory which always occur in anharmonic systems and are contained only in quantum back-reaction. (Some numerical studies of state evolution have been done for anharmonic cosmologies with holonomy-type variables; see, e.g., [27]. However, studies so far have not attempted an analysis of the generic behavior free of bias toward the initial state.)

The effective procedure has several advantages compared with traditional techniques of solving quantum constrained systems: Physical states can be implemented by reality conditions; no integral form of the physical inner product or a complete physical Hilbert space are required. The procedure is then much more tractable especially for non-deparameterizable systems [44–46], where alternatives for the construction of Hilbert spaces [97–99] encounter difficulties in the evolution of observables, freezing at turning points of internal times [100]. Moreover, if there are several constraints $\hat{C}^i$ one can address the anomaly problem by computing Poisson brackets for C_Q^i and higher-order constraints. If quantum corrections in the constraints are sufficiently parameterized, to take into account quantization ambiguities and also properties of general semi-classical states in which to take expectation values, one can solve for parameter choices that make the system anomaly-free. (Parameterizations usually require counterterms [53] not seen directly in simple correction functions such as (7.22).) These calculations can be quite involved, but they are still much easier than what would be required at an operator level.

7.4.4 Isotropic Cosmology

Methods of effective equations and constraints can be applied in quite some detail to cosmology in isotropic ADM or Ashtekar–Barbero variables, not only illustrating more general applications to inhomogeneity but also providing relevant information on background evolution and coherent states. With spatial flatness, we have an ADM canonical pair $(a, 3a\dot{a}/4\pi G)$ with the scale factor a, transformed canonically (up to a constant) to a pair $c = \gamma\dot{a}$, $|p| = a^2$, with $\{c, p\} = 8\pi\gamma G/3$. These variables are realized as the components of an isotropic connection $A_a^i = c\delta_a^i$ and densitized triad $E_i^a = p\delta_i^a$. (The triad component p can take both signs thanks to its orientation freedom, and $\gamma > 0$ is the Barbero–Immirzi parameter as in Sect. 7.3.)

In these variables, we have the Hamiltonian constraint

$$C := -\frac{c^2\sqrt{|p|}}{\gamma^2} + \frac{8\pi G}{3}H_{\text{matter}} = 0,$$

which when it vanishes is equivalent to the Friedmann equation. It generates the Raychaudhuri equation in proper time by $\dot{p} = \{p, C\}$, $\dot{c} = \{c, C\}$.

In loop quantum cosmology, we represent the constraint as a state equation $\hat{C}|\psi\rangle = 0$, but with $\exp(if_0|p|^xc)$ used for c as a version of holonomies:

$$-\frac{\sin^2(f_0|p|^xc)\sqrt{|p|}}{\gamma^2 f_0^2|p|^{2x}} + \frac{8\pi G}{3}H_{\text{matter}} = 0. \tag{7.47}$$

The two parameters f_0 and x parameterize the dependence of the discreteness scale $L = f_0|p|^{x+1/2}/\gamma$ on the geometry due to lattice refinement [62, 63], for instance with $x = -1/2$ as used in [17].

Writing the equation for physical states in the triad representation, for coefficients ψ_μ of $|\psi\rangle = \sum_\mu \psi_\mu(\varphi)|\mu\rangle$ expanded in triad eigenstates $|\mu\rangle$ (with an extra label φ for a possible matter fields) we obtain a difference equation [9, 15] for a wavefunction of the universe:

$$C_+(\mu)\psi_{\mu+1}(\varphi) - C_0(\mu)\psi_\mu(\varphi) + C_-(\mu)\psi_{\mu-1}(\varphi) = \hat{H}_\varphi(\mu)\psi_\mu(\varphi). \tag{7.48}$$

This is a recurrence relation for the wavefunction on mini-superspace, but we can interpret it as an 'evolution' equation in 'internal time' μ, the quantized triad component. With or without an evolution picture, the dynamics is non-singular because the wavefunction can be extended uniquely across the classical singularity at $\mu = 0$ [36]. As a geometrical picture, one may view evolution as a bounce: For negative μ, evolving forward in μ implies that the volume $V_\mu \propto |\mu|^{3/2}$ shrinks, turning into expansion after crossing $\mu = 0$ [101]. But in general, a simple geometrical picture is not available in the strong quantum regime surrounding the classical singularity.

With effective equations, intuitive geometrical pictures may be derived more systematically. Calculations are tractable in harmonic systems in which the dynamics of expectation values and moments decouples to finite sets. However, quantum cosmology is not close to the harmonic oscillator, the eponymous harmonic system. More generally, for the absence of quantum back-reaction we need a linear dynamical algebra of basic variables with the Hamiltonian. Such a system can be arranged by coupling spatially flat isotropic geometry to a free, massless scalar, so that the loop-modified Hamiltonian constraint becomes

$$-\frac{\sin^2(f_0|p|^x c)\sqrt{|p|}}{\gamma^2 f_0^2|p|^{2x}} + \frac{4\pi G}{3}\frac{p_\varphi^2}{|p|^{3/2}} = 0. \tag{7.49}$$

Now choosing φ as internal time, evolution is generated by

$$p_\varphi \propto H(c,p) = \left|\sin\left(f_0|p|^x c\right)|p|^{1-x}\right|,$$

again parameterizing lattice refinement by f_0 and x.

After a canonical transformation to $V = |p|^{1-x}/f_0(1-x)$ and $P := -f_0|p|^x c$, we realize the system as a harmonic one: We have a linear dynamical algebra of $sl(2,\mathbb{R})$-type with operators $\hat{V}$, $\hat{J} = \widehat{V\exp(iP)}$ and $\hat{H} = |\text{Im}\,\hat{J}|$:

$$[\hat{V},\hat{J}] = \hbar\hat{H}, \qquad [\hat{V},\hat{H}] = -\hbar\hat{J}, \qquad [\hat{J},\hat{H}] = \hbar\hat{V}. \tag{7.50}$$

(The absolute value in $\hat{H}$ might seem problematic at first, but is harmless [31, 95].) Solving the resulting Hamiltonian equations of motion and imposing reality conditions to ensure that $\exp(iP)$ is quantized to a unitary operator shows bouncing behavior, with $V(\varphi) \propto \cosh(\varphi)$ (provided the evolving state is semi-classical at least once) [30, 31].

For anharmonic cosmological systems, the effective dynamics is governed by the quantum Hamiltonian $\langle\hat{H}\rangle((\cdot), \Delta(\cdot))$ with moment couplings from interaction terms. As an example for the low-curvature behavior ignoring the holonomy modification, with a cosmological constant we have $H = V\sqrt{P^2 - \Lambda}$ classically and

$$\frac{d\langle\hat{P}\rangle}{d\varphi} \propto -\sqrt{\langle\hat{P}\rangle^2 - \Lambda} + \frac{1}{2}\Lambda\frac{(\Delta P)^2}{(\langle\hat{P}\rangle^2 - \Lambda)^{3/2}} + \cdots, \tag{7.51}$$

$$\frac{d\langle\hat{V}\rangle}{d\varphi} \propto \frac{\langle\hat{V}\rangle\langle\hat{P}\rangle}{\sqrt{\langle\hat{P}\rangle^2 - \Lambda}} + \frac{3}{2}\Lambda\frac{\langle\hat{V}\rangle\langle\hat{P}\rangle(\Delta P)^2}{(\langle\hat{P}\rangle^2 - \Lambda)^{5/2}} - \Lambda\frac{\Delta(VP)}{(\langle\hat{P}\rangle^2 - \Lambda)^{3/2}} + \cdots, \tag{7.52}$$

with moments up to second order [89] but no appearance of ΔV since H is linear in V. We can make ΔV large without affecting the dynamics much (until other moments coupled to ΔV increase). Large fluctuations of a certain kind therefore do not necessarily imply strong quantum back-reaction, in contrast to the usual statistical properties of fluctuations. The case of a positive cosmological constant is of interest because here some moments can become large even in low-curvature regimes. The evolution of states has been analyzed with different methods [89, 102].

Finally, the most general case of isotropic models with quantum back-reaction is obtained for a general, possibly self-interacting scalar. The simple version is again the free, massless scalar, or a stiff fluid with $\rho = p$, producing bouncing background solutions. Initially, we have coupled equations such as (7.51) and (7.52) but with holonomy modifications. One can eliminate $\langle \hat{P} \rangle$ in (7.52) or $\langle \hat{J} \rangle$ for holonomy-type variable, using the modified Friedmann constraint, then parameterize correlations (even to higher orders) by one parameter η, define the quantum-gravity scale $\rho_{QG} = 3/(8\pi GL^2) + \cdots$ with extra terms from fluctuations, and write the equation for $d\langle \hat{V} \rangle / d\varphi$ as one for

$$\dot{a} = \frac{da}{d\tau} = \frac{d\varphi}{d\tau} \frac{d(f_0(1-x)\langle \hat{V} \rangle)^{1/(2(1-x))}}{d\varphi} \propto \frac{p_\varphi}{2a^3} \langle \hat{V} \rangle^{\frac{2x-1}{2(1-x)}} \frac{d\langle \hat{V} \rangle}{d\varphi}.$$

Evolution is then governed by the effective Friedmann equation

$$\left(\frac{\dot{a}}{a} \right)^2 = \frac{8\pi G}{3} \left[\rho \left(1 - \frac{\rho}{\rho_{QG}} \right) \right.$$
$$\left. \pm \frac{1}{2} \sqrt{1 - \frac{\rho}{\rho_{QG}}} \eta (\rho - p) + \frac{(\rho - p)^2}{\rho + p} \eta^2 \right] \tag{7.53}$$

for the scale factor a via $\langle \hat{V} \rangle$ [34, 35]. If only the top line is present, as in the free scalar model with $\rho = p$ [32, 33], it follows immediately that a bounce is possible when the density reaches the quantum-gravity scale $\rho_{QG} = 3/(8\pi GL^2)$. The second line, however, makes it impossible, at the present stage, to tell what happens in general without kinetic domination (with dynamics not close to a harmonic model). Is there a bounce, or not? The question of whether the singularity is always avoided remains open, even in isotropic models. Alternatively to a bounce, there could be an asymptotic approach to smaller volume, or oscillations at a small level. (With a self-interacting or massive scalar, we can no longer deparameterize. However, effective equations can be applied even to constrained systems with local internal times, valid only for a finite range [44–46].)

As discussed in Sects. 7.1 and 7.3.1, holonomy corrections are significant in regimes where higher-curvature corrections are also strong. In canonical quantizations, higher-curvature terms and the higher time derivatives they contain follow from quantum back-reaction. Therefore, the moment terms in (7.53) with their correlation parameter should, for anharmonic models, be as significant as holonomy modifications that give rise to the ρ^2 / ρ_{QG}-term in the first line of (7.53). With dynamical states, however, η is time dependent and difficult to predict in high-curvature and strong quantum regimes. No easy conclusions about bounces are possible.

7.4.5 Beginning

Further results show several crucial properties of relevance for general dynamical aspects and the meaning but also indeterminism of 'bounces'. The bounce in the harmonic model is a consequence of modified classical dynamics, replacing P^2 by a

Fig. 7.8 Harmonic
cosmology displayed in
analogy with the harmonic
oscillator in Fig. 7.4: the
expectation value is in the
middle, and spreads
according to fluctuations. For
an unsqueezed Gaussian
state, the fluctuations are
symmetric. The squeezing
determines the asymmetry of
the state, but remains largely
unrestricted in quantum
cosmology, implying cosmic
forgetfulness

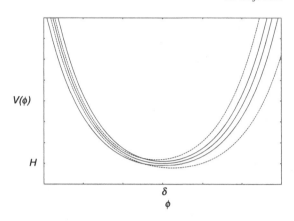

periodic function. The modification is motivated by quantum geometry, in particular
the use of holonomy operators in background-independent quantizations, but it does
not result from quantum dynamics. (See also [103] and [104].)

To shed light on state dependence, we may go back to the harmonic model and
find properties of solutions for state parameters. One general consequence is cosmic
forgetfulness: there are upper bounds

$$\left| \left(\frac{\Delta V}{\langle \hat{V} \rangle} \right)_{\varphi \to \infty} - \left(\frac{\Delta V}{\langle \hat{V} \rangle} \right)_{\varphi \to -\infty} \right| \leq \frac{\Delta p_\varphi}{\langle \hat{p}_\varphi \rangle}$$

for the change of fluctuations in a dynamical coherent [105] or semi-classical state
[106], with $\varphi \to \pm\infty$ at the two asymptotic ends of large volume, separated by the
bounce regime. (In the free, massless model, $\langle \hat{p}_\varphi \rangle$ and Δp_φ are constant.) Semi-
classicality of the state is preserved, not altogether surprisingly in this harmonic
model, but the moments do vary, perhaps considerably so. Rewriting the inequality
as

$$\left| 1 - \frac{\Delta V_{\varphi \to -\infty}}{\Delta V_{\varphi \to \infty}} \right| \leq \frac{\Delta p_\varphi / \langle \hat{p}_\varphi \rangle}{(\Delta V / \langle \hat{V} \rangle)_{\varphi \to \infty}} \tag{7.54}$$

shows that the change of volume fluctuations can be quite dramatic if matter is more
quantum than geometry at large volume: $\Delta p_\varphi / \langle \hat{p}_\varphi \rangle \gg (\Delta V / \langle \hat{V} \rangle)|_{\varphi \to \infty}$, a usual as-
sumption in quantum field theory on curved space-time. (One may always find states
for which the values of fluctuations are more symmetric, for instance unsqueezed
Gaussian ones [31]. But this does not disprove cosmic forgetfulness as a statement
about general classes of states free of prejudice on their form.)

Harmonicity does not rule out significant changes of the quantum behavior
through the bounce. Cosmic forgetfulness [105, 107] shows that a strong sensitivity
to initial values of the moments makes it practically impossible to reconstruct the
pre-big bang state from knowledge of the state after the big bang, as illustrated in
Fig. 7.8. Also in the bounce regime, state parameters are largely unrestricted. We
cannot estimate well what value the correlation parameter η in (7.53) might be at
high density. With this limitation, resulting from a combination of quantum back-

reaction and cosmic forgetfulness, generic bounce statements are extremely difficult to obtain.

Moreover, we have to keep in mind that non-singular solutions for expectation values do not necessarily imply a classical space-time structure. We must embed the model in a consistent deformation with holonomy corrections. Partial implementations are available [38], but the classicality of the pre-bounce space-time structure remains unclear until a complete picture for general holonomy corrections is obtained. What has been seen already is that the bounce phase itself is not at all dynamical, because holonomy corrections also make the space-time signature turn Euclidean [39]; see Sect. 7.3.5. (An evolution picture is not required for cosmic forgetfulness and the inequality (7.54): we are just comparing values of variables along gauge orbits of the Hamiltonian constraint.) The universe does not evolve through a bounce; it emerges from Euclidean space.

All these results show that we cannot reconstruct the pre-bounce phase; instead, a non-singular beginning is derived in loop quantum cosmology.

7.5 Conclusions

The evaluation of the dynamics of loop quantum gravity greatly benefits from a general framework for effective equations. It allows us to expand around excited or general semi-classical states, not just the ground state. General semi-classical states beyond Gaussians (or even density states) are possible in terms of the moments, significantly going beyond other analyses in which a Gaussian form is often assumed. One can deal with effective constraints, and perform an anomaly analysis. Technically, calculations are much more tractable and less redundant than working with wavefunctions.

The deep quantum regime remains hard, but even here effective equations do provide independent reliable information. In semi-classical regimes, we can address long-standing issues such as the anomaly problem, or the problem of time. We learn that quantum gravity may have drastic effects on the space-time structure, including signature change at high densities. Perturbation theory derived from effective equations is crucial for semi-classical or low-energy aspects—just what is needed for potentially observable cosmology. Thanks largely to the presence of inverse-triad corrections, the theory becomes falsifiable in this systematic framework.

Acknowledgements   This work was supported in part by NSF grant PHY0748336.

References

1. M. Bojowald, Class. Quantum Gravity **17**, 1489 (2000). gr-qc/9910103
2. M. Bojowald, Living Rev. Relativ. **11**, 4 (2008). http://www.livingreviews.org/lrr-2008-4. gr-qc/0601085

3. M. Bojowald, *Quantum Cosmology: A Fundamental Theory of the Universe* (Springer, New York, 2011)
4. C. Rovelli, *Quantum Gravity* (Cambridge University Press, Cambridge, 2004)
5. T. Thiemann, *Introduction to Modern Canonical Quantum General Relativity* (Cambridge University Press, Cambridge, 2007)
6. A. Ashtekar, J. Lewandowski, Class. Quantum Gravity **21**, R53 (2004). gr-qc/0404018
7. K.V. Kuchař, M.P. Ryan, Phys. Rev. D **40**, 3982 (1989)
8. M. Bojowald, H.A. Kastrup, Class. Quantum Gravity **17**, 3009 (2000). hep-th/9907042
9. M. Bojowald, Class. Quantum Gravity **19**, 2717 (2002). gr-qc/0202077
10. A. Ashtekar, M. Bojowald, J. Lewandowski, Adv. Theor. Math. Phys. **7**, 233 (2003). gr-qc/0304074
11. B.S. DeWitt, Phys. Rev. **160**, 1113 (1967)
12. D.L. Wiltshire, in *Cosmology: The Physics of the Universe*, ed. by B. Robson, N. Visvanathan, W.S. Woolcock (World Scientific, Singapore, 1996), pp. 473–531. gr-qc/0101003
13. M. Bojowald, Class. Quantum Gravity **17**, 1509 (2000). gr-qc/9910104
14. M. Bojowald, Class. Quantum Gravity **18**, 1055 (2001). gr-qc/0008052
15. M. Bojowald, Class. Quantum Gravity **18**, 1071 (2001). gr-qc/0008053
16. A. Ashtekar, T. Pawłowski, P. Singh, Phys. Rev. D **73**, 124038 (2006). gr-qc/0604013
17. A. Ashtekar, T. Pawłowski, P. Singh, Phys. Rev. D **74**, 084003 (2006). gr-qc/0607039
18. W.F. Blyth, C.J. Isham, Phys. Rev. D **11**, 768 (1975)
19. W. Kaminski, J. Lewandowski, Class. Quantum Gravity **25**, 035001 (2008). arXiv:0709.3120
20. W. Kaminski, T. Pawłowski, Phys. Rev. D **81**, 024014 (2010). arXiv:0912.0162
21. M. Martín-Benito, L.J. Garay, G.A. Mena Marugán, Phys. Rev. D **78**, 083516 (2008). arXiv:0804.1098
22. D. Brizuela, G.A. Mena Marugán, T. Pawłowski, Class. Quantum Gravity **27**, 052001 (2010). arXiv:0902.0697
23. A. Ashtekar, M. Campiglia, A. Henderson, Phys. Lett. B **681**, 347 (2009). arXiv:0909.4221
24. A. Ashtekar, M. Campiglia, A. Henderson, Class. Quantum Gravity **27**, 135020 (2010). arXiv:1001.5147
25. C. Rovelli, F. Vidotto, Class. Quantum Gravity **27**, 145005 (2010). arXiv:0911.3097
26. A. Henderson, C. Rovelli, F. Vidotto, E. Wilson-Ewing, Class. Quantum Gravity **28**, 025003 (2011). arXiv:1010.0502
27. E. Bentivegna, T. Pawłowski, Phys. Rev. D **77**, 124025 (2008). arXiv:0803.4446
28. M. Martín-Benito, G.A. Mena Marugán, T. Pawłowski, Phys. Rev. D **80**, 084038 (2009). arXiv:0906.3751
29. G.A. Mena Marugán, J. Olmedo, T. Pawłowski, Phys. Rev. D **84**, 064012 (2011). arXiv:1108.0829
30. M. Bojowald, Phys. Rev. D **75**, 081301(R) (2007). gr-qc/0608100
31. M. Bojowald, Phys. Rev. D **75**, 123512 (2007). gr-qc/0703144
32. K. Vandersloot, Phys. Rev. D **71**, 103506 (2005). gr-qc/0502082
33. P. Singh, Phys. Rev. D **73**, 063508 (2006). gr-qc/0603043
34. M. Bojowald, Phys. Rev. Lett. **100**, 221301 (2008). arXiv:0805.1192
35. M. Bojowald, Gen. Relativ. Gravit. **40**, 2659 (2008). arXiv:0801.4001
36. M. Bojowald, Phys. Rev. Lett. **86**, 5227 (2001). gr-qc/0102069
37. M. Novello, S.E.P. Bergliaffa, Phys. Rep. **463**, 127 (2008). arXiv:0802.1634
38. T. Cailleteau, J. Mielczarek, A. Barrau, J. Grain, arXiv:1111.3535
39. M. Bojowald, G.M. Paily, arXiv:1112.1899
40. M. Bojowald, in *Foundations of Space and Time: Reflections on Quantum Gravity*, ed. by J. Murugan, A. Weltman, G.F.R. Ellis (Cambridge University Press, Cambridge, 2012). arXiv:1101.5592
41. K. Banerjee, G. Calcagni, M. Martín-Benito, SIGMA **8**, 016 (2012). arXiv:1109.6801
42. A. Perez, Class. Quantum Gravity **20**, R43 (2003). gr-qc/0301113
43. D. Oriti, *Approaches to Quantum Gravity* (Cambridge University Press, Cambridge, 2009)

44. M. Bojowald, P.A. Höhn, A. Tsobanjan, Class. Quantum Gravity **28**, 035006 (2011). arXiv:1009.5953
45. M. Bojowald, P.A. Höhn, A. Tsobanjan, Phys. Rev. D **83**, 125023 (2011). arXiv:1011.3040
46. P.A. Höhn, E. Kubalova, A. Tsobanjan, arXiv:1111.5193
47. P.A.M. Dirac, Proc. R. Soc. A **246**, 333 (1958)
48. K.V. Kuchař, J. Math. Phys. **15**, 708 (1974)
49. S.A. Hojman, K. Kuchař, C. Teitelboim, Ann. Phys. **96**, 88 (1976)
50. J.F. Donoghue, Phys. Rev. D **50**, 3874 (1994). gr-qc/9405057
51. C.P. Burgess, Living Rev. Relativ. **7**, 5 (2004). http://www.livingreviews.org/lrr-2004-5. gr-qc/0311082
52. N. Kaloper, M. Kleban, A. Lawrence, S. Shenker, Phys. Rev. D **66**, 123510 (2002). hep-th/0201158
53. M. Bojowald, G. Hossain, M. Kagan, S. Shankaranarayanan, Phys. Rev. D **78**, 063547 (2008). arXiv:0806.3929
54. G. Immirzi, Class. Quantum Gravity **14**, L177 (1997). gr-qc/9612030
55. J.F. Barbero, Phys. Rev. D **51**, 5507 (1995). gr-qc/9410014
56. A. Ashtekar, Phys. Rev. D **36**, 1587 (1987)
57. C. Rovelli, L. Smolin, Nucl. Phys. B **331**, 80 (1990)
58. C. Rovelli, L. Smolin, Nucl. Phys. B **442**, 593 (1995). Erratum: Nucl. Phys. B **456**, 753 (1995). gr-qc/9411005
59. A. Ashtekar, J. Lewandowski, Class. Quantum Gravity **14**, A55 (1997). gr-qc/9602046
60. A. Perez, Phys. Rev. D **73**, 044007 (2006). gr-qc/0509118
61. A. Henderson, A. Laddha, C. Tomlin, arXiv:1204.0211 [gr-qc]
62. M. Bojowald, Gen. Relativ. Gravit. **38**, 1771 (2006). gr-qc/0609034
63. M. Bojowald, Gen. Relativ. Gravit. **40**, 639 (2008). arXiv:0705.4398
64. M. Bojowald, A. Skirzewski, Rev. Math. Phys. **18**, 713 (2006). math-ph/0511043
65. M. Bojowald, A. Skirzewski, Int. J. Geom. Methods Mod. Phys. **4**, 25 (2007). hep-th/0606232
66. T. Thiemann, Class. Quantum Gravity **15**, 1281 (1998). gr-qc/9705019
67. M. Bojowald, Class. Quantum Gravity **26**, 075020 (2009). arXiv:0811.4129
68. M. Bojowald, G. Calcagni, S. Tsujikawa, Phys. Rev. Lett. **107**, 211302 (2011). arXiv:1101.5391
69. M. Bojowald, G. Hossain, M. Kagan, S. Shankaranarayanan, Phys. Rev. D **79**, 043505 (2009). arXiv:0811.1572
70. M. Bojowald, G. Calcagni, J. Cosmol. Astropart. Phys. **1103**, 032 (2011). arXiv:1011.2779
71. M. Bojowald, G. Hossain, Phys. Rev. D **77**, 023508 (2008). arXiv:0709.2365
72. G. Calcagni, G. Hossain, Adv. Sci. Lett. **2**, 184 (2009). arXiv:0810.4330
73. M. Bojowald, G. Calcagni, S. Tsujikawa, J. Cosmol. Astropart. Phys. **1111**, 046 (2011). arXiv:1107.1540
74. E.J. Copeland, D.J. Mulryne, N.J. Nunes, M. Shaeri, Phys. Rev. D **77**, 023510 (2008). arXiv:0708.1261
75. E.J. Copeland, D.J. Mulryne, N.J. Nunes, M. Shaeri, Phys. Rev. D **79**, 023508 (2009). arXiv:0810.0104
76. G. Calcagni, M.V. Cortês, Class. Quantum Gravity **24**, 829 (2007). gr-qc/0607059
77. M. Shimano, T. Harada, Phys. Rev. D **80**, 063538 (2009). arXiv:0909.0334
78. A. Barrau, J. Grain, Phys. Rev. Lett. **102**, 081301 (2009). arXiv:0902.0145
79. J. Grain, A. Barrau, A. Gorecki, Phys. Rev. D **79**, 084015 (2009). arXiv:0902.3605
80. J. Grain, T. Cailleteau, A. Barrau, A. Gorecki, Phys. Rev. D **81**, 024040 (2010). arXiv:0910.2892
81. J. Mielczarek, T. Cailleteau, J. Grain, A. Barrau, Phys. Rev. D **81**, 104049 (2010). arXiv:1003.4660
82. J. Mielczarek, M. Szydłowski, Phys. Lett. B **657**, 20 (2007). arXiv:0705.4449
83. M. Bojowald, J.D. Reyes, R. Tibrewala, Phys. Rev. D **80**, 084002 (2009). arXiv:0906.4767
84. J.D. Reyes, Ph.D. thesis, The Pennsylvania State University, 2009

85. A. Perez, D. Pranzetti, Class. Quantum Gravity **27**, 145009 (2010). arXiv:1001.3292
86. T. Cailleteau, A. Barrau, arXiv:1111.7192
87. F. Cametti, G. Jona-Lasinio, C. Presilla, F. Toninelli, in *Proceedings of the International School of Physics 'Enrico Fermi', Course CXLIII* (IOS Press, Amsterdam, 2000), pp. 431–448. quant-ph/9910065
88. M. Bojowald, A. Tsobanjan, Phys. Rev. D **80**, 125008 (2009). arXiv:0906.1772
89. M. Bojowald, D. Brizuela, H.H. Hernandez, M.J. Koop, H.A. Morales-Técotl, Phys. Rev. D **84**, 043514 (2011). arXiv:1011.3022
90. M. Bojowald, T. Strobl, Rev. Math. Phys. **15**, 663 (2003). hep-th/0112074
91. M. Bojowald, *Canonical Gravity and Applications: Cosmology, Black Holes, and Quantum Gravity* (Cambridge University Press, Cambridge, 2010)
92. M. Bojowald, H. Hernández, A. Skirzewski, Phys. Rev. D **76**, 063511 (2007). arXiv:0706.1057
93. M. Bojowald, R. Tavakol, Phys. Rev. D **78**, 023515 (2008). arXiv:0803.4484
94. M. Bojowald, D. Mulryne, W. Nelson, R. Tavakol, Phys. Rev. D **82**, 124055 (2010). arXiv:1004.3979
95. M. Bojowald, B. Sandhöfer, A. Skirzewski, A. Tsobanjan, Rev. Math. Phys. **21**, 111 (2009). arXiv:0804.3365
96. M. Bojowald, A. Tsobanjan, Class. Quantum Gravity **27**, 145004 (2010). arXiv:0911.4950
97. R.M. Wald, Phys. Rev. D **48**, 2377 (1993). gr-qc/9305024
98. T. Thiemann, Class. Quantum Gravity **23**, 2211 (2006). gr-qc/0305080
99. B. Dittrich, T. Thiemann, Class. Quantum Gravity **23**, 1025 (2006). gr-qc/0411138
100. A. Higuchi, R.M. Wald, Phys. Rev. D **51**, 544 (1995). gr-qc/9407038
101. M. Bojowald, Gen. Relativ. Gravit. **35**, 1877 (2003). gr-qc/0305069
102. J. Mielczarek, W. Piechocki, Phys. Rev. D **83**, 104003 (2011). arXiv:1011.3418
103. R. Helling, arXiv:0912.3011
104. J. Haro, E. Elizalde, arXiv:0901.2861
105. M. Bojowald, Proc. R. Soc. A **464**, 2135 (2008). arXiv:0710.4919
106. W. Kaminski, T. Pawłowski, Phys. Rev. D **81**, 084027 (2010). arXiv:1001.2663
107. M. Bojowald, Nat. Phys. **3**, 523 (2007)

Chapter 8
Asymptotic Safety, Fractals, and Cosmology

Martin Reuter and Frank Saueressig

Abstract These lecture notes introduce the basic ideas of the asymptotic safety approach to quantum Einstein gravity (QEG). In particular they provide the background for recent work on the possibly multi-fractal structure of the QEG spacetimes. Implications of asymptotic safety for the cosmology of the early Universe are also discussed.

8.1 Introduction

Finding a consistent and fundamental quantum theory for gravity is still one of the most challenging open problems in theoretical high-energy physics to date. As is well known, the perturbative quantization of the classical description for gravity, general relativity, results in a non-renormalizable quantum theory [1–3]. One possible lesson drawn from this result may assert that gravity constitutes an effective field theory valid at low energies, whose ultraviolet (UV) completion requires the introduction of new degrees of freedom and symmetries. This is the path followed, e.g., by string theory. In a less radical approach, one retains the fields and symmetries known from general relativity and conjectures that gravity constitutes a fundamental theory at the non-perturbative level. One proposal along this line is the asymptotic safety scenario [4–6] initially put forward by Weinberg [7–10]. The key ingredient in this scenario is a non-Gaussian fixed point (NGFP) of the gravitational renormalization group (RG) flow, which controls the behavior of the theory at high energies and renders physical quantities safe from unphysical divergences. Given that the NGFP comes with a finite number of unstable (or relevant) directions, this construction is as predictive as a 'standard' perturbatively renormalizable quantum field theory.

M. Reuter (✉) · F. Saueressig
Institute of Physics, University of Mainz, Staudingerweg 7, 55099 Mainz, Germany
e-mail: reuter@thep.physik.uni-mainz.de

F. Saueressig
e-mail: saueressig@thep.physik.uni-mainz.de

G. Calcagni et al. (eds.), *Quantum Gravity and Quantum Cosmology*,
Lecture Notes in Physics 863, DOI 10.1007/978-3-642-33036-0_8,
© Springer-Verlag Berlin Heidelberg 2013

(1) The primary tool for investigating this scenario is the functional renormalization group equation (FRGE) for gravity [11], which constitutes the spring-board for the detailed investigations of the non-perturbative renormalization group behavior of quantum Einstein gravity [11–60]. The FRGE defines a Wilsonian RG flow on a theory space which consists of all diffeomorphism-invariant functionals of the metric $g_{\mu\nu}$, and turned out to be ideal for investigating the asymptotic safety conjecture [4–8]. In fact, it yielded substantial evidence for the non-perturbative renormalizability of quantum Einstein gravity. The theory emerging from this construction (henceforth denoted 'QEG') is not a quantization of classical general relativity. Instead, its bare action corresponds to a non-trivial fixed point of the RG flow and is a prediction therefore.

The approach of [11] employs the effective average action Γ_k [61–72] which has crucial advantages as compared to other continuum implementations of the Wilsonian RG flow [73–76]. The scale dependence of Γ_k is governed by the FRGE [61]

$$k\partial_k \Gamma_k[\Phi, \bar{\Phi}] = \frac{1}{2}\mathrm{STr}\left[\left(\frac{\delta^2\Gamma_k}{\delta\Phi^A\delta\Phi^B} + \mathscr{R}_k\right)^{-1} k\partial_k\mathscr{R}_k\right]. \tag{8.1}$$

Here Φ^A is the collection of all dynamical fields considered, $\bar{\Phi}^A$ denotes their background counterparts and STr denotes a generalized functional trace carrying a minus sign for fermionic fields and a factor 2 for complex fields. Moreover $\mathscr{R}_k$ is a matrix-valued infrared cutoff, which provides a k-dependent mass-term for fluctuations with momenta $p^2 \ll k^2$, while vanishing for $p^2 \gg k^2$. Solutions of the flow equation give rise to families of effective field theories $\{\Gamma_k[g_{\mu\nu}], 0 \leq k < \infty\}$ labeled by the coarse-graining scale k. The latter property opens the door to a rather direct extraction of physical information from the RG flow, at least in single-scale cases: If the physical process under consideration involves a single typical momentum scale p_0 only, it can be described by a tree-level evaluation of $\Gamma_k[g_{\mu\nu}]$, with $k = p_0$.

(2) Already soon after the asymptotic safety program had taken its modern form, various indications pointed in the direction that in QEG space-time should have certain features in common with a fractal. In ref. [13] the four-dimensional graviton propagator has been studied in the regime of asymptotically large momenta and it has been found that near the Planck scale a kind of dynamical dimensional reduction occurs. As a consequence of the NGFP controlling the UV behavior of the theory, the four-dimensional graviton propagator essentially behaves as two-dimensional on microscopic scales.

Subsequently, the "finger prints" of the NGFP on the fabric of the effective QEG space-times have been discussed in [15], where it was shown that asymptotic safety induces a characteristic self-similarity of space-time on length scales below the Planck length ℓ_{Pl}. The graviton propagator becomes scale-invariant in this regime [13]. Based on this observation it was argued that, in a cosmological context, the geometry fluctuations it describes can give rise to a scale-free spectrum of primordial density perturbations responsible for structure formation [77–81]. Thus the overall picture of the space-time structure in asymptotically safe gravity as it emerged about ten years ago comprises a smooth classical manifold on large distance scales, while on small scales one encounters a low-dimensional effective fractal [13, 15].

The characteristic feature at the heart of these results is that the effective field equations derived from the gravitational average action equip every given smooth space-time manifold with, in principle, infinitely many different (pseudo-) Riemannian structures, one for each coarse-graining scale [82, 83]. Thus, very much like in the famous example of the coast line of England [84], the proper length on a QEG space-time depends on the 'length of the yardstick' used to measure it. Earlier on similar fractal properties had already been found in other quantum gravity theories, in particular near dimension 2 [85], in a non-asymptotically safe model [86] and by analyzing the conformal anomaly [87].

Along a different line of investigations, the causal dynamical triangulation (CDT) approach has been developed and first Monte-Carlo simulations were performed [88–95]; see [96] for a recent review. In this framework one attempts to compute quantum gravity partition functions by numerically constructing the continuum limit of an appropriate statistical mechanics system. This limit amounts to a second-order phase transition. If CDT and its counterpart QEG, formulated in the continuum by means of the average action, belong to the same universality class, one may expect that the phase transition of the former is described by the non-trivial fixed point underlying the asymptotic safety of the latter.

Remarkably, ref. [90–92] reported results which indicated that the four-dimensional CDT space-times, too, undergo a dimensional reduction from four to two dimensions as one 'zooms' in on short distances. In particular, it had been demonstrated that the spectral dimension d_s measured in the CDT simulations has the very same limiting behaviors, $4 \to 2$, as in QEG [97]. Therefore it was plausible to assume that both approaches indeed 'see' the same continuum physics.

This interpretation became problematic when ref. [94] carried out CDT simulations for $d = 3$ macroscopic dimensions, which favor a value near $d_s = 2$ on the shortest length-scale probed since, in this case, the QEG prediction for the fixed point region is the value $d_s = 3/2$ [97]. Furthermore, the authors of ref. [98] reported simulations within the *Euclidean* dynamical triangulation (EDT) approach in $d = 4$, which favor a drop of the spectral dimension from 4 to about 1.5; this is again in conflict with the QEG expectations if one interprets the latter dimension as the value in the continuum limit.

Later on we will present several types of scale-dependent effective dimensions, specifically the spectral dimension d_s and the walk dimension d_w for the effective QEG space-times. We shall see that on length scales slightly *larger* than ℓ_{Pl} there exists a further regime which exhibits the phenomenon of dynamical dimensional reduction. There the spectral dimension is even smaller than near the fixed point, namely $d_s = 4/3$ in the case of 4 dimensions classically. Moreover, we shall argue that the (3-dimensional) results reported in [94] are in perfect accord with QEG, but that the shortest possible length scale achieved in the simulations is not yet close to the Planck length. Rather, the Monte Carlo data probe the transition between the classical and the newly discovered 'semi-classical' regime [99].

For similar work on fractal features in different approaches we must refer to the literature [100–122].

(3) As for possible physics implications of the RG flow predicted by QEG, ideas from particle physics, in particular the 'RG improvement', have been employed in order to study the leading quantum gravity effects in black holes [123–126], cosmological space-times [77–81, 127–136] or possible observable signatures from asymptotic safety at the LHC [137–140]. Among other results, it was found [123–125] that the quantum effects tend to decrease the Hawking temperature of black holes, and that their evaporation process presumably stops completely once the black holes mass is of the order of the Planck mass. In cosmology it turned out that inflation can occur without the need of an inflaton, and that the running of the cosmological constant might be responsible for the observed entropy of the present Universe [79–81].

These lectures are intended to provide the necessary background for these developments. They consist of three main parts, dealing with the basic ideas of asymptotic safety, the fractal QEG space-times, and possible implications of asymptotic safety for cosmology, respectively.

8.2 Theory Space and Its Truncation

We start by reviewing the basic ideas underlying asymptotic safety, referring to [4–6] for a more detailed discussion. The arena in which the Wilsonian RG dynamics takes place is 'theory space'. Albeit a somewhat formal notion, it helps in visualizing various concepts related to functional renormalization group equations; see Fig. 8.1. To describe it, we shall consider an arbitrary set of fields $\Phi(x)$. Then the corresponding theory space consists of all (action) functionals $A : \Phi \mapsto A[\Phi]$ depending on this set, possibly subject to certain symmetry requirements (a $\mathbb{Z}_2$-symmetry for a single scalar, or diffeomorphism invariance if Φ denotes the space-time metric, for instance). So the theory space $\{A[\cdot]\}$ is completely determined once the field content and the symmetries are fixed. Let us assume we can find a set of 'basis functionals' $\{P_\alpha[\cdot]\}$ so that every point of theory space has an expansion of the form

$$A[\Phi, \bar{\Phi}] = \sum_{\alpha=1}^{\infty} \bar{u}_\alpha P_\alpha[\Phi, \bar{\Phi}]. \tag{8.2}$$

The basis $\{P_\alpha[\cdot]\}$ will include both local field monomials and non-local invariants and we may use the 'generalized couplings' $\{\bar{u}_\alpha, \alpha = 1, 2, \ldots\}$ as local coordinates. More precisely, the theory space is coordinatized by the subset of 'essential couplings', i.e., those coordinates which cannot be absorbed by a field reparameterization.

Geometrically speaking the FRGE for the effective average action, Eq. (8.1), defines a vector field β on theory space. The integral curves along this vector field are the 'RG trajectories' $k \mapsto \Gamma_k$ parameterized by the scale k. They start, for $k \to \infty$, at the microscopic action S and terminate at the ordinary effective action at

Fig. 8.1 The points of theory space are the action functionals $A[\cdot]$. The RG equation defines a vector field β on this space; its integral curves are the RG trajectories $k \mapsto \Gamma_k$. They emanate from the fixed point action $\Gamma_* \equiv \Gamma_\infty$, which might differ from the bare action by a simple explicitly known functional, and end at the standard effective action Γ

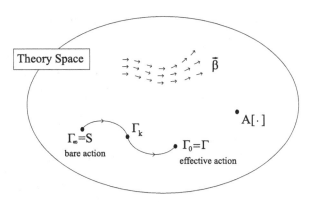

$k = 0$. The natural orientation of the trajectories is from higher to lower scales k, the direction of increasing 'coarse graining'. Expanding Γ_k as in (8.2),

$$\Gamma_k[\Phi, \bar{\Phi}] = \sum_{\alpha=1}^{\infty} \bar{u}_\alpha(k) P_\alpha[\Phi, \bar{\Phi}], \tag{8.3}$$

the trajectory is described by infinitely many 'running couplings' $\bar{u}_\alpha(k)$. Inserting (8.3) into the FRGE we obtain a system of infinitely many coupled differential equations for the $\bar{u}_\alpha$'s:

$$k \partial_k \bar{u}_\alpha(k) = \overline{\beta}_\alpha(\bar{u}_1, \bar{u}_2, \ldots; k), \quad \alpha = 1, 2, \ldots. \tag{8.4}$$

Here the 'beta functions' $\overline{\beta}_\alpha$ arise by expanding the trace on the right-hand side of the FRGE in terms of $\{P_\alpha[\cdot]\}$, i.e., $\frac{1}{2} \mathrm{Tr}[\cdots] = \sum_{\alpha=1}^{\infty} \overline{\beta}_\alpha(\bar{u}_1, \bar{u}_2, \ldots; k) P_\alpha[\Phi, \bar{\Phi}]$. The expansion coefficients $\overline{\beta}_\alpha$ have the interpretation of beta functions similar to those of perturbation theory, but not restricted to relevant couplings. In standard field theory jargon one would refer to $\bar{u}_\alpha(k = \infty)$ as the 'bare' parameters and to $\bar{u}_\alpha(k = 0)$ as the 'renormalized' or 'dressed' parameters.

The notation with the bar on $\bar{u}_\alpha$ and $\overline{\beta}_\alpha$ is to indicate that we are still dealing with dimensionful couplings. Usually the flow equation is reexpressed in terms of the dimensionless couplings

$$u_\alpha \equiv k^{-d_\alpha} \bar{u}_\alpha, \tag{8.5}$$

where d_α is the canonical mass dimension of $\bar{u}_\alpha$. Correspondingly, the essential u_α's are used as coordinates of theory space. The resulting RG equations

$$k \partial_k u_\alpha(k) = \beta_\alpha(u_1, u_2, \ldots) \tag{8.6}$$

are a coupled system of autonomous differential equations. The β_α's have no explicit k-dependence and define a 'time independent' vector field on theory space.

In this language, the basic idea of renormalization can be understood as follows. The boundary of theory space depicted in Fig. 8.1 is meant to separate points with coordinates $\{u_\alpha, \alpha = 1, 2, \ldots\}$ with all the essential couplings u_α well defined, from points with undefined, divergent couplings. The basic task of renormalization theory consists in constructing an 'infinitely long' RG trajectory which lies entirely within this theory space, i.e., a trajectory which neither leaves theory space (that is, develops divergences) in the UV limit $k \to \infty$ nor in the infrared (IR) limit $k \to 0$. Every such trajectory defines one possible quantum theory.

The consistent UV behavior can be ensured by performing the limit $k \to \infty$ at a fixed point $\{u_\alpha^*, \alpha = 1, 2, \ldots\} \equiv u^*$ of the RG flow. The fixed point is a zero of the vector field $\beta \equiv (\beta_\alpha)$, i.e., $\beta_\alpha(u^*) = 0$ for all $\alpha = 1, 2, \ldots$. The RG trajectories, solutions of $k\partial_k u_\alpha(k) = \beta_\alpha[u(k)]$, have a low 'velocity' near a fixed point because the β_α's are small there and directly at the fixed point the running stops completely. As a result, one can 'use up' an infinite amount of RG time near/at the fixed point if one bases the quantum theory on a trajectory which runs into such a fixed point for $k \to \infty$. This construction ensures that in the UV limit the trajectory ends at an 'inner point' of theory space giving rise to a well-behaved action functional. Thus we can be sure that, for $k \to \infty$, the trajectory does not develop pathological properties such as divergent couplings. The resulting quantum theory is 'safe' from unphysical divergences.

At this stage it is natural to distinguish two classes of fixed points. First, the UV limit may be performed at a Gaussian fixed point (GFP) where $u_\alpha^* = 0$, $\forall \alpha = 1, 2, \ldots$. In this case, the fixed point functional does not contain interactions and the theory becomes asymptotically free in the UV. This is the construction underlying perturbatively renormalizable quantum field theories. More general, one can also use a non-Gaussian fixed point (NGFP) for letting $k \to \infty$, where, by definition, not all of the coordinates u_α^* vanish. In the context of gravity, Weinberg [7, 8] proposed that the UV limit of the theory is provided by such a NGFP.

Note that at the NGFP it is the *dimensionless* essential couplings (8.5) which assume constant values. Therefore, even directly at a NGFP where $u_\alpha(k) \equiv u_\alpha^*$, the dimensionful couplings keep running according to a power law

$$\bar{u}_\alpha(k) = u_\alpha^* k^{d_\alpha}. \tag{8.7}$$

Furthermore, non-essential dimensionless couplings are not required to attain fixed point values.

Given a fixed point, an important concept is its *UV critical hypersurface* $\mathscr{S}_{\text{UV}}$, or synonymously, its *unstable manifold*. By definition, it consists of all points of theory space which are pulled into the fixed point by the inverse RG flow, i.e., for *increasing* k. Its dimensionality $\dim(\mathscr{S}_{\text{UV}}) \equiv \Delta_{\text{UV}}$ is given by the number of attractive (for *increasing* cutoff k) directions in the space of couplings.

For the RG equations (8.6), the linearized flow near the fixed point is governed by the Jacobi matrix $\mathbf{B} = (B_{\alpha\gamma})$, $B_{\alpha\gamma} \equiv \partial_\gamma \beta_\alpha(u^*)$:

$$k\partial_k u_\alpha(k) = \sum_\gamma B_{\alpha\gamma}\left(u_\gamma(k) - u_\gamma^*\right). \tag{8.8}$$

The general solution to this equation reads

$$u_\alpha(k) = u_\alpha^* + \sum_I C_I V_\alpha^I \left(\frac{k_0}{k}\right)^{\theta_I}, \qquad (8.9)$$

where the V^I's are the right-eigenvectors of $\mathbf{B}$ with eigenvalues $-\theta_I$, i.e., $\sum_\gamma B_{\alpha\gamma} V_\gamma^I = -\theta_I V_\alpha^I$. Since $\mathbf{B}$ is not symmetric in general the θ_I's are not guaranteed to be real. We assume that the eigenvectors form a complete system though. Furthermore, k_0 is a fixed reference scale, and the C_I's are constants of integration. The quantities θ_I are referred to as *critical exponents* since when the renormalization group is applied to critical phenomena (second-order phase transitions) the traditionally defined critical exponents are related to the θ_I's in a simple way [68, 69].

If $u_\alpha(k)$ is to describe a trajectory in $\mathscr{S}_{\mathrm{UV}}$, $u_\alpha(k)$ must approach u_α^* in the limit $k \to \infty$ and therefore we must set $C_I = 0$ for all I with $\mathrm{Re}\,\theta_I < 0$. Hence the dimensionality Δ_{UV} equals the number of $\mathbf{B}$-eigenvalues with a negative real part, i.e., the number of θ_I's with $\mathrm{Re}\,\theta_I > 0$. The corresponding eigenvectors span the tangent space to $\mathscr{S}_{\mathrm{UV}}$ at the NGFP. If we *lower* the cutoff for a generic trajectory with all C_I nonzero, only Δ_{UV} 'relevant' parameters corresponding to the eigendirections tangent to $\mathscr{S}_{\mathrm{UV}}$ grow ($\mathrm{Re}\,\theta_I > 0$), while the remaining 'irrelevant' couplings pertaining to the eigendirections normal to $\mathscr{S}_{\mathrm{UV}}$ decrease ($\mathrm{Re}\,\theta_I < 0$). Thus near the NGFP a generic trajectory is attracted towards $\mathscr{S}_{\mathrm{UV}}$.

Coming back to the asymptotic safety construction, let us now use this fixed point in order to take the limit $k \to \infty$. The trajectories which define an infinite cutoff limit are special in the sense that all irrelevant couplings are set to zero: $C_I = 0$ if $\mathrm{Re}\,\theta_I < 0$. These conditions place the trajectory exactly on $\mathscr{S}_{\mathrm{UV}}$. There is a Δ_{UV}-parameter family of such trajectories, and the experiment must decide which one is realized in Nature. Therefore the predictive power of the theory increases with decreasing dimensionality of $\mathscr{S}_{\mathrm{UV}}$, i.e., number of UV attractive eigendirections of the NGFP. If $\Delta_{\mathrm{UV}} < \infty$, the quantum field theory thus constructed is comparable to and as predictive as a perturbatively renormalizable model with Δ_{UV} 'renormalizable couplings'. Summarizing, we call a theory *asymptotically safe* if its UV behavior is controlled by a non-Gaussian fixed point with a finite number of relevant directions. The former condition ensures that the theory is safe from unphysical UV divergences while the latter requirement guarantees the predictivity of the construction.

Up to this point our discussion did not involve any approximation. A method which gives rise to non-perturbative approximate solutions is to truncate the theory space $\{A[\cdot]\}$. The basic idea is to project the RG flow onto a finite-dimensional subspace of theory space. The subspace should be chosen in such a way that the projected flow encapsulates the essential physical features of the exact flow on the full space.

Concretely, the projection onto a truncation subspace is performed as follows. One makes an ansatz of the form $\Gamma_k[\Phi, \bar{\Phi}] = \sum_{i=1}^N \bar{u}_i(k) P_i[\Phi, \bar{\Phi}]$, where the k-independent functionals $\{P_i[\cdot], i = 1, \ldots, N\}$ form a 'basis' on the subspace selected. For a scalar field ϕ, say, examples include pure potential terms $\int d^d x \phi^m(x), \int d^d x \phi^n(x) \ln \phi^2(x), \ldots$, a standard kinetic term $\int d^d x (\partial \phi)^2$, higher order derivative terms $\int d^d x \, \phi (\partial^2)^n \phi, \int d^d x \, f(\phi)(\partial^2)^n \phi (\partial^2)^m \phi, \ldots$, and non-local terms like $\int d^d x \, \phi \ln(-\partial^2)\phi, \ldots$. Even if Γ_∞ is simple, a standard ϕ^4 action, say, the evolution from $k = \infty$ downwards will generate such terms.

The projected RG flow is described by a set of ordinary (if $N < \infty$) differential equations for the couplings $\bar{u}_i(k)$. They arise as follows. Let us assume we expand the Φ-dependence of $\frac{1}{2}\text{Tr}[\cdots]$ (with the ansatz for $\Gamma_k[\Phi, \bar{\Phi}]$ inserted) in a basis $\{P_\alpha[\cdot]\}$ of the *full* theory space which contains the P_i's spanning the truncated space as a subset:

$$\frac{1}{2}\text{Tr}[\cdots] = \sum_{\alpha=1}^{\infty} \overline{\beta}_\alpha(\bar{u}_1, \ldots, \bar{u}_N; k) P_\alpha[\Phi, \bar{\Phi}]$$

$$= \sum_{i=1}^{N} \overline{\beta}_i(\bar{u}_1, \ldots, \bar{u}_N; k) P_i[\Phi, \bar{\Phi}] + \text{rest}. \tag{8.10}$$

Here the 'rest' contains all terms outside the truncated theory space; the approximation consists in neglecting precisely those terms. Thus, equating (8.10) to the LHS of the flow equation, $\partial_t \Gamma_k = \sum_{i=1}^{N} \partial_t \bar{u}_i(k) P_i$, the linear independence of the P_i's implies the coupled system of ordinary differential equations

$$\partial_t \bar{u}_i(k) = \overline{\beta}_i(\bar{u}_1, \ldots, \bar{u}_N; k), \quad i = 1, \ldots, N. \tag{8.11}$$

Solving (8.11) one obtains an *approximation* to the exact RG trajectory projected onto the chosen subspace. Note that this approximate trajectory does, in general, not coincide with the projection of the exact trajectory, but if the subspace is well chosen, it will not be very different from it.

8.3 The Effective Average Action for Gravity

The effective average action for gravity which has been introduced in ref. [11] is a concrete implementation of the general ideas outlined above. The ultimate goal is to give meaning to an integral over 'all' metrics $\gamma_{\mu\nu}$ of the form $\int \mathscr{D}\gamma_{\mu\nu} \exp\{-S[\gamma_{\mu\nu}] + \text{source terms}\}$ whose bare action $S[\gamma_{\mu\nu}]$ is invariant under general coordinate transformations. The first step consists in splitting the quantum metric according to

$$\gamma_{\mu\nu} = \overline{g}_{\mu\nu} + h_{\mu\nu} \tag{8.12}$$

where $\overline{g}_{\mu\nu}$ is a fixed, but unspecified, background metric and $h_{\mu\nu}$ are the quantum fluctuations around this background which are not necessarily small. This allows the formal construction of the gauge-fixed (Euclidean) gravitational path integral

$$\int \mathscr{D}h \mathscr{D}C^\mu \mathscr{D}\bar{C}_\mu \exp\left\{-S[\bar{g}+h]-S^{\mathrm{gf}}[h;\bar{g}]-S^{\mathrm{ghost}}[h,C,\bar{C};\bar{g}]-\Delta_k S[h,C,\bar{C};\bar{g}]\right\}.$$

(8.13)

Here $S[\bar{g}+h]$ is a generic action, which depends on $\gamma_{\mu\nu}$ only, while the background gauge fixing $S^{\mathrm{gf}}[h;\bar{g}]$ and ghost contribution $S^{\mathrm{ghost}}[h,C,\bar{C};\bar{g}]$ contain $\bar{g}_{\mu\nu}$ and $h_{\mu\nu}$ in such a way that they do not combine into a full $\gamma_{\mu\nu}$. We take $S^{\mathrm{gf}}[h;\bar{g}]$ to be a gauge fixing 'of the background type' [141], i.e., it is invariant under diffeomorphisms acting on both $h_{\mu\nu}$ and $\bar{g}_{\mu\nu}$.

The key ingredient in the construction of the FRGE is the coarse graining term $\Delta_k S[h,C,\bar{C};\bar{g}]$. It is quadratic in the fluctuation field,

$$\int d^d x \sqrt{\bar{g}} h_{\mu\nu} \mathscr{R}_k^{\mu\nu\rho\sigma}(-\bar{D}^2) h_{\rho\sigma},$$

plus a similar term for the ghosts. The kernel $\mathscr{R}_k^{\mu\nu\rho\sigma}(p^2)$ provides a k-dependent mass term which separates the fluctuations into high momentum modes $p^2 \gg k^2$ and low momentum modes $p^2 \ll k^2$ with respect to the scale set by the covariant Laplacian of the background metric. The profile of $\mathscr{R}_k^{\mu\nu\rho\sigma}(p^2)$ ensures that the high momentum modes are integrated out unsuppressed while the contribution of the low momentum modes to the path integral is suppressed by the k-dependent mass term. Varying k then naturally realizes Wilson's idea of coarse graining by integrating out the quantum fluctuations shell by shell.

The k-derivative of Eq. (8.13) with $h_{\mu\nu}$ and the ghosts coupled to appropriate sources, provides the starting point for the construction of the functional renormalization group equation for the effective average action Γ_k [61–67]. (See [68, 69] for reviews.) For gravity this flow equation takes the form [11]

$$\partial_t \Gamma_k[\bar{h},\xi,\bar{\xi};\bar{g}] = \frac{1}{2}\mathrm{STr}\left[\left(\Gamma_k^{(2)}+\mathscr{R}_k\right)^{-1}\partial_t \mathscr{R}_k\right].$$

(8.14)

Here $t = \log(k/k_0)$, STr is a functional supertrace which includes a minus sign for the ghosts $\xi \equiv \langle C \rangle$, $\bar{\xi} \equiv \langle \bar{C} \rangle$, $\mathscr{R}_k$ is the matrix-valued (in field space) IR cutoff introduced above, and $\Gamma_k^{(2)}$ is the second variation of Γ_k with respect to the *fluctuation fields*. Notably, $\Gamma_k[\bar{h},\xi,\bar{\xi};\bar{g}]$ depends on *two* metrics, $\bar{g}_{\mu\nu}$ and

$$g_{\alpha\beta} \equiv \langle \gamma_{\alpha\beta} \rangle = \bar{g}_{\alpha\beta} + \bar{h}_{\alpha\beta}, \qquad \bar{h}_{\alpha\beta} \equiv \langle h_{\alpha\beta} \rangle.$$

(8.15)

In this sense, Γ_k is of an intrinsically *bimetric* nature, and therefore we often write $\Gamma_k[g,\bar{g},\xi,\bar{\xi}] \equiv \Gamma_k[\bar{h}=g-\bar{g},\xi,\bar{\xi};\bar{g}]$. This functional is invariant under background gauge transformations acting on all four fields simultaneously. It is a k-dependent generalization of the standard effective action $\Gamma \equiv \Gamma_0$ to which it reduces in the limit $k \to 0$. It can also be shown that Γ_k in the limit $k \to \infty$ is essentially equivalent to the bare action S. (For further details about Γ_k for gravity we refer to [11].)

8.4 The Einstein–Hilbert Truncation

Solving the FRGE (8.14) is equivalent to (and as difficult as) calculating the functional integral over $\gamma_{\mu\nu}$. It is therefore important to devise efficient approximation methods. The truncation of theory space is the one which makes maximum use of the FRGE reformulation of the quantum field theory problem at hand.

The first truncation for which the RG flow has been worked out [11] is the 'Einstein–Hilbert truncation' which retains in Γ_k only the terms $\int d^d x \sqrt{g}$ and $\int d^d x \sqrt{g} R$, already present in the in the classical action, with k-dependent coupling constants, as well as the classical gauge fixing and ghost terms:

$$\Gamma_k = \frac{1}{16\pi G_k} \int d^d x \sqrt{g} \{-R + 2\bar{\lambda}_k\} + \text{class. gf- and gh-terms.} \quad (8.16)$$

In this case the truncation subspace is 2-dimensional. The ansatz (8.16) contains two free functions of the scale, the running cosmological constant $\bar{\lambda}_k$ and the running Newton constant G_k.

Upon inserting the ansatz (8.16) into the flow equation (8.14) it boils down to a system of two ordinary differential equations. We shall display them here in terms of the dimensionless running cosmological constant and Newton constant, respectively:

$$\lambda_k \equiv k^{-2} \bar{\lambda}_k, \qquad g_k \equiv k^{d-2} G_k. \quad (8.17)$$

Using λ_k and g_k the RG equations become autonomous

$$k \partial_k g(k) = \beta_g \big[g(k), \lambda(k)\big], \qquad k \partial_k \lambda(k) = \beta_\lambda \big[g(k), \lambda(k)\big], \quad (8.18)$$

with

$$\beta_g(g_k, \lambda_k) = \big[d - 2 + \eta_N(g_k, \lambda_k)\big] g_k. \quad (8.19)$$

Here $\eta_N \equiv \partial_t \ln G_k$ is the anomalous dimension of the operator $\sqrt{g} R$. The explicit form of the beta functions β_g and β_λ for arbitrary cutoff $\mathcal{R}_k$ and dimension can be found in ref. [11]. Here we only display the result for $d = 4$ and a sharp cutoff:

$$\partial_t \lambda_k = -(2 - \eta_N)\lambda_k - \frac{g_k}{\pi}\left[5\ln(1 - 2\lambda_k) - 2\zeta(3) + \frac{5}{2}\eta_N\right], \quad (8.20a)$$

$$\partial_t g_k = (2 + \eta_N)g_k, \quad (8.20b)$$

$$\eta_N = -\frac{2g_k}{6\pi + 5g_k}\left[\frac{18}{1 - 2\lambda_k} + 5\ln(1 - 2\lambda_k) - \zeta(2) + 6\right]. \quad (8.20c)$$

In [14] this system has been analyzed in detail, using both analytical and numerical methods. In particular all RG trajectories have been classified, and examples have been computed numerically. The most important classes of trajectories in the phase portrait on the g–λ-plane are shown in Fig. 8.2.

The RG flow is found to be dominated by two fixed points (g^*, λ^*): the GFP at $g^* = \lambda^* = 0$, and a NGFP with $g^* > 0$ and $\lambda^* > 0$. There are three classes of

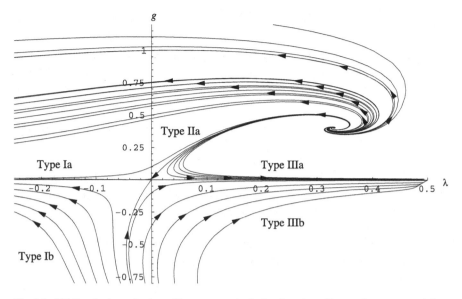

Fig. 8.2 RG flow in the g–λ-plane. The arrows point in the direction of increasing coarse graining, i.e., of decreasing k. (From [14])

trajectories emanating from the NGFP: trajectories of Type Ia and IIIa run towards negative and positive cosmological constants, respectively, and the single trajectory of Type IIa ('separatrix') hits the GFP for $k \to 0$. The high momentum properties of QEG are governed by the NGFP; for $k \to \infty$, in Fig. 8.2 all RG trajectories on the half-plane $g > 0$ run into this point. The two critical exponents are a complex conjugate pair $\theta_{1,2} = \theta' \pm i\theta''$ with $\theta' > 0$. The fact that at the NGFP the dimensionless coupling constants g_k, λ_k approach constant, non-zero values then implies that the dimensionful quantities run according to

$$G_k = g^* k^{2-d}, \qquad \bar{\lambda}_k = \lambda^* k^2. \tag{8.21}$$

Hence for $k \to \infty$ and $d > 2$ the dimensionful Newton constant vanishes while the cosmological constant diverges.

So, the Einstein–Hilbert truncation does indeed predict the existence of a NGFP with exactly the properties needed for the asymptotic safety construction. Clearly the crucial question is whether the NGFP found is the projection of an exact fixed point in the full theory or merely the artifact of an insufficient approximation. This question has been analyzed during the past decade within truncations of ever increasing complexity. All investigations performed to date support the existence of a NGFP in the exact theory, and without exception they predict a projected RG flow on the g–λ-plane which is qualitatively similar to that of the Einstein–Hilbert truncation. In fact, the phase portrait in Fig. 8.2 has survived substantial generalizations of the truncation ansatz for the average action. Furthermore, clear evidence for a small, finite dimensionality of $\mathscr{S}_{UV}$ was found, first in $2 + \varepsilon$ dimensions [15] and then by an impressively complex calculation in $d = 4$ also [31, 32].

Beside its successes in describing gravity at high energies, QEG also recovers classical general relativity at low energies. Concretely, it was shown in [123–125] that Fig. 8.2 contains Type IIIa trajectories which are in agreement with observational data. This analysis is fairly robust and clear-cut; it does not involve the NGFP. All that is needed is the RG flow linearized about the GFP. In its vicinity one has [11]

$$\bar{\lambda}(k) = \bar{\lambda}_0 + v\bar{G}k^d + \cdots, \qquad G(k) = \bar{G} + \cdots, \qquad (8.22)$$

i.e., $\bar{\lambda}$ displays a running $\propto k^d$ and G is approximately constant. Here v is a positive constant of order unity [11, 14]. These equations are valid if $\lambda(k) \ll 1$ and $g(k) \ll 1$. They describe a 2-parameter family of RG trajectories labeled by the pair $(\bar{\lambda}_0, \bar{G})$. It will prove convenient to use an alternative labeling (λ_T, k_T) with $\lambda_T \equiv (4v\bar{\lambda}_0\bar{G})^{1/2}$ and $k_T \equiv (\bar{\lambda}_0/v\bar{G})^{1/4}$. The old labels are expressed in terms of the new ones as $\bar{\lambda}_0 = \frac{1}{2}\lambda_T k_T^2$ and $\bar{G} = \lambda_T/(2vk_T^2)$. It is furthermore convenient to introduce the abbreviation $g_T \equiv \lambda_T/(2v)$. When parameterized by the pair (λ_T, k_T) the trajectories assume the form

$$\bar{\lambda}(k) = \frac{1}{2}\lambda_T k_T^2 \big[1 + (k/k_T)^4\big] \equiv \bar{\lambda}_0\big[1 + (k/k_T)^4\big],$$
$$G(k) = \frac{\lambda_T}{2vk_T^2} \equiv \frac{g_T}{k_T^2}, \qquad (8.23)$$

or, in dimensionless form,

$$\lambda(k) = \frac{1}{2}\lambda_T\left[\left(\frac{k_T}{k}\right)^2 + \left(\frac{k}{k_T}\right)^2\right], \qquad g(k) = g_T\left(\frac{k}{k_T}\right)^2. \qquad (8.24)$$

As for the interpretation of the new variables, it is clear that $\lambda_T \equiv \lambda(k \equiv k_T)$ and $g_T \equiv g(k = k_T)$, while k_T is the scale at which β_λ (but not β_g) vanishes according to the linearized running: $\beta_\lambda(k_T) \equiv k d\lambda(k)/dk|_{k=k_T} = 0$. Thus we see that (g_T, λ_T) are the coordinates of the turning point T of the Type IIIa trajectory considered, and k_T is the scale at which it is passed. The regimes $k > k_T$ ($k < k_T$) are conveniently referred to as the 'UV regime' ('IR regime').

Let us now hypothesize that, within a certain range of k-values, the RG trajectory realized in Nature can be approximated by (8.24). In order to determine its parameters $(\bar{\lambda}_0, \bar{G})$ or (λ_T, k_T) we must perform a measurement of G and $\bar{\lambda}$. If we interpret the observed values $G_{observed} = m_{Pl}^{-2}$, $m_{Pl} \approx 1.2 \times 10^{19}$ GeV, and $\bar{\lambda}_{observed} \approx 10^{-120}m_{Pl}^2$ as the running $G(k)$ and $\bar{\lambda}(k)$ evaluated at a scale $k \ll k_T$, then we get from (8.23) that $\bar{\lambda}_0 = \bar{\lambda}_{observed}$ and $\bar{G} = G_{observed}$. Using the definitions of λ_T and k_T along with $v = O(1)$ this leads to the order-of-magnitude estimates $g_T \approx \lambda_T \approx 10^{-60}$ and $k_T \approx 10^{-30}m_{Pl} \approx (10^{-3}$ cm$)^{-1}$. Because of the tiny values of g_T and λ_T the turning point lies in the linear regime of the GFP. Going beyond the linear regime, the k-dependence of G and $\bar{\lambda}$ is plotted schematically in Fig. 8.3.

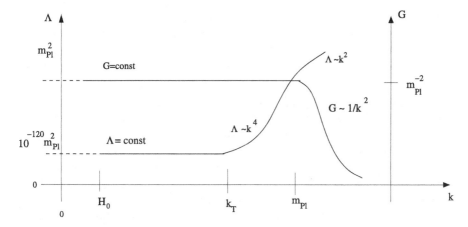

Fig. 8.3 The dimensionful $\Lambda(k) \equiv \bar{\lambda}(k)$ and $G(k)$ for a Type IIIa trajectory with realistic parameters

8.5 The Multi-fractal Properties of QEG Space-Times

We now proceed by discussing an intriguing consequence arising from the scale-dependence of the gravitational effective action, namely that the QEG space-time at short distances develops fractal properties [13, 15, 97]. As we have seen, the effective average action $\Gamma_k[g_{\mu\nu}]$ defines an infinite set of effective field theories, valid near a variable mass scale k. Intuitively speaking, the solution $\langle g_{\mu\nu}\rangle_k$ of the scale dependent field equation

$$\frac{\delta \Gamma_k}{\delta g_{\mu\nu}(x)}\left[\langle g\rangle_k\right] = 0 \qquad (8.25)$$

can be interpreted as the metric averaged over (Euclidean) space-time volumes of a linear extension ℓ which typically is of the order of $1/k$. Knowing the scale dependence of Γ_k, i.e., the renormalization group trajectory $k \mapsto \Gamma_k$, we can in principle follow the solution $\langle g_{\mu\nu}\rangle_k$ from the ultraviolet ($k \to \infty$) to the infrared ($k \to 0$).

(1) Quantum space-times. It is an important feature of this approach that the infinitely many equations of (8.25), one for each scale k, are valid *simultaneously*. They all refer *to the same* physical system, the 'quantum space-time', but describe its effective metric structure on different scales. An observer using a 'microscope' with a resolving power $\ell \approx k^{-1}$ will perceive the Universe to be a Riemannian manifold with metric $\langle g_{\mu\nu}\rangle_k$.[1] At every fixed k, $\langle g_{\mu\nu}\rangle_k$ is a smooth classical metric. But since the quantum space-time is characterized by the infinity of Eqs. (8.25) with $k = 0, \ldots, \infty$ it can acquire very non-classical and in particular fractal features. In

---

[1]The 'resolving power' ℓ of the microscope is in general a complicated function of k. It can be found by an algorithm outlined in [97]. For the purposes of the present discussion it is sufficient to think of this relationship as $\ell \approx 1/k$, like on flat space.

particular, it was concluded in [13, 15] that the effective dimensionality of space-time is scale dependent. It equals 4 at macroscopic distances ($\ell \gg \ell_{\mathrm{Pl}}$) but, near $\ell \approx \ell_{\mathrm{Pl}}$, it gets dynamically reduced to the value 2. For $\ell \ll \ell_{\mathrm{Pl}}$ space-time resembles a 2-dimensional fractal. In the following we review the arguments that led to this conclusion.

(2) Self-similarity in the fixed point regime. For simplicity we use the Einstein–Hilbert truncation to start with, and we consider space-times with classical dimensionality $d = 4$. The corresponding RG trajectories are shown in Fig. 8.2. The physically relevant ones, for $k \to \infty$, all approach the NGFP at (g_*, λ_*) so that the dimensionful quantities run according to (8.21). This scaling behavior is realized in the asymptotic scaling regime $k \gg m_{\mathrm{Pl}}$. Near $k = m_{\mathrm{Pl}}$ the trajectories cross over towards the GFP at $g = \lambda = 0$, and then run towards negative, vanishing, and positive values of λ, respectively. For our present purpose, it suffices to consider the limiting cases of very small and very large distances of a RG trajectory. We assume that G_k and $\bar{\lambda}_k$ behave as in (8.21) for $k \gg m_{\mathrm{Pl}}$, and that they are constant for $k \ll m_{\mathrm{Pl}}$. The precise interpolation between the two regimes will not be needed here.

The argument of ref. [15] concerning the fractal nature of the QEG space-times was as follows. Within the Einstein–Hilbert truncation of theory space, the effective field equations (8.25) happen to coincide with the ordinary Einstein equation, but with G_k and $\bar{\lambda}_k$ replacing the classical constants. Without matter,

$$R_{\mu\nu}(\langle g \rangle_k) = \frac{2}{2 - d} \bar{\lambda}_k \langle g_{\mu\nu} \rangle_k. \tag{8.26}$$

Since in absence of dimensionful constants of integration $\bar{\lambda}_k$ is the only quantity in this equation which sets a scale, every solution to (8.26) has a typical radius of curvature $r_c(k) \propto 1/\sqrt{\bar{\lambda}_k}$. (For instance, the maximally symmetric S^4-solution has the radius $r_c = r = \sqrt{3/\bar{\lambda}_k}$.) If we want to explore the space-time structure at a fixed length scale ℓ we should use the action $\Gamma_k[g_{\mu\nu}]$ at $k = 1/\ell$ because with this functional a tree level analysis is sufficient to describe the essential physics at this scale, including the relevant quantum effects. Hence, when we observe the space-time with a microscope of resolution ℓ, we will see an average radius of curvature given by $r_c(\ell) \equiv r_c(k = 1/\ell)$. Once ℓ is smaller than the Planck length $\ell_{\mathrm{Pl}} \equiv m_{\mathrm{Pl}}^{-1}$ we are in the fixed point regime where $\bar{\lambda}_k \propto k^2$ so that $r_c(k) \propto 1/k$, or

$$\boxed{r_c(\ell) \propto \ell.} \tag{8.27}$$

Thus, when we look at the structure of space-time with a microscope of resolution $\ell \ll \ell_{\mathrm{Pl}}$, the average radius of curvature which we measure is proportional to the resolution itself. If we want to probe finer details and decrease ℓ we automatically decrease r_c and hence *increase* the average curvature. Space-time seems to be more strongly curved at small distances than at larger ones. The scale-free relation (8.27) suggests that at distances below the Planck length the QEG space-time is a special kind of fractal with a self-similar structure. It has no intrinsic scale because in the fractal regime, i.e., when the RG trajectory is still close to the NGFP, the parameters which usually set the scales of the gravitational interaction, G and $\bar{\lambda}$, are not yet 'frozen out'. This happens only later on, somewhere half way between the non-Gaussian and the Gaussian fixed point, at a scale of the order of m_{Pl}.

Below this scale, G_k and $\bar{\lambda}_k$ stop running and, as a result, $r_c(k)$ becomes independent of k so that $r_c(\ell) = \mathrm{const}$ for $\ell \gg \ell_{\mathrm{Pl}}$. In this regime $\langle g_{\mu\nu} \rangle_k$ is k-independent, indicating that the macroscopic space-time is describable by a single smooth, classical Riemannian manifold.

(3) Anomalous dimension and graviton propagator. An independent argument supporting the assertion that the QEG space-time has an effective dimensionality which is k-dependent and non-integer in general based upon the *anomalous dimension* $\eta_N \equiv \partial_t \ln G_k$ has been put forward in ref. [13]. In a sense which we shall make more precise in a moment, the effective dimensionality of space-time equals $4 + \eta_N$. The RG trajectories of the Einstein–Hilbert truncation (within its domain of validity) have $\eta_N \approx 0$ for $k \to 0$ and $\eta_N \approx -2$ for $k \to \infty$, the smooth change by two units occurring near $k \approx m_{\mathrm{Pl}}$. As a consequence, the effective dimensionality is 4 for $\ell \gg \ell_{\mathrm{Pl}}$ and 2 for $\ell \ll \ell_{\mathrm{Pl}}$.

In fact, the UV fixed point has an anomalous dimension $\eta \equiv \eta_N(g_*, \lambda_*) = -2$. We can use this information in order to determine the momentum dependence of the dressed graviton propagator for momenta $p^2 \gg m_{\mathrm{Pl}}^2$. Expanding (8.16) about flat space and omitting the standard tensor structures we find the inverse propagator $\widetilde{\mathscr{G}}_k(p)^{-1} \propto G_k^{-1} p^2$. The conventional dressed propagator $\widetilde{\mathscr{G}}(p)$ contained in $\Gamma \equiv \Gamma_{k=0}$ is obtained from the *exact* $\widetilde{\mathscr{G}}_k$ in the limit $k \to 0$. For $p^2 > k^2 \gg m_{\mathrm{Pl}}^2$ the actual cutoff scale is the physical momentum p^2 itself so that the k-evolution of $\widetilde{\mathscr{G}}_k(p)$ stops at the threshold $k = \sqrt{p^2}$. Therefore,

$$\widetilde{\mathscr{G}}(p)^{-1} \propto p^2 G_k^{-1}\big|_{k=\sqrt{p^2}} \propto \left(p^2\right)^{1-\frac{\eta}{2}} \qquad (8.28)$$

because $G_k^{-1} \propto k^{-\eta}$ when η is (approximately) constant. In d flat dimensions, and for $\eta \neq 2 - d$, the Fourier transform of $\widetilde{\mathscr{G}}(p) \propto 1/(p^2)^{1-\eta/2}$ yields the following propagator in position space:

$$\mathscr{G}(x; y) \propto \frac{1}{|x - y|^{d-2+\eta}}. \qquad (8.29)$$

This form of the propagator is well known from the theory of critical phenomena, for instance. (In the latter case it applies to large distances.) Equation (8.29) is not valid directly at the NGFP. For $d = 4$ and $\eta = -2$ the dressed propagator is $\widetilde{\mathscr{G}}(p) = 1/p^4$, which has the following representation in position space:

$$\mathscr{G}(x; y) = -\frac{1}{8\pi^2} \ln\left(\mu |x - y|\right). \qquad (8.30)$$

Here μ is an arbitrary constant with the dimension of a mass. Obviously (8.30) has the same form as a $1/p^2$-propagator in 2 dimensions.

Slightly away from the NGFP, before other physical scales intervene, the propagator is of the familiar type (8.29) which shows that the quantity η_N has the standard interpretation of an anomalous dimension in the sense that fluctuation effects modify the decay properties of $\mathscr{G}$ so as to correspond to a space-time of effective dimensionality $4 + \eta_N$.

Thus the properties of the RG trajectories imply a remarkable dimensional reduction: *Space-time, probed by a 'graviton' with $p^2 \ll m_{\mathrm{Pl}}^2$ is 4-dimensional, but it appears to be 2-dimensional for a graviton with $p^2 \gg m_{\mathrm{Pl}}^2$* [13]. More generally, in d classical dimensions, where the macroscopic space-time is d-dimensional, the anomalous dimension at the fixed point is $\eta_N = 2 - d$. Therefore, for any d, the dimensionality of the fractal as implied by η_N is $d + \eta_N = 2$ [13, 15].

8.6 Spectral, Walk, and Hausdorff Dimension

The fractal properties of the QEG space-time can be further quantified by investigating random walks and diffusion processes on fractals. In this course one is led to introduce various notions of fractal dimensions, such as the spectral or walk dimension [142]. A priori they have no reason to equal the effective dimension $d_{\mathrm{eff}} = d + \eta$ implied by the running Newton constant and the graviton propagator.

(1) The spectral dimension. Consider the diffusion process where a spinless test particle performs a Brownian random walk on an ordinary Riemannian manifold with a fixed classical metric $g_{\mu\nu}(x)$. It is described by the heat-kernel $K_g(x, x'; T)$ which gives the probability density for a transition of the particle from x to x' during the fictitious time T. It satisfies the heat equation

$$\partial_T K_g(x, x'; T) = -\Delta_g K_g(x, x'; T), \tag{8.31}$$

where $\Delta_g = -D^2$ denotes the Laplace operator. In flat space, this equation is easily solved by

$$K_g(x, x'; T) = \int \frac{d^d p}{(2\pi)^d} e^{ip \cdot (x-x')} e^{-p^2 T}. \tag{8.32}$$

In general, the heat-kernel is a matrix element of the operator $\exp(-T\Delta_g)$. In the random walk picture its trace per unit volume,

$$P_g(T) = V^{-1} \int d^d x \sqrt{g(x)} K_g(x, x; T) \equiv V^{-1} \operatorname{Tr} \exp(-T\Delta_g), \tag{8.33}$$

has the interpretation of an average return probability. Here $V \equiv \int d^d x \sqrt{g(x)}$ denotes the total volume. It is well known that P_g possesses an asymptotic early time expansion (for $T \to 0$) of the form $P_g(T) = (4\pi T)^{-d/2} \sum_{n=0}^{\infty} A_n T^n$, with A_n denoting the Seeley–DeWitt coefficients. From this expansion one can motivate the definition of the spectral dimension d_s as the T-independent logarithmic derivative

$$d_s \equiv -2\frac{d \ln P_g(T)}{d \ln T}\bigg|_{T=0}. \tag{8.34}$$

On smooth manifolds, where the early-time expansion of $P_g(T)$ is valid, the spectral dimension agrees with the topological dimension d of the manifold.

Given $P_g(T)$, it is natural to define an, in general T-dependent, generalization of the spectral dimension by

$$\mathscr{D}_s(T) \equiv -2\frac{d \ln P_g(T)}{d \ln T}. \tag{8.35}$$

According to (8.34), we recover the true spectral dimension of the space-time by considering the shortest possible random walks, i.e., by taking the limit $d_s = \lim_{T \to 0} \mathscr{D}_s(T)$. Note that in view of a possible comparison with other (discrete) approaches to quantum gravity the generalized, scale-dependent version (8.35) will play a central role later on.

(2) The walk dimension. Regular Brownian motion in flat space has the celebrated property that the random walker's average square displacement increases linearly with time: $\langle r^2 \rangle \propto T$. Indeed, performing the integral (8.32) we obtain the familiar probability density

$$K(x, x'; T) = \frac{1}{(4\pi T)^{d/2}} \exp\left(-\frac{|x - x'|^2}{4T}\right). \tag{8.36}$$

Using (8.36) yields the expectation value $\langle r^2 \rangle \equiv \langle x^2 \rangle = \int d^d x \, x^2 K(x, 0; T) \propto T$.

Many diffusion processes of physical interest (such as diffusion on fractals) are anomalous in the sense that this linear relationship is generalized to a power law $\langle r^2 \rangle \propto T^{2/d_w}$ with $d_w \neq 2$. The interpretation of the so-called walk dimension d_w is as follows. The trail left by the random walker is a random object, which is interesting in its own right. It has the properties of a fractal, even in the 'classical' case when the walk takes place on a regular manifold. The quantity d_w is precisely the fractal dimension of this trail. Diffusion processes are called regular if $d_w = 2$, and anomalous when $d_w \neq 2$.

(3) The Hausdorff dimension. Finally, we introduce the Hausdorff dimension d_H. Instead of working with its mathematically rigorous definition in terms of the Hausdorff measure and all possible covers of the metric space under consideration, the present, simplified definition may suffice for our present purposes. On a smooth set, the scaling law for the volume $V(r)$ of a d-dimensional ball of radius r takes the form

$$V(r) \propto r^{d_H}. \tag{8.37}$$

The Hausdorff dimension is then obtained in the limit of infinitely small radius,

$$d_H \equiv \lim_{r \to 0} \frac{\ln V(r)}{\ln r}. \tag{8.38}$$

Contrary to the spectral or walk dimension whose definitions are linked to dynamical diffusion processes on space-time, no such dynamics is associated with d_H.

8.7 Fractal Dimensions Within QEG

Upon introducing various concepts for fractal dimensions in the last section, we now proceed with their evaluation for the QEG effective space-times, following refs. [97] and [99]. Our discussion will mostly be based on the Einstein–Hilbert truncation. As we shall see this restriction is actually unnecessary in the asymptotic scaling regime, i.e., when the RG trajectory is close to the NGFP. In this case we can derive *exact* results for the spectral and walk dimension by exploiting the scale invariance of the theory at the fixed point.

8.7.1 Diffusion Processes on QEG Space-Times

Since in QEG one integrates over all metrics, the central idea is to replace $P_g(T)$ by its expectation value

$$P(T) \equiv \langle P_\gamma(T) \rangle \equiv \int \mathscr{D}\gamma \, \mathscr{D}C \, \mathscr{D}\bar{C} \, P_\gamma(T) \exp\left(-S_{\text{bare}}[\gamma, C, \bar{C}]\right). \tag{8.39}$$

Here $\gamma_{\mu\nu}$ denotes the microscopic metric and S_{bare} is the bare action related to the UV fixed point, with the gauge-fixing and the pieces containing the ghosts C and $\bar{C}$ included. For the untraced heat-kernel, we define likewise $K(x, x'; T) \equiv \langle K_\gamma(x, x'; T) \rangle$. These expectation values are most conveniently calculated from the effective average action Γ_k, which equips the d-dimensional smooth manifolds underlying the QEG effective space-times with a family of metric structures $\{\langle g_{\mu\nu} \rangle_k, 0 \le k < \infty\}$, one for each coarse-graining scale k [82, 97]. These metrics are solutions to the effective field equations implied by Γ_k.

We shall again approximate the latter by the Einstein–Hilbert truncation (8.16). The corresponding effective field equation is given by (8.26). Based on this equation, we can easily find the k-dependence of the corresponding solution $\langle g_{\mu\nu} \rangle_k$ by rewriting (8.26) as $[\bar{\lambda}_{k_0}/\bar{\lambda}_k] R^\mu{}_\nu(\langle g \rangle_k) = \frac{2}{2-d} \bar{\lambda}_{k_0} \delta^\mu{}_\nu$ for some fixed reference scale k_0, and exploiting that $R^\mu{}_\nu(cg) = c^{-1} R^\mu{}_\nu(g)$ for any constant $c > 0$. This shows that the metric and its inverse scale according to, for any d,

$$\langle g_{\mu\nu}(x) \rangle_k = \frac{\bar{\lambda}_{k_0}}{\bar{\lambda}_k} \langle g_{\mu\nu}(x) \rangle_{k_0}, \qquad \langle g^{\mu\nu}(x) \rangle_k = \frac{\bar{\lambda}_k}{\bar{\lambda}_{k_0}} \langle g^{\mu\nu}(x) \rangle_{k_0}. \tag{8.40}$$

Denoting the Laplace operators corresponding to the metrics $\langle g_{\mu\nu} \rangle_k$ and $\langle g_{\mu\nu} \rangle_{k_0}$ by $\Delta(k)$ and $\Delta(k_0)$, respectively, these relations imply

$$\Delta(k) = \frac{\bar{\lambda}_k}{\bar{\lambda}_{k_0}} \Delta(k_0). \tag{8.41}$$

At this stage, the following remark is in order. In the asymptotic scaling regime associated with the NGFP, the scale-dependence of the couplings is fixed by the fixed-point condition (8.21). This implies in particular

$$\langle g_{\mu\nu}(x)\rangle_k \propto k^{-2} \quad (k \to \infty). \tag{8.42}$$

This asymptotic relation is actually an *exact* consequence of asymptotic safety, which solely relies on the scale-independence of the theory at the fixed point.

We can evaluate the expectation value (8.39) by exploiting the effective field theory properties of the effective average action. Since Γ_k defines an effective field theory at the scale k we know that $\langle \mathcal{O}(\gamma_{\mu\nu})\rangle \approx \mathcal{O}(\langle g_{\mu\nu}\rangle_k)$ provided the observable $\mathcal{O}$ involves only momentum scales of the order of k. We apply this rule to the RHS of the diffusion equation, $\mathcal{O} = -\Delta_\gamma K_\gamma(x, x'; T)$. The subtle issue here is the correct identification of k. If the diffusion process involves (approximately) only a small interval of scales near k over which $\bar{\lambda}_k$ does not change much, the corresponding heat equation contains the operator $\Delta(k)$ for this specific, fixed value of k: $\partial_T K(x, x'; T) = -\Delta(k)K(x, x'; T)$. Denoting the eigenvalues of $\Delta(k_0)$ by $\mathcal{E}_n$ and the corresponding eigenfunctions by ϕ_n, this equation is solved by

$$K(x, x'; T) = \sum_n \phi_n(x)\phi_n(x') \exp\left[-F(k^2)\mathcal{E}_n T\right]. \tag{8.43}$$

Here we introduced the convenient notation $F(k^2) \equiv \bar{\lambda}_k/\bar{\lambda}_{k_0}$. Knowing the propagation kernel, we can time-evolve any initial probability distribution $p(x; 0)$ according to

$$p(x; T) = \int d^d x' \sqrt{g_0(x')} K(x, x'; T) p(x'; 0), \tag{8.44}$$

with g_0 the determinant of $\langle g_{\mu\nu}\rangle_{k_0}$. If the initial distribution has an eigenfunction expansion of the form $p(x; 0) = \sum_n C_n\phi_n(x)$, we obtain

$$p(x; T) = \sum_n C_n\phi_n(x) \exp\left[-F(k^2)\mathcal{E}_n T\right]. \tag{8.45}$$

If the C_n's are significantly different from zero only for a single eigenvalue $\mathcal{E}_N$, we are dealing with a single-scale problem and would identify $k^2 = \mathcal{E}_N$ as the relevant scale at which the running couplings are to be evaluated. In general the C_n's are different from zero over a wide range of eigenvalues. In this case we face a multiscale problem where different modes ϕ_n probe the space-time on different length scales. If $\Delta(k_0)$ corresponds to flat space, say, the eigenfunctions $\phi_n = \phi_p$ are plane waves with momentum p^μ, and they resolve structures on a length scale ℓ of order $1/|p|$. Hence, in terms of the eigenvalue $\mathcal{E}_n \equiv \mathcal{E}_p = p^2$ the resolution is $\ell \approx 1/\sqrt{\mathcal{E}_n}$. This suggests that when the manifold is probed by a mode with eigenvalue $\mathcal{E}_n$ it 'sees' the metric $\langle g_{\mu\nu}\rangle_k$ for the scale $k = \sqrt{\mathcal{E}_n}$. Actually, the identification $k = \sqrt{\mathcal{E}_n}$ is correct also for curved space since, in the construction of Γ_k, the parameter k is introduced precisely as a cutoff in the spectrum of the covariant Laplacian.

As a consequence, under the spectral sum of (8.45), we must use the scale $k^2 = \mathscr{E}_n$ which depends explicitly on the resolving power of the corresponding mode. Likewise, in Eq. (8.43), $F(k^2)$ is to be interpreted as $F(\mathscr{E}_n)$:

$$K(x, x'; T) = \sum_n \phi_n(x)\phi_n(x') \exp\left[-F(\mathscr{E}_n)\mathscr{E}_n T\right]$$

$$= \sum_n \phi_n(x) \exp\left\{-F[\Delta(k_0)]\Delta(k_0)T\right\}\phi_n(x'). \tag{8.46}$$

As in [97], we choose k_0 as a macroscopic scale in the classical regime, and we assume that at k_0 the cosmological constant is small, so that $\langle g_{\mu\nu}\rangle_{k_0}$ can be approximated by the flat metric on $\mathbb{R}^d$. The eigenfunctions of $\Delta(k_0)$ are plane waves then and Eq. (8.46) becomes

$$K(x, x'; T) = \int \frac{d^d p}{(2\pi)^d} e^{ip\cdot(x-x')} e^{-p^2 F(p^2)T}, \tag{8.47}$$

where the scalar products are performed with respect to the flat metric, $\langle g_{\mu\nu}\rangle_{k_0} = \delta_{\mu\nu}$. The kernel (8.47) satisfies the relation $K(x, x'; 0) = \delta^d(x - x')$ and, provided that $\lim_{p\to 0} p^2 F(p^2) = 0$, also $\int d^d x K(x, x'; T) = 1$.

Taking the trace of (8.47) within this 'flat-space approximation' yields [97]

$$P(T) = \int \frac{d^d p}{(2\pi)^d} e^{-p^2 F(p^2)T}. \tag{8.48}$$

Introducing $z = p^2$, the final result for the average return probability reads

$$P(T) = \frac{1}{(4\pi)^{d/2}\Gamma(d/2)} \int_0^\infty dz\, z^{d/2-1} \exp\left[-zF(z)T\right], \tag{8.49}$$

where $F(z) \equiv \bar{\lambda}(k^2 = z)/\bar{\lambda}_{k_0}$. In the classical case, $F(z) = 1$, the relation (8.49) reproduces the familiar result $P(T) = 1/(4\pi T)^{d/2}$, whence $\mathscr{D}_s(T) = d$ independently of T. We shall now discuss the spectral dimension for several other illustrative and important examples.

8.7.2 The Spectral Dimension in QEG

(A) Let us evaluate the average return probability (8.49) for a simplified RG trajectory where the scale dependence of the cosmological constant is given by a power law, with the same exponent δ for all values of k:

$$\bar{\lambda}_k \propto k^\delta \quad \Longrightarrow \quad F(z) \propto z^{\delta/2}. \tag{8.50}$$

By rescaling the integration variable in (8.49) we see that in this case

$$P(T) = \frac{\text{const}}{T^{d/(2+\delta)}}. \tag{8.51}$$

Hence (8.35) yields the important result

$$\boxed{\mathscr{D}_s(T) = \frac{2d}{2+\delta}.}$$

(8.52)

It happens to be T-independent, so that for $T \to 0$ trivially $d_s = 2d/(2+\delta)$.

(B) Next, let us be slightly more general and assume that the power law (8.50) is valid only for squared momenta in a certain interval, $p^2 \in [z_1, z_2]$, but $\bar{\lambda}_k$ remains unspecified otherwise. In this case we can obtain only partial information about $P(T)$, namely for T in the interval $[z_2^{-1}, z_1^{-1}]$. The reason is that for $T \in [z_2^{-1}, z_1^{-1}]$ the integral in (8.49) is dominated by momenta for which approximately $1/p^2 \approx T$, i.e., $z \in [z_1, z_2]$. This leads us again to the formula (8.52), which now, however, is valid only for a restricted range of diffusion times T; in particular the spectral dimension of interest may not be given by extrapolating (8.52) to $T \to 0$.

(C) Let us consider an arbitrary asymptotically safe RG trajectory so that its behavior for $k \to \infty$ is controlled by the NGFP. In this case the running of the cosmological constant for $k \gtrsim M$, with M a characteristic mass scale of the order of the Planck mass, is given by a quadratic scale-dependence $\bar{\lambda}_k = \lambda_* k^2$, independently of d. This corresponds to a power law with $\delta = 2$, which entails in the *NGFP regime*, i.e., for $T \lesssim 1/M^2$,

$$\mathscr{D}_s(T) = \frac{d}{2} \quad \text{(NGFP regime)}.$$

(8.53)

This dimension, again, is locally T-independent. It coincides with the $T \to 0$ limit:

$$d_s = \frac{d}{2}.$$

(8.54)

This is the result first derived in ref. [97]. As it was explained there, it is actually an exact consequence of asymptotic safety which relies solely on the existence of the NGFP and does not depend on the Einstein–Hilbert truncation.

(D) Returning to the Einstein–Hilbert truncation, let us consider the piece of the Type IIIa RG trajectory depicted in Fig. 8.4 which lies inside the linear regime of the GFP. Newton's constant is approximately k-independent there and the cosmological constant evolves according to (8.22). When k is not too small, so that $\bar{\lambda}_0$ can be neglected relative to $\nu \bar{G} k^d$, we are in what we shall call the 'k^d regime'; it is characterized by a pure power law $\bar{\lambda}_k \approx k^\delta$ with $\delta = d$. The physics behind this scale dependence is simple and well-known: It represents the vacuum energy density obtained by summing up the zero-point energies of all field modes integrated out. For T in the range of scales pertaining to the k^d regime we find

$$\mathscr{D}_s(T) = \frac{2d}{2+d} \quad (k^d \text{ regime}).$$

(8.55)

8.7.3 The Walk Dimension in QEG

In order to determine the walk dimension for the diffusion on the effective QEG space-times, we return to Eq. (8.47) for the untraced heat-kernel. We restrict ourselves to a regime with a power-law running of $\bar{\lambda}_k$, whence $F(p^2) = (Lp)^\delta$ with some constant length-scale L.

Introducing $q_\mu \equiv p_\mu T^{1/(2+\delta)}$ and $\xi_\mu \equiv (x_\mu - x'_\mu)/T^{1/(2+\delta)}$, we can rewrite (8.47) in the form

$$K(x, x'; T) = \frac{1}{T^{d/(2+\delta)}} \Phi\left(\frac{|x - x'|}{T^{1/(2+\delta)}}\right) \tag{8.56}$$

with the function

$$\Phi(|\xi|) \equiv \int \frac{d^d q}{(2\pi)^d} e^{iq \cdot \xi} e^{-L^\delta q^{2+\delta}}. \tag{8.57}$$

For $\delta = 0$, this obviously reproduces (8.36). From the argument of Φ in (8.56) we infer that $r = |x - x'|$ scales as $T^{1/(2+\delta)}$ so that the walk dimension can be read off as

$$\boxed{\mathscr{D}_w(T) = 2 + \delta.} \tag{8.58}$$

In analogy with the spectral dimension, we use the notation $\mathscr{D}_w(T)$ rather than d_w to indicate that it might refer to an approximate scaling law which is valid for a finite range of scales only.

For $\delta = 0, 2$, and d we find, in particular, for any topological dimension d,

$$\mathscr{D}_w = \begin{cases} 2 & \text{classical regime,} \\ 4 & \text{NGFP regime,} \\ 2 + d & k^d \text{ regime.} \end{cases} \tag{8.59}$$

Regimes with all three walk dimensions of (8.59) can be realized along a single RG trajectory. Again, the result for the NGFP regime, $\mathscr{D}_w = 4$, is exact in the sense that it does not rely on the Einstein–Hilbert truncation.

8.7.4 The Hausdorff Dimension in QEG

The smooth manifold underlying QEG has *per se* no fractal properties whatsoever. In particular, the volume of a d-ball $\mathscr{B}^d$ covering a patch of the smooth manifold of QEG space-time scales as $V(\mathscr{B}^d) = \int_{\mathscr{B}^d} d^d x \sqrt{g_k} \propto (r_k)^d$. Thus, by comparing to Eq. (8.37), we read off that the Hausdorff dimension is strictly equal to the topological one:

$$\boxed{d_H = d.} \tag{8.60}$$

8.7.5 Relations Between Dimensions

(1) The Alexander-Orbach relation. For standard fractals the quantities d_s, d_w, and d_H are not independent but are related by [143]

$$\frac{d_s}{2} = \frac{d_H}{d_w}. \tag{8.61}$$

By combining Eqs. (8.52), (8.58), and (8.60) we see that the same relation holds true for the effective QEG space-times, at least within the Einstein–Hilbert approximation and when the underlying RG trajectory is in a regime with power-law scaling of $\bar{\lambda}_k$. For every value of the exponent δ we have

$$\frac{\mathscr{D}_s(T)}{2} = \frac{d_H}{\mathscr{D}_w(T)}. \tag{8.62}$$

(2) (Non-) Recurrence. The results $d_H = d$, $\mathscr{D}_w = 2 + \delta$ imply that, as soon as $\delta > d - 2$, we have $\mathscr{D}_w > d_H$ and the random walk is *recurrent* then [142]. Classically ($\delta = 0$) this condition is met only in low dimensions $d < 2$, but in the case of the QEG space-times it is always satisfied in the k^d regime ($\delta = d$), for example. So also from this perspective the QEG space-times, due to the specific quantum gravitational dynamics to which they owe their existence, appear to have a dimensionality smaller than their topological one.

(3) Four dimensions are special. It is intriguing that, in the NGFP regime, $\mathscr{D}_w = 4$ independently of d. Hence the walk is recurrent ($\mathscr{D}_w > d_H$) for $d < 4$, non-recurrent for $d > 4$, and the marginal case $\mathscr{D}_w = d_H$ is realized if and only if $d = 4$, making $d = 4$ a distinguished value.

Notably, there is another feature of the QEG space-times which singles out $d = 4$: It is the only dimensionality for which $\mathscr{D}_s$(NGFP regime) $= d/2$ coincides with the effective dimension $d_{\mathrm{eff}} = d + \eta_* = 2$ obtained from the scale-dependent graviton propagator (see Sect. 8.5.)

8.8 The RG Running of $\mathscr{D}_s$ and $\mathscr{D}_w$

Let us consider an arbitrary RG trajectory $k \mapsto (g_k, \lambda_k)$, where $g_k \equiv G_k k^{d-2}$ and $\lambda_k \equiv \bar{\lambda}_k k^{-2}$ are the dimensionless Newton constant and cosmological constant, respectively. Along such a RG trajectory there might be isolated intervals of k-values where the cosmological constant evolves according to a power law, $\bar{\lambda}_k \propto k^\delta$, for some constant exponents δ which are not necessarily the same on different such intervals. If the intervals are sufficiently long, it is meaningful to ascribe a spectral and walk dimension to them since $\delta = $ const implies k-independent values $\mathscr{D}_s = 2d/(2 + \delta)$ and $\mathscr{D}_w = 2 + \delta$.

In between the intervals of approximately constant $\mathscr{D}_s$ and $\mathscr{D}_w$, where the k-dependence of $\bar{\lambda}_k$ is not a power law, the notion of a spectral or walk dimension might not be meaningful. The concept of a *scale-dependent* dimension $\mathscr{D}_s$ or $\mathscr{D}_w$ is to some extent arbitrary with respect to the way it interpolates between the 'plateaus' on which $\delta = \text{const}$ for some extended period of RG time. While RG methods allow the computation of the $\mathscr{D}_s$ and $\mathscr{D}_w$ values on the various plateaus, it is a matter of convention how to combine them into continuous functions $k \mapsto \mathscr{D}_s(k)$, $\mathscr{D}_w(k)$ which interpolate between the respective values.

(1) The exponent δ as a function on theory space. Next we describe a special proposal for a k-dependent $\mathscr{D}_s(k)$ and $\mathscr{D}_w(k)$ which is motivated by technical simplicity and the general insights it allows. We retain Eqs. (8.52) and (8.58), but promote $\delta \to \delta(k)$ to a k-dependent quantity

$$\delta(k) \equiv k\partial_k \ln(\bar{\lambda}_k). \tag{8.63}$$

When $\bar{\lambda}_k$ satisfies a power law, $\bar{\lambda}_k \propto k^\delta$, this relation reduces to the case of constant δ. If not, δ has its own scale dependence, but no direct physical interpretation should be attributed to it. The particular definition (8.63) has the special property that it actually can be evaluated without first solving for the RG trajectory. The function $\delta(k)$ can be seen as arising from a certain scalar function on theory space, $\delta = \delta(g, \lambda)$, whose k-dependence results from inserting an RG trajectory: $\delta(k) \equiv \delta(g_k, \lambda_k)$. In fact, (8.63) implies $\delta(k) = k\partial_k \ln(k^2\lambda_k) = 2 + \lambda_k^{-1}k\partial_k\lambda_k$ so that $\delta(k) = 2 + \lambda_k^{-1}\beta_\lambda(g_k, \lambda_k)$ upon using the RG-equation $k\partial_k\lambda_k = \beta_\lambda(g, \lambda)$. Thus when we consider δ as a function on theory space, coordinatized by g and λ, it reads

$$\delta(g, \lambda) = 2 + \frac{1}{\lambda}\beta_\lambda(g, \lambda). \tag{8.64}$$

Substituting this relation into (8.52) and (8.58), the spectral and the walk dimensions become functions on the g–λ-plane,

$$\mathscr{D}_s(g, \lambda) = \frac{2d}{4 + \lambda^{-1}\beta_\lambda(g, \lambda)}, \tag{8.65}$$

and

$$\mathscr{D}_w(g, \lambda) = 4 + \lambda^{-1}\beta_\lambda(g, \lambda). \tag{8.66}$$

As we discussed already, the scaling regime of a NGFP has the exponent $\delta = 2$. From Eq. (8.64) we learn that this value is realized at all points (g, λ) where $\beta_\lambda = 0$. The second condition for the NGFP, $\beta_g = 0$, is not required here, so that we have $\delta = 2$ along the entire line in theory space:

$$\mathscr{B} = \{(g, \lambda) \mid \beta_\lambda(g, \lambda) = 0\}. \tag{8.67}$$

For $d = 4$ the curve $\mathscr{B}$ is shown as the dashed line in Fig. 8.4. Both the GFP $(g, \lambda) = (0, 0)$ and the NGFP, $(g, \lambda) = (g^*, \lambda^*)$, are located on this curve. Furthermore, the turning points T of all Type IIIa trajectories are also situated on $\mathscr{B}$, and the same holds for all the higher-order turning points which occur when the trajectory spirals around the NGFP. The line $\mathscr{B}$ divides the (g, λ)-plane in three domains:

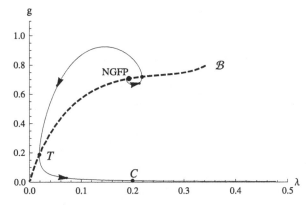

Fig. 8.4 The (g, λ) theory space with the line of turning points, $\mathscr{B}$, and a typical trajectory of Type IIIa. The *arrows* point in the direction of decreasing k. The *big black dot* indicates the NGFP while the *smaller dots* represent points at which the RG trajectory switches from increasing to decreasing λ or vice versa. The point T is the lowest turning point, and C is a typical point within the classical regime. For $\lambda \gtrsim 0.4$, the RG flow leaves the classical regime and is no longer reliably captured by the Einstein–Hilbert truncation

(i) Above $\mathscr{B}$: $\beta_\lambda > 0$, $\delta > 2 \Rightarrow \mathscr{D}_s < d/2$, $\mathscr{D}_w > 4$. (ii) Below $\mathscr{B}$: $\beta_\lambda < 0$, $\delta < 2 \Rightarrow \mathscr{D}_s > d/2$, $\mathscr{D}_w < 4$. (iii) On $\mathscr{B}$: $\beta_\lambda = 0$, $\delta = 2 \Rightarrow \mathscr{D}_s = d/2$, $\mathscr{D}_w = 4$. This observation leads us to an important conclusion: The values $\delta = 2 \Longleftrightarrow \mathscr{D}_s = d/2$, $\mathscr{D}_w = 4$ which (without involving any truncation) are found in the NGFP regime, actually also apply to all points $(g, \lambda) \in \mathscr{B}$, provided the Einstein–Hilbert truncation is reliable and no matter is included.

(2) Running dimensions along a RG trajectory. We proceed by investigating how the spectral and walk dimension of the effective QEG space-times changes along a given RG trajectory. As discussed above, our interest is in scaling regimes where $\mathscr{D}_s$ and $\mathscr{D}_w$ remain (approximately) constant for a long interval of k-values. For the remainder of this section, we will restrict ourselves to the Einstein–Hilbert truncation in $d = 4$.

We start by numerically solving the coupled differential equations (8.18) with the β-functions from [11] for a series of initial conditions keeping $\lambda_{\text{init}} = \lambda(k_0) = 0.2$ fixed and successively lowering $g_{\text{init}} = g(k_0)$. The result is a family of RG trajectories where the classical regime becomes more and more pronounced. Subsequently, these solutions are substituted into (8.65) and (8.66), which give $\mathscr{D}_s(t; g_{\text{init}}, \lambda_{\text{init}})$ and $\mathscr{D}_w(t; g_{\text{init}}, \lambda_{\text{init}})$ in dependence of the RG-time $t \equiv \ln(k)$ and the RG trajectory. One can verify explicitly that substituting the RG trajectory into the return probability (8.49) and computing the spectral dimension from (8.34) by carrying out the resulting integrals numerically gives rise to the same picture.

Figure 8.5 shows the resulting spectral dimension and the localization of the plateau-regimes on the RG trajectory. In the left diagram, g_{init} decreases by one order of magnitude for each shown trajectory, starting with the highest value to the very left. As a central result, Fig. 8.5 establishes that the RG flow gives rise to *three* plateaus where $\mathscr{D}_s(t)$ and $\mathscr{D}_w(t)$ are approximately constant:

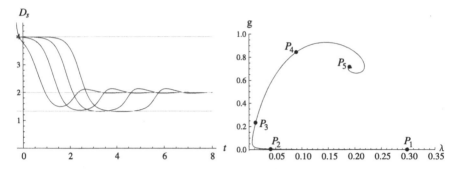

Fig. 8.5 The $t \equiv \ln(k)$-dependent spectral dimension along illustrative solutions of the RG-equations (8.18) in $d = 4$. The trajectories develop three plateaus: the classical plateau with $\mathscr{D}_s = 4$, $\mathscr{D}_w = 2$, the semi-classical plateau where $\mathscr{D}_s = 4/3$, $\mathscr{D}_w = 6$ and the NGFP plateau with $\mathscr{D}_s = 2$, $\mathscr{D}_w = 4$. (Recall that $\mathscr{D}_w = 2d/\mathscr{D}_s = 8/\mathscr{D}_s$.) The plateau values are indicated by the horizontal lines. The second figure shows the location of these plateaus on the RG trajectory: the classical, k^4, and NGFP regime appear between the points P_1 and P_2, P_3 and P_4, and above P_5, respectively

(i) For small values k, below $t \simeq 1.8$, say, one finds a *classical plateau* where $\mathscr{D}_s = 4$, $\mathscr{D}_w = 2$ for a long range of k-values. Here $\delta = 0$, indicating that the cosmological constant is indeed constant.

(ii) Following the RG flow towards the UV (larger values of t) one next encounters the *semi-classical plateau* where $\mathscr{D}_s = 4/3$, $\mathscr{D}_w = 6$. In this case $\delta(k) = 4$ so that $\bar{\lambda}_k \propto k^4$ on the corresponding part of the RG trajectory.

(iii) Finally, the *NGFP plateau* is characterized by $\mathscr{D}_s = 2$, $\mathscr{D}_w = 4$, which results from the scale-dependence of the cosmological constant at the NGFP $\bar{\lambda}_k \propto k^2 \Longleftrightarrow \delta = 2$.

The plateaus become more and more extended the closer the trajectory's turning point T gets to the GFP, i.e., the smaller the IR value of the cosmological constant.

8.9 Matching the Spectral Dimensions of QEG and CDT

The key advantage of the spectral dimension $\mathscr{D}_s(T)$ is that it may be defined and computed within various *a priori* unrelated approaches to quantum gravity. In particular, it is easily accessible in Monte Carlo simulations of the causal dynamical triangulations (CDT) approach in $d = 4$ [90–92] and $d = 3$ [94] as well as in Euclidean dynamical triangulations (EDT) [98]. This feature allows a direct comparison between $\mathscr{D}_s^{\text{CDT}}(T)$ and $\mathscr{D}_s^{\text{EDT}}(T)$ obtained within the discrete approaches and $\mathscr{D}_s^{\text{QEG}}(T)$ capturing the fractal properties of the QEG effective space-times. In [99] we carried out this analysis for $d = 3$, using the Monte Carlo data obtained in [94] according to the following scheme:

(i) First, we numerically construct a RG trajectory $g_k(g_0, \lambda_0)$, $\lambda_k(g_0, \lambda_0)$ depending on the initial conditions g_0, λ_0, by solving the flow equations (8.18).

Table 8.1 Initial conditions g_0^{fit}, λ_0^{fit} for the RG trajectory providing the best fit to the Monte Carlo data [94]. The fit-quality $(\Delta \mathcal{D}_s)^2$, given by the sum of the squared residues, improves systematically when increasing the number of simplices in the triangulation

	g_0^{fit}	λ_0^{fit}	$(\Delta \mathcal{D}_s)^2$
70k	0.7×10^{-5}	7.5×10^{-5}	0.680
100k	8.8×10^{-5}	39.5×10^{-5}	0.318
200k	13×10^{-5}	61×10^{-5}	0.257

(ii) We evaluate the resulting spectral dimension $\mathcal{D}_s^{\text{QEG}}(T; g_0, \lambda_0)$ of the corresponding effective QEG space-time. This is done by first finding the return probability $P(T; g_0, \lambda_0)$, Eq. (8.49), for the RG trajectory under consideration and then substituting the resulting expression into (8.35). The spectral dimension constructed in this way depends not only on the length of the random walk but also on the initial conditions of the RG trajectory.

(iii) We determine the RG trajectory underlying the CDT-simulations by fitting the parameters g_0, λ_0 to the Monte Carlo data. The corresponding best-fit values are obtained via an ordinary least-squares fit, minimizing the squared Euclidean distance

$$(\Delta \mathcal{D}_s)^2 \equiv \sum_{T=20}^{500} \left[\mathcal{D}_s^{\text{QEG}}(T; g_0^{\text{fit}}, \lambda_0^{\text{fit}}) - \mathcal{D}_s^{\text{CDT}}(T) \right]^2, \tag{8.68}$$

between the (continuous) function $\mathcal{D}_s^{\text{QEG}}(T; g_0, \lambda_0)$ and the points $\mathcal{D}_s^{\text{CDT}}(T)$. We thereby restrict ourselves to the random walks with discrete, integer length $20 \leq T \leq 500$, which constitute the 'reliable' part of the data.

The resulting best-fit values $g_0^{\text{fit}}, \lambda_0^{\text{fit}}$ for triangulations with $N = 70{,}000$, $N = 100{,}000$, and $N = 200{,}000$ simplices are collected in Table 8.1. Notably, the sum over the squared residuals in the third column of the table improves systematically with an increasing number of simplices. By integrating the flow equation for $g(k), \lambda(k)$ for the best-fit initial conditions one furthermore observes that the points $g_0^{\text{fit}}, \lambda_0^{\text{fit}}$ are actually located on *different* RG trajectories. Increasing the size of the simulation N leads to a mild but systematic increase of the distance between the turning point T and the GFP of the corresponding best-fit trajectories.

Figure 8.6 then shows the direct comparison between the spectral dimensions obtained by the simulations (continuous curves) and the best-fit QEG trajectories (dashed curves) for 70k, 100k and 200k in the upper left, upper right and lower left panel, respectively. This data is complemented by the relative error

$$\varepsilon \equiv -\frac{\mathcal{D}_s^{\text{QEG}}(T; g_0^{\text{fit}}, \lambda_0^{\text{fit}}) - \mathcal{D}_s^{\text{CDT}}(T)}{\mathcal{D}_s^{\text{QEG}}(T; g_0^{\text{fit}}, \lambda_0^{\text{fit}})} \tag{8.69}$$

for the three fits in the lower right panel. The 70k data still shows a systematic deviation from the classical value $\mathcal{D}_s(T) = 3$ for long random walks, which is not present

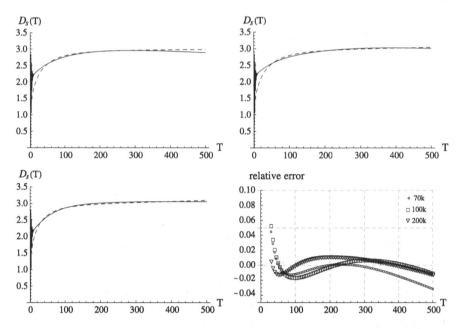

Fig. 8.6 Comparison between the spectral dimension measured in 3-dimensional CDT space–times build from 70k (*upper left*), 100k (*upper right*), and 200k simplices (*lower left*) obtained in [94] (*continuous curves*) and the best fit values for $\mathscr{D}_s^{\mathrm{QEG}}(T; g_0^{\mathrm{fit}}, \lambda_0^{\mathrm{fit}})$ (*dashed curves*). The relative errors for the fits to the CDT-datasets with $N = 70,000$ (*circles*), $N = 100,000$ (*squares*) and $N = 200,000$ (*triangles*) simplices are shown in the *lower right*. The residuals growth for very small and very large durations T of the random walk, consistent with discreteness effects at small distances and the compactness of the simulation for large values of T, respectively. The quality of the fit improves systematically for triangulations containing more simplices. For the $N = 200k$ data the relative error is $\approx 1\%$

in the QEG results. This mismatch decreases systematically for larger triangulations where the classical regime becomes more and more pronounced. Nevertheless and most remarkably we find that for the 200k-triangulation $\varepsilon \lesssim 1\%$, throughout.

We conclude this section by extending $\mathscr{D}_s^{\mathrm{QEG}}(T; g_0^{\mathrm{fit}}, \lambda_0^{\mathrm{fit}})$ obtained from the 200k data to the region of very short random walks $T < 20$. The result is depicted in Fig. 8.7 which displays $\mathscr{D}_s^{\mathrm{CDT}}(T)$ (continuous curve) and $\mathscr{D}_s^{\mathrm{QEG}}(T; g_0^{\mathrm{fit}}, \lambda_0^{\mathrm{fit}})$ (dashed curve) as a function of $\log(T)$. Similarly to the four-dimensional case discussed in Fig. 8.5, the function $\mathscr{D}_s^{\mathrm{QEG}}(T; g_0^{\mathrm{fit}}, \lambda_0^{\mathrm{fit}})$ obtained for $d = 3$ develops three plateaus where the spectral dimension is approximately constant over a long T-interval. For successively decreasing duration of the random walks, these plateaus correspond to the classical regime $\mathscr{D}_s^{\mathrm{QEG}}(T) = 3$, the semi-classical regime where $\mathscr{D}_s^{\mathrm{QEG}}(T) \approx 1$ and the NGFP regime where $\mathscr{D}_s^{\mathrm{QEG}}(T) = 3/2$. The figure illustrates that $\mathscr{D}_s^{\mathrm{CDT}}(T)$ probes the classical regime and part of the first crossover towards the semi-classical regime only. This is in perfect agreement with the assertion [94] that the present simulations do not yet probe structures below the Planck scale. This assessment resolves the apparent contradiction between the extrapolation re-

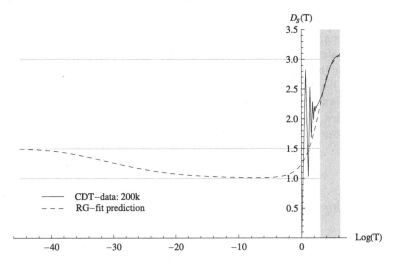

Fig. 8.7 Comparison between the spectral dimensions obtained from the dynamical triangulation with 200k simplices (*continuous curve*) and the corresponding $\mathscr{D}_s^{\mathrm{QEG}}(T; g_0^{\mathrm{fit}}, \lambda_0^{\mathrm{fit}})$ predicted by QEG (*dashed curve*). In the latter case, the scaling regime corresponding to the NGFP is reached for $\log(T) < -40$, which is well below the distance scales probed by the Monte Carlo simulation

sult $\lim_{T \to 0} \mathscr{D}_s^{\mathrm{CDT}}(T) \approx 2$ and the QEG prediction $\lim_{T \to 0} \mathscr{D}_s^{\mathrm{QEG}}(T) = 3/2$. Performing the extrapolation of $\lim_{T \to 0} \mathscr{D}_s^{\mathrm{CDT}}(T)$ based on the leading corrections to the classical regime does not reliably identify the signature of a non-Gaussian fixed point in $\mathscr{D}_s(T)$.

A similar conclusion also holds true in four dimensions. Comparing the profiles of $\mathscr{D}_s^{\mathrm{QEG}}(T)$ shown in Fig. 8.5 with the fitting functions used in the CDT [90–92] or EDT [98] simulations shows that all the Monte Carlo data points obtained are positioned on the *infrared* side of the turning point of the RG trajectories underlying the QEG effective space-times. They neither probe the semi-classical plateau nor the scaling regime of the NGFP. Depending on where the data are cut off, one obtains different tangents to the first crossover, which lead to widely different extrapolations for the value $d_s = \mathscr{D}_s(T)|_{T=0}$. We believe that this is actually at the heart of the apparent mismatch in the spectral dimension for infinitesimal random walks reported from the CDT and EDT computations.

8.10 Asymptotic Safety in Cosmology

At this point we switch to another field where QEG effects might be relevant, the cosmology of the early Universe. As we discussed at the end of Sect. 8.4, the Type IIIa trajectories displayed in Fig. 8.2 possess all the qualitative properties one would expect from the RG trajectory describing gravitational phenomena in the real Universe. They can have a long classical regime and a small, positive cosmological constant in the infrared. Remarkably, along the RG trajectory realized by Nature [79–

81, 133], the dimensionful running cosmological constant $\bar{\lambda}(k)$ changes by about 120 orders of magnitude between k-values of the order of the Planck mass and macroscopic scales, while the dimensionful Newton constant $G(k)$ has no strong k-dependence in this regime. For $k > m_{\text{Pl}}$, the scale dependence of $G(k)$ and $\bar{\lambda}(k)$ is governed by the NGFP, implying that $\bar{\lambda}(k)$ diverges and $G(k)$ approaches zero, see Eq. (8.21). An immediate question is whether there is any experimental or observational evidence that would hint at this enormous scale dependence of the gravitational parameters. Clearly, the natural place to search for such phenomena is cosmology.

8.10.1 RG Improved Einstein Equations

The computational setting for investigating the signatures arising from the scale-dependent couplings are the RG improved Einstein equations: By means of a suitable cutoff identification $k = k(t)$ we turn the scale dependence of $G(k)$ and $\bar{\lambda}(k)$ into a time dependence, and then substitute the resulting $G(t) \equiv G[k(t)]$ and $\bar{\lambda}(t) \equiv \bar{\lambda}[k(t)]$ into the Einstein equations $G_{\mu\nu} = -\bar{\lambda}(t)g_{\mu\nu} + 8\pi G(t)T_{\mu\nu}$. We specialize $g_{\mu\nu}$ to describe a spatially flat ($K = 0$) Robertson–Walker (FRW) metric with scale factor $a(t)$, and we take $T_\mu{}^\nu = \text{diag}[-\rho, p, p, p]$ to be the energy-momentum tensor of an ideal fluid with equation of state $p = w\rho$, where $w > -1$ is constant. Then the improved Einstein equation boils down to the modified Friedmann equation and a continuity equation:

$$H^2 = \frac{8\pi}{3}G(t)\rho + \frac{1}{3}\bar{\lambda}(t), \qquad (8.70a)$$

$$\dot{\rho} + 3H(\rho + p) = -\frac{\dot{\bar{\lambda}} + 8\pi\rho\dot{G}}{8\pi G}. \qquad (8.70b)$$

The modified continuity equation (8.70b) is the integrability condition for the improved Einstein equation implied by Bianchi identity, $D^\mu[\bar{\lambda}(t)g_{\mu\nu} - 8\pi G(t)T_{\mu\nu}] = 0$. It describes the energy exchange between the matter and gravitational degrees of freedom (geometry). For later use let us note that upon defining the critical density $\rho_{\text{crit}}(t) \equiv 3H(t)^2/[8\pi G(t)]$, the relative density $\Omega_{\text{M}} \equiv \rho/\rho_{\text{crit}}$ and $\Omega_{\bar{\lambda}} = \rho_{\bar{\lambda}}/\rho_{\text{crit}}$ the modified Friedmann equation (8.70a) can be written as $\Omega_{\text{M}}(t) + \Omega_{\bar{\lambda}}(t) = 1$.

8.10.2 Solving the RG Improved Einstein Equations

The general strategy for solving Eqs. (8.70a), (8.70b) is as follows. First we obtain $G(k)$ and $\bar{\lambda}(k)$ by solving the flow equation in the Einstein–Hilbert truncation before constructing the cosmologies by numerically solving the RG improved evolution equations. We shall employ the cutoff identification $k(t) = \xi H(t)$, where ξ is a fixed positive constant of order unity. This is a natural choice since in a Robertson–Walker geometry the Hubble parameter measures the curvature of space-time; its inverse H^{-1} defines the size of the 'Einstein elevator'.

The very early part of the cosmology can be described analytically. For $k \to \infty$ the trajectory approaches the NGFP so that $G(k) = g^*/k^2$ and $\bar{\lambda}(k) = \lambda^* k^2$. In this case the differential equation can be solved analytically, with the result

$$
H(t) = \frac{\alpha}{t}, \qquad a(t) = A t^\alpha, \qquad \alpha = \left[\frac{1}{2}(3 + 3w)\left(1 - \Omega_{\bar{\lambda}}^*\right) \right]^{-1}, \qquad (8.71)
$$

and $\rho(t) = \widehat{\rho} t^{-4}$, $G(t) = \widehat{G} t^2$, $\bar{\lambda}(t) = \widehat{\bar{\lambda}}/t^2$. Here A, $\widehat{\rho}$, $\widehat{G}$, and $\widehat{\bar{\lambda}}$ are positive constants. They parametrically depend on the relative vacuum energy density in the fixed point regime, $\Omega_{\bar{\lambda}}^*$, which assumes values in the interval $(0, 1)$. If $\alpha > 1$ the deceleration parameter $q = \alpha^{-1} - 1$ is negative and the Universe is in a phase of *power-law inflation*. Furthermore, it has *no particle horizon* if $\alpha \geq 1$, but does have a horizon of radius $d_H = t/(1 - \alpha)$ if $\alpha < 1$. In the case of $w = 1/3$ this means that there is a horizon for $\Omega_{\bar{\lambda}}^* < 1/2$, but none if $\Omega_{\bar{\lambda}}^* \geq 1/2$.

8.10.3 Inflation in the Fixed-Point Regime

Next we discuss in more detail the cosmologies originating from the epoch of power-law inflation which is realized in the NGFP regime if $\Omega_{\bar{\lambda}}^* > 1/2$. Since the transition from the fixed point to the classical FRW regime is rather sharp, it will be sufficient to approximate the RG improved UV cosmologies by the following caricature: For $0 < t < t_{\mathrm{tr}}$, the scale factor behaves as $a(t) \propto t^\alpha$, $\alpha > 1$. Here $\alpha = (2 - 2\Omega_{\bar{\lambda}}^*)^{-1}$ since $w = 1/3$ will be assumed. Thereafter, for $t > t_{\mathrm{tr}}$, we have a classical, entirely matter-driven expansion $a(t) \propto t^{1/2}$. Clearly this is a very attractive scenario: *neither to trigger inflation nor to stop it one needs any ad hoc ingredients such as an inflaton field or a special potential.* It suffices to include the leading quantum effects in the gravity + matter system. Following [79–81], the RG improved cosmological evolution for the RG trajectory realized by Nature is characterized as follows:

(A) The transition time t_{tr} is dictated by the RG trajectory. It leaves the asymptotic scaling regime near $k \approx m_{\mathrm{Pl}}$. Hence $H(t_{\mathrm{tr}}) \approx m_{\mathrm{Pl}}$ and since $\xi = O(1)$ and $H(t) = \alpha/t$, we find the estimate

$$
t_{\mathrm{tr}} = \alpha t_{\mathrm{Pl}}. \qquad (8.72)
$$

Here, as always, the Planck mass, time, and length are defined in terms of the value of Newton's constant in the classical regime: $t_{\mathrm{Pl}} = \ell_{\mathrm{Pl}} = m_{\mathrm{Pl}}^{-1} = \bar{G}^{1/2} = G_{\mathrm{observed}}^{1/2}$. Let us now assume that $\Omega_{\bar{\lambda}}^*$ is very close to 1 so that α is large: $\alpha \gg 1$. Then (8.72) implies that the transition takes place at a cosmological time which is much later than the Planck time. At the transition the *Hubble parameter* is of order m_{Pl}, but

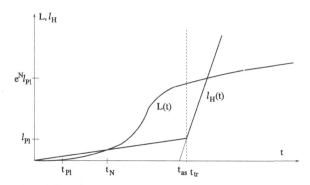

Fig. 8.8 The proper length L and the Hubble radius as a function of time. The NGFP and FRW cosmologies are valid for $t < t_{tr}$ and $t > t_{tr}$, respectively. The classical cosmology has an apparent initial singularity at t_{as} outside its domain of validity. Structures of size $e^N \ell_{Pl}$ at t_{tr} cross the Hubble radius at t_N, a time which can be larger than the Planck time

the *cosmological time* is in general not of the order of t_{Pl}. Stated differently, the 'Planck time' is *not* the time at which H and the related physical quantities assume Planckian values. The Planck time as defined above is well within the NGFP regime: $t_{Pl} = t_{tr}/\alpha \ll t_{tr}$.

In the NGFP regime $0 < t < t_{tr}$ the Hubble radius $\ell_H(t) \equiv 1/H(t)$, i.e., $\ell_H(t) = t/\alpha$, increases linearly with time but, for $\alpha \gg 1$, with a very small slope. At the transition, $t = t_{tr}$, the NGFP solution is to be matched continuously with a FRW cosmology (with vanishing cosmological constant). We may use the classical formula $a \propto \sqrt{t}$ for the scale factor, but we must shift the time axis on the classical side such that a, H, and then as a result of (8.70a), also ρ are continuous at t_{tr}. Therefore, $a(t) \propto (t - t_{as})^{1/2}$ and $H(t) = \frac{1}{2}(t - t_{as})^{-1}$ for $t > t_{tr}$. Equating this Hubble parameter at $t = t_{tr}$ to $H(t) = \alpha/t$, valid in the NGFP regime, we find that the shift t_{as} must be chosen as $t_{as} = (\alpha - \frac{1}{2})t_{Pl} = (1 - \frac{1}{2\alpha})t_{tr} < t_{tr}$. Here the subscript 'as' stands for 'apparent singularity'. This is to indicate that if one continues the classical cosmology to times $t < t_{tr}$, it has an initial singularity ('big bang') at $t = t_{as}$. Since, however, the FRW solution is not valid there, nothing special happens at t_{as}; the true initial singularity is located at $t = 0$ in the NGFP regime; see Fig. 8.8.

(B) We now consider some structure of comoving length Δx, a single wavelength of a density perturbation, for instance. The corresponding physical, i.e., proper length is $L(t) = a(t)\Delta x$ then. In the NGFP regime it has the time dependence $L(t) = (t/t_{tr})^\alpha L(t_{tr})$. The ratio of $L(t)$ and the Hubble radius evolves according to $L(t)/\ell_H(t) = (t/t_{tr})^{\alpha-1} L(t_{tr})/\ell_H(t_{tr})$. For $\alpha > 1$, i.e., $\Omega_\lambda^* > 1/2$, the proper length of any object grows faster than the Hubble radius. So objects which are of 'sub-Hubble' size at early times can cross the Hubble radius and become 'super-Hubble' at later times; see Fig. 8.8.

Let us focus on a structure which, at $t = t_{tr}$, is e^N times larger than the Hubble radius. Before the transition we have $L(t)/\ell_H(t) = e^N (t/t_{tr})^{\alpha-1}$. Assuming $e^N > 1$, there exists a time $t_N < t_{tr}$ at which $L(t_N) = \ell_H(t_N)$, so that the structure considered 'crosses' the Hubble radius at the time t_N. It is given by

$$t_N = t_{\text{tr}} \exp\left(-\frac{N}{\alpha - 1}\right). \tag{8.73}$$

What is remarkable about this result is that, even with rather moderate values of α, one can easily 'inflate' structures to a size which is by many e-folds larger than the Hubble radius *during a very short time interval at the end of the NGFP epoch*.

The largest structures in the present Universe, evolved backward in time by the classical equations to the point where $H = m_{\text{Pl}}$, have a size of about $e^{60} \ell_{\text{Pl}}$ there. We can use (8.73) with $N = 60$ to find the time t_{60} at which those structures crossed the Hubble radius. With $\alpha = 25$ the result is $t_{60} = 2.05 t_{\text{Pl}} = t_{\text{tr}}/12.2$. Remarkably, t_{60} is smaller than t_{tr} by one order of magnitude only. As a consequence, the physical conditions prevailing at the time of the crossing are not overly 'exotic' yet. The Hubble parameter, for instance, is only one order of magnitude larger than at the transition: $H(t_{60}) \approx 12 m_{\text{Pl}}$. The same is true for the temperature; one can show that $T(t_{60}) \approx 12 T(t_{\text{tr}})$ where $T(t_{\text{tr}})$ is of the order of m_{Pl}. Note that t_{60} is larger than t_{Pl}.

(C) QEG offers a natural mechanism for generating primordial fluctuations during the NGFP epoch. The idea is that the NGFP amounts to a kind of 'critical phenomenon' with characteristic fluctuations on all scales. They turn out to have a scale-free spectrum with a spectral index close to $n = 1$. For a detailed discussion of this mechanism the reader is referred to [15, 77–81]. Suffice it to say that the quantum mechanical generation of the primordial fluctuations makes essential use of the dimensionally reduced form of the graviton propagator; it happens on sub-Hubble distance scales. However, thanks to the inflationary NGFP era the modes relevant to cosmological structure formation were indeed smaller than the Hubble radius at a sufficiently early time, for $t < t_{60}$, say. (See the $L(t)$ curve in Fig. 8.8.)

8.10.4 Entropy and the Renormalization Group

In standard Friedmann–Robertson–Walker cosmology where the expansion is adiabatic, the entropy (within a comoving volume) is constant. It has always been somewhat puzzling therefore where the huge amount of entropy contained in the present Universe comes from. Presumably it is dominated by the cosmic microwave background radiation (CMBR) photons which contribute an amount of about 10^{88} to the entropy within the present Hubble sphere.

The observation that the value of the cosmological constant decreases during the expansion of the universe hints at another mechanism at work within the RG improved cosmologies: the dynamical creation of entropy. Following [79–81] we shall argue that in principle the entire entropy of the massless fields in the present Universe can be understood as arising from this effect. If energy can be exchanged freely between the cosmological constant and the matter degrees of freedom, the entropy observed today is obtained precisely if the initial entropy at the big bang vanishes. The assumption that the matter system must allow for an unhindered energy exchange with $\bar{\lambda}$ is essential; see refs. [77–81].

To make the argument as transparent as possible, let us first consider a Universe without matter, but with a positive $\bar{\lambda}$. Assuming maximal symmetry, this is nothing but de Sitter space, of course. In static coordinates its metric is given by $ds^2 = -[1 + 2\Phi_N(r)]dt^2 + [1 + 2\Phi_N(r)]^{-1}dr^2 + r^2(d\theta^2 + \sin^2\theta d\phi^2)$ with $\Phi_N(r) = -\frac{1}{6}\bar{\lambda}r^2$. In the weak field and slow motion limit $\Phi_N(r)$ has the interpretation of a Newtonian potential; for $\bar{\lambda} > 0$ it is an upside-down parabola. Point particles in this space-time 'roll down the hill' and are rapidly driven away from the origin $r = 0$ and from any other particle. Now assume that the magnitude of $|\bar{\lambda}|$ is slowly ('adiabatically') decreased. This will cause the potential $\Phi_N(r)$ to move upward as a whole, its slope decreases. So the change in $\bar{\lambda}$ increases the particle's potential energy. This is the simplest way of understanding that a *positive decreasing* cosmological constant has the effect of 'pumping' energy into the matter degrees of freedom.

We are thus led to suspect that, because of the decreasing cosmological constant, there is a continuous inflow of energy into the cosmological fluid contained in an expanding Universe. It will 'heat up' the fluid or, more exactly, lead to a slower decrease of the temperature than in standard cosmology. Furthermore, by elementary thermodynamics, it will *increase* the entropy of the fluid. If during the time dt an amount of heat $dQ > 0$ is transferred into a volume V at the temperature T the entropy changes by an amount $dS = dQ/T > 0$. To be as conservative (i.e., close to standard cosmology) as possible, we assume that this process is reversible. If not, dS is even larger.

In order to quantify this argument, we model the matter in the early Universe by a gas with n_b bosonic and n_f fermionic massless degrees of freedom, all at the same temperature. *In equilibrium* its energy density, pressure, and entropy density are given by the usual relations (here $n_{\text{eff}} = n_b + \frac{7}{8}n_f$)

$$\rho = 3p = \frac{\pi^2}{30}n_{\text{eff}}T^4, \tag{8.74a}$$

$$s = \frac{2\pi^2}{45}n_{\text{eff}}T^3, \tag{8.74b}$$

so that in terms of $U \equiv \rho V$ and $S \equiv sV$,

$$T\,dS = dU + p\,dV. \tag{8.74c}$$

In an out-of-equilibrium process of entropy generation the question arises how the various thermodynamical quantities are related then. To be as conservative as possible, we make the assumption that the irreversible inflow of energy destroys thermal equilibrium as little as possible in the sense that the equilibrium relations (8.74a)–(8.74c) continue to be (approximately) valid. Such minimally non-adiabatic processes were termed 'adiabatic' (with the quotation marks) in ref. [144, 145].

8.10.5 Primordial Entropy Generation

Let us return to the modified continuity equation (8.70b). After multiplication by a^3 it reads

$$\left[\dot{\rho} + 3H(\rho + p)\right]a^3 = \widetilde{\mathscr{P}}(t),\tag{8.75}$$

where we defined

$$\widetilde{\mathscr{P}} \equiv -\left(\frac{\dot{\bar{\lambda}} + 8\pi\rho\dot{G}}{8\pi G}\right)a^3.\tag{8.76}$$

Without assuming any particular equation of state, Eq. (8.75) can be rewritten as

$$\frac{d}{dt}\left(\rho a^3\right) + p\frac{d}{dt}\left(a^3\right) = \widetilde{\mathscr{P}}(t).\tag{8.77}$$

The interpretation of this equation is as follows. Let us consider a unit *coordinate*, i.e., comoving volume in the Robertson–Walker space-time. Its corresponding *proper* volume is $V = a^3$ and its energy contents is $U = \rho a^3$. The rate of change of these quantities is subject to (8.77):

$$\frac{dU}{dt} + p\frac{dV}{dt} = \widetilde{\mathscr{P}}(t).\tag{8.78}$$

In classical cosmology where $\widetilde{\mathscr{P}} \equiv 0$ this equation together with the standard thermodynamic relation $dU + pdV = TdS$ is used to conclude that the expansion of the Universe is adiabatic, i.e., the entropy inside a comoving volume does not change as the Universe expands, $dS/dt = 0$.

When $\bar{\lambda}$ and G are time dependent, $\widetilde{\mathscr{P}}$ is non-zero and we interpret (8.78) as describing the process of energy (or 'heat') exchange between the scalar fields $\bar{\lambda}$ and G and the ordinary matter. This interaction causes S to change,

$$T\frac{dS}{dt} = T\frac{d}{dt}\left(sa^3\right) = \widetilde{\mathscr{P}}(t),\tag{8.79}$$

where here and in the following we write $S \equiv sa^3$ for the entropy carried by the matter inside a unit comoving volume and s for the corresponding proper entropy density. The actual rate of change of the comoving entropy is

$$\frac{dS}{dt} = \frac{d}{dt}\left(sa^3\right) = \mathscr{P}(t),\tag{8.80}$$

where $\mathscr{P} \equiv \widetilde{\mathscr{P}}/T$. If T is known as a function of t we can integrate (8.79) to obtain $S = S(t)$. In the RG improved cosmologies the entropy production rate per comoving volume

$$\mathscr{P}(t) = -\left[\frac{\dot{\bar{\lambda}} + 8\pi\rho\dot{G}}{8\pi G}\right]\frac{a^3}{T}\tag{8.81}$$

is non-zero because the gravitational 'constants' $\bar{\lambda}$ and G have acquired a time dependence.

Clearly we can convert the heat exchanged, TdS, to an entropy change only if the dependence of the temperature T on the other thermodynamical quantities, in particular ρ and p is known. For this reason we shall now make the following assumption about the matter system and its (non-equilibrium!) dynamics:

The matter system is assumed to consist of n_{eff} species of effectively massless degrees of freedom which all have the same temperature T. The equation of state is $p = \rho/3$, i.e., $w = 1/3$, and ρ depends on T as

$$\rho(T) = \kappa^4 T^4, \qquad \kappa \equiv \left(\frac{\pi^2}{30} n_{\mathrm{eff}}\right)^{1/4}. \tag{8.82}$$

No assumption is made about the relation $s = s(T)$.

The first assumption, radiation dominance and equal temperature, is plausible since we shall find that there is no significant entropy production any more once $H(t)$ has dropped substantially below m_{Pl}. The second assumption, Eq. (8.82), amounts to the hypothesis formulated above, the approximate validity of the *equilibrium* relations among ρ, p, and T.

Note that while we used (8.74c) in relating $\mathscr{P}(t)$ to the entropy production and also postulated Eq. (8.74a), we do not assume the validity of the formula for the entropy density, Eq. (8.74b), *a priori*. We shall see that the latter is an automatic consequence of the cosmological equations. To make the picture as clear as possible we shall neglect in the following all ordinary dissipative processes in the cosmological fluid.

Using $p = \rho/3$ and (8.82) the entropy production rate can be seen to be a total time derivative, $\mathscr{P}(t) = d/dt[(4/3)\kappa a^3 \rho^{3/4}]$. Therefore we can immediately integrate (8.79) and obtain

$$S(t) = \frac{4}{3}\kappa a^3 \rho^{3/4} + S_c, \tag{8.83}$$

or, in terms of the proper entropy density, $s(t) = (4/3)\kappa \rho(t)^{3/4} + S_c/a(t)^3$. Here S_c is a constant of integration. In terms of T, using (8.82) again,

$$s(t) = \frac{2\pi^2}{45} n_{\mathrm{eff}} T(t)^3 + \frac{S_c}{a(t)^3}. \tag{8.84}$$

The final result (8.84) is very remarkable for at least two reasons. First, for $S_c = 0$, Eq. (8.84) has exactly the form (8.74b) which is valid for radiation in equilibrium. Note that we did not postulate this relationship, only the $\rho(T)$-law was assumed. The equilibrium formula $s \propto T^3$ was *derived* from the cosmological equations, i.e., the modified conservation law. This result makes the hypothesis 'non-adiabatic, but as little as possible' self-consistent. Second, if $\lim_{t\to 0} a(t)\rho(t)^{1/4} = 0$, which is actually the case for the most interesting class of cosmologies we shall find, then $S(t \to 0) = S_c$ by Eq. (8.83). As we mentioned in the introduction, the most plausible initial value of S is $S = 0$ which means a vanishing constant of integration

S_c here. But then, with $S_c = 0$, Eq. (8.83) tells us that *the entire entropy carried by the massless degrees of freedom today (CMBR photons) is due to the RG running*.

8.10.6 Entropy Production for RG Trajectory Realized by Nature

Substituting the NGFP solution (8.71) for $w = 1/3$ the entropy production rate (8.81) reads $\mathscr{P}(t) = 4\kappa(\alpha - 1)A^3\hat{\rho}^{3/4}t^{3\alpha-4}$. For the entropy per unit comoving volume we find, if $\alpha \neq 1$, $S(t) = S_c + (4/3)\kappa A^3\hat{\rho}^{3/4}t^{3(\alpha-1)}$, and the corresponding proper entropy density is $s(t) = S_c/(A^3 t^{3\alpha}) + 4\kappa\hat{\rho}^{3/4}/(3t^3)$. For the discussion of the entropy we must distinguish three qualitatively different cases.

(i) The case $\alpha > 1$, i.e., $1/2 < \Omega^*_{\bar{\lambda}} < 1$: Here $\mathscr{P}(t) > 0$ so that the entropy and energy content of the matter system increases with time. By Eq. (8.81), $\mathscr{P} > 0$ implies $\dot{\bar{\lambda}} + 8\pi\rho\dot{G} < 0$. Since $\dot{\bar{\lambda}} < 0$ but $\dot{G} > 0$ in the NGFP regime, the energy exchange is predominantly due to the decrease of $\bar{\lambda}$ while the increase of G is subdominant in this respect. The comoving entropy $S(t)$ has a finite limit for $t \to 0$, $S(t \to 0) = S_c$, and $S(t)$ grows monotonically for $t > 0$. If $S_c = 0$, which would be the most natural value in view of the discussion above, *all* of the entropy carried by the matter fields is due to the energy injection from $\bar{\lambda}$.

(ii) The case $\alpha < 1$, i.e., $0 < \Omega^*_{\bar{\lambda}} < 1/2$: Here $\mathscr{P}(t) < 0$ so that the energy and entropy of matter decreases. Since $\mathscr{P} < 0$ amounts to $\dot{\bar{\lambda}} + 8\pi\rho\dot{G} > 0$, the dominant physical effect is the increase of G with time, the counteracting decrease of $\bar{\lambda}$ is less important. The comoving entropy starts out from an infinitely positive value at the initial singularity, $S(t \to 0) \to +\infty$. This case is unphysical probably.

(iii) The case $\alpha = 1$, $\Omega^*_{\bar{\lambda}} = 1/2$: Here $\mathscr{P}(t) \equiv 0$, $S(t) = $ const. The effect of a decreasing $\bar{\lambda}$ and increasing G cancels exactly.

At lower scales the RG trajectory leaves the NGFP and very rapidly 'crosses over' to the GFP. This is most clearly seen in the behavior of the anomalous dimension $\eta_N(k) \equiv k\partial_k \ln G(k)$ which changes from its NGFP value $\eta_* = -2$ to the classical $\eta_N = 0$. This transition happens near $k \approx m_{Pl}$ or, since $k(t) \approx H(t)$, near a cosmological 'transition' time t_{tr} defined by the condition $k(t_{tr}) = \xi H(t_{tr}) = m_{Pl}$. (Recall that $\xi = O(1)$.) The complete solution to the improved equations can be found with numerical methods only. It proves convenient to use logarithmic variables normalized with respect to their respective values at the turning point. Beside the 'RG time' $\tau \equiv \ln(k/k_T)$, we use $x \equiv \ln(a/a_T)$, $y \equiv \ln(t/t_T)$, and $\mathscr{U} \equiv \ln(H/H_T)$.

Summarizing the numerical results, one can say that for any value of $\Omega^*_{\bar{\lambda}}$ the UV cosmologies consist of two scaling regimes and a relatively sharp crossover region near $k, H \approx m_{Pl}$ corresponding to $x \approx -34.5$ which connects them. At higher k-scales the fixed point approximation is valid, at lower scales one has a classical FRW cosmology in which $\bar{\lambda}$ can be neglected.

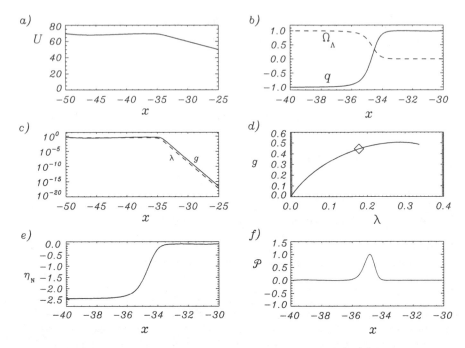

Fig. 8.9 The crossover epoch of the cosmology for $\Omega^*_{\tilde{\lambda}} = 0.98$. The plots (**a**), (**b**), (**c**) display the logarithmic Hubble parameter $\mathcal{U}$, as well as q, $\Omega_{\tilde{\lambda}}$, g and λ as a function of the logarithmic scale factor x. A crossover is observed near $x \approx -34.5$. The *diamond* in plot (**d**) indicates the point on the RG trajectory corresponding to this x-value. (The lower horizontal part of the trajectory is not visible on this scale.) The plots (**e**) and (**f**) show the x-dependence of the anomalous dimension and entropy production rate, respectively

As an example, Fig. 8.9 shows the crossover cosmology with $\Omega^*_{\tilde{\lambda}} = 0.98$ and $w = 1/3$. The entropy production rate $\mathcal{P}$ has a maximum at t_{tr} and quickly goes to zero for $t > t_{\mathrm{tr}}$; it is non-zero for all $t < t_{\mathrm{tr}}$. By varying the $\Omega^*_{\tilde{\lambda}}$-value one can check that the early cosmology is indeed described by the NGFP solution (8.71). For the logarithmic H vs. a plot, for instance, it predicts $\mathcal{U} = -2(1 - \Omega^*_{\tilde{\lambda}})x$ for $x < -34.4$. The left part of the plot in Fig. 8.9a and its counterparts with different values of $\Omega^*_{\tilde{\lambda}}$ indeed comply with this relation. If $\Omega^*_{\tilde{\lambda}} \in (1/2, 1)$ we have $\alpha = (2 - 2\Omega^*_{\tilde{\lambda}})^{-1} > 1$ and $a(t) \propto t^{\alpha}$ describes a phase of accelerated power-law inflation.

When $\Omega^*_{\tilde{\lambda}} \nearrow 1$ the slope of $\mathcal{U}(x) = -2(1 - \Omega^*_{\tilde{\lambda}})x$ decreases and finally vanishes at $\Omega^*_{\tilde{\lambda}} = 1$. This limiting case corresponds to a constant Hubble parameter, i.e., to de Sitter space. For values of $\Omega^*_{\tilde{\lambda}}$ smaller than, but close to 1 this de Sitter limit is approximated by an expansion $a \propto t^{\alpha}$ with a very large exponent α. The phase of power-law inflation automatically comes to a halt once the RG running has reduced $\tilde{\lambda}$ to a value where the resulting vacuum energy density no longer can overwhelm the matter energy density.

8.11 Conclusions

In these lectures we reviewed the basic ideas of asymptotic safety and explained why we believe that quantum Einstein gravity is likely to be renormalizable in the modern non-perturbative sense. We argued that the scale-dependence of the gravitational couplings intrinsic to asymptotic safety gives rise to multi-fractal features of the effective space-times and should also have an impact on the cosmological evolution of the Universe we live in. In the latter context, we proposed three possible candidate signatures: a period of automatic, cosmological constant-driven inflation that requires no *ad hoc* inflaton, the entropy carried by the radiation which fills the Universe today, and the primordial density perturbations necessary for structure formation. If these perturbations are an imprint of the metric fluctuations in the NGFP regime, the 'critical phenomenon' properties of the latter might be the origin of the observed scale free spectrum of the former. It is indeed an exciting idea that what we see when we look at the starry sky, during a clear summer night on the Cycladic Islands, for instance, might actually be a snapshot of the geometry fluctuations governed by the short-distance limit of QEG, and tremendously magnified by the cosmic expansion.

Acknowledgements   M.R. would like to thank the organizers of the 6th Aegean Summer School for the opportunity to present this material and their cordial hospitality at Naxos. We are also grateful to D. Benedetti and J. Henson for sharing their Monte Carlo data with us, and A. Nink for a careful reading of the manuscript. The research of F.S. is supported by the Deutsche Forschungsgemeinschaft (DFG) within the Emmy-Noether program (Grant SA/1975 1-1).

References

1. G. 't Hooft, M.J.G. Veltman, Ann. Henri Poincaré Phys. Theor. A **20**, 69 (1974)
2. M.H. Goroff, A. Sagnotti, Phys. Lett. B **160**, 81 (1985)
3. A.E.M. van de Ven, Nucl. Phys. B **378**, 309 (1992)
4. M. Niedermaier, M. Reuter, Living Rev. Relativ. **7**, 9 (2006). http://www.livingreviews.org/lrr-2006-5
5. M. Reuter, F. Saueressig, in *Geometric and Topological Methods for Quantum Field Theory*, ed. by H. Ocampo, S. Paycha, A. Vargas (Cambridge University Press, Cambridge, 2010). arXiv:0708.1317
6. R. Percacci, in *Approaches to Quantum Gravity*, ed. by D. Oriti (Cambridge University Press, Cambridge, 2009). arXiv:0709.3851
7. S. Weinberg, in *General Relativity, an Einstein Centenary Survey*, ed. by S.W. Hawking, W. Israel (Cambridge University Press, Cambridge, 1979)
8. S. Weinberg, in *Conceptual Foundations of Quantum Field Theory*, ed. by T.Y. Cao (Cambridge University Press, Cambridge, 1999), pp. 241–251. hep-th/9702027
9. S. Weinberg, arXiv:0903.0568
10. S. Weinberg, PoS CD **09**, 001 (2009). arXiv:0908.1964
11. M. Reuter, Phys. Rev. D **57**, 971 (1998). hep-th/9605030
12. D. Dou, R. Percacci, Class. Quantum Gravity **15**, 3449 (1998). hep-th/9707239
13. O. Lauscher, M. Reuter, Phys. Rev. D **65**, 025013 (2002). hep-th/0108040
14. M. Reuter, F. Saueressig, Phys. Rev. D **65**, 065016 (2002). hep-th/0110054
15. O. Lauscher, M. Reuter, Phys. Rev. D **66**, 025026 (2002). hep-th/0205062

16. O. Lauscher, M. Reuter, Class. Quantum Gravity **19**, 483 (2002). hep-th/0110021
17. O. Lauscher, M. Reuter, Int. J. Mod. Phys. A **17**, 993 (2002). hep-th/0112089
18. W. Souma, Prog. Theor. Phys. **102**, 181 (1999). hep-th/9907027
19. M. Reuter, F. Saueressig, Phys. Rev. D **66**, 125001 (2002). hep-th/0206145
20. M. Reuter, F. Saueressig, Fortschr. Phys. **52**, 650 (2004). hep-th/0311056
21. A. Bonanno, M. Reuter, J. High Energy Phys. **0502**, 035 (2005). hep-th/0410191
22. R. Percacci, D. Perini, Phys. Rev. D **67**, 081503 (2003). hep-th/0207033
23. R. Percacci, D. Perini, Phys. Rev. D **68**, 044018 (2003). hep-th/0304222
24. D. Perini, Nucl. Phys. Proc. Suppl. **127C**, 185 (2004). hep-th/0305053
25. A. Codello, R. Percacci, Phys. Rev. Lett. **97**, 221301 (2006). hep-th/0607128
26. D. Litim, Phys. Rev. Lett. **92**, 201301 (2004). hep-th/0312114
27. P. Fischer, D. Litim, Phys. Lett. B **638**, 497 (2006). hep-th/0602203
28. R. Percacci, D. Perini, Class. Quantum Gravity **21**, 5035 (2004). hep-th/0401071
29. R. Percacci, J. Phys. A **40**, 4895 (2007). hep-th/0409199
30. A. Codello, R. Percacci, C. Rahmede, Int. J. Mod. Phys. A **23**, 143 (2008). arXiv:0705.1769
31. P.F. Machado, F. Saueressig, Phys. Rev. D **77**, 124045 (2008). arXiv:0712.0445
32. A. Codello, R. Percacci, C. Rahmede, Ann. Phys. **324**, 414 (2009). arXiv:0805.2909
33. M. Reuter, H. Weyer, Phys. Rev. D **79**, 105005 (2009). arXiv:0801.3287
34. M. Reuter, H. Weyer, Gen. Relativ. Gravit. **41**, 983 (2009). arXiv:0903.2971
35. M. Reuter, H. Weyer, Phys. Rev. D **80**, 025001 (2009). arXiv:0804.1475
36. P.F. Machado, R. Percacci, Phys. Rev. D **80**, 024020 (2009). arXiv:0904.2510
37. J.E. Daum, M. Reuter, Adv. Sci. Lett. **2**, 255 (2009). arXiv:0806.3907
38. D. Benedetti, P.F. Machado, F. Saueressig, Mod. Phys. Lett. A **24**, 2233 (2009). arXiv:0901.2984
39. D. Benedetti, P.F. Machado, F. Saueressig, Nucl. Phys. B **824**, 168 (2010). arXiv:0902.4630
40. D. Benedetti, P.F. Machado, F. Saueressig, arXiv:0909.3265
41. J.-E. Daum, U. Harst, M. Reuter, J. High Energy Phys. **1001**, 084 (2010). arXiv:0910.4938
42. U. Harst, M. Reuter, J. High Energy Phys. **1105**, 119 (2011). arXiv:1101.6007
43. A. Eichhorn, H. Gies, M.M. Scherer, Phys. Rev. D **80**, 104003 (2009). arXiv:0907.1828
44. K. Groh, F. Saueressig, J. Phys. A **43**, 365403 (2010). arXiv:1001.5032
45. A. Eichhorn, H. Gies, Phys. Rev. D **81**, 104010 (2010). arXiv:1001.5033
46. E. Manrique, M. Reuter, Ann. Phys. **325**, 785 (2010). arXiv:0907.2617
47. E. Manrique, M. Reuter, F. Saueressig, Ann. Phys. **326**, 440 (2011). arXiv:1003.5129
48. E. Manrique, M. Reuter, F. Saueressig, Ann. Phys. **326**, 463 (2011). arXiv:1006.0099
49. E. Manrique, M. Reuter, Phys. Rev. D **79**, 025008 (2009). arXiv:0811.3888
50. E. Manrique, S. Rechenberger, F. Saueressig, Phys. Rev. Lett. **106**, 251302 (2011). arXiv:1102.5012
51. J.-E. Daum, M. Reuter, Phys. Lett. B **710**, 215 (2012). arXiv:1012.4280
52. J.-E. Daum, M. Reuter, PoS **CNCFG 2010**, 003 (2010). arXiv:1111.1000
53. D. Benedetti, K. Groh, P.F. Machado, F. Saueressig, J. High Energy Phys. **1106**, 079 (2011). arXiv:1012.3081
54. M. Niedermaier, Phys. Rev. Lett. **103**, 101303 (2009)
55. K. Groh, S. Rechenberger, F. Saueressig, O. Zanusso, PoS EPS -HEP2011, 124 (2011). arXiv:1111.1743
56. D. Benedetti, New J. Phys. **14**, 015005 (2012). arXiv:1107.3110
57. P. Forgács, M. Niedermaier, hep-th/0207028
58. M. Niedermaier, J. High Energy Phys. **0212**, 066 (2002). hep-th/0207143
59. M. Niedermaier, Nucl. Phys. B **673**, 131 (2003). hep-th/0304117
60. M. Niedermaier, Class. Quantum Gravity **24**, R171 (2007). gr-qc/0610018
61. C. Wetterich, Phys. Lett. B **301**, 90 (1993)
62. M. Reuter, C. Wetterich, Nucl. Phys. B **417**, 181 (1994)
63. M. Reuter, C. Wetterich, Nucl. Phys. B **427**, 291 (1994)
64. M. Reuter, C. Wetterich, Nucl. Phys. B **391**, 147 (1993)
65. M. Reuter, C. Wetterich, Nucl. Phys. B **408**, 91 (1993)

66. M. Reuter, Phys. Rev. D **53**, 4430 (1996). hep-th/9511128
67. M. Reuter, Mod. Phys. Lett. A **12**, 2777 (1997). hep-th/9604124
68. J. Berges, N. Tetradis, C. Wetterich, Phys. Rep. **363**, 223 (2002). hep-ph/0005122
69. C. Wetterich, Int. J. Mod. Phys. A **16**, 1951 (2001). hep-ph/0101178
70. M. Reuter, hep-th/9602012
71. J. Pawlowski, Ann. Phys. **322**, 2831 (2007). hep-th/0512261
72. H. Gies, hep-ph/0611146
73. C. Bagnuls, C. Bervillier, Phys. Rep. **348**, 91 (2001). hep-th/0002034
74. T.R. Morris, Prog. Theor. Phys. Suppl. **131**, 395 (1998). hep-th/9802039
75. J. Polonyi, Cent. Eur. J. Phys. **1**, 1 (2004)
76. O.J. Rosten, Phys. Rep. **511**, 177 (2012). arXiv:1003.1366
77. A. Bonanno, M. Reuter, Phys. Rev. D **65**, 043508 (2002). hep-th/0106133
78. M. Reuter, F. Saueressig, J. Cosmol. Astropart. Phys. **0509**, 012 (2005). hep-th/0507167
79. A. Bonanno, M. Reuter, J. Phys. Conf. Ser. **140**, 012008 (2008). arXiv:0803.2546
80. A. Bonanno, M. Reuter, J. Cosmol. Astropart. Phys. **0708**, 024 (2007). arXiv:0706.0174
81. A. Bonanno, M. Reuter, Entropy **13**, 274 (2011). arXiv:1011.2794
82. M. Reuter, J. Schwindt, J. High Energy Phys. **0601**, 070 (2006). hep-th/0511021
83. M. Reuter, J. Schwindt, J. High Energy Phys. **0701**, 049 (2007). hep-th/0611294
84. B. Mandelbrot, *The Fractal Geometry of Nature* (Freeman, New York, 1977)
85. H. Kawai, M. Ninomiya, Nucl. Phys. B **336**, 115 (1990)
86. R. Floreanini, R. Percacci, Nucl. Phys. B **436**, 141 (1995). hep-th/9305172
87. I. Antoniadis, P.O. Mazur, E. Mottola, Phys. Lett. B **444**, 284 (1998). hep-th/9808070
88. J. Ambjørn, J. Jurkiewicz, R. Loll, Phys. Rev. Lett. **93**, 131301 (2004). hep-th/0404156
89. J. Ambjørn, J. Jurkiewicz, R. Loll, Phys. Lett. B **607**, 205 (2005). hep-th/0411152
90. J. Ambjørn, J. Jurkiewicz, R. Loll, Phys. Rev. Lett. **95**, 171301 (2005). hep-th/0505113
91. J. Ambjørn, J. Jurkiewicz, R. Loll, Phys. Rev. D **72**, 064014 (2005). hep-th/0505154
92. J. Ambjørn, J. Jurkiewicz, R. Loll, Contemp. Phys. **47**, 103 (2006). hep-th/0509010
93. J. Ambjørn, S. Jordan, J. Jurkiewicz, R. Loll, Phys. Rev. Lett. **107**, 211303 (2011). arXiv:1108.3932
94. D. Benedetti, J. Henson, Phys. Rev. D **80**, 124036 (2009). arXiv:0911.0401
95. R. Kommu, arXiv:1110.6875
96. J. Ambjørn, J. Jurkiewicz, R. Loll, Lect. Notes Phys. **807**, 59 (2010). arXiv:0906.3947
97. O. Lauscher, M. Reuter, J. High Energy Phys. **0510**, 050 (2005). hep-th/0508202
98. J. Laiho, D. Coumbe, Phys. Rev. Lett. **107**, 161301 (2011). arXiv:1104.5505
99. M. Reuter, F. Saueressig, J. High Energy Phys. **1112**, 012 (2011). arXiv:1110.5224
100. L. Modesto, Class. Quantum Gravity **26**, 242002 (2009). arXiv:0812.2214
101. L. Modesto, arXiv:0905.1665
102. F. Caravelli, L. Modesto, arXiv:0905.2170
103. E. Magliaro, C. Perini, L. Modesto, arXiv:0911.0437
104. S. Carlip, AIP Conf. Proc. **1196**, 72 (2009). arXiv:0909.3329
105. S. Carlip, in *Foundations of Space and Time*, ed. by G. Ellis, J. Murugan, A. Weltman (Cambridge University Press, Cambridge, 2012). arXiv:1009.1136
106. S. Carlip, D. Grumiller, Phys. Rev. D **84**, 084029 (2011). arXiv:1108.4686
107. A. Connes, J. High Energy Phys. **0611**, 081 (2006). hep-th/0608226
108. A.H. Chamseddine, A. Connes, M. Marcolli, Adv. Theor. Math. Phys. **11**, 991 (2007). hep-th/0610241
109. D. Guido, T. Isola, J. Funct. Anal. **203**, 362 (2003). math.OA/0202108
110. D. Guido, T. Isola, in *Advances in Operator Algebras and Mathematical Physics*, ed. by F. Boca, O. Bratteli, R. Longo, H. Siedentop (Theta, Bucharest, 2005). math.OA/0404295
111. C. Antonescu, E. Christensen, math.OA/0309044
112. D. Benedetti, Phys. Rev. Lett. **102**, 111303 (2009). arXiv:0811.1396
113. T.P. Sotiriou, M. Visser, S. Weinfurtner, Phys. Rev. Lett. **107**, 131303 (2011). arXiv:1105.5646
114. G. Calcagni, Phys. Rev. Lett. **104**, 251301 (2010). arXiv:0912.3142

115. G. Calcagni, J. High Energy Phys. **1003**, 120 (2010). arXiv:1001.0571
116. G. Calcagni, Phys. Lett. B **697**, 251 (2011). arXiv:1012.1244
117. M. Arzano, G. Calcagni, D. Oriti, M. Scalisi, Phys. Rev. D **84**, 125002 (2011). arXiv:1107.5308
118. G. Calcagni Adv. Theor. Math. Phys. (to be published). arXiv:1106.5787
119. G. Calcagni, J. High Energy Phys. **1201**, 065 (2012). arXiv:1107.5041
120. E. Akkermans, G.V. Dunne, A. Teplyaev, Phys. Rev. Lett. **105**, 230407 (2010). arXiv:1010.1148
121. E. Akkermans, G.V. Dunne, A. Teplyaev, Europhys. Lett. **88**, 40007 (2009). arXiv:0903.3681
122. C.T. Hill, Phys. Rev. D **67**, 085004 (2003). hep-th/0210076
123. A. Bonanno, M. Reuter, Phys. Rev. D **62**, 043008 (2000). hep-th/0002196
124. A. Bonanno, M. Reuter, Phys. Rev. D **73**, 083005 (2006). hep-th/0602159
125. A. Bonanno, M. Reuter, Phys. Rev. D **60**, 084011 (1999). gr-qc/9811026
126. M. Reuter, E. Tuiran, Phys. Rev. D **83**, 044041 (2011). arXiv:1009.3528
127. A. Bonanno, M. Reuter, Phys. Lett. B **527**, 9 (2002). astro-ph/0106468
128. A. Bonanno, M. Reuter, Int. J. Mod. Phys. D **13**, 107 (2004). astro-ph/0210472
129. E. Bentivegna, A. Bonanno, M. Reuter, J. Cosmol. Astropart. Phys. **0401**, 001 (2004). astro-ph/0303150
130. S. Weinberg, Phys. Rev. D **81**, 083535 (2010). arXiv:0911.3165
131. M. Reuter, H. Weyer, Phys. Rev. D **69**, 104022 (2004). hep-th/0311196
132. M. Reuter, H. Weyer, Phys. Rev. D **70**, 124028 (2004). hep-th/0410117
133. M. Reuter, H. Weyer, J. Cosmol. Astropart. Phys. **0412**, 001 (2004). hep-th/0410119
134. B.F.L. Ward, Mod. Phys. Lett. A **23**, 3299 (2008). arXiv:0808.3124
135. A. Bonanno, A. Contillo, R. Percacci, Class. Quantum Gravity **28**, 145026 (2011). arXiv:1006.0192
136. M. Hindmarsh, D. Litim, C. Rahmede, J. Cosmol. Astropart. Phys. **1107**, 019 (2011). arXiv:1101.5401
137. J. Hewett, T. Rizzo, J. High Energy Phys. **0712**, 009 (2007). arXiv:0707.3182
138. D.F. Litim, T. Plehn, Phys. Rev. Lett. **100**, 131301 (2008). arXiv:0707.3983
139. B. Koch, Phys. Lett. B **663**, 334 (2008). arXiv:0707.4644
140. K. Falls, D.F. Litim, A. Raghuraman, Int. J. Mod. Phys. A **27**, 1250019 (2012). arXiv:1002.0260
141. B.S. DeWitt, *The Global Approach to Quantum Field Theory* (Clarendon Press, Oxford, 2003)
142. D. ben-Avraham, S. Havlin, *Diffusion and Reactions in Fractals and Disordered Systems* (Cambridge University Press, Cambridge, 2004)
143. S. Alexander, R. Orbach, J. Phys. Lett. (Paris) **43**, L625 (1982)
144. J.A.S. Lima, Phys. Rev. D **54**, 2571 (1996). gr-qc/9605055
145. J.A.S. Lima, Gen. Relativ. Gravit. **29**, 805 (1997). gr-qc/9605056

Chapter 9
Holography for Inflationary Cosmology

Paul McFadden

Abstract We review the construction of a holographic framework for cosmology enabling four-dimensional inflationary universes to be described in terms of three-dimensional dual quantum field theories. We show how cosmological observables are encoded in the correlation functions of the dual QFT, and obtain precise holographic formulae for the primordial power spectra. Through perturbative QFT calculations, we compute holographically the observational signatures of a universe emerging from a non-geometric phase in which the gravitational description is strongly coupled at early times. A custom fit to WMAP7 and other astrophysical data allows us to estimate the parameters of the holographic model and assess its performance relative to standard power-law ΛCDM.

9.1 Introduction

The notion of holography first emerged from considerations of black hole physics. In [49],'t Hooft observed that the onset of gravitational collapse imposes an upper bound on the entropy of any given region of spacetime. Through simple scaling arguments, he showed that the configuration with the maximum possible entropy consists of a single black hole completely filling the region, whose entropy then scales as the area of the region's boundary in Planck units. Since entropy is a measure of the number of (Boolean) degrees of freedom, it follows that the number of degrees of freedom in any gravitational theory scales as the area of the boundary. In particular, the gravitational entropy scales in the same way as the ordinary extensive entropy of a non-gravitational QFT living in one dimension less. Driving this argument to its ultimate conclusion [47, 49], one arrives at the *holographic principle*: that any quantum gravitational system should admit an equivalent dual description in terms of a non-gravitational QFT living in one dimension less.

Astonishingly, a concrete realisation of this conjecture—the AdS/CFT correspondence—was subsequently found in string theory [30], and a precise holographic

P. McFadden (✉)
Perimeter Institute for Theoretical Physics, Waterloo, ON, Canada N2J 2Y5
e-mail: pmcfadden@perimeterinstitute.ca

G. Calcagni et al. (eds.), *Quantum Gravity and Quantum Cosmology*,
Lecture Notes in Physics 863, DOI 10.1007/978-3-642-33036-0_9,
© Springer-Verlag Berlin Heidelberg 2013

dictionary linking bulk and boundary quantities was established soon thereafter [16, 53]. In spite of the generality of 't Hooft's original argument, however, almost all explicit realisations of holography to date necessarily involve spacetimes with a negative cosmological constant. The extent to which holographic dualities may be extended to encompass more general spacetimes is an unanswered question of the greatest importance.

One promising strategy is to start with spacetimes closely related to those already in possession of a well understood holographic description. Inflationary cosmologies provide just such an example, and of course are of great interest in their own right. In a recent series of papers [7, 14, 33–37] we proposed how to set up a holographic framework for cosmology, enabling cosmological evolution to be described through the physics of a dual three-dimensional non-gravitational QFT. While our chief focus is the primordial inflationary epoch, the framework we propose may also be applied to the late-time de Sitter epoch our universe is currently entering. In this article, we will review this basic holographic framework and discuss the holographic calculation of the cosmological power spectrum, expanding and updating our earlier account[1] [35].

Aside from the conceptual interest attached to a holographic description of inflationary cosmology, there are a number of more pragmatic reasons motivating such a development. Firstly, uncovering the structure of three-dimensional QFT in cosmological observables brings in new intuition about their structure and may lead to more efficient computational techniques, particularly for the calculation of non-Gaussianities. Secondly, the standard inflationary scenario, despite its successes, is still unsatisfactory in a number of ways: it generically requires fine tuning, and there are trans-Planckian issues and questions about the initial conditions for inflation, see for instance [6, 50]. A key feature of known holographic dualities is their strong/weak coupling nature, meaning that in the regime where the one description is weakly coupled the other is strongly coupled. A holographic framework for cosmology thus provides a natural arena for constructing new models with intrinsic strong-coupling gravitational dynamics at early times—a holographic non-geometric phase—that have only a weakly coupled three-dimensional QFT description. Such models lie beyond the scope of the conventional inflationary paradigm, and may potentially be free from the problems besetting conventional inflationary models. Moreover, as we will see, models of this nature generically lead to qualitatively different predictions for cosmological observables that will be measured in the near future.

Any holographic proposal for cosmology should specify what the dual QFT is and how to use it to compute cosmological observables. The holographic description we propose uses the one-to-one correspondence between cosmologies and so-called 'domain-wall' spacetimes discussed in [9, 45], and assumes that standard holographic dualities (also known as gauge/gravity dualities, since the dual QFT

---

[1]The slides for the talk accompanying these proceedings may also be found online at http://www.physics.ntua.gr/cosmo11/Naxos2011/MorningLectures/McFadden.pdf.

is typically a large-N gauge theory) are valid. More precisely, the steps involved are illustrated in Fig. 9.1. The first step is to map any given inflationary spacetime to a domain-wall spacetime. For cosmologies that at late times approach either a de Sitter spacetime or a power-law scaling solution, the corresponding domain-wall solutions describe holographic renormalisation group flows (i.e., spacetimes in which the radial evolution of the bulk geometry holographically encodes RG flow in the dual QFT). For these cases there is an operational gauge/gravity duality, meaning one has a dual description in terms of a three-dimensional QFT. Crucially, the map between cosmologies and domain-walls may equivalently be expressed entirely in terms of QFT variables, and amounts to a certain analytic continuation of parameters and momenta. Applying this analytic continuation to the regular QFT dual to the domain-wall spacetime, we obtain the QFT dual of the original cosmological spacetime.

We call this latter theory a 'pseudo'-QFT because we currently only have an operational definition for it, namely, we do the computations in the regular QFT dual to the corresponding domain-wall spacetime and then apply the analytic continuation. From the standard holographic dictionary, to compute tree-level cosmological correlators we need continue only the large-N correlators of the dual QFT, where N is the rank of the gauge group of the dual QFT. (Loop corrections to cosmological correlators then correspond to $1/N^2$ corrections in the dual QFT.) In the large-N limit, this analytic continuation is well defined and simply amounts to the insertion of a few minus signs in specific formulae. Thus, from a strictly pragmatic point of view, our operational definition of the pseudo-QFT is sufficient to compute all observable quantities of interest. Nonetheless, one might ultimately hope for a more fundamental definition, in particular one that is valid beyond large-N perturbation theory. Interesting progress along these lines (albeit for a very different bulk gravitational theory) was recently made in [1], where the pseudo-QFT dual to Vasiliev higher spin gravity on de Sitter space was identified as a specific $Sp(N)$ gauge theory.

The remainder of this article is organised as follows. In Sect. 9.2, we discuss the general features of domain-walls and cosmologies: the form of the background metric and perturbations, their dynamics, the domain-wall/cosmology correspondence, and the primordial cosmological power spectrum. In Sect. 9.3, we discuss holography for cosmology: the background solutions of interest, the basics of holography including the radial Hamiltonian formulation, properties of the stress tensor 2-point function, and the analysis required to derive holographic formulae for the cosmological power spectra. In Sect. 9.4, we introduce holographic phenomenology for cosmology, and discuss the computation of the holographic power spectra up to 2-loop order in perturbation theory. Finally, in Sect. 9.5, we discuss the observational compatibility of the predicted holographic power spectrum in the light of WMAP7 and other astrophysical data.

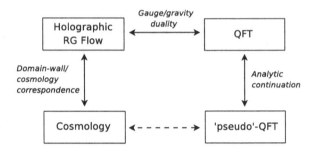

Fig. 9.1 The 'pseudo'-QFT dual to inflationary cosmology is operationally defined using the correspondence of cosmologies to domain-walls and standard gauge/gravity duality

9.2 Domain-Walls and Cosmologies

Let us begin by introducing the various objects appearing on the l.h.s. of Fig. 9.1, which represents the bulk gravitational physics. For simplicity, we will focus throughout on spatially flat universes with a single inflaton field, Φ, which we take to be minimally coupled and equipped with a potential $V(\Phi)$. All our results may straightforwardly be extended to more general cases if desired, (e.g., non-flat, multi-scalar, non-canonical kinetic terms, etc.).

9.2.1 Defining the Perturbations

The metric and scalar field for the unperturbed background solution take the form

$$ds^2 = \sigma dz^2 + a^2(z)dx_i dx^i, \qquad \Phi = \varphi(z), \tag{9.1}$$

where the spatial index i runs from 1 to 3, and σ is a sign taking values $\sigma = \pm 1$. If $\sigma = -1$, the metric describes a flat FRW cosmology with z the proper time co-ordinate. The remaining case, where $\sigma = +1$, we will refer to as a domain-wall spacetime.[2] In this latter case, z now plays the role of a radial coordinate. Note also that we have chosen the domain-wall to be Euclidean. A Lorentzian domain-wall may easily be obtained by continuing one of the x^i coordinates to become a time coordinate [45]. The continuation to a Euclidean domain-wall will turn out to be convenient, however, since the QFT vacuum state implicit in the Euclidean formulation maps to the Bunch-Davies vacuum on the cosmology side. (Other choices of cosmological vacuum require considering the boundary QFT in different states, which may be accomplished using the real-time formalism of [46].)

---

[2]The name 'domain-wall' spacetime dates back to earlier work featuring solutions of this form that interpolate between two stationary points of the scalar field potential, one at $z = +\infty$ and another at $z = -\infty$. Unfortunately in the present context the name is slightly misleading, since we consider only the $z > 0$ part of the geometry, i.e., there is no actual domain wall. We will nevertheless stick with this terminology as it is standard issue in high-energy physics.

Turning now to include perturbations, the inflaton may be decomposed into a background piece φ and a perturbation $\delta\varphi$,

$$\Phi(z, \mathbf{x}) = \varphi(z) + \delta\varphi(z, \mathbf{x}),$$

while the perturbed metric may be written in the ADM form

$$ds^2 = \sigma N^2 dz^2 + g_{ij}(dx^i + N^i dz)(dx^j + N^j dz), \qquad (9.2)$$

where the perturbed lapse and shift functions

$$N = 1 + \delta N(z, \mathbf{x}), \qquad N_i = g_{ij}N^j = \delta N_i(z, \mathbf{x}), \qquad g_{ij} = a^2(z)\big(\delta_{ij} + h_{ij}(z, \mathbf{x})\big).$$

We may then further decompose the perturbations into scalar, vector and tensor pieces according to

$$\delta N_i = a^2(\nu_{,i} + v_i), \qquad h_{ij} = -2\psi\delta_{ij} + 2\chi_{,ij} + 2\omega_{(i,j)} + \gamma_{ij}, \qquad (9.3)$$

where the vector perturbations v_i and ω_i are transverse, $v_{i,i} = 0$ and $\omega_{i,i} = 0$, while the tensor perturbation γ_{ij} is transverse traceless, $\gamma_{ij,i} = 0$ and $\gamma_{ii} = 0$. (Here, and in the remainder of this article, we adopt the convention that repeated covariant indices are summed using the Kronecker delta. In contrast, an index is raised or lowered using the full metric.)

Gauge-invariant variables may be defined by relating the perturbations in a general gauge to those in some fully-fixed gauge. We will see shortly that the dynamics comprise only a single scalar degree of freedom, plus one tensor mode. To parametrise this scalar degree of freedom, a particularly useful choice is the gauge-invariant variable $\zeta(z, \mathbf{x})$ encoding the curvature perturbation on uniform energy density slices. More precisely, ζ is defined so that in comoving gauge, where $\delta\varphi$ vanishes, the spatial part of the perturbed metric reads

$$g_{ij} = a^2 e^{2\zeta}\big[e^{\hat{\gamma}}\big]_{ij},$$

where $\hat{\gamma}_{ij}$ is transverse traceless and the exponential is to be expanded out (i.e., $[e^{\hat{\gamma}}]_{ij} = \delta_{ij} + \hat{\gamma}_{ij} + \cdots$). This prescription fixes the gauge completely, thereby defining the gauge-invariant variables ζ and $\hat{\gamma}_{ij}$ to all orders in perturbation theory. In our forthcoming discussion of the holographic power spectrum, however, we will only need to work to linear order in perturbation theory. In this case, upon transforming to a general gauge, we find ζ and $\hat{\gamma}_{ij}$ correspond to the gauge-invariant combinations

$$\zeta = -\psi - \frac{H}{\dot{\varphi}}\delta\varphi, \qquad \hat{\gamma}_{ij} = \gamma_{ij}.$$

The corresponding expressions at quadratic order in perturbation theory, as required for the treatment of non-Gaussianities, may be found in [36, 37]. (Note in particular that γ_{ij} is no longer gauge-invariant at quadratic order.)

Finally, when working in momentum space, it is useful to decompose the transverse traceless tensors in a helicity basis according to

$$\hat{\gamma}_{ij}(z, \mathbf{q}) = \sum_{s=\pm} \hat{\gamma}^{(s)}(z, \mathbf{q}) \varepsilon_{ij}^{(s)}(\mathbf{q}), \tag{9.4}$$

where $\mathbf{q}$ is the spatial 3-momentum and the helicity tensors $\varepsilon_{ij}^{(s)}(\mathbf{q})$ are normalised so that

$$\varepsilon_{ij}^{(s)}(\mathbf{q})\varepsilon_{ij}^{(s')}(-\mathbf{q}) = 2\delta^{ss'}, \qquad \sum_{s=\pm} \varepsilon_{ij}^{(s)}(\mathbf{q})\varepsilon_{kl}^{(s)}(-\mathbf{q}) = 2\Pi_{ijkl}, \tag{9.5}$$

with the transverse traceless projector Π_{ijkl} and the transverse projector π_{ij} defined by

$$\Pi_{ijkl} = \frac{1}{2}(\pi_{ik}\pi_{jl} + \pi_{il}\pi_{jk} - \pi_{ij}\pi_{kl}), \qquad \pi_{ij} = \delta_{ij} - \frac{q_i q_j}{q^2}.$$

Note that with these normalisation conventions, $\hat{\gamma}^{(s)}(z, \mathbf{q}) = (1/2)\varepsilon_{ij}^{(s)}(-\mathbf{q})\hat{\gamma}_{ij}(z, \mathbf{q})$. Explicit expressions for the helicity tensors may be found in [37, 51]. Under complex conjugation, $\varepsilon_{ij}^{(s)*}(\mathbf{q}) = \varepsilon_{ij}^{(s)}(-\mathbf{q}) = \varepsilon_{ij}^{(-s)}(\mathbf{q})$.

9.2.2 Dynamics

Having written the metric in the ADM form (9.2), we may now write the action for both domain-walls and cosmologies in the combined form

$$S = \frac{1}{2\kappa^2} \int d^4x N \sqrt{g} \left[K_{ij} K^{ij} - K^2 + N^{-2}(\dot{\Phi} - N^i \Phi_{,i})^2 \right.$$
$$\left. + \sigma \left(-R + g^{ij} \Phi_{,i} \Phi_{,j} + 2\kappa^2 V(\Phi) \right) \right]. \tag{9.6}$$

Here, $\kappa^2 = 8\pi G$, R is the scalar curvature of the spatial metric g_{ij} and $K_{ij} = [(1/2)\dot{g}_{ij} - \nabla_{(i}N_{j)}]/N$ is the extrinsic curvature of constant-z slices. While this expression might seem unfamiliar, it is simply the action of gravity minimally coupled to a scalar field with a potential $V(\Phi)$ in disguise. Setting $\sigma = -1$ for example, the action (9.6) is equivalent to the familiar inflationary action

$$S_C = \frac{1}{2\kappa^2} \int d^4x \sqrt{g^{(4)}} \left[R^{(4)} - (\partial \Phi)^2 - 2\kappa^2 V(\Phi) \right].$$

Our reason for preferring the ADM form (9.6) is simply that it neatly encompasses both *Lorentzian* cosmologies and *Euclidean* domain-walls: the spatial gradient and potential terms on the second line appear with positive sign for Euclidean domain-walls and with negative sign for Lorentzian cosmologies, while the kinetic terms on the first line take the same sign for both.

In the following, we will restrict our consideration to background solutions in which the evolution of the scalar field $\varphi(z)$ is (piece-wise) monotonic in z, as appropriate for describing holographic RG flows. For such solutions, $\varphi(z)$ can be inverted to $z(\varphi)$, allowing the Hubble rate $H \equiv \dot{a}/a$ to be expressed as a function of φ, say as $H(z) = -(1/2)W(\varphi)$. The complete equations of motion for the background then take the simple form

$$H = -\frac{1}{2}W, \qquad \dot{\varphi} = W_{,\varphi}, \qquad 2\sigma\kappa^2 V = (W_{,\varphi})^2 - \frac{3}{2}W^2. \tag{9.7}$$

In cosmology, this first-order formalism dates back to the work of [42], where it was obtained by application of the Hamilton-Jacobi method. For domain-walls, this formalism has been discussed from variety of standpoints in [10, 12, 15, 44, 45].

An action for the perturbations may be obtained by solving the Hamiltonian and momentum constraints for the comoving-gauge lapse and shift in terms of ζ and $\hat{\gamma}_{ij}$, then backsubstituting into (9.6). Keeping track of the sign σ, at quadratic order we find

$$S = \frac{1}{\kappa^2} \int d^4x \left[a^3 \varepsilon \dot{\zeta}^2 + \sigma a \varepsilon (\partial \zeta)^2 + \frac{a^3}{8}\dot{\hat{\gamma}}_{ij}\dot{\hat{\gamma}}_{ij} + \frac{\sigma a}{8}\hat{\gamma}_{ij,k}\hat{\gamma}_{ij,k} \right], \tag{9.8}$$

where $\varepsilon = -\dot{H}/H^2 = \dot{\phi}^2/2H^2 = 2(W_{,\varphi}/W)^2$. (In standard inflation ε would be the usual slow-roll parameter, however we have no need to assume slow roll here.) The action at cubic order may be derived by the same method, albeit with more work; the result may be found in [31] (or including the sign σ, in [36]).

In momentum space, the corresponding linear equations of motion are

$$0 = \ddot{\zeta} + (3H + \dot{\varepsilon}/\varepsilon)\dot{\zeta} - \sigma a^{-2}q^2\zeta, \qquad 0 = \ddot{\gamma}^{(s)} + 3H\dot{\gamma}^{(s)} - \sigma a^{-2}q^2\hat{\gamma}^{(s)}. \tag{9.9}$$

From the first of these equations one finds that ζ tends to a constant on superhorizon scales for which $q \ll aH$. This property accounts for the utility of ζ in inflationary cosmology: perturbations exit the horizon during inflation, after which they remain constant until their eventual re-entry in the subsequent radiation- or matter-dominated eras.

Finally, in preparation for our holographic analysis to follow, it is useful to define *response functions* relating the canonical momenta to the perturbations. From the quadratic action (9.8), the canonical momenta (times an overall factor of κ^2) are

$$\Pi = \kappa^2 \frac{\partial \mathcal{L}}{\partial \dot{\zeta}} = 2\varepsilon a^3 \dot{\zeta}, \qquad \Pi_{ij} = \kappa^2 \frac{\partial \mathcal{L}}{\partial \dot{\hat{\gamma}}_{ij}} = \frac{1}{4}a^3 \dot{\hat{\gamma}}_{ij}. \tag{9.10}$$

In momentum space, we may further decompose Π_{ij} in a helicity basis,

$$\Pi_{ij}(z, \mathbf{q}) = \sum_{s=\pm} \Pi^{(s)}(z, \mathbf{q})\varepsilon_{ij}^{(s)}(\mathbf{q}), \qquad \Pi^{(s)}(z, \mathbf{q}) = \frac{1}{4}a^3 \dot{\hat{\gamma}}^{(s)}(z, \mathbf{q}). \tag{9.11}$$

The linear response functions Ω and E are then defined by

$$\Pi(z, \mathbf{q}) = \Omega(z, q)\zeta(z, \mathbf{q}), \qquad \Pi^{(s)}(z, \mathbf{q}) = E(z, q)\hat{\gamma}^{(s)}(z, \mathbf{q}). \qquad (9.12)$$

Note that Ω and E are perfectly well-defined once a given solution of (9.9) has been specified (the normalisation of the solution does not matter for this purpose). From (9.9), the response functions satisfy

$$0 = \dot{\Omega} + \frac{1}{2a^3\varepsilon}\Omega^2 - 2\sigma a\varepsilon q^2, \qquad 0 = \dot{E} + \frac{4}{a^3}E^2 - \frac{\sigma a}{4}q^2. \qquad (9.13)$$

To deal with holographic non-Gaussianities one must extend the above definition of response functions to quadratic order in perturbation theory [36, 37]. The linear response functions Ω and E are however sufficient to derive the holographic power spectra, as is our goal here.

9.2.3 The Domain-Wall/Cosmology Correspondence

Defining the analytically continued variables $\bar{\kappa}$ and $\bar{q}$ according to

$$\bar{\kappa}^2 = -\kappa^2, \qquad \bar{q} = -iq, \qquad (9.14)$$

where $q = +\sqrt{q^2}$ and $\bar{q} = +\sqrt{\bar{q}^2}$ denote *magnitudes* of spatial 3-momenta, it is easy to see that a perturbed cosmological solution written in terms of the variables κ and q continues to a perturbed Euclidean domain-wall solution expressed in terms of the variables $\bar{\kappa}$ and $\bar{q}$. The first continuation is equivalent to reversing the sign of the potential in the background equation of motion (9.7). (We will see shortly, however, that the continuation we have chosen has a clearer interpretation in terms of the variables of the dual QFT.) The second of these analytic continuations generates the necessary sign change $\bar{q}^2 = -q^2$ in the linear equations of motion (9.9). The choice of branch cut we made (i.e., $\bar{q} = -iq$ rather than $\bar{q} = +iq$) stems from the necessity of mapping the cosmological Bunch-Davies vacuum behaviour at early times,

$$\zeta, \hat{\gamma}^{(s)} \sim \exp(-iq\tau) \quad \text{as } \tau = \int^z \frac{dz'}{a(z')} \to -\infty, \qquad (9.15)$$

to the domain-wall solution that decays smoothly in the domain-wall interior,

$$\zeta, \hat{\gamma}^{(s)} \sim \exp(\bar{q}\tau) \quad \text{as } \tau \to -\infty,$$

as required in the calculation of holographic correlation functions.

Turning now to the linear response functions, if we choose $\Omega(z, q)$ and $E(z, q)$ to be cosmological response functions solving (9.13) with $\sigma = -1$ and Bunch-Davies initial conditions, then the corresponding domain-wall response functions $\bar{\Omega}(z, \bar{q})$ and $\bar{E}(z, \bar{q})$ are given by the analytic continuation

$$\bar{\Omega}(z, -iq) = \Omega(z, q), \qquad \bar{E}(z, -iq) = E(z, q). \tag{9.16}$$

(Note we have defined our response functions so that they are independent of κ^2; this was in fact our motivation for introducing the extra factor of κ^2 in (9.10).)

We have thus established that the correspondence between cosmologies and domain-walls holds, not only for the background solutions, but also for linear perturbations around them. This is the basis for the relation between power spectra and holographic 2-point functions, to be discussed shortly. The correspondence also holds at higher order in perturbation theory, allowing cosmological non-Gaussianities to be related to holographic higher-point functions. This may be established in a straightforward fashion by working out the momentum-space Lagrangian for ζ and $\hat{\gamma}^{(s)}$ at higher order in perturbation theory (for explicit results at cubic order, see [36, 37]). One finds that the sign σ is always associated with factors of momenta such that the continuation (9.14) indeed maps perturbed cosmological solutions to perturbed domain-wall solutions.

Finally, let us note the analytic continuations (9.14) may equivalently be expressed in terms of dual QFT variables as

$$\bar{N} = -iN, \qquad \bar{q} = -iq, \tag{9.17}$$

where $\bar{N}$ is the rank of the gauge group of the QFT dual to the domain-wall spacetime, and N is the rank of the gauge group of the *pseudo-QFT* dual to the corresponding cosmology. These relations follow directly from (9.14) noting that in the standard holographic dictionary $\bar{\kappa}^{-2} \propto \bar{N}^2$, working in units where the AdS radius has been set to unity. (Indeed, in our later results, we will see explicitly that holographic correlation functions calculated from the gravity side of the correspondence appear with an overall prefactor of $\bar{\kappa}^{-2}$. On the QFT side of the correspondence, this prefactor corresponds to the overall prefactor of $\bar{N}^2$ in correlators arising from the trace over gauge indices.) Our choice of branch cut in the continuation of $\bar{N}$ has been chosen so that the dimensionless effective QFT coupling, $g_{\text{eff}}^2 = g_{\text{YM}}^2 \bar{N}/\bar{q} = g_{\text{YM}}^2 N/q$, does not change when we analytically continue from QFT to pseudo-QFT. As we will see later in Sect. 9.4.2, this will turn out to be important because the QFT correlators are in general non-analytic functions of g_{eff}^2 at large N [2, 19].

9.2.4 Cosmological Power Spectra

In the inflationary paradigm, cosmological perturbations originate on sub-horizon scales as quantum fluctuations of the vacuum. Quantising the interaction picture fields ζ and $\hat{\gamma}_{ij}$ in standard fashion (see, e.g., [28]),

$$\zeta(z, \mathbf{q}) = a(\mathbf{q})\zeta_q(z) + a^\dagger(-\mathbf{q})\zeta_q^*(z), \tag{9.18}$$

$$\hat{\gamma}^{(s)}(z, \mathbf{q}) = b^{(s)}(\mathbf{q})\hat{\gamma}_q(z) + b^{(s)\dagger}(-\mathbf{q})\hat{\gamma}_q^*(z), \tag{9.19}$$

where the creation and annihilation operators obey the usual commutation relations

$$[a(\mathbf{q}), a^\dagger(\mathbf{q}')] = (2\pi)^3\delta(\mathbf{q} - \mathbf{q}'), \qquad [b^{(s)}(\mathbf{q}), b^{(s')\dagger}(\mathbf{q}')] = (2\pi)^3\delta(\mathbf{q} - \mathbf{q}')\delta^{ss'}, \tag{9.20}$$

and the mode functions $\zeta_q(z)$ and $\hat{\gamma}_q(z)$ are solutions of the linearised equations of motion (9.9), with initial conditions corresponding to the Bunch-Davies vacuum (9.15). The normalisation of the mode functions is fixed by imposing the canonical commutation relations,[3]

$$[\zeta(z, \mathbf{q}), \kappa^{-2}\Pi(z, \mathbf{q}')] = i(2\pi)^3\delta(\mathbf{q} + \mathbf{q}'), \tag{9.21}$$

$$[\hat{\gamma}_{ij}(z, \mathbf{q}), \kappa^{-2}\Pi_{kl}(z, \mathbf{q}')] = i(2\pi)^3\delta(\mathbf{q} + \mathbf{q}')\Pi_{ijkl}, \tag{9.22}$$

where the latter, upon converting to the helicity basis using (9.5), reads

$$[\hat{\gamma}^{(s)}(z, \mathbf{q}), \kappa^{-2}\Pi^{(s')}(z, \mathbf{q}')] = \frac{i}{2}(2\pi)^3\delta(\mathbf{q} + \mathbf{q}')\delta^{ss'}. \tag{9.23}$$

Using (9.10)–(9.11) and the mode decompositions (9.18)–(9.19), the canonical commutation relations are equivalent to the Wronskian relations

$$i = 2\varepsilon a^3\kappa^{-2}\left(\zeta_q(z)\dot{\zeta}_q^*(z) - \dot{\zeta}_q(z)\zeta_q^*(z)\right), \tag{9.24}$$

$$\frac{i}{2} = \frac{1}{4}a^3\kappa^{-2}\left(\hat{\gamma}_q(z)\dot{\hat{\gamma}}_q^*(z) - \dot{\hat{\gamma}}_q(z)\hat{\gamma}_q^*(z)\right). \tag{9.25}$$

As well as fixing the normalisation of the mode functions, these relations imply

$$|\zeta_q(z)|^2 = \frac{-\kappa^2}{2\,\mathrm{Im}[\Omega(z, q)]}, \qquad |\hat{\gamma}_q(z)|^2 = \frac{-\kappa^2}{4\,\mathrm{Im}[E(z, q)]}, \tag{9.26}$$

where we have used the definition of the response functions Ω and E given in (9.12). Computing now the 2-point functions, we find

$$\langle\langle\zeta(z, q)\zeta(z, -q)\rangle\rangle = |\zeta_q(z)|^2, \qquad \langle\langle\hat{\gamma}^{(s)}(z, q)\hat{\gamma}^{(s')}(z, -q)\rangle\rangle = |\hat{\gamma}_q(z)|^2\delta^{ss'}, \tag{9.27}$$

---

[3]Recall that $\kappa^{-2}\Pi$ and $\kappa^{-2}\Pi^{(s)}$, rather than Π and $\Pi^{(s)}$, are the actual canonical momenta.

where our double bracket notation indicates dropping the delta function associated with momentum conservation, e.g.,

$$\langle z(z, \mathbf{q})\zeta(z, \mathbf{q}')\rangle = (2\pi)^3 \delta(\mathbf{q} + \mathbf{q}')\langle\langle \zeta(z, q)\zeta(z, -q)\rangle\rangle. \tag{9.28}$$

According to convention, the late-time scalar and tensor power spectra are then defined as

$$\Delta_S^2(q) \equiv \frac{q^3}{2\pi^2}\langle\langle \zeta(q)\zeta(-q)\rangle\rangle = \frac{q^3}{2\pi^2}|\zeta_{q(0)}|^2, \tag{9.29}$$

$$\Delta_T^2(q) \equiv \frac{q^3}{2\pi^2}\langle\langle \hat{\gamma}_{ij}(q)\hat{\gamma}_{ij}(-q)\rangle\rangle = \frac{2q^3}{\pi^2}|\hat{\gamma}_{q(0)}|^2, \tag{9.30}$$

where $\zeta_{q(0)}$ and $\hat{\gamma}_{q(0)}$ are the constant late-time values of the cosmological mode functions. Physically, the power spectra represent the contribution to the field variance in position space per logarithmic interval in wavenumbers, e.g.,

$$\langle \zeta(\mathbf{x})^2\rangle = \int \frac{1}{(2\pi)^3}(4\pi q^2 dq)\langle\langle \zeta(q)\zeta(-q)\rangle\rangle = \int \Delta_S^2(q)\mathrm{d}\ln q.$$

Using (9.26), the cosmological power spectra may also be expressed in terms of the late-time values $\Omega_{(0)}$ and $E_{(0)}$ of the response functions:

$$\Delta_S^2(q) = -\frac{\kappa^2 q^3}{4\pi^2 \,\mathrm{Im}[\Omega_{(0)}(q)]}, \qquad \Delta_T^2(q) = -\frac{\kappa^2 q^3}{2\pi^2 \,\mathrm{Im}[E_{(0)}(q)]}. \tag{9.31}$$

We will see shortly that the holographic 2-point functions for the corresponding domain-wall spacetime may similarly be expressed in terms of the domain-wall linear response functions. Since the domain-wall response functions are related to the cosmological response functions via (9.16), we will therefore be able to relate the cosmological power spectra above to the holographic 2-point functions.

9.3 Holography for Cosmology

In the present section we turn our attention to the top half of Fig. 9.1, depicting standard gauge/gravity duality. We begin by enumerating the well understood classes of holographic RG flows and discussing some basic features of holographic dualities. We then review Hamiltonian holographic renormalisation and a useful decomposition of the stress tensor 2-point function. Proceeding with our main holographic analysis, we obtain expressions for the stress tensor 2-point function in terms of the domain-wall response functions. Ultimately, our purpose is to derive the holographic formulae for the cosmological power spectra given in (9.66). Readers not concerned with the more intricate aspects of holographic analysis may prefer to begin with Sects. 9.3.1 and 9.3.2 then skip to Sect. 9.3.6.

9.3.1 Background Solutions

At present, there are two general classes of domain-wall solutions for which a well understood holographic description exists. We list these classes below: it is for these backgrounds that our holographic framework for cosmology is most readily applicable.

(i) *Asymptotically AdS domain-walls.* In this case the solution behaves asymptotically as

$$a(z) \sim e^z, \qquad \varphi \sim 0 \quad \text{as } z \to \infty.$$

The boundary theory has a UV fixed point which corresponds to the bulk AdS critical point. Depending on the rate at which φ approaches zero as $z \to \infty$, the QFT is either a deformation of the conformal field theory (CFT), or else the CFT in a state in which the dual scalar operator acquires a nonvanishing vacuum expectation value (see [43] for details). Under the domain-wall/cosmology correspondence, these solutions are mapped to cosmologies that are asymptotically de Sitter at late times.

(ii) *Asymptotically power-law solutions.* In this case the solution behaves asymptotically as

$$a(z) \sim (z/z_0)^n, \qquad \varphi \sim \sqrt{2n} \log(z/z_0) \quad \text{as } z \to \infty, \qquad (9.32)$$

where $z_0 = n-1$. Examples of such dualities are provided by considering the near-horizon limit of the non-conformal branes [5, 18]. In particular, for the case $n = 7$, the asymptotic geometry corresponds to the near-horizon limit of a stack of D2 brane solutions. The detailed holographic dictionary for these theories has been worked out only relatively recently [23, 24, 52]. These theories are characterised by the fact that they have a 'generalised conformal structure' [20–22, 24] (see also Sect. 9.4.1). Under the domain-wall/cosmology correspondence, asymptotically power-law domain walls are mapped to cosmologies that are asymptotically power-law at late times.

9.3.2 Basics of Holography

Gauge/gravity duality is an exact equivalence between a bulk gravitational theory and a boundary QFT. Typically, the boundary QFT is a gauge theory that admits a large-N expansion. The N here denotes the rank of the gauge group: an example of such theory, with gauge group $SU(N)$, is discussed in Sect. 9.4.1. The large-N limit consists of taking $N \to \infty$ while keeping the 't Hooft coupling $\lambda = g_{YM}^2 N$ fixed. One can show that in this limit only planar diagrams survive [48]. On the bulk side, taking the large-N limit means that one suppresses loop effects. The value of λ then controls whether the supergravity approximation is valid or not.

Under the duality, bulk fields are related to local gauge-invariant operators of the boundary QFT. In particular, the bulk metric is related to the boundary stress tensor, T_{ij}, while bulk scalar fields, such as the inflaton, correspond to boundary scalar operators (e.g., tr $F_{ij}F^{ij}$ where F_{ij} is the gauge field strength and the trace runs over the gauge group indices). More precisely, the map is specified as follows. First, recall that in order to define a quantum theory we must specify the behaviour of the fields at infinity. In a gravitational theory, this means in particular that the spacetime asymptotics must be prescribed. In gauge/gravity duality, the fields that specify the boundary conditions on the bulk side are identified with the sources of the boundary QFT operators [16, 53]. Correlation functions for these gauge-invariant operators may then be extracted from the asymptotics of bulk solutions. Conversely, given the correlation functions of dual operators, one may reconstruct the bulk asymptotics.

Thus, to define the bulk theory, we need to specify appropriate boundary conditions. These boundary conditions must involve an arbitrary boundary metric, since this will act as a source for the stress tensor. Such boundary conditions are supplied by giving an asymptotically locally AdS metric, which in four dimensions takes the form,

$$
\mathrm{d}s^2 = \mathrm{d}r^2 + g_{ij}(r, x)\mathrm{d}x^i \mathrm{d}x^j,
$$
$$
g_{ij}(r, x) = e^{2r}\left(g_{(0)ij}(x) + e^{-2r}g_{(2)ij}(x) + \cdots + e^{-2mr}g_{(2m)ij}(x) + \cdots\right).
$$
(9.33)

This encompasses the boundary conditions for the bulk metric, both for asymptotically AdS domain-walls and for asymptotically power-law solutions. In the former case, the radial coordinate r may be identified with z, and $2m = 3$. For asymptotically power-law solutions, one may perform a conformal transformation to the *dual frame* [5] defined by $\tilde{g}_{ij} = \exp(-\lambda\Phi)g_{ij}$, where $\lambda = \sqrt{2/n}$. The asymptotic solution above then describes the most general asymptotics for the dual frame metric $\tilde{g}_{ij}$, where now $2m = (3n-1)/(n-1) > 3$ and $r = \int \exp(-\lambda\Phi/2)\mathrm{d}z$ (see [24] for details). In general, much of the holographic analysis for spacetimes with power-law asymptotics may be obtained from that for asymptotically AdS$_{2m+1}$ spacetimes, which are related to power-law spacetimes via dimensional reduction on a T^{2m-3} torus followed by an analytic continuation in m [23].

In the asymptotic expansion (9.33), the leading coefficient $g_{(0)ij}(x)$ is an arbitrary (non-degenerate) three-dimensional metric on the conformal boundary of the bulk spacetime. Since this is the metric on which the dual QFT lives, $g_{(0)ij}$ acts as the source for the dual stress tensor T_{ij}. The subleading coefficients $g_{(2k)ij}(x)$, with $k < m$, are then locally determined in terms of $g_{(0)ij}$ via an asymptotic analysis of the field equations. The coefficient $g_{(2m)ij}(x)$, however, is only partially constrained by this asymptotic analysis. (On the QFT side, these constraints correspond to the QFT Ward identities.) In fact, one finds that the coefficient $g_{(2m)ij}(x)$ is directly related to the expectation value of the boundary stress tensor [11, 24]:

$$\langle T_{ij} \rangle = \frac{1}{2\bar{\kappa}^2}(2mg_{(2m)ij}).   \tag{9.34}$$

An analogous relation also exists for the expectation value of the dual scalar operator in terms of the asymptotic behaviour of the bulk scalar field (see [11, 24] for details). We emphasize that this result only requires that Einstein equations hold asymptotically.

Here, we focused our discussion on the stress tensor, but an analogous discussion holds for all operators provided we specify appropriate boundary conditions for the corresponding bulk fields. If one includes such additional fields, then the holographic formulae such as (9.34) will in general acquire additional terms [3, 4], but the structure described above remains the same. More importantly for our purposes, since we are only interested in correlation functions of the stress tensor, we only need to turn on a source for the stress tensor, in which case the formulae above hold unchanged.[4]

The relation (9.34) may be read in two ways: (i) given a bulk gravitational solution we may read off the dual QFT data encoded by the solution; (ii) given QFT data we may reconstruct the bulk asymptotic solution. We stress that this asymptotic reconstruction is possible even when gravity is strongly coupled in the interior (corresponding to a weakly coupled boundary QFT). The coefficients up to $g_{(2m)ij}$ simply encode the boundary conditions, i.e., the fact that we are considering asymptotically locally AdS configurations (in the dual frame for the power-law case). In gauge/gravity duality, these terms encode the fact that we have turned on a source for the dual operator (the stress tensor for the case at hand), and this is unrelated to whether the dual QFT is at weak or strong coupling. The first term to depend on the bulk dynamics is $g_{(2m)ij}$. When gravity is weakly coupled, this coefficient is determined by the behaviour of the gravitational solution deep in the interior. When gravity is strongly coupled, this coefficient should be obtained by solving the full stringy dynamics in the interior. Gauge/gravity duality requires that the value obtained this way *must agree* with the $g_{(2m)ij}$ determined via (9.34) from the weakly coupled dual QFT.

9.3.3 Hamiltonian Holographic Renormalisation

In the following, rather than using (9.34) directly, we will instead employ the radial Hamiltonian formulation of [38, 39]. Here, the radial direction plays a role equivalent to that of time in the usual Hamiltonian formalism. The radial Hamiltonian formulation has a number of advantages for our present purposes; in particular, it

---

[4]Modulo contributions to (9.34) from condensates of low-dimension operators, cf. the discussion of the Coulomb branch flow in [3, 4]. Such cases can be analysed along similar lines but we will not discuss this here.

leads to a universal formula for the 1-point function that is independent of any of the issues (additional fields, etc.) discussed in the previous subsection. It further permits us to work with an arbitrary potential for the scalar field, so long as this potential admits background solutions of either the asymptotically AdS or the asymptotically power-law form. (In contrast, the formula (9.34) must be established on a case by case basis for different potentials, as in [3, 4, 11].)

A key feature of spacetimes of the form (9.33) is that, to leading order as $r \to \infty$, the radial derivative is equal to the dilatation operator δ_D, i.e.,

$$\partial_r = \delta_D \left(1 + O\left(e^{-2r}\right)\right), \tag{9.35}$$

where the δ_D acts on the metric as $\delta_D g_{ij}(x,r) = 2g_{ij}(x,r)$. (In particular, this means the scale factor a transforms as $\delta_D a = a$.) The bulk scalar field also transforms with a specific conformal weight. Equation (9.35) is a sharp version of the oft-quoted relation between the radial direction and the energy scale of the dual QFT. This equivalence allows one to trade the asymptotic radial expansion (9.33) for a covariant expansion in eigenfunctions of the dilatation operator. By definition, an eigenfunction $A_{(n)}$ of weight n satisfies

$$\delta_D A_{(n)} = -n A_{(n)},$$

hence, for example, the scale factor a has weight minus one. From (9.35), $A_{(n)} \sim e^{-nr}(1 + O(e^{-2r}))$, so the radial expansion and the expansion in eigenfunctions of the dilatation operator are closely related. The latter expansion is manifestly covariant, however, whereas expanding in the bulk radial coordinate is not a covariant operation.

In the radial Hamiltonian formalism then, the expectation value of the dual stress tensor is given by

$$\langle T^i_j \rangle = \left(\frac{-2}{\sqrt{g}} \Pi^i_j \right)_{(3)}, \tag{9.36}$$

where Π^i_j is the radial canonical momentum in Fefferman-Graham gauge where $N_i = 0$ and $N = 1$, and the subscript indicates taking the piece with overall dilatation weight three.[5] Indeed, one might have anticipated this on general grounds, since in three dimensions the conformal dimension of the stress tensor is three. Equation (9.36) is the universal formula we mentioned above.[6] To extract the piece with

---

[5] In odd bulk dimensions the transformation of this specific coefficient also has an additional anomalous contribution due to the conformal anomaly [17]. In four bulk dimensions there is no anomaly, however, and this coefficient is a true eigenfunction of δ_D.

[6] Strictly speaking, while (9.36) holds universally, expressing Π^i_j in terms of the coefficients in the asymptotic expansion of the bulk fields depends on the details of theory under consideration (field content, interactions, etc.). Fortunately, however, we will not need this information here.

dilatation weight three, Π^i_j may first be decomposed in eigenfunctions of the dilatation operator. In general, the radial canonical momentum will contain pieces with weight less than three: the process of holographic renormalisation then amounts to determining these terms through the asymptotic analysis and subtracting them. In [38, 39], it is shown that removing these pieces is equivalent to adding local boundary covariant counterterms to the on-shell action.

For asymptotically AdS domain-walls, the radial canonical momentum is

$$\Pi^i_j = \frac{1}{2\bar{\kappa}^2}\sqrt{g}\big(K^i_j - K\delta^i_j\big), \tag{9.37}$$

where $K_{ij} = (1/2)\partial_z g_{ij}$ is the extrinsic curvature of constant-z slices. (Recall for domain-walls, the z coordinate is a radial variable.) In the case of asymptotically power-law domain-walls, the relevant radial canonical momentum is instead that of the dual frame [24], namely

$$\tilde{\Pi}^i_j = \frac{1}{2\bar{\kappa}^2}\sqrt{\tilde{g}}e^{\lambda\Phi}\big(\tilde{K}^i_j - (\tilde{K} + \lambda\Phi_{,r})\delta^i_j\big). \tag{9.38}$$

Here, all tilded quantities belong to the dual frame and $\partial_r = e^{\lambda\varphi/2}\partial_z$. (Note the r.h.s. of (9.36) should also be evaluated in the dual frame.)

9.3.3.1 Constraint Equations

In our later analysis, we will need to make use of the Hamiltonian and momentum constraint equations, and so it is convenient to first present these here. As mentioned above, for asymptotically AdS domain-walls, the holographic analysis is performed in Fefferman-Graham gauge with $N = 1$ and $N_i = 0$. In the case of asymptotically power-law domain-walls, the holographic analysis also requires the choice of Fefferman-Graham gauge, but in the *dual frame*. The corresponding Einstein frame metric $g_{ij} = e^{\lambda\Phi}\tilde{g}_{ij}$ then has vanishing shift N_i but a nonzero lapse perturbation $\delta N = (\lambda/2)\delta\varphi$. In the following, to cover both cases, we will assume the shift has been gauged to zero but allow for a nonzero lapse perturbation.

Differentiating the Lagrangian (9.6) with respect to N and N_i, we obtain the domain-wall Hamiltonian and momentum constraints

$$0 = -R + K^2 - K_{ij}K^{ij} + 2\bar{\kappa}^2 V - N^{-2}\dot{\Phi}^2 + g^{ij}\Phi_{,i}\Phi_{,j}, \tag{9.39}$$

$$0 = \nabla_j\big(K^j_i - \delta^j_i K\big) - N^{-1}\dot{\Phi}\Phi_{,i}, \tag{9.40}$$

where $K_{ij} = (1/2N)\dot{g}_{ij}$. Expanding to linear order then yields

$$0 = -4a^{-2}\partial^2\psi + 2H\dot{h} + 4\bar{\kappa}^2 V\delta N - 2\dot{\varphi}\delta\dot{\varphi} + 2\bar{\kappa}^2 V'\delta\varphi, \tag{9.41}$$

$$0 = \frac{1}{2}\dot{h}_{ij,j} - \frac{1}{2}\dot{h}_{,i} + 2H\delta N_{,i} - \dot{\varphi}\delta\varphi_{,i}. \tag{9.42}$$

Acting on the latter equation with $\partial^{-2}\partial_i$ we may extract the scalar part

$$0 = 2\dot{\psi} - \dot{\varphi}\delta\varphi + 2H\delta N. \tag{9.43}$$

9.3.4 The Stress Tensor 2-Point Function

Prior to commencing our holographic calculation, let us briefly discuss the QFT correlator that will be of interest to us, the stress tensor 2-point function. As we have in mind some regular three-dimensional QFT dual to a domain-wall spacetime, we will denote momenta using $\bar{\mathbf{q}}$ rather than $\mathbf{q}$. The boundary metric on which the QFT lives will moreover be flat in the absence of sources.

Quite generally, the diffeomorphism Ward identity implies that the stress tensor 2-point function is transverse, i.e.,

$$0 = \bar{q}_i \langle\!\langle T_{ij}(\bar{q}) T_{kl}(-\bar{q}) \rangle\!\rangle.$$

Our double bracket notation here once again suppresses the delta function associated with momentum conservation, as in (9.28). It is then a simple exercise to show that only two transverse tensors with the correct symmetries can be built from the momentum $\bar{q}_i$ and the background metric δ_{ij}. Through this argument, we find the stress tensor 2-point function admits the general decomposition

$$\langle\!\langle T_{ij}(\bar{q}) T_{kl}(-\bar{q}) \rangle\!\rangle = A(\bar{q}) \Pi_{ijkl} + B(\bar{q}) \pi_{ij} \pi_{kl}. \tag{9.44}$$

Here, $A(\bar{q})$ encodes the transverse traceless piece of the 2-point function while $B(\bar{q})$ encodes the trace piece, since

$$\langle\!\langle T^{(s)}(\bar{q}) T^{(s')}(-\bar{q}) \rangle\!\rangle = \frac{1}{2} A(\bar{q}) \delta^{ss'}, \qquad \langle\!\langle T(\bar{q}) T(-\bar{q}) \rangle\!\rangle = 4B(\bar{q}),$$

where $T^{(s)}(\bar{\mathbf{q}}) = (1/2)\varepsilon_{ij}^{(s)}(-\bar{\mathbf{q}}) T_{ij}(\bar{\mathbf{q}})$ in parallel with our earlier treatment of $\hat{\gamma}^{(s)}(\bar{\mathbf{q}})$.

At a more formal level, the 2-point function encodes the variation of the 1-point function in the presence of sources, $\delta \langle T_j^i \rangle_s$, under a linear variation of the appropriate source, in this case the metric $g_{(0)ij}$ on which the QFT lives. Setting $g_{(0)ij} = \delta_{ij} + \delta g_{(0)ij}$, we therefore have

$$\delta \langle T_j^i(\mathbf{x}) \rangle_s = \delta^{im} \delta \langle T_{mj}(\mathbf{x}) \rangle_s = -\frac{1}{2} \int d\mathbf{x}' \delta^{im} \langle T_{mj}(\mathbf{x}) T_{kl}(\mathbf{x}') \rangle \delta g_{(0)}^{kl}(\mathbf{x}'),$$

where in the first equality we used the fact that the 1-point function $\langle T_{ij}(\mathbf{x})\rangle$ vanishes on a flat background (i.e., with the source set to zero). In momentum space, this becomes

$$\delta\langle T^i_j(\bar{\mathbf{q}})\rangle = -\frac{1}{2}\delta^{im}\langle\langle T_{mj}(\bar{q})T_{kl}(-\bar{q})\rangle\rangle\delta g^{kl}_{(0)}(\bar{\mathbf{q}}).$$

In particular, inserting (9.44) and decomposing the metric variation as in (9.3), we find

$$\delta\langle T^{(s)}(\bar{\mathbf{q}})\rangle_s = \frac{1}{2}A(\bar{q})\gamma^{(s)}_{(0)}(\bar{\mathbf{q}}), \qquad \delta\langle T(\bar{\mathbf{q}})\rangle_s = -4B(\bar{q})\psi_{(0)}(\bar{\mathbf{q}}). \qquad (9.45)$$

9.3.5 Holographic Analysis

Our goal is now to evaluate the stress tensor 2-point function in terms of the domain-wall response functions. To do so, we will expand (9.36) to linear order in the sources then compare with (9.45). We will deal first of all with the case of asymptotically AdS domain-walls before turning to the case of power-law asymptotics.

Working in Fefferman-Graham gauge where $N = 1$ and $N_i = 0$, expanding out (9.37) in momentum space to linear order, we find

$$\delta\langle T^{(s)}\rangle_s = -\frac{1}{2}\bar{\kappa}^{-2}\dot{\gamma}^{(s)}_{(3)}, \qquad \delta\langle T\rangle_s = \bar{\kappa}^{-2}\dot{h}_{(3)}. \qquad (9.46)$$

Substituting for $\dot{\gamma}^{(s)}$ using (9.11)–(9.12), the first of these equations reads

$$\delta\langle T^{(s)}\rangle_s = -2\bar{\kappa}^{-2}\left[a^{-3}\bar{E}(\bar{q})\gamma^{(s)}\right]_{(3)}.$$

As the factor a^{-3} has dilation weight three, the coefficient of the source $\gamma^{(s)}_{(0)}$ is $-2\bar{\kappa}^{-2}\bar{E}_{(0)}(\bar{q})$. Comparing with (9.45), we may then identify the transverse traceless piece of the 2-point function,

$$A(\bar{q}) = -4\bar{\kappa}^{-2}\bar{E}_{(0)}(\bar{q}). \qquad (9.47)$$

In this formula, the zero subscript indicates taking the piece of the response function that has zero weight under dilatations, i.e., the piece that is independent of r as $r \to \infty$. In general, $\bar{E}$ diverges as $r \to \infty$, and so to extract $\bar{E}_{(0)}$ correctly requires first determining the terms with eigenvalue less than zero and subtracting these from $\bar{E}$, before taking the limit $r \to \infty$ (see the example in the next subsection, and also [39]). The issue here is that the subtraction of the infinite pieces may induce a change in the finite part as well, which may happen if the local covariant counterterms needed to cancel the infinities necessarily have a finite part as well.

To identify the trace piece $B(\bar{q})$ of the stress tensor 2-point function we need to express $\dot{h}$ in terms of ψ, which requires use of the constraint equations. Setting $\delta N = 0$, the Hamiltonian and momentum constraints (9.41) and (9.43) take the form

$$\dot{h} = -\frac{2\bar{q}^2}{a^2 H}\psi + \frac{\dot{\varphi}}{H}\delta\dot{\varphi} + (\ldots)\delta\varphi, \qquad \dot{\psi} = (\ldots)\delta\varphi. \tag{9.48}$$

Here, and in the following, we will ignore terms proportional to $\delta\varphi$ since these do not contribute to $B(\bar{q})$. (Instead, since $\delta\phi_{(0)}$ sources the dual scalar operator O, they contribute to the correlator $\langle T\,O\rangle$.) Now, on the one hand, we have

$$\dot{\zeta} = \left(-\dot{\psi} - \frac{H}{\dot{\varphi}}\delta\dot{\varphi}\right) = -\frac{H}{\dot{\varphi}}\delta\dot{\varphi} + (\ldots)\delta\varphi, \tag{9.49}$$

while on the other hand,

$$\dot{\zeta} = \frac{1}{2a^3\varepsilon}\Pi = \frac{1}{2a^3\varepsilon}\bar{\Omega}(\bar{q})\zeta = -\frac{1}{2a^3\varepsilon}\bar{\Omega}(\bar{q})\psi + (\ldots)\delta\varphi. \tag{9.50}$$

Thus, at linear order,

$$\delta\dot{\varphi} = \frac{H}{a^3\dot{\varphi}}\bar{\Omega}(\bar{q})\psi + (\ldots)\delta\varphi, \qquad \dot{h} = \left(\frac{\bar{\Omega}(\bar{q})}{a^3} - \frac{2\bar{q}^2}{a^2 H}\right)\psi + (\ldots)\delta\varphi. \tag{9.51}$$

From (9.46) and (9.45), we then identify

$$B(\bar{q}) = -\frac{1}{4}\bar{\kappa}^{-2}\bar{\Omega}_{(0)}. \tag{9.52}$$

Note we have dropped the contribution to $B(\bar{q})$ from the term in (9.51) proportional to $\bar{q}^2$: this contribution simply amounts to a scheme-dependent contact term which may be removed through the addition of a finite local counterterm. In extracting the zero-dilatation weight piece of the response function $\bar{\Omega}_{(0)}$, similar considerations apply as discussed above for the case of $\bar{E}_{(0)}$.

Having seen how the stress tensor 2-point function for asymptotically AdS domain-walls is given by the zero-dilatation weight pieces of the appropriate response functions, let us now turn to the case of asymptotically power-law domain-walls. Fortunately the analysis is very closely related to that above. We start by writing the perturbed dual frame metric in Fefferman-Graham gauge as

$$d\tilde{s}^2 = e^{-\lambda\Phi}ds^2 = dr^2 + \tilde{a}^2[\delta_{ij} + \tilde{h}_{ij}]dx^i dx^j, \tag{9.53}$$

$$\tilde{h}_{ij} = -2\tilde{\psi}\delta_{ij} + 2\tilde{\chi}_{,ij} + 2\tilde{\omega}_{(i,j)} + \tilde{\gamma}_{ij}, \tag{9.54}$$

where $\tilde{a} = ae^{-\lambda\varphi/2}$ and $\mathrm{d}r = e^{-\lambda\varphi/2}\mathrm{d}z$. These dual frame perturbations are then related to their Einstein frame counterparts by

$$\tilde{\psi} = \psi + (\lambda/2)\delta\varphi, \qquad \tilde{\chi} = \chi, \qquad \tilde{\omega}_i = \omega_i, \qquad \tilde{\gamma}_{ij} = \gamma_{ij}.$$

In addition, we have a nonzero Einstein frame lapse perturbation $\delta N = (\lambda/2)\delta\varphi$ as noted previously.

The 1-point function in the presence of sources is given by (9.36), using the dual frame canonical momentum (9.38). Expanding (9.38) to linear order and converting dual frame perturbations to Einstein frame perturbations (as well as r-derivatives to z-derivatives), we obtain

$$\delta\langle T^{(s)}\rangle_s = -\frac{1}{2}\bar{\kappa}^{-2}\big[e^{3\lambda\varphi/2}\dot{\gamma}^{(s)}\big]_{(3)}, \qquad \delta\langle T\rangle_s = \bar{\kappa}^{-2}\big[e^{3\lambda\varphi/2}\dot{h} + (\ldots)\delta\varphi\big]_{(3)}.$$

One may now proceed as in the asymptotically AdS case, using the response functions to substitute for the radial derivatives of metric perturbations, the only difference being that there is now a nonzero lapse perturbation. Since the constraint equations (9.41) and (9.43) only involve δN and its spatial derivative $\delta N_{,i}$, but never its radial derivative $\delta\dot{N}$, the new terms involving the lapse perturbation can only contribute to the $(\ldots)\delta\varphi$ piece and never to the piece of interest proportional to ψ. (If there *were* a piece proportional to $\delta\dot{N} = (\lambda/2)\delta\dot{\varphi}$, this would contribute a term proportional to ψ via (9.51).) We may therefore recycle our analysis above giving

$$\delta\langle T^{(s)}\rangle_s = -2\bar{\kappa}^{-2}\big[e^{3\lambda\varphi/2}a^{-3}\bar{E}(\bar{q})\gamma^{(s)}\big]_{(3)}, \tag{9.55}$$

$$\delta\langle T\rangle_s = \bar{\kappa}^{-2}\left[e^{-3\lambda\varphi/2}\left(\frac{\bar{\Omega}(\bar{q})}{a^3} - \frac{2\bar{q}^2}{a^2H}\right)\psi + (\ldots)\delta\varphi\right]_{(3)}. \tag{9.56}$$

Returning to the dual frame,

$$\delta\langle T^{(s)}\rangle_s = -2\bar{\kappa}^{-2}\big[\tilde{a}^{-3}\bar{E}(\bar{q})\tilde{\gamma}^{(s)}\big]_{(3)}, \tag{9.57}$$

$$\delta\langle T\rangle_s = \bar{\kappa}^{-2}\left[\left(\frac{\bar{\Omega}(\bar{q})}{\tilde{a}^3} - \frac{2\bar{q}^2e^{\lambda\varphi/2}}{\tilde{a}^2H}\right)\tilde{\psi} + (\ldots)\delta\varphi\right]_{(3)}. \tag{9.58}$$

Since the dilatation weight of $\tilde{a}$ in the dual frame is minus one, examining the above we in fact recover precisely our previous results (9.47) and (9.52). These results are thus valid for both asymptotically AdS domain-walls as well as for asymptotically power-law domain-walls. In the latter case, however, the subtraction of terms with negative dilatation weight before sending $r \to \infty$ should be performed in the dual frame, as we will see in the following example.

9.3.5.1 An Example: Exact Power-Law Inflation

To illustrate the above discussion, let us consider the domain-wall backgrounds exactly equal (rather than merely asymptotic) to (9.32), namely

$$a = (z/z_0)^n, \qquad \varphi = \sqrt{2n}\ln(z/z_0), \qquad z_0 = n - 1 > 0.$$

Under the domain-wall/cosmology correspondence, these solutions are mapped to cosmologies undergoing exact power-law inflation. While this particular model is strongly constrained by the WMAP data [25], this need not concern us here since our purpose is simply to illustrate the steps involved in the holographic computation. Furthermore, we will see in Sect. 9.4 that the strong coupling version of these models (i.e., where gravity is strongly coupled at early times but the dual three-dimensional QFT is weakly coupled) are compatible with observations.

Referring back to the background equations of motion (9.7), we find the function $W = -(2n/z_0)\exp(-\varphi/\sqrt{2n})$. It then follows that $\varepsilon = 1/n$ and both mode functions $\hat{\gamma}_q$ and ζ_q obey the same equation of motion, which for the domain-wall spacetime reads

$$0 = \ddot{\zeta}_{\bar{q}} + (3n/z)\dot{\zeta}_{\bar{q}} - (z/z_0)^{-2n}\bar{q}^2\zeta_{\bar{q}}. \tag{9.59}$$

Imposing regularity in the interior, the solution is

$$\zeta_{\bar{q}} = C_{\bar{q}}\rho^\sigma K_\sigma(\rho),$$

where K_σ is a modified Bessel function of the second kind of order $\sigma = (3n - 1)/2(n - 1) > 3/2$, the radial coordinate $\rho = \bar{q}(z/z_0)^{1-n}$, and $C_{\bar{q}}$ is an arbitrary function of $\bar{q}$. The boundary $z \to \infty$ corresponds to $\rho = 0$ while the domain-wall interior corresponds to $\rho \to \infty$. The corresponding radial canonical momentum (times $\bar{\kappa}^2$) is equal to

$$\Pi_{\bar{q}} = \frac{2\varepsilon}{a^3}\dot{\zeta}_{\bar{q}} = -\frac{2C_{\bar{q}}}{n}\left(\frac{\rho}{\bar{q}}\right)^{-2\sigma}\rho\partial_\rho\left(\rho^\sigma K_\sigma(\rho)\right). \tag{9.60}$$

Expanding about $\rho = 0$, we find

$$\zeta_{\bar{q}} = C_{\bar{q}}\left(1 + \frac{1}{4(1-\sigma)}\rho^2 + \cdots - \frac{\Gamma(1-\sigma)}{4^\sigma\Gamma(1+\sigma)}\rho^{2\sigma} + \cdots\right),$$

$$\Pi_{\bar{q}} = -C_{\bar{q}}\frac{2\bar{q}^{2\sigma}}{n}\left(\frac{1}{2(1-\sigma)}\rho^{2(1-\sigma)} + \cdots - \frac{2\sigma\Gamma(1-\sigma)}{4^\sigma\Gamma(1+\sigma)} + \cdots\right), \tag{9.61}$$

and thus

$$\Omega(\bar{q}) = \frac{\Pi_{\bar{q}}}{\zeta_{\bar{q}}} = -\frac{2\bar{q}^{2\sigma}}{n}\left(\frac{1}{2(1-\sigma)}\rho^{2(1-\sigma)} + \cdots - \frac{2\sigma\Gamma(1-\sigma)}{4^\sigma\Gamma(1+\sigma)} + \cdots\right). \tag{9.62}$$

As expected, this diverges as $\rho \to 0$. To compute the 2-point function we need to identify the parts that have negative dilatation eigenvalue, subtract them from (9.62), and then take $\rho \to 0$.

To do this, we first transform to the dual frame via $\tilde{g}_{ij} = e^{-\sqrt{(2/n)}\varphi} g_{ij}$ and then change radial variable, $r = z_0 \ln(z/z_0) = -\ln(\rho/\bar{q})$. The metric is now that of AdS,

$$ds^2 = dr^2 + e^{2r} dx^2,$$

and the dilatation operator is exactly equal to the radial derivative,

$$\delta_D = \partial_r = -\rho \partial_\rho.$$

(This reflects the fact that the AdS isometry group is the same as the conformal group in one dimension less.) It follows that any monomial in ρ is an eigenfunction of δ_D,

$$\delta_D \rho^n = -n\rho^n,$$

and one can simply identify in (9.62) all terms with negative eigenvalue; for example, $\bar{\Omega}_{(-2\sigma+2)} = -\bar{q}^{2\sigma} \rho^{-2(\sigma-1)}/(n(1-\sigma))$. We then have

$$\bar{\Omega}_{(0)} = \frac{4\sigma \Gamma(1-\sigma)}{n4^\sigma \Gamma(1+\sigma)} \bar{q}^{2\sigma}.$$

In this example, the identification of the terms with negative eigenvalues could be accomplished by inspection. In more complicated examples, however, this is no longer the case, so we briefly indicate here how one could compute them (see [39] for a more complete discussion). Starting from (9.13) with $\sigma = +1$ and changing the radial coordinate from z to r, one obtains

$$\partial_r \bar{\Omega} + \frac{n}{2} \bar{\Omega}^2 e^{-2\sigma r} - \frac{2}{n} \bar{q}^2 e^{2(\sigma-1)r} = 0. \tag{9.63}$$

This equation may now be solved asymptotically by expanding Ω in dilatation eigenvalues,

$$\bar{\Omega} = \sum_{k \geq 1} \bar{\Omega}_{(-2\sigma+2k)},$$

making use[7] of $\partial_r = \delta_D$ and collecting all terms with the same weight. For example, to leading order, at weight $(-2\sigma+2)$, only the first and last term in (9.63) can have this weight, and one obtains $\bar{\Omega}_{(-2\sigma+2)} = -(\bar{q}^2/n(1-\sigma)) \exp(2(\sigma-1)r)$ in agreement with our earlier result. Through iteration, one may obtain all coefficients with negative eigenvalue.

---

[7] In examples where the background solution is only asymptotically AdS, the relation between the dilatation operator and the radial derivative contains subleading terms (see (9.35)) that must be taken into account. For a full discussion, see [39].

Having obtained $\bar{\Omega}_{(0)}$, we finally compute $B(\bar{q})$:

$$B(\bar{q}) = -\frac{1}{4}\bar{\kappa}^{-2}\bar{\Omega}_{(0)} = -\frac{\sigma\,\Gamma(1-\sigma)}{n4^\sigma\,\Gamma(1+\sigma)}\bar{\kappa}^{-2}\bar{q}^{2\sigma} = -\frac{\pi}{4^\sigma\,\Gamma^2(\sigma)n\sin\pi\sigma}\bar{\kappa}^{-2}\bar{q}^{2\sigma}.$$

A near-identical argument holds for the tensors $\hat{\gamma}^{(s)}$ yielding $\bar{\Omega}_{(0)} = (8/n)\bar{E}_{(0)}$, and hence $A(\bar{q}) = 2nB(\bar{q})$. Via the domain-wall/cosmology correspondence, applying the continuations (9.14), the imaginary parts of the cosmological response functions are

$$\mathrm{Im}\,\Omega_{(0)} = (8/n)\,\mathrm{Im}\,E_{(0)} = -\frac{4\pi}{n4^\sigma\,\Gamma^2(\sigma)}\kappa^{-2}q^{2\sigma}. \tag{9.64}$$

From (9.31), we then recover the expected cosmological power spectra:

$$\Delta_S^2(q) = \frac{n}{16}\Delta_T^2(q) = \frac{n4^{\sigma-2}\Gamma^2(\sigma)}{\pi^3}\kappa^2 q^{3-2\sigma}. \tag{9.65}$$

Note that we could equally well have obtained (9.64) by applying the continuations (9.14) to the *unrenormalised* domain-wall response function (9.62), then taking the imaginary part followed by the limit $z \to \infty$. This is because the divergent terms one subtracts to obtain the renormalised response functions are all analytic functions of $\bar{q}^2$ (as may be seen from (9.62), where the leading term is proportional to $\bar{q}^2$) and hence under the continuation $\bar{q}^2 = -q^2$, these terms remain real and do not contribute to the imaginary part of the cosmological response functions. Only the leading *non-analytic* piece of the domain-wall response functions contributes to the late-time imaginary part of the cosmological response functions: this leading non-analytic piece is finite and is simply $\bar{\Omega}_{(0)}$. In fact, the late-time values of the imaginary parts of the cosmological response functions have to be finite as a consequence of the Wronskian relations (9.24)–(9.25) and the fact that ζ and $\hat{\gamma}_{ij}$ tend to finite constants at late times.

9.3.6 Holographic Formulae for the Power Spectra

After the detailed arguments of the preceding subsections, let us summarise our progress thus far. Firstly, in (9.31), we expressed the cosmological power spectra in terms of the imaginary pieces of the cosmological response functions at late times. Secondly, in (9.47) and (9.52), we saw how the stress tensor 2-point function of the dual QFT is given by the zero-dilatation weight pieces of the corresponding domain-wall response functions. To extract these zero-dilatation weight pieces, the domain-wall response functions had first to be renormalised by subtracting counterterms with negative dilatation weight, before sending $z \to \infty$. As we saw in the previous subsection, however, these counterterms are necessarily analytic functions of $\bar{q}^2$, and so do not contribute to the imaginary part of the corresponding cosmological response function at late times. The latter is therefore precisely given by analytically continuing the zero-dilatation weight piece of the domain-wall response function according to (9.16) and taking the imaginary part. Putting all this together, we arrive at our principal result: that the cosmological power spectra are directly related to the stress tensor 2-point function of the dual QFT via the holographic formulae

$$\Delta_S^2(q) = \frac{-q^3}{16\pi^2 \, \text{Im} \, B(-iq)}, \qquad \Delta_T^2(q) = \frac{-2q^3}{\pi^2 \, \text{Im} \, A(-iq)}. \tag{9.66}$$

In these formulae, as well as the analytic continuation of momentum indicated, one must also continue $\bar{N} = -iN$. As one might expect, the scalar power spectrum is related to the trace piece of the stress-tensor 2-point function, while the tensor power spectrum is related to the transverse traceless piece of the 2-point function.

9.4 Holographic Phenomenology for Cosmology

As noted in the introduction, one of the most striking features of holographic dualities is that they are strong/weak coupling dualities, meaning that when one description is weakly coupled, the other is strongly coupled, and vice versa. In the regime where the dual QFT is strongly coupled then, the gravitational description is weakly coupled and our holographic formulae should (and indeed they do) reproduce the results of standard single-field inflation. In this situation, application of the holographic framework offers a fresh perspective, and may lead to new insights, but offers no new predictions.

In the regime in which the dual QFT is weakly coupled, however, the corresponding gravitational description is instead *strongly coupled* at very early times. Let us emphasize that by 'strongly coupled' gravity we do *not* mean that the perturbative fluctuations around the background FRW spacetime are strongly coupled, but rather, that the description in terms of metric fluctuations is itself not valid. This is a non-geometric 'stringy' phase. A geometric description emerges only asymptotically, and at late times one recovers a specific accelerating FRW spacetime (to be matched to conventional hot big bang cosmology), along with a specific set of inhomogeneities. Crucially, these inhomogeneities are not linked with a perturbative quantisation around the FRW spacetime as in conventional inflation, but rather, they originate from the dynamics of the dual weakly coupled QFT. Holography thus suggests a natural generalisation of the inflationary mechanism to strongly coupled gravity, in which the properties of cosmological perturbations may be determined through three-dimensional perturbative QFT calculations. To follow these late-time inhomogeneities through the reheating transition to the post-inflationary universe, just as in conventional inflation, one then makes use of the conservation of ζ and $\hat{\gamma}_{ij}$ on superhorizon scales (or more generally, the 'separate universes' argument, see e.g., [8, 42]).

In order to compute the observational predictions of such a scenario, it is necessary to specify more precisely the nature of the dual QFT. Ideally, one would be able to deduce this from first principles via some string/M-theoretic construction. In the absence of such a construction, we will instead pursue a (holographic) phenomenological approach. As with other known holographic dualities, the dual QFT will in general involve scalars, fermions and gauge fields, and it should admit a large N limit. The question is then whether one can find a theory which is compatible with current observations. A further guiding principle is to consider QFTs of the type featured earlier in Sect. 9.3.1, for which the holographic dual is well understood. One might thus consider either deformations of CFTs (dual to asymptotically de Sitter cosmologies) or else QFTs with a generalised conformal structure (dual to asymptotically power-law cosmologies). In the following we will focus on the latter class of QFTs, leaving exploration of the former to future work.

9.4.1 A Prototype Dual QFT

Any QFT dual to an asymptotically power-law cosmology is required to satisfy quite a restrictive set of properties [24]. Specifically, (i) it should admit a large-N limit, (ii) all fields should be massless, (iii) it should have a dimensionful coupling constant, and (iv) all terms in the Lagrangian should have the same scaling dimension, which should be different from three. The properties (ii)–(iv) imply that the theory admits a *generalised conformal structure* [22], i.e., the theory would be conformal if the coupling constant is promoted to a background field transforming non-trivially under conformal transformations.

A simple class of models exhibiting these properties is given by three-dimensional $SU(\bar{N})$ Yang-Mills theory coupled to a number of massless scalars and fermions, all transforming in the adjoint of $SU(\bar{N})$, and with interactions consisting of Yukawa and quartic scalar terms. (We write the rank of the QFT gauge group as $\bar{N}$ here, since we will first be performing calculations using the QFT dual to the domain-wall spacetime before analytically continuing to the pseudo-QFT.) Theories of this type are typical in holography where they appear as the worldvolume theories of D-branes. In three dimensions, the Yang-Mills coupling g_{YM}^2 has dimension one and so the theory is super-renormalisable. Moreover, by rescaling the fields appropriately, one may arrange that the coupling appears only as an overall constant multiplying the action. Assigning scaling dimension one to scalars and gauge fields, and $3/2$ to fermions, one finds that kinetic terms and the interactions all have dimension four. Allowing $\mathcal{N}_A$ gauge fields A^I ($I = 1, \ldots, \mathcal{N}_A$); $\mathcal{N}_\phi$ minimal scalars ϕ^J ($J = 1, \ldots, \mathcal{N}_\phi$); $\mathcal{N}_\chi$ conformal scalars χ^K ($K = 1, \ldots, \mathcal{N}_\chi$) and $\mathcal{N}_\psi$ fermions ψ^L ($L = 1, \ldots, \mathcal{N}_\psi$), the Lagrangian then takes the form

$$S = \frac{1}{g_{YM}^2} \int d^3x \, \text{tr} \left[\frac{1}{2} F_{ij}^I F^{Iij} + \frac{1}{2} \left(D\phi^J \right)^2 + \frac{1}{2} \left(D\chi^K \right)^2 + \bar{\psi}^L D\psi^L \right]$$

$$+ \lambda_{M_1 M_2 M_3 M_4} \Phi^{M_1} \Phi^{M_2} \Phi^{M_3} \Phi^{M_4} + \mu_{M L_1 L_2}^{\alpha\beta} \Phi^M \psi_\alpha^{L_1} \psi_\beta^{L_2} \Big]. \qquad (9.67)$$

Here, the couplings $\lambda_{M_1 M_2 M_3 M_4}$ and $\mu_{M L_1 L_2}^{\alpha\beta}$ (where α and β are spinor indices) are dimensionless, and we have grouped the scalars appearing in the interaction terms as $\Phi^M = (\{\phi^J\}, \{\chi^K\})$. When we couple the theory to gravity, the conformal scalars acquire an additional $R\chi^2$ coupling; on a flat background this means the conformal scalars have a different stress tensor to their minimally coupled counterparts. Specifically, the stress tensor on a flat background is given by

$$T_{ij} = \frac{1}{g_{YM}^2} \, \text{tr} \Big[2 F_{ik}^I F_j^{Ik} + D_i \phi^J D_j \phi^J + D_i \chi^K D_j \chi^K$$

$$- \frac{1}{8} D_i D_j (\chi^K)^2 + \frac{1}{2} \bar\psi^L \gamma_{(i} \overleftrightarrow{D}_{j)} \psi^L$$

$$- \delta_{ij} \Big(\frac{1}{2} F_{kl}^I F^{Ikl} + \frac{1}{2} (D\phi^J)^2 + \frac{1}{2} (D\chi^K)^2 - \frac{1}{8} D^2 (\chi^K)^2$$

$$+ \lambda_{M_1 M_2 M_3 M_4} \Phi^{M_1} \Phi^{M_2} \Phi^{M_3} \Phi^{M_4} + \mu_{M L_1 L_2}^{\alpha\beta} \Phi^M \psi_\alpha^{L_1} \psi_\beta^{L_2} \Big) \Big]. \qquad (9.68)$$

9.4.2 Calculating the Holographic Power Spectra

To extract predictions, we need to compute the coefficients $A(\bar q)$ and $B(\bar q)$ appearing in the general decomposition (9.44) of the stress tensor 2-point function, analytically continue the results, and then insert them in the holographic formulae (9.66) for the power spectra. This task is made somewhat simpler by the generalised conformal structure and large-$\bar N$ counting, which together imply that the general form of the 2-point function at large $\bar N$ is

$$A(\bar q) = \bar q^3 \bar N^2 f_A(g_{eff}^2), \qquad B(\bar q) = \bar q^3 \bar N^2 f_B(g_{eff}^2), \qquad (9.69)$$

where $f_A(g_{eff}^2)$ and $f_B(g_{eff}^2)$ are general functions[8] of the dimensionless effective 't Hooft coupling

$$g_{eff}^2 = g_{YM}^2 \bar N / \bar q.$$

Under the QFT analytic continuations (9.17),

$$\bar N^2 \bar q^3 \to -i N^2 q^3, \qquad g_{eff}^2 \to g_{eff}^2,$$

---

[8]If instead we had imposed only the generalised conformal structure and not the large-$\bar N$ counting, then the r.h.s. of (9.69) would be modified as $\bar N^2 f(g_{eff}^2) \to f(\bar N^2, g_{eff}^2)$, where $f(\bar N^2, g_{eff}^2)$ is a general function of two variables.

Fig. 9.2 1-loop contribution to the stress tensor 2-point function. We sum over the contributions from gauge fields, scalars and fermions, with each diagram yielding a contribution of order $\sim \bar{N}^2 \bar{q}^3$

$T_{ij}(q)$ $T_{kl}(-q)$

hence $A(\bar{q})$ and $B(\bar{q})$ continue very simply in theories with generalised conformal invariance. (Recall here that the invariance of g_{eff}^2 was our original reason for continuing $\bar{N} = -iN$ and not $\bar{N} = +iN$ in (9.17). The invariance of g_{eff}^2 is required since $f_A(g_{\text{eff}}^2)$ and $f_B(g_{\text{eff}}^2)$ are in general non-analytic functions of g_{eff}^2.) Inserting (9.69) into the holographic formulae (9.66) then, the cosmological power spectra are

$$\Delta_S^2(q) = \frac{1}{16\pi^2 N^2} \frac{1}{f_A(g_{\text{eff}}^2)}, \qquad \Delta_T^2(q) = \frac{2}{\pi^2 N^2} \frac{1}{f_B(g_{\text{eff}}^2)}. \qquad (9.70)$$

In principle these formulae receive subleading $1/N^2$ corrections, however, as we shall see shortly, the observational data favour $N \sim 10^4$ rendering such terms negligible in practice. In the following, we now turn to evaluate the functions $f_A(g_{\text{eff}}^2)$ and $f_B(g_{\text{eff}}^2)$ in the perturbative limit where g_{eff}^2 is small.

9.4.2.1 1-Loop Calculation

The leading contribution to the 2-point function of the stress tensor is at one loop (see Fig. 9.2). Since the stress tensor has dimension three, and the only dimensionful quantity that can appear to this order is $\bar{q}$ (1-loop amplitudes are independent of g_{YM}^2), it follows that

$$A(\bar{q}) = f_A^{(0)} \bar{N}^2 \bar{q}^3 + O(g_{\text{eff}}^2), \qquad B(\bar{q}) = f_B^{(0)} \bar{N}^2 \bar{q}^3 + O(g_{\text{eff}}^2), \qquad (9.71)$$

i.e., $f_{A/B} = f_{A/B}^{(0)} + O(g_{\text{eff}}^2)$ where $f_{A/B}^{(0)}$ are numerical coefficients whose value depends only on the field content. Explicit calculation then reveals that

$$f_A^{(0)} = (\mathcal{N}_A + \mathcal{N}_\phi + \mathcal{N}_\chi + 2\mathcal{N}_\psi)/256, \qquad f_B^{(0)} = (\mathcal{N}_A + \mathcal{N}_\phi)/256. \qquad (9.72)$$

Inserting this into our holographic formulae, we find

$$\Delta_S^2(q) = \frac{1}{16\pi^2 N^2 f_B^{(0)}} + O(g_{\text{eff}}^2), \qquad \Delta_T^2(q) = \frac{2}{\pi^2 N^2 f_A^{(0)}} + O(g_{\text{eff}}^2). \qquad (9.73)$$

From an observational perspective, the cosmological power spectra are known to be well fitted by the empirical parametrisations

$$\Delta_S^2(q) = \Delta_S^2(q_*)\left(\frac{q}{q_*}\right)^{n_S(q)-1}, \qquad \Delta_T^2(q) = \Delta_T^2(q_*)\left(\frac{q}{q_*}\right)^{n_T(q)} \qquad (9.74)$$

where $\Delta_{S/T}^2(q_*)$ is the scalar/tensor amplitude at some chosen pivot scale q_*, and $n_{S/T}(q)$ is the scalar/tensor spectral tilt. Comparing with (9.73), we see immediately that the power spectra are *scale-invariant* to leading order (i.e. $n_S = 1 + O(g_{\text{eff}}^2)$, $n_T = O(g_{\text{eff}}^2)$), regardless of the precise field content of the model. To estimate the value of N we may compare with the observed amplitude of the scalar power spectrum. From the WMAP data [25] we have $\Delta_S^2(q_*) \sim O(10^{-9})$, hence $N \sim O(10^4)$, justifying our use of the large N limit.

The observational data also serve to provide an upper bound on the ratio of tensor to scalar power spectra. From (9.73), we find

$$r = \Delta_T^2/\Delta_S^2 = 32 f_B^{(0)}/f_A^{(0)} + O(g_{\text{eff}}^2),$$

and hence an upper bound on r translates into a constraint on the field content of the dual QFT through (9.72). A smaller upper bound on r requires increasing the number of conformal scalars and massless fermions and/or decreasing the number of gauge fields and minimal scalars.

9.4.2.2 2-Loop Corrections

Corrections to the stress tensor 2-point function at 2-loop order give rise to small deviations from scale invariance. In the following we will focus on the case of the scalar power spectrum, since this is the more tightly constrained by observational data. (The behaviour of the tensor power spectrum is essentially identical, however, with only the values of the coefficients being different.) At 2-loop order then, either by inspection or from direct calculation of some of the contributing diagrams depicted in Fig. 9.3, the function $f_B(g_{\text{eff}}^2)$ takes the form

$$f_B(g_{\text{eff}}^2) = f_B^{(0)}\left(1 - f_B^{(1)}g_{\text{eff}}^2 \ln g_{\text{eff}}^2 + f_B^{(2)}g_{\text{eff}}^2 + O(g_{\text{eff}}^4)\right), \qquad (9.75)$$

where $f_B^{(1)}$ and $f_B^{(2)}$ are numerical coefficients depending on the QFT field content, as well as the Yukawa and the quartic couplings.

As is well known, in perturbation theory super-renormalisable theories with massless fields display severe infrared divergences. Indeed, each of the 2-loop diagrams listed in Fig. 9.3 evaluates to an overall factor of $\bar{N}^3 g_{\text{YM}}^2$ multiplying an integral with superficial degree of (infrared) divergence two. Imposing an infrared

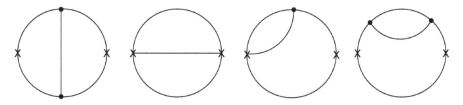

Fig. 9.3 Diagram topologies contributing at 2-loop order

cut-off, $\bar{q}_{\mathrm{IR}}$, one may evaluate the integrals to obtain $\sim \bar{q}^2 \ln(\bar{q}/\bar{q}_{\mathrm{IR}})$. Altogether, one finds a 2-loop contribution to the stress tensor 2-point function of the order

$$\bar{N}^2 \bar{q}^3 g_{\mathrm{eff}}^2 \ln(\bar{q}/\bar{q}_{\mathrm{IR}}) = \bar{N}^2 \bar{q}^3 \left(-g_{\mathrm{eff}}^2 \ln g_{\mathrm{eff}}^2 + g_{\mathrm{eff}}^2 \ln(g_{\mathrm{YM}}^2 \bar{N}/\bar{q}_{\mathrm{IR}})\right). \tag{9.76}$$

Thus, $f_B^{(1)}$ is determined by the full 2-loop calculation but $f_B^{(2)}$ remains undetermined, since $\bar{q}_{\mathrm{IR}}$ is so far arbitrary. It was argued in [19], however, that this infrared divergence is an artefact of perturbation theory and that instead the theory develops a physical scale that acts as a cut-off. To compute this scale generally requires nonperturbative information. For a specific class of models, it was shown in [2] that a large-$\bar{N}$ resummation leads to a finite answer with $\bar{q}_{\mathrm{IR}} \sim g_{\mathrm{YM}}^2 \bar{N}$. Similar behaviour is expected for the class of QFTs we consider here, but a precise determination of the infrared scale $\bar{q}_{\mathrm{IR}}$ (and hence $f_B^{(2)}$) is not yet to hand.

Instead, we will simply assume that all the cosmological scales relevant to the CMB lie far above the infrared scale $\bar{q}_{\mathrm{IR}}$, allowing the effects of the latter to be neglected. The validity of this assumption may then be cross-checked through comparison with the observational data. To this end, we rearrange (9.75) in the form

$$f\left(g_{\mathrm{eff}}^2\right) = f_B^{(0)} \left(1 + f_B^{(1)} g_{\mathrm{eff}}^2 \ln\left(1/\left(f_B^{(3)} g_{\mathrm{eff}}^2\right)\right) + O\left(g_{\mathrm{eff}}^4\right)\right), \tag{9.77}$$

where $f_B^{(3)} = \exp(-f_B^{(2)}/f_B^{(1)})$. Thus, as long as we probe the theory at momentum scales far above $\bar{q}_{\mathrm{IR}}$, the specific value of $f_B^{(3)}$ should only provide a small correction since $|\ln g_{\mathrm{eff}}^2| \gg |\ln f_B^{(3)}|$. We will thus write $f_B^{(3)} = \beta |f_B^{(1)}|$ and take $\beta = 1$ in the following. (For the effects of allowing the parameter β to vary, see [14].)

To simplify our notation, we set

$$f_B^{(1)} g_{\mathrm{YM}}^2 \bar{N} = g\bar{q}_*, \tag{9.78}$$

where $\bar{q}_*$ is the pivot scale. Substituting back into (9.70), we obtain the following 2-loop approximation to the power spectrum[9]

---

[9]Note that in previous treatments [33–35] we chose to Taylor expand the result (9.79); here, we retain the full form to provide better accuracy in the case that gq_*/q is not so small.

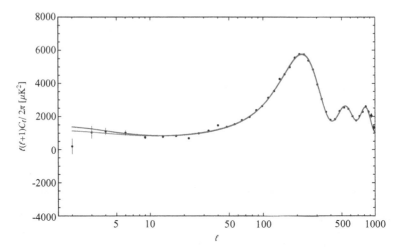

Fig. 9.4 Perturbative theoretical prediction for the power spectrum of the holographic model. The *lower curve* corresponds to $g > 0$ while the *upper* corresponds to $g < 0$. The perturbative calculation is reliable for $g_{\mathrm{eff}}^2 \sim gq_*/q \ll 1$, corresponding to large momenta $q/gq_* \gg 1$ far from the peak/trough feature at $\ln|q/gq_*| = 1$. At sufficiently high momenta, the power spectrum becomes nearly scale invariant, with $g > 0$ corresponding to a blue tilt and $g < 0$ to a red tilt

$$\Delta_S^2(q) = \Delta_S^2(q_*) \frac{1}{1 + (gq_*/q)\ln|q/gq_*|}, \qquad (9.79)$$

where $\Delta_S^2(q_*) = 1/(4\pi^2 N^2 f_B^{(0)})$.

The power spectrum (9.79) is plotted in Fig. 9.4 for both positive and negative g. At sufficiently large momenta the spectrum rapidly becomes nearly scale invariant, with positive values of g resulting in a slight blue tilt and negative values of g yielding a slight red tilt. This behaviour reflects the fact that the dual QFT becomes asymptotically free at high momenta, with the free theory itself corresponding to an exact Harrison-Zel'dovich spectrum.

At lower momenta, the existence of the non-perturbative infrared scale q_{IR} becomes apparent, resulting in the peak/trough feature in the spectrum at $q = egq_*$. Note, however, that the perturbative calculation of $f_B(g_{\mathrm{eff}}^2)$ underpinning the power spectrum (9.79) breaks down when $g_{\mathrm{eff}}^2 \sim gq_*/q$ becomes of order unity (recalling that $f_B^{(1)}$ is a constant of order unity). This means that the perturbative result (9.79) becomes unreliable at low momenta close to the peak/trough feature in Fig. 9.4. Moreover, our approximation $\beta = 1$ is no longer justified in this regime and one should retain β as an independent parameter. Since the smallest momentum scale appearing in the CMB is of the order 10^{-4} Mpc^{-1}, if the power spectrum (9.79) is to reliably fit the entire range of CMB scales, then we conclude that the maximum value of g is restricted to be of the order $|g|_{\mathrm{max}} \sim 2 \times 10^{-3}$.

9.5 Confronting Observations

Having obtained the 2-loop approximation (9.79) to the holographic power spectrum, in this section we discuss its compatibility with the current observational data [14] (see also [13]). In addition to specifying the primordial power spectrum (9.79), we need to specify the matter content of the post-inflationary universe: for simplicity we will assume a six-parameter 'holographic-ΛCDM' model describing a flat universe with radiation, baryons, cold dark matter and a cosmological constant. Four of the six parameters thus describe the composition and expansion of the universe, namely the Hubble rate $H_0 = 100h$ km/s/Mpc, the physical baryon and dark matter densities $\Omega_b h^2$ and $\Omega_c h^2$, and the optical depth due to re-ionisation τ. (Given that we do not need spatial curvature to fit the data, the current dark energy contribution then follows from the requirement that the overall density of the universe is equal to the critical value.) The remaining two parameters are those featuring in the holographic power spectrum (9.79), namely the amplitude $\Delta_S^2(q_*)$ and the holographic coupling g. (The pivot scale is arbitrary and we will take it to be $\bar{q}_* = 0.05$ Mpc^{-1}.)

As a benchmark for the performance of the holographic model, we will also evaluate the performance the conventional power-law ΛCDM model. This latter model may be obtained by replacing the holographic power spectrum (9.79) with a spectrum of the power-law form (9.74), with the spectral index n_s assumed to be constant. (Such a power spectrum provides a good approximation to the predictions of simple conventional inflationary models, for which the running $\alpha_s = \mathrm{d}n_s/\mathrm{d}\ln q$ is of higher order in slow roll than the departure from scale invariance $n_s - 1$ [27].) Both ΛCDM and the holographic model thus have six parameters: in place of the holographic coupling g, ΛCDM has the spectral index n_s, with the other five parameters $\Omega_b h^2$, $\Omega_c h^2$, h, τ and $\Delta_S^2(q_*)$ being common to both models.

The best-fit values for the six parameters of the holographic-ΛCDM model are summarised[10] in Table 9.1, based on the analysis of [14]. Analogous results for power-law ΛCDM may be found in Table 9.2. The results are quoted both for the seven-year WMAP data [26], as well as for the combined data sets WMAP+BAO+H_0 and WMAP+CMB also introduced in [26]. (The former is a combination of WMAP7 with priors on the Hubble constant [41] and angular diameter distances [40], while the latter is a combination of WMAP7 with small-scale CMB experiments.)

Comparing Tables 9.1 and 9.2, we see that the estimated values of those parameters common to both models are essentially overlapping, with only $\Omega_b h^2$ differing by about one standard deviation. The best-fit value of the holographic coupling g is as expected small, indicating a nearly scale-invariant spectrum. The best-fit value of g is not so small, however, that we can be fully comfortable with our approximation

---

[10]Note we have exchanged the dimensionless Hubble parameter h for the parameter θ denoting the ratio between the sound horizon at the time of last scattering and the angular diameter distance of the surface of last scattering. Physically, this ratio fixes the position of the acoustic peaks and is tightly constrained by the data. Theoretically, θ is a function of $\Omega_b h^2$, $\Omega_c h^2$ and h, hence given $\Omega_b h^2$ and $\Omega_c h^2$, θ may be expressed in terms of h (for more details see, e.g., Sect. 7.2 of [51]).

Table 9.1 Parameters of holographic-ΛCDM and their uncertainties at the 68 % confidence level

	WMAP7	WMAP+BAO+H_0	WMAP+CMB
$\Omega_b h^2$	0.02310 ± 0.00045	0.02312 ± 0.00043	0.02326 ± 0.00045
$\Omega_c h^2$	0.1077 ± 0.0051	0.1120 ± 0.0036	0.1076 ± 0.0042
100θ	1.0407 ± 0.0026	1.0406 ± 0.0026	1.0423 ± 0.0022
τ	0.087 ± 0.015	0.084 ± 0.015	0.088 ± 0.016
Δ_S^2	$(2.146 \pm 0.088) \times 10^{-9}$	$(2.172 \pm 0.086) \times 10^{-9}$	$(2.151 \pm 0.084) \times 10^{-9}$
g	-0.00127 ± 0.00093	-0.00136 ± 0.00094	-0.00114 ± 0.00088

Table 9.2 Parameters of power-law ΛCDM and their uncertainties at the 68 % confidence level

	WMAP7	WMAP+BAO+H_0	WMAP+CMB
$\Omega_b h^2$	0.02252 ± 0.00056	0.02257 ± 0.00053	0.02265 ± 0.00051
$\Omega_c h^2$	0.1116 ± 0.0054	0.1127 ± 0.0035	0.1124 ± 0.0048
100θ	1.0394 ± 0.0027	1.0400 ± 0.0026	1.0411 ± 0.0022
τ	0.088 ± 0.014	0.088 ± 0.014	0.088 ± 0.014
$\Delta_S^2(q_*)$	$(2.183 \pm 0.073) \times 10^{-9}$	$(2.191 \pm 0.075) \times 10^{-9}$	$(2.190 \pm 0.068) \times 10^{-9}$
n_s	0.969 ± 0.014	0.970 ± 0.012	0.969 ± 0.013

Table 9.3 Best-fit log likelihood values $-\ln\mathscr{L}$ for both the holographic model and ΛCDM, as well as the difference $\Delta \ln\mathscr{L} = \ln\mathscr{L}_{\Lambda\mathrm{CDM}} - \ln\mathscr{L}_{\mathrm{hol}}$ between them. The errors on the best-fit log likelihoods are estimated to be around 0.1

	Holographic Model	ΛCDM	$\Delta \ln\mathscr{L}_{\mathrm{best}}$
WMAP7	3735.5	3734.3	1.2
WMAP+BAO+H_0	3737.3	3735.7	1.6
WMAP+CMB	3815.0	3812.5	2.5

$|g|q_*/q \ll 1$ used to derive the holographic power spectrum. At the lower end of the range of momentum scales contributing to the CMB, $q \approx 10^{-4}$ Mpc^{-1}, we find $|g|q_*/q \approx 0.65$, indicating that the higher-order loop corrections and the effects of the infrared scale are potentially becoming important. Understanding the magnitude and significance of these effects is an important goal for future work.

The best-fit log likelihoods for both models are summarised in Table 9.3. The likelihood function $\mathscr{L}(\alpha_M) \equiv P(D|\alpha_M)$ encodes the probability of obtaining the data D, given the model M with some choice of parameters α_M. Thus, from Table 9.3, the probability of obtaining the observed WMAP7 data is approximately three times as likely given the power-law ΛCDM model with all parameters set to their best-fit values as for the holographic model, also with its parameters set to their best-fit values. Power-law ΛCDM is therefore slightly better at fitting the data, as illustrated in Fig. 9.5.

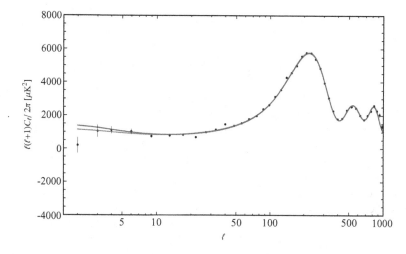

Fig. 9.5 Best-fit angular power spectra for power-law ΛCDM (*lower curve*, coloured red online) and holographic model (*upper curve*, coloured green online), versus the WMAP7 data

In performing a comparison of two models, however, the relevant quantity is *not* goodness of fit as measured by the best-fit log likelihood, but rather, the *Bayesian evidence* (see, e.g., [29] for an extended discussion). Given two models M_1 and M_2, we wish to evaluate which model is most probable given the data, i.e., the ratio

$$\frac{P(M_1|D)}{P(M_2|D)} = \frac{P(D|M_1)}{P(D|M_2)}\frac{P(M_1)}{P(M_2)},$$

where from Bayes' theorem

$$P(M_1|D) = \frac{P(D|M_1)P(M_1)}{P(D)}$$

and similarly for M_2. (The unconditional probability for the data $P(D)$ is a model-independent constant and so drops out of the ratio $P(M_1|D)/P(M_2|D)$.) Assuming that each model is a priori equally as likely so $P(M_1)/P(M_2)$ is unity, the relevant quantity to compute is then the *evidence ratio* E_1/E_2, where

$$E_1 \equiv P(D|M_1) = \int \mathrm{d}\alpha_{M_1}\, P(\alpha_{M_1})\mathscr{L}(\alpha_{M_1})$$

and similarly for M_2. The evidence thus naturally takes into account our uncertainty regarding the parameters of the model by integrating the likelihood over the entire parameter space, weighted by the prior probability $P(\alpha_M)$. (In contrast, the best-fit likelihood is simply the maximum value attained by the likelihood at any single point in this parameter space.)

To compute the evidence then, we need to assign prior probability distributions $P(\alpha_M)$ for the parameters of each model. If we assume flat priors, namely, a prior probability that is constant over some defined region and zero outside, the evidence reduces to the integral

$$E = \frac{1}{\mathrm{Vol}_M} \int d\alpha_M \mathscr{L}(\alpha_M), \qquad (9.80)$$

where Vol_M is the volume of the region in parameter space over which the prior probability distribution is non-zero. If the likelihood function is strongly peaked with support over only a relatively small region inside Vol_M, changing the prior region can strongly affect the computed evidence. Provided the changes to the overall volume of the parameter space do not add or exclude regions where $\mathscr{L}$ is large, the integral will be unaffected while Vol_M can change substantially, with the computed evidence being inversely proportion to Vol_M.

With the exception of n_s and g, both models have the same parameters. By using the same priors for the variables shared by the holographic and standard ΛCDM scenarios the ambiguity in the evidence associated with Vol_M is minimised. The situation with g and n_s is however more problematic. For the holographic model, we should restrict g to values where perturbative expansion used to derive (9.79) is valid. As we estimated at the end of Sect. 9.4.2.2, this corresponds to restricting $|g| < |g|_{\max} \approx 2 \times 10^{-3}$. (In fact the computed value of the evidence is only mildly dependent on the value of $|g|_{\max}$, as we will see in Figs. 9.6 and 9.7, see also [14].)

The choice prior for n_s is less straightforward since, unlike $|g|_{\max}$, the spectral index is a purely empirical parameter and we cannot restrict it by appealing to the internal consistency of some underlying theory. Moreover, our best information about n_s is derived from the WMAP data we are using to compute the evidence, and it would be inappropriately circular to set the prior on n_s directly from a parameter estimate derived from the WMAP data itself! To illustrate the consequences of this dilemma, we will consider two different choices of prior, $0.92 < n_s < 1.0$ and $0.9 < n_s < 1.1$. The first choice includes only the range over which the likelihood is appreciably different from zero, maximising the evidence for ΛCDM at the risk of being circular. The second choice is centered symmetrically on the scale-invariant Harrison-Zel'dovich spectrum, and hence does not provide any information about the sign of n_s. (In this sense the second choice is fairer, since we do not provide the holographic model with the sign of the tilt, corresponding to the sign of g, either.)

The result of the evidence calculation for the WMAP7 data set is presented in Fig. 9.6, and the results including the other data sets are given in Fig. 9.7. (Details of the numerical implementation of the computation and the choice of priors for the parameters common to both models may be found in [14].) Examining the plots, if we assume the narrow prior of $0.92 < n_2 < 1$, we find the difference in $-\ln E$ is of order 1.2 to 1.6, which corresponds to weak evidence in favour of ΛCDM. The difference in evidence is slightly more pronounced for the combined data sets WMAP7+BAO+H_0 and WMAP+CMB when compared to the pure WMAP7 data set. On the other hand, for the wider prior of $0.9 < n_s < 1.1$, the difference in evidence is less than unity, and as such is not considered statistically significant. Regardless of which prior is used, we do not find any strong evidence in favour of ΛCDM, where 'strong' evidence is generally taken to mean differences in $-\ln E$

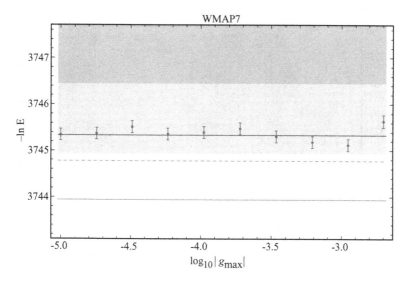

Fig. 9.6 The Bayesian evidence for the WMAP7 data set. The ΛCDM model with the narrow and broad priors corresponds to the *lowermost solid line* and the *middle dashed line* respectively (coloured orange online). The *uppermost solid line* (*red*) is the evidence for the pure Harrison-Zel'dovich spectrum with $n_s = 1$. The data points represent evidence computed for the holographic model, as a function of $|g|_{max}$, as indicated on the horizontal axis. As a guide to the eye, the shading indicates differences in evidence $\Delta \ln E$ of 1, 2.5 and 5 relative to ΛCDM with the narrow prior (i.e., relative to the lowermost solid line)

of greater than 2.5. At this stage then, the only firm conclusion that can be drawn is that better data is required. Fortunately, with the imminent release of data from the Planck satellite we will not have long to wait.

An important clue to the theoretical issues at stake is provided by comparing the evidence for the holographic model with that of the exactly scale-invariant Harrison-Zel'dovich spectrum. From Figs. 9.6 and 9.7, we find that the computed evidence for both models is roughly identical. With hindsight, this is perhaps not too surprising since the holographic power spectrum coincides with the Harrison-Zel'dovich spectrum when the holographic coupling $g = 0$ (i.e., when the dual QFT is free). Moreover, to ensure the validity of the perturbative calculation underpinning the holographic power spectrum (9.79) across the entire range of CMB momentum scales, we restricted the maximum value of the holographic coupling to $|g| < |g|_{max}$. In effect, this restriction limits the amount of scale-dependence that may be obtained from the holographic model, accounting for its similarity in performance to the Harrison-Zel'dovich spectrum.

A number of potential approaches to this problem present themselves: the most conservative would be to improve our determination of g_{max} by computing the unknown coefficient $f_B^{(1)}$ at 2-loop order (the result will in general depend on the Yukawa and quartic couplings, as well as the field content of the dual QFT). It may be that the simple order-of-magnitude estimate of $|g|_{max}$ used here is too small, meaning that our perturbative calculation in fact allows more scale-dependence than

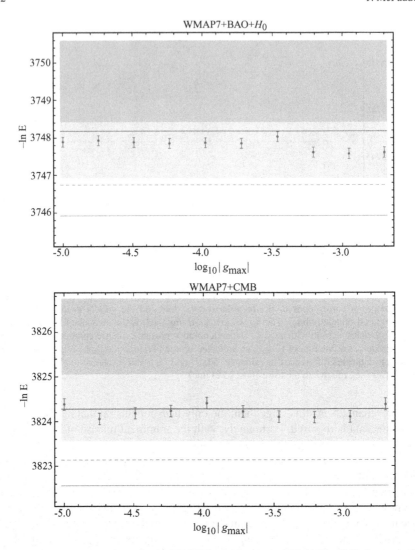

Fig. 9.7 The Bayesian evidence for the WMAP7+BAO+H_0 and WMAP7+CMB data sets, displayed using the same conventions as in Fig. 9.6

we have permitted here. An improved understanding of the infrared scale is also required to determine how far we may push our perturbative calculation of the power spectrum. A more radical approach would be to relax the assumption that the dual QFT should be perturbative over the entire range of CMB momentum scales, and instead permit g_{eff}^2 to become large at the lower end of the CMB range. To pursue this connection, it would be interesting to see if $f_B(g_{eff}^2)$, and hence the holographic power spectrum, could be computed numerically via lattice simulations.

9.6 Conclusion

In this article, we presented a holographic framework for inflationary cosmology, based on standard holography in combination with the domain-wall/cosmology correspondence. Cosmological observables are related by this framework to correlation functions of a dual three-dimensional QFT. The correlation functions of this dual QFT may be obtained by a straightforward analytic continuation of the correlators of the regular QFT dual to the corresponding domain-wall spacetime, at least within the context of large-N perturbation theory. Analysing the behaviour of linearised fluctuations, we obtained precise holographic formulae relating the scalar and tensor cosmological power spectra to the 2-point function of the stress tensor of the dual QFT.

When the dual QFT is strongly coupled, the gravitational description is weakly coupled and we recover the predictions of conventional inflationary scenarios, albeit from a holographic perspective. When the dual QFT is weakly coupled, however, we obtain a scenario in which the gravitational description is strongly coupled at early times, describing an early universe which is in a non-geometric phase. In general, there are two classes of dual QFT for which the holographic description is comparatively well understood. The first class comprises deformations of conformal field theories, and corresponds to universes which are asymptotically de Sitter towards the end of the inflationary epoch. The properties and phenomenology of these models is relatively unexplored and is a major direction for future research. The second class, on which we focused here, describe QFTs with generalised conformal symmetry and correspond to universes whose geometry is asymptotically power-law towards the end of the inflationary era. Dual QFTs of this form are super-renormalisable and are moreover free from infrared divergences, offering in principle a complete description of the corresponding cosmological evolution. For this class of models, we saw how to compute the scalar and tensor cosmological power spectra explicitly up to 2-loops in perturbation theory.

The overall amplitude of the power spectrum is proportional to $1/N^2$, implying that $N \sim O(10^4)$ in accordance with the large-N limit. At leading 1-loop order, the power spectrum is moreover scale-invariant on simple dimensional grounds, with small deviations from scale invariance arising only from corrections at 2-loop order in perturbation theory. The validity of perturbation theory (or equivalently, the assumed strong coupling of the gravitational description at early times) is therefore directly linked to the near scale-invariance of the observed power spectrum. A custom fit of the predicted 2-loop holographic power spectrum to WMAP7 and other astrophysical data sets confirms that these predictions are indeed fully compatible with current observational data. Nevertheless, there are strong prospects for observationally distinguishing the holographic-ΛCDM model from conventional power-law ΛCDM following the release of data from the Planck satellite and other forthcoming observational probes. In particular, the holographic power spectrum rapidly becomes scale invariant at high momenta, essentially as a consequence of asymptotic freedom in the dual QFT. This generates a relatively strong running of the power spectrum, in which successive logarithmic derivatives $d^k n_s(q)/d\ln q^k$ are of comparable order, in sharp contrast to conventional inflationary models for which successive logarithmic derivatives are of higher order in slow roll.

Some theoretical issues concerning the holographic power spectrum nonetheless remain. One of the most pressing is the need to determine more accurately the range of validity of the 2-loop approximation to the power spectrum, motivating a full 2-loop calculation as well as a detailed investigation of the effects of the nonperturbative infrared scale. Needless to say, should the future observational data definitively rule out a primordial power spectrum of the form we have considered here, only the specific class of dual QFTs possessing generalised conformal symmetry would be ruled out, rather than the notion of holography for cosmology per se. In this eventuality, one would then be able to focus exclusively on QFTs describing deformations of conformal field theories.

Another important issue for models assuming a strongly coupled gravitational description at early times is the exact nature of the post-inflationary transition to a conventional weakly coupled hot big bang phase. It would be very interesting to develop a detailed theory for this transition period, the analogue of the reheating period in conventional scenarios. In order to exit the holographic period we would need to modify the UV structure of the dual QFT (since the UV of the QFT corresponds to late times), which may be achieved by adding irrelevant operators to the QFT. At momenta far below the momentum scale q_{UV} set by the lowest dimension irrelevant operator, the computation of the 2-point function (and therefore of the power spectrum) is however well approximated by the computation we performed here. Thus, as long as q_{UV} is much larger that the largest momentum scale seen by CMB (i.e., $q_{UV} \gg 10^{-1}$ Mpc^{-1}), the error incurred by omitting this exit period is very small. In principle, though, one could compute corrections to the holographic formulae due to such irrelevant operators and extract from the data the best-fit value for q_{UV}. We leave such a study for future work, but note that the ability to fit the data well without these corrections suggests they are indeed small.

A final topic we have not discussed here is the predictions of holographic models for primordial non-Gaussianity. Up to the level of the 3-point function, these predictions have been calculated in great detail in [7, 36, 37]. As one might expect, cosmological 3-point functions are related via holographic formulae to 3-point functions of the stress tensor of the dual QFT. The analysis is of essentially the same character as that studied here for the holographic power spectrum, though the necessity of working at quadratic order in perturbation theory leads to a somewhat greater technical complexity. A consideration of special importance is the appearance of 'semi-local' contact terms in the holographic formulae: these terms contribute when two of the three points in a 3-point function are coincident. From a purely QFT perspective one might be tempted to discard terms of this form, however when inserted into the relevant holographic formulae they contribute to 'local'-type cosmological non-Gaussianity and so must be retained. Intriguingly, for model QFTs of the type considered here, the cosmological scalar bispectrum is predicted [37] to be of *exactly* the equilateral template (to our knowledge, the only model known to do so), with the nonlinearity parameter $f_{NL} = 5/36$. Remarkably, this value for f_{NL} is a concrete prediction fully independent of the field content of the dual QFT. The analogous predictions for 3-point functions involving tensors were determined in [7]. In particular, for the case of three gravitons, one may recover the *exact* 3-point function of conventional slow-roll inflation, extending the result recently reported in [32].

To conclude, let us remind ourselves of some of the many pressing questions that remain. Can we use the holographic description to enhance our understanding of inflationary fine tunings? Is there a holographic description for the late-time de Sitter epoch we find ourselves entering, and if so, what are the consequences? Can we understand the entropy of de Sitter space holographically?

Acknowledgements   I am grateful to the organisers of the Sixth Aegean Summer School for a stimulating conference and an exciting opportunity to present this work, as well as to my collaborators Adam Bzowski, Richard Easther, Raphael Flauger and Kostas Skenderis with whom this work was developed. This research was funded through a VENI grant from NWO, the Netherlands Organisation for Scientific Research.

References

1. D. Anninos, T. Hartman, A. Strominger, Higher spin realization of the DS/CFT correspondence (2011). arXiv:1108.5735
2. T. Appelquist, R.D. Pisarski, High-temperature Yang-Mills theories and three-dimensional quantum chromodynamics. Phys. Rev. D **23**, 2305 (1981). doi:10.1103/PhysRevD.23.2305
3. M. Bianchi, D.Z. Freedman, K. Skenderis, How to go with an RG flow. J. High Energy Phys. **0108**, 041 (2001)
4. M. Bianchi, D.Z. Freedman, K. Skenderis, Holographic renormalization. Nucl. Phys. B **631**, 159–194 (2002)
5. H.J. Boonstra, K. Skenderis, P.K. Townsend, The domain wall/QFT correspondence. J. High Energy Phys. **01**, 003 (1999)
6. R.H. Brandenberger, Inflationary cosmology: progress and problems (1999). hep-ph/9910410
7. A. Bzowski, P. McFadden, K. Skenderis, Holographic predictions for cosmological 3-point functions (2011). arXiv:1112.1967
8. P. Creminelli, A. Nicolis, M. Zaldarriaga, Perturbations in bouncing cosmologies: dynamical attractor versus scale invariance. Phys. Rev. D **71**, 063505 (2005). doi:10.1103/PhysRevD.71.063505
9. M. Cvetic, H.H. Soleng, Naked singularities in dilatonic domain wall space times. Phys. Rev. D **51**, 5768–5784 (1995). doi:10.1103/PhysRevD.51.5768
10. J. de Boer, E.P. Verlinde, H.L. Verlinde, On the holographic renormalization group. J. High Energy Phys. **08**, 003 (2000)
11. S. de Haro, S.N. Solodukhin, K. Skenderis, Holographic reconstruction of spacetime and renormalization in the AdS/CFT correspondence. Commun. Math. Phys. **217**, 595–622 (2001). doi:10.1007/s002200100381
12. O. DeWolfe, D.Z. Freedman, S.S. Gubser, A. Karch, Modeling the fifth dimension with scalars and gravity. Phys. Rev. D **62**, 046008 (2000). doi:10.1103/PhysRevD.62.046008
13. M. Dias, Cosmology at the boundary of de Sitter using the dS/QFT correspondence. Phys. Rev. D **84**, 023512 (2011). doi:10.1103/PhysRevD.84.023512
14. R. Easther, R. Flauger, P. McFadden, K. Skenderis, Constraining holographic inflation with WMAP. JCAP **1109**(030) (2011). doi:10.1088/1475-7516/2011/09/030
15. D.Z. Freedman, C. Nunez, M. Schnabl, K. Skenderis, Fake supergravity and domain wall stability. Phys. Rev. D **69**, 104027 (2004). doi:10.1103/PhysRevD.69.104027
16. S. Gubser, I.R. Klebanov, A.M. Polyakov, Gauge theory correlators from noncritical string theory. Phys. Lett. B **428**, 105–114 (1998). doi:10.1016/S0370-2693(98)00377-3

17. M. Henningson, K. Skenderis, The holographic Weyl anomaly. J. High Energy Phys. **9807**, 023 (1998)
18. N. Itzhaki, J.M. Maldacena, J. Sonnenschein, S. Yankielowicz, Supergravity and the large N limit of theories with sixteen supercharges. Phys. Rev. D **58**, 046004 (1998). doi:10.1103/PhysRevD.58.046004
19. R. Jackiw, S. Templeton, How superrenormalizable interactions cure their infrared divergences. Phys. Rev. D **23**, 2291 (1981). doi:10.1103/PhysRevD.23.2291
20. A. Jevicki, T. Yoneya, Space-time uncertainty principle and conformal symmetry in D-particle dynamics. Nucl. Phys. B **535**, 335–348 (1998). doi:10.1016/S0550-3213(98)00578-1
21. A. Jevicki, Y. Kazama, T. Yoneya, Quantum metamorphosis of conformal transformation in D3-brane Yang-Mills theory. Phys. Rev. Lett. **81**, 5072–5075 (1998). doi:10.1103/PhysRevLett.81.5072
22. A. Jevicki, Y. Kazama, T. Yoneya, Generalized conformal symmetry in D-brane matrix models. Phys. Rev. D **59**, 066001 (1999). doi:10.1103/PhysRevD.59.066001
23. I. Kanitscheider, K. Skenderis, Universal hydrodynamics of non-conformal branes. J. High Energy Phys. **04**, 062 (2009). doi:10.1088/1126-6708/2009/04/062
24. I. Kanitscheider, K. Skenderis, M. Taylor, Precision holography for non-conformal branes. J. High Energy Phys. **09**, 094 (2008). doi:10.1088/1126-6708/2008/09/094
25. E. Komatsu et al., Five-year Wilkinson Microwave Anisotropy Probe (WMAP) observations: cosmological interpretation. Astrophys. J. Suppl. **180**, 330–376 (2009). doi:10.1088/0067-0049/180/2/330
26. E. Komatsu et al., Seven-year Wilkinson Microwave Anisotropy Probe (WMAP) observations: cosmological interpretation. Astrophys. J. Suppl. **192**, 18 (2011). doi:10.1088/0067-0049/192/2/18
27. A. Kosowsky, M.S. Turner, CBR anisotropy and the running of the scalar spectral index. Phys. Rev. D **52**, 1739–1743 (1995). doi:10.1103/PhysRevD.52.R1739
28. D. Langlois, Lectures on inflation and cosmological perturbations. Lect. Notes Phys. **800**, 1–57 (2010). doi:10.1007/978-3-642-10598-21
29. D. MacKay, *Information Theory, Inference, and Learning Algorithms* (Cambridge University Press, Cambridge, 2003). Available online at http://www.inference.phy.cam.ac.uk/mackay/itila/book.html
30. J.M. Maldacena, The large N limit of superconformal field theories and supergravity. Adv. Theor. Math. Phys. **2**, 231–252 (1998). doi:10.1023/A:1026654312961
31. J.M. Maldacena, Non-Gaussian features of primordial fluctuations in single field inflationary models. J. High Energy Phys. **05**, 013 (2003)
32. J.M. Maldacena, G.L. Pimentel, On graviton non-Gaussianities during inflation. J. High Energy Phys. **1109**, 045 (2011). doi:10.1007/JHEP09(2011)045
33. P. McFadden, K. Skenderis, Holography for cosmology. Phys. Rev. D **81**, 021301 (2010). doi:10.1103/PhysRevD.81.021301
34. P. McFadden, K. Skenderis, Observational signatures of holographic models of inflation, in *Proceedings of 12th Marcel Grossmann Meeting*, (2010) arXiv:1010.0244
35. P. McFadden, K. Skenderis, The holographic Universe. J. Phys. Conf. Ser. **222**, 012007 (2010)
36. P. McFadden, K. Skenderis, Cosmological 3-point correlators from holography. J. Cosmol. Astropart. Phys. **1106**, 030 (2011). doi:10.1088/1475-7516/2011/06/030
37. P. McFadden, K. Skenderis, Holographic non-Gaussianity. J. Cosmol. Astropart. Phys. **1105**, 013 (2011). doi:10.1088/1475-7516/2011/05/013
38. I. Papadimitriou, K. Skenderis, AdS/CFT correspondence and geometry, in *Proceedings of the Strasburg Meeting on AdS/CFT* (2004). hep-th/0404176
39. I. Papadimitriou, K. Skenderis, Correlation functions in holographic RG flows. J. High Energy Phys. **10**, 075 (2004). doi:10.1088/1126-6708/2004/10/075
40. B.A. Reid et al., Baryon acoustic oscillations in the Sloan digital sky survey data release 7 galaxy sample. Mon. Not. R. Astron. Soc. **401**, 2148–2168 (2010). doi:10.1111/j.1365-2966.2009.15812.x

41. A.G. Riess, L. Macri, S. Casertano, M. Sosey, H. Lampeitl et al., A redetermination of the Hubble constant with the Hubble space telescope from a differential distance ladder. Astrophys. J. **699**, 539–563 (2009). doi:10.1088/0004-637X/699/1/539
42. D.S. Salopek, J.R. Bond, Nonlinear evolution of long wavelength metric fluctuations in inflationary models. Phys. Rev. D **42**, 3936–3962 (1990). doi:10.1103/PhysRevD.42.3936
43. K. Skenderis, Lecture notes on holographic renormalization. Class. Quantum Gravity **19**, 5849–5876 (2002). doi:10.1088/0264-9381/19/22/306
44. K. Skenderis, P.K. Townsend, Gravitational stability and renormalization-group flow. Phys. Lett. B **468**, 46–51 (1999). doi:10.1016/S0370-2693(99)01212-5
45. K. Skenderis, P.K. Townsend, Hidden supersymmetry of domain walls and cosmologies. Phys. Rev. Lett. **96**, 191301 (2006). doi:10.1103/PhysRevLett.96.191301
46. K. Skenderis, B.C. van Rees, Real-time gauge/gravity duality. Phys. Rev. Lett. **101**, 081,601 (2008). doi:10.1103/PhysRevLett.101.081601
47. L. Susskind, The World as a hologram. J. Math. Phys. **36**, 6377–6396 (1995). doi:10.1063/1.531249
48. G. 't Hooft, A planar diagram theory for strong interactions. Nucl. Phys. B **72**, 461 (1974). doi:10.1016/0550-3213(74)90154-0
49. G. 't Hooft, Dimensional reduction in quantum gravity (1993). gr-qc/9310026
50. N. Turok, A critical review of inflation. Class. Quantum Gravity **19**, 3449–3467 (2002). doi:10.1088/0264-9381/19/13/305
51. S. Weinberg, *Cosmology* (Oxford University Press, Oxford, 2008)
52. T. Wiseman, B. Withers, Holographic renormalization for coincident Dp-branes. J. High Energy Phys. **10**, 037 (2008). doi:10.1088/1126-6708/2008/10/037
53. E. Witten, Anti-de Sitter space and holography. Adv. Theor. Math. Phys. **2**, 253–291 (1998)

Part III
Observational Status

Chapter 10
Observational Status of Dark Matter

Joseph Silk

Abstract Identification of dark matter is one of the most urgent problems in cosmology. I describe the astrophysical case for dark matter, from both an observational and a theoretical perspective. I also review the current status of direct and indirect detection of dark matter, and review the prospects for future advances.

10.1 Introduction

Identification of dark matter is one of the most urgent problems in cosmology. It is most likely a weakly interacting particle that is yet to be discovered. One cannot eliminate exotic scalar fields as a model for dark matter or even alternative theories of gravity that dispense with dark matter. However theory favours a weakly interacting particle, to the extent that models such as SUSY provide a plethora of potential dark matter candidates. Moreover SUSY is highly motivated, so it behooves us to examine its predictions carefully. Of course should evidence for SUSY fail to emerge in the near future from the LHC one would have to reconsider a much wider range of dark matter models. These are not lacking. However because the SUSY LSP is such an appealing candidate on theoretical grounds, almost all dark matter searches are designed around the LSP. This overview will therefore focus on the observational motivations rather than the particle physics aspects of dark matter constraints on SUSY dark matter candidates such as the LSP, or NLSP, or even on non-SUSY candidates.

J. Silk
Institut d'Astrophysique de Paris, UMR 7095 CNRS, Université Pierre et Marie Curie, 98 bis Boulevard Arago, Paris 75014, France

J. Silk
Department of Physics and Astronomy, The Johns Hopkins University, 3701 San Martin Drive, Baltimore, MD 21218, USA

J. Silk (✉)
Beecroft Institute of Particle Astrophysics and Cosmology, University of Oxford, 1 Keble Road, Oxford OX1 3RH, UK
e-mail: silk@astro.ox.ac.uk

G. Calcagni et al. (eds.), *Quantum Gravity and Quantum Cosmology*,
Lecture Notes in Physics 863, DOI 10.1007/978-3-642-33036-0_10,
© Springer-Verlag Berlin Heidelberg 2013

10.2 The Observational Case

The first evidence for dark matter emerged from studies of galaxy clusters in the 1930s [1], on megaparsec scales. There is now overwhelming evidence for dark matter from kiloparsec scales to scales of hundreds of megaparsecs. Our best laboratories for dark matter are dwarf spheroidal galaxies. Most of these, a kiloparsec or less across, are almost pure dark matter. The ratio of dark matter to baryonic matter is an order of magnitude larger than the canonical value of 15 from the Big Bang. In the Milky Way, within say the orbit of the sun 8 kpc from the galactic center, there are approximately equal masses of ordinary matter and dark matter. Only on much larger scales does the dark matter to ordinary matter ratio approach the canonical value.

In fact this convergence to the primordial value is a function of the mass of the system. The Milky Way in its entirety, halo included, is deficient in ordinary matter by about a factor of 2. This is on a scale of 100 kpc. One has to go to galaxy groups and clusters, on a scale of order a Mpc, before the asymptotic value is attained, From here onto the horizon, the dark matter dominance amounts to a factor of 15. I conclude that dark matter is ubiquitous.

In addition, large-scale structure simulations demonstrate unambiguously that the dark matter is cold. Theory favours the idea that dark matter most likely is a weakly interacting massive particle (WIMP), with a favoured candidate being the LSP found in the theory of supersymmetry, in the mass range 0.001–10 TeV. The motivation for a WIMP arises from the so-called WIMP miracle: the relic abundance of dark matter arises naturally from production followed by thermal freeze-out of generic Majorana particle candidates with generically weak-like interactions if

$$1 \sim < n\sigma v > \sim \left(3 \times 10^{-26} \text{ cm}^3/\text{s}\right)(\Omega_\chi/0.3),$$

where σ is the self-annihilation cross-section. Of course there are numerous non-WIMP dark matter candidates ranging from very light particle such as axions (mass $\sim 10^{-6}$ eV) to GUT or even Planck-scale mass particles, as well as exotic scalar fields. However physicists are far from identifying the specific particle.

In this review I will focus on the astrophysics. I will describe the observational evidence for dark matter and illustrate how the field has evolved in recent years.

10.3 From Galaxies to Clusters

10.3.1 Galaxy Rotation Curves

Perhaps the best studied galaxy for dark matter is the Milky Way Galaxy. A new rotation curve model leads to estimates of the local dark matter density near the Sun at 8 kpc from the centre of the Galaxy of 0.235 ± 0.030 GeV cm^{-3}, and the total mass inside the Galaxy at 385 kpc, halfway to M31, of $(7.03 \pm 1.01) \times 10^{11} M_\odot$. This leads to a stellar baryon fraction of 0.072 ± 0.018, or about half of the primordial value [2].

Disk galaxies are generally dominated by dark matter. The dark matter problem assumed a central position in cosmology for two reasons. New developments in optical and in radio astronomy allowed dynamical measurements in the outer regions of individual spiral galaxies.

In the 1970s, Rubin and Roberts, among others, pioneered observations of extended flat rotation curves in the optical and 21 cm wave bands respectively. The first discussion of the need for unseen matter seems to be by Roberts and Rots (1973) [3] who argue that "The shapes of the rotation curves at large radii indicate a significant amount of matter at these distances and imply that spiral galaxies are larger than found from photometric measurements." Indeed an important paper that establishes the systematic flatness of rotation curves from optical data [4] builds on the earlier study led by Rubin [5]. They state that "Roberts and his collaborators deserve credit for first calling attention to flat rotation curves."

However uncertainty remained about the interpretation because of possible gradients in the disk mass-to-light (M/L) ratios (reviewed in [6] who state that "By the 1970s, flat rotation curves were routinely detected (Rogstad and Shostak 1972) [7] but worries about side bands still persisted, and a variation in M/L across the disk was a possible explanation."

At the same time there was new theoretical insight. This occurred in 1970. The first convincing dark matter inference was made by Freeman, [8] who modelled a self-gravitating exponential disk and demonstrated that the predicted decline of the rotation curve requires addition of dark matter to match the flat rotation curves known at the time. His transformational 1970 paper was the first indication from rotation curve analysis that the rotation curve is not determined by the mass distribution in the disk alone, but requires a contribution to its amplitude from an extended distribution of dark matter. This insight led to the concept of individual galaxies embedded in dark halos.

This was followed by a dynamical argument advanced by [9] that dark halos are required by global stability arguments in order to avoid non-axisymmetric instabilities and bar formation, A similar argument was given independently by [10]. This argument is now partly discounted because bulges stabilise and bars are virtually universal. Global stability requires a halo containing 60 % of the disk mass at the disk edge [11], but the presence of bulges may reduce this requirement. There does remain the issue of bulgeless galaxies however. Some massive galaxies are bulgeless and exceedingly flat [14], requiring a dark matter-dominated halo.

Freeman's argument was further refined by the notion of maximum disks, introduced in 1985 [12] because of the unknown disk M/L value. Maximum disks provide the maximum contribution of the disk mass to the rotation curve. Dark matter is required to account for 15 % of the rotation curve or 30 % of the mass within the scale of maximum rotation velocity [13], and dominates further out where the rotation curve flattens. Kinematical data demonstrates that most disks are indeed sub-maximal. Dark matter is universally accepted as required in disk galaxy halos, unless recourse is made to alternative theories of gravity such as MOND or TEVES, cf. [15].

Dwarf spheroidal galaxies are dark matter laboratories, dominated by dark matter. However the numbers defy interpretation. Feedback is readily adjusted to reduce the numbers of low mass dwarfs [16], but the most massive dwarfs predicted by LCDM simulations should be observed: they are not. Unorthodox feedback (AGN) may be a solution [17].

Most dwarfs have cores rather than cusps as predicted by CDM-only simulations. Supernova feedback may turn cusps into cores by gas sloshing [18]. Baryon feedback reconciles data with simulations that include baryon feedback and associated gas outflows [19].

The local dark matter density is poorly known. It is important for direct detection experiments. A disk component is predicted from dragging and disruption of satellites [20]. For an isothermal population of old tracers (A and F stars) [21], one find $\rho_{dm} = 0.003 \pm 0.008 M_\odot/\mathrm{pc}^3$ (90 per cent confidence level). However, the vertical dispersion profile of these tracers is poorly known. For a non-isothermal profile (similar to the blue disc stars from SDSS DR-7), the local density increases to $\rho_{dm} = 0.033 \pm 0.008 M_\odot/\mathrm{pc}^3$.

Galaxy clusters are a promising venue for testing dark matter predictions. The central dark matter cusp, if it exists, can be constrained by combining measurements of the stellar kinematics of the central galaxy with a strong lensing analysis of radial and tangential arcs near the cluster center (e.g., [22]). Outside the cluster core, the cluster mass profile can be measured through weak lensing (see [23]). The inferred cluster concentrations probe the cluster formation epoch. There is no consensus on whether the results are consistent with LCDM, or require additional large-scale power such as might be provided by non-gaussianity or by dynamical dark energy. X-ray studies of the hot intracluster medium (ICM) provide the gas pressure gradient. By assuming hydrostatic equilibrium, this yields the cluster mass (e.g., [24]). One can avoid assumptions about hydrostatic equilibrium via weak lensing, and also probe the ICM gas in a complementary fashion via the Sunyaev-Zeldovich effect on the cosmic microwave background (see, e.g., [25]).

Cluster counts are sensitive to the universal dark matter value, and in particular to the growth rate of density fluctuations. This is partially suppressed at recent epochs as dark energy dominates, and hence number counts of clusters are reduced [26].

10.4 Large-Scale Structure

10.4.1 Redshift Space Distortions

Galaxy redshift surveys have historically been the main probe of dark matter on large scales, those of clusters and of superclusters of galaxies. Redshift space distortions measure Ω_m. On smaller scales, these provide virial estimators [27] and are sensitive on quasilinear scales to the growth rate of density fluctuations. The new surveys (2DF, SDSS, WiggleZ) are able to probe the power spectrum of galaxies over $0.1 < z < 0.9$. Redshift-space distortions are measured on large scales, to over

$k < 0.3h/$Mpc. The growth rate is strongly dependent on Ω_m which is found to be 0.27 to about 5 % and is well probed over this redshift range [28].

10.4.2 Baryon Acoustic Oscillations

The acoustic imprint of the matter-radiation plasma prior to decoupling leaves baryon as well as radiation acoustic oscillations in the residual power spectra. The acoustic wavelength is a geometrical probe of the curvature of the universe. The baryon acoustic oscillations are especially powerful as a probe because one can slice the universe by redshift. Assuming the dark energy is a cosmological constant and allowing the spatial curvature to vary, recent studies of large galaxy samples find that this geometrical measurement of the curvature of universe yields $\Omega_k = 0.0035 \pm 0.0054$ [29, 30].

10.4.3 Cosmic Microwave Background

The radiation power spectrum has a high significance detection of the acoustic peaks. However projection onto the last scattering surface introduces additional degeneracies. These arise because the distance to last scatter (equivalently the age of the universe) is degenerate with respect to curvature. The spectral index adds further uncertainty. The situation has been improved with fine-scale measurements of the CMB anisotropies that probe the damping tail. Both the ACT and SPT experiments are able to measure the damping of the primordial primary CMB fluctuations and reconstruct the BAOs. To make progress one has to remove degeneracies that limit independent determinations of Ω_{DM}, Ω_b. For a Hubble constant prior ($h = 0.74$), one obtains $\Omega_b = 0.023 \pm 0.0012$, a more constraining result than obtained from primordial nucleosynthesis $\Omega_b = 0.022 \pm 0.002$ [31].

Other canonical parameters that have less than a few percent uncertainty are the scalar spectral index $n_s = 0.965$ and the normalisation to unit variance of galaxy count to mass fluctuations, $\sigma_8 = 0.8$, on mass scale $2.5 \times 10^{14} h^{-1} M_\odot$.

10.5 Future Prospects in Observation

Dark matter and dark energy surveys are complementary. The four leading methods in dark energy measurements are supernovae, BAO, weak lensing, and counts of clusters of galaxies. These measure various nearly orthogonal combinations of dark matter and dark energy, and are primarily being developed to constrain theories of cosmic acceleration. However improved dark matter diagnostics are an inevitable corollary. These methods are reviewed in [32]. The conclusions are that:

(a) Type Ia supernovae provide immense precision for measuring distances relative to local calibrators (i.e., distances in h^{-1} Mpc) at $z \sim 0.5$, with future surveys designed to achieve statistical errors of 0.01 mag or less (or $\sim$0.5 % in distance). However systematic uncertainties may be dominant, including imperfect photometric calibration, redshift evolution in the population of SNe, and the effects of dust extinction.

(b) The BAO method augments the SN method by measuring absolute distances (in Mpc), assuming a calibration of the sound horizon. Spectroscopic BAO measurements cover a greater comoving volume and measure $H(z)$ directly in addition to the distance-redshift relation. Cosmic variance-limited BAO surveys provide sensitivity to dark energy over the range $1 < z < 3$, independently of supernovae. However if the universe were more inhomogeneous than usually assumed over 100 Mpc scales, there would be considerable uncertainty in BAO approaches.

(c) Weak lensing measurements probe both the distance-redshift relation and the linear growth rate of structure. One challenge is to obtain an accurate PSF which affects galaxy images and must be determined to very high accuracy ($\sim$0.001). Other major challenges are calibration of photometric redshift distributions to a similar level of accuracy, and correction for the intrinsic alignment of galaxies.

(d) Cluster abundance measurements measure the growth rate of structure and can thereby probe alternative gravity models. A major challenge is obtain the calibration of the cluster mass scale to better than 1 %. The combination of new x-ray and SZ surveys should help refine cluster mass determinations.

10.6 Future Prospects in Astrophysical Theory

Theory lacks adequate resolution and physics. Of course these issues are intricately connected. One needs to tackle baryon physics and the associated possibilities for feedback. At this point in time, the leading simulations, such as the ERIS cosmological simulation of the MWG, provide at best 10 pc resolution in a state of the art simulation with gas and star formation. The gas and star formation physics is included in an ad hoc way, because of the resolution limitation. For example, a star formation threshold in density is adopted. and varied to explore possible sensitivity of the results. However in reality it is the unresolved subgrid physics that determines the actual threshold, if one even exists. Mastery of the required subparsec-scale physics will take time, but there is no obvious reason why with orders of magnitude improvement in computing power we cannot achieve this goal.

For the moment, phenomenology drives all modelling. This is true especially for local star formation. A serious consequence is that physics honed on local star-forming regions, where one has high resolution probes of star-forming clouds and of ongoing feedback, may not necessarily apply in the more extreme conditions of the early universe.

One issue that arises frequently is whether the perceived challenges to LCDM justify a new theory of gravity. From MOND onwards, there are any number of alternative theories that are designed to explain certain observations. However none can explain all observations, as is often said to be the case for LCDM. But to the extent that any unexplained anomalies exist, these are invariably at no more than the 2-sigma level of significance. It seems to me that such "evidence" is not adequate motivation for abandoning Einstein-Newton gravity. While it is overwhelmingly clear that there are many potential discrepancies with LCDM, we have certainly not developed the optimal LCDM theory of galaxy formation. Current theory does not adequately include the baryons nor do we reliably understand star formation, let alone feedback.

Here is a summary of some of the key reasons that LCDM does not provide a robust explanation of the following observations: I list 10 examples.

- (a) Massive bulgeless galaxies with thin disks are reasonably common [14]. Simulations invariably make thick disks and bulges. Indeed the bulges are typically overly massive relative to the disks for all galaxies other than S0s. Massive thin disks are especially hard to simulate unless very fine-tuned feedback is applied. A consensus is that the feedback prescriptions are far from unique. One appealing solution involves supernova feedback. This drives a galactic fountain that feeds the bulge. A wind is driven from the bulge where star formation is largely suppressed for sufficiently high feedback [33]. Another proposal includes radiation pressure from massive stars as well as supernovae. The combined feedback helps expand the halo expansion, thereby limiting dynamical friction and bulge formation [34].
- (b) Dark matter cores are generally inferred in dwarf spheroidal galaxies, whereas LCDM theory predicts a cusp, the NFW profile. Strong supernova feedback can eject enough baryons from the innermost region to create a core [35].
- (c) The excessive predicted numbers of dwarf galaxies are one of the most cited problems with LCDM. The discrepancy amounts to orders of magnitude. The issue of dwarf visibility is addressed by feedback that ejects most of the baryons and thereby renders the dwarfs invisible, at least in the optical bands. There are three commonly discussed mechanisms for dwarf feedback: reionization of the universe at early epochs, supernovae and tidal stripping. AGN-driven outflows via intermediate mass black holes provide another alternative to which relatively little attention has been paid [36].

  Reionization only works for the lowest mass dwarfs. The ultrafaint dwarfs in the MWG may be fossils of these first galaxies [37]. It is argued that supernova feedback solves the problem for the more massive dwarfs [38]. However this conclusion is disputed by [17] for whom prediction in simulations of massive dwarfs is a problem. These authors argue that the relatively massive dwarfs should form stars, and we see no counterparts of these systems, apart possibly from rare massive dwarfs such as the Magellanic Clouds.

  One can also appeal to a lower star formation efficiency (SFE) in dwarfs, plausibly associated with low metallicities and hence low dust and H_2 content. Models based on metallicity-regulated star formation can account for the numbers and radial distribution of the dwarfs by a decreasing SFE [39]. This explanation is disputed by [40], who infer a range in SFEs for the dwarfs of some two orders of magnitude. A similar result appeals to the halo mass threshold below which star formation must be suppressed to account for the dwarf luminosity function, with the stellar masses of many observed dwarfs violating this condition [41]. Finally, tidal stripping may provide a solution [42], at least for the inner dwarfs.

- (d) Another long-standing problem relates to downsizing. Massive galaxies are in place before lower mass galaxies as measured by stellar mass assembly, and their star formation time-scales and chemical evolution time-scales at their formation/assembly epoch are shorter. It is possible to develop galaxy formation models with suitable degrees and modes of feedback that address these issues. However a major difficulty confronted by all semi-analytical models (SAMs) is that the evolution of the galaxy luminosity function contradicts the data, either at high or at low redshift. The SAMs that are normalised to low redshift and tuned to account for the properties of local galaxies fail at high redshift by generating too many red galaxies [43]. Too few blue galaxies are predicted at $z = 0.3$. This problem has been addressed by including AGB stars in the stellar populations. This fix results in a more rapid reddening time-scale by speeding up the evolution of the rest-frame near-infrared galaxy luminosity function [44]. There is a price to be paid however: now there are excess numbers of blue galaxies predicted at $z = 0.5$.

- (e) The luminosity function problem is most likely related to another unexplained property of high redshift galaxies.The SSFR evolution at high z is very different from that at low z. Essentially, it saturates. One finds an infrared main sequence of galactic star formation rates: SFR versus M_* [45].

- (f) Much has been made of nearby rotation curve wiggles that trace similar dips in the stellar surface density that seemingly reduce the significance of any dark matter contribution. Maximum disks optimise the contribution of stars to the rotation curve, and these wiggles are most likely associated with spiral density waves. A similar result may be true for low surface brightness gas-rich dwarf galaxies [46]. High mass-to-light ratios are sometimes required, but these are easily accommodated if the IMF is somewhat bottom-heavy. The case for IMF variations has been made for several data sets, primarily for early-type galaxies (e.g. [47]). The LSB dwarfs are plausible relics of the building blocks expected in hierarchical formation theories.

- (g) Spiral arms are seen in the HI distribution in the outer regions of some disks. This tells us that significant angular momentum transfer is helping feed in the optical inner disk. The baron self-gravity is large enough that one does not for example need to appeal to a flattened halo, which might otherwise be problematic for the DM model [48].

- (h) The slope and normalisation of the baryon Tully-Fisher relation does not agree with the simplest LCDM prediction. The observed slope is approximately 4, similar to what is found for MOND [49]. LCDM (without feedback) gives a slope of 3 [50], but fails to account for the observed dispersion and possible curvature.

- (i) The baryon fraction in galaxies is some 50 % of the primordial value predicted by light element nucleosynthesis. These baryons are not in hot gaseous halos [51]. Convergence to the universal value on cluster scales is controversial: convergence to the WMAP value is seen for x-ray clusters above a temperature of 5 keV [52], but could be as large as 30 % even for massive clusters [53, 54]. If the latter discrepancy were to be confirmed, one would need significant bias of baryons relative to dark matter, presumably due to feedback, on unprecedentedly large scales.

• (j) Bulk flows are found over 100 Mpc scales that are up to several deviations larger than expected in LCDM [55]. The technique primarily uses Tully-Fisher and fundamental plane galaxy calibrators of the distance scale. An x-ray approach, calibrating via kSZ, claims the existence of a bulk flow out to 800 Mpc [56]. However the discrepancies with LCDM are controversial because of possible systematics.

10.7 Direct Detection

Many weakly interacting massive elementary particles, if dark matter, must pass through us every second, about 10^6 m^{-2} s^{-1}. Detection techniques involve large masses of some suitable material that is studied for weak signals from the rare WIMP interactions. The detectors are located deep underground or under mountains, to avoid spurious cosmic-ray induced events. The nuclear recoil signatures include ionisation, phonons and scintillation, and ideally require all of these effects.

Event detections have been reported by several experiments. These include CDMS2 (X kg germanium), CoGeNT and CRESST-II. However none of these have sufficient significance to be attributed to dark matter. The one exception is the NaI scintillation experiment, DAMA/LIBRA, now running for 14 years at Gran Sasso. This experiment uses solar modulation to enhance the direct detection signal and reports a 8.9 σ detection. The report of an almost 3 sigma detection of annual modulation in CoGeNT has produced considerable excitement, but tension remains with the other experiments in both amplitude of the modulation and scattering cross-section. The competing experiments rule out most explanations, including incoherent spin-independent scatterings. However windows remaining are via coherent spin-dependent scatterings by light WIMPs on protons, or via spin-dependent scatterings with isospin suppression of neutron scatterings. Alternatively, allowance for streams in the local dark matter density adds sufficient uncertainty to reduce these tensions [57]. The allowed WIMP mass range is 5–20 GeV. Discounting DAMA/LIBRA, the allowed window for neutralinos extends up to several TeV.

10.8 Indirect Detection

Halo Majorana fermion WIMPs occasionally annihilate today into energetic particles: $\nu, \gamma, \bar{p}, e^+$. They are also trapped by the sun and other stars. All of these lead to possible signals. Introduction of a primordial asymmetry reduces the annihilation signal relative to the direct detection signal, at the expense of increasing the annihilation rate for the subdominant symmetric component [58].

10.8.1 Helioseismology

WIMP scattering on protons modifies the solar temperature profile. Low mass ($m_\chi \gtrsim 5$ GeV) WIMPS are trapped and fill the solar core and modify $T(r)$. This leads to a detectable signal from solar physics-motivated experiments. Helioseismology has successfully studied p-modes from the outer regions of the sun. These measurements are sensitive to the temperature profile. The predicted signal probes solar structure. The revised solar opacities have thrown this field into disarray, since the totality of solar data, including solar neutrinos and helioseismology, can no longer be fit by the solar standard model. Addition of low mass WIMPs adds a new degree of freedom, and affects the helioseismology signal because of the modified solar temperature profile. The effect is especially strong for 5 GeV WIMPs that interact via spin-dependent scatterings. If their abundance is high enough, e.g. if annihilations are partially or totally suppressed, one can even eliminate them as a DM candidate. Annihilation suppression in favour of a built-in asymmetry is reasonably natural for WIMPs in the mass range 5–10 GeV. Asymmetric dark matter (aDM) provides a compelling explanation for the observed baryon fraction $\sim m_p/m_\chi$, admittedly at the price of losing the perhaps less "natural" SUSY LSP-motivated explanation for Ω_χ. Collider constraints on the large annihilation cross-sections required for the Majorana component require a light mediator particle that allows new annihilation channels that are weakly coupled to the standard model [59], although these limits are only restrictive for 10 GeV WIMPs if elliptical galaxy halo shapes are introduced as a constraint on the self-interaction dark matter cross-section.

10.8.2 High Energy Cosmic Rays

Rare particles in cosmic rays, most notably $\bar{p}$ and e^+, are a unique signature of dark matter annihilations. The search for high energy antiprotons has led to no surprises so far, although in principle because secondary $\bar{p}$ from cosmic ray spallations are Lorentz-boosted, there is a potential signal to be sought below 1 GeV. However solar modulation effects make this a difficult measurement.

Cosmic ray positrons have provided a far more productive target. Hints of a signal came with the HEAT balloon-borne experiment that detected a rise in the positron fraction $e^+/(e^+ + e^-)$ above ~ 10 GeV. This result has been confirmed by the PAMELA satellite to ~ 100 GeV, and most recently by FERMI to ~ 200 GeV [60], and cannot easily be attributed to cosmic ray secondary production of e^+. Additional sources are needed. The associated cosmic ray electron flux has been measured by FERMI to ~ 1 TeV , and to ~ 3 TeV by HESS and most recently by MAGIC, [61] The spectrum shows a drop at a few TeV.

Possible explanations include nearby astrophysical positron sources, dark matter decays or dark matter annihilations. The most likely sources are nearby pair-wind pulsars by Milagro at a median gamma ray energy of 20 TeV. More distant pulsars will also contribute, but the nearest sources dominate in typical cosmic ray diffusion models. Supernova remnant acceleration models also present a viable option [62]. Such astrophysical solutions will be tested by the predicted anisotropy, which in the pulsar explanation already is close to the FERMI one-year upper limit [63].

The dark matter explanation of the positron excess requires a TeV particle. In the case of annihilations, considerable local substructure is required to give a boost to the annihilation rate. A halo dark matter clumpiness factor as large as 10^3 is usually invoked in order to boost the signal, since at a specified dark matter density (determined by the galactic rotation curve), the annihilation flux is inversely proportional to the square of the neutralino mass.

Theory struggles to generate such large clumpiness factors. One solution is via a Sommerfeld enhancement for ultracold dark matter. This might be expected for substructure in cold dense clumps (of order solar mass or below) in CDM. In this case, one achieves a local annihilation cross-section as required of order 10^{-23} cm^3 s^{-1}. Production of excessive gamma rays from the inner galaxy is avoided if tidal destruction of substructure destroys most of the boost in the bulge region [64]. Extragalactic constraints are constraining but are unable to definitively eliminate the annihilation interpretation of the essentially local positron/electron fluxes. The strongest constraints include the effects of prolonging the decoupling of the CMB as well as diffuse gamma ray signals from dwarfs, but are insensitive for TeV WIMPs.

10.8.3 Gamma Rays

Recent data from the Fermi satellite has constrained dark matter models. The FERMI energy range spans 0.02–300 GeV, with angular resolution of 5 degrees to 5 arcmin, depending on the energy, and energy resolution of around 10 %. Theory of dark matter annihilations (and decays) predicts several gamma ray smoking guns. These include a harder spectrum than expected via π^0 decay channels, spectral bumps and lines, and inverse Compton gammas, as well as radio synchrotron photons from high energy electrons and positrons. The ideal laboratory for dark matter detection via annihilations is to look at dark matter laboratories such as gamma rays from nearby dark matter-dominated dwarf galaxies. Hitherto only upper limits have been set on gamma ray emission, with Fermi setting stronger limits at lower particle masses, and the ACT arrays at higher masses. For thermal decoupling, the neutralino mass must exceed $\sim$30 GeV from Fermi dwarf [65] and CMB [66] constraints.

10.8.4 The WMAP Microwave Haze

Dark matter annihilations in the galactic bulge lead to a possible radio synchrotron signal. The WMAP quasi-spherical haze residuals in the lowest frequency WMAP channels has been interpreted as such a signal [67], and led to the prediction that the same high energy electrons would lead to an inverse Compton gamma ray flux, produced by Compton scattering of e+e- on the interstellar radiation field. This leads to an expected Fermi haze, once known templates were subtracted [68]. Analysis of the diffuse gamma ray emission in the inner bulge, once known templates were subtracted, revealed the presence of enormous bubble-like features, north and south of the Galactic Center [69]. These clearly are not due to dark matter injection but rather arise from an immense explosion some tens of millions of years ago that requires local reacceleration over tens of degrees (at least a kpc) in order to account for the short electron lifetimes. The dark matter contribution has been recently revived. In addition to this large-scale diffuse emission, there is an unexplained spectral distortion within the central degree where part of the Fermi haze is unexplained by known sources or foregrounds. A second diffuse component seems to be required in addition to cosmic ray-induced gammas in the lower energy channels. A reasonable spectral and morphological fit is attained with neutralinos in the mass range 7–45 GeV for different annihilation channels with leptonic or hadronic final states [70]. The Fermi collaboration remains agnostic on these results, having produced significant unexplained residuals when all known sources are subtracted out in the GC region [71]. The origin of the possibly associated WMAP haze, also confirmed as a new CMB foreground component [72], still remains a mystery. Indeed the same electron component postulated for the Fermi spectral excess generates a synchrotron component that has been interpreted as contributing to the WMAP haze signal [73].

10.8.5 Decaying Dark Matter

Another dark matter option is via decays of massive neutralinos. The required decay time is $\sim 10^{26}$ sec [74]. The morphological differences between annihilating and decaying dark matter provides a distinguishable characteristic [75]. Decaying dark matter in galaxy clusters turns out to be the best probe since the nearest clusters just fill the Fermi beam and thereby give optimal sensitivity to a possible diffuse signal from the cluster. FERMI constraints effectively eliminate decaying dark matter as an option [76].

10.9 The Future

10.9.1 The Sun

As the sun orbits the galaxy, it traps massive neutralinos that scatter off protons. These accumulate in the solar core where they annihilate, producing energetic neutrinos that may induce signals via muon production in experiments under ice such as IceCube, or under water such as ANTARES. Future scaled-up experiments should be capable of imaging the sun if neutralinos indeed annihilate at masses up to a TeV.

If WIMPs do not self-annihilate, as would be the case for asymmetric WIMPs, the numbers build up in the sun and lead to another signal. At low masses, WIMPs fill the core of the sun and WIMP recoils redistribute the solar temperature profile. This effect is optimised at the lowest masses that do not evaporate from the sun ($\sim$5 GeV) but still gives a helioseismological signal for WIMPs below $\sim$20 GeV. This effect will be especially relevant once solar g-modes are detected [77]. There is also a potentially detectable solar neutrino signal [78] if WIMPs are allowed to accumulate and scatter via spin-dependent couplings where direct detection limits are weak.

10.9.2 Direct Detection

How low do we need to go in direct detection in order to eliminate SUSY-motivated WIMPs? Tonne-scale detectors are under construction [79] and should be able to go well beyond the LHC benchmark models in terms of sensitivity to dark matter.

10.9.3 Air Cerenkov Telescopes

Another technique that allows sensitive determinations of gamma rays measures atmospheric Cerenkov radiation from muon-poor air showers. These are induced by TeV gamma rays and have adequate resolution to resolve out identifiable discrete sources. An ultimate Cerenkov telescope array with 10 km^2 area can probe down to 10 GeV and achieve SUSY-model sensitivities comparable and complementary to those of ton-scale direct detection experiments [80]. ACTs provide the most promising avenue for complementing direct detection.

10.9.4 Strange Stars

A neutron star is a dark matter collector. If neutron matter is metastable, the energy from WIMP annihilations may trigger the conversion of a neutron star to a quark star [81]. The rest mass energy of the neutron star is liberated in high energy particles, neutrinos and photons. One might be able to observe such an event, in a region of high dark matter density, as a gamma ray burst of unusual characteristics. The explosion is intrinsically off-centre because of the thermal distribution of WIMPS that spans the inner part of the neutron star core. The resulting anisotropic ejection can provide a momentum kick to the surviving quark star.

10.9.5 The Galactic Centre

There is a black hole of mass $4 \times 10^6 M_\odot$ identified with the radio source SagA*
at the Galactic Centre. Theoretical arguments suggest that when it formed it may
have acquired a steep dark matter cusp that would yield an enhanced annihilation
signal in gamma rays. The characteristic features of this spectrum are an exponential
plus flat power-law, and no variability. HESS data confirms the exponential cut-off
above a few TeV and no detectable variability [82], but the power-law is too steep
for an annihilating particle with a unique mass. Addition of Fermi data confirms a
complex inflected spectrum [83]. There are two possible interpretations: an astrophysical source, with novel spectral characteristics, or dark matter annihilations of
a TeV particle together with a steep power-law contribution from an astrophysical
source (and/or a lower mass annihilating particle).

10.9.6 LHC

The LHC reach overlaps with indirect dark matter detection experiments. The SUSY
benchmark models for direct detection are accessible at the LHC and there is com-
plementarity with indirect searches [84]. However the ultimate sensitivity to these
models will come from combining direct detection with ACT array telescopes.

10.10 Summary

The case for dark matter is powerful. Alternative theories of gravity are far more
complex than Einstein gravity. For example, both vector and tensor degrees of freedom are invoked in TEVES in addition to the usual scalar potential. And even with
this extra freedom, a vigorous debate rages as to whether there remain observations
that defy explanation. Motivation for exploring alternative gravity requires more
than the need to test Einstein's theory, since there are a vast variety of alternatives
waiting in the wings. Indeed Einstein gravity awaits its first major confrontation with
the hopefully imminent detection of gravity waves. Rather, one needs a discrepancy
of significance comparable to the precession of Mercury's perihelion advance that
motivated Einstein to go beyond Newtonian gravity. The astronomical data show no
such evidence. This is certainly true for galaxies and galaxy clusters. To reconcile
with LCDM, there is a price to pay, namely that of astrophysical complexity. But
this is hardly headline news. We do not invoke new physics to account for unusual
weather patterns.

On the largest scales, there are intriguing hints of possible anomalies. These
range from bulk flows to CMB features. However the data is too compromised by
possible systematics to reach any robust conclusions. The greatest weakness in the
dark matter saga is that we have not identified the nature of the dark matter itself.

This is a serious issue. But patience is counselled. We live at a moment when the new discipline of particle astrophysics is flourishing. Many experiments are underway or being planned to search for direct and indirect traces of dark matter, generally on the assumption that it is a weakly interacting elementary particle. The LHC is searching for hints of particle candidates for dark matter, motivated by supersymmetry. These arguments may be wrong. Theorists may be guilty of hubris. But as we finally approach the ability to probe large swathes of SUSY-motivated parameter space, the tantalizing claims of "discoveries" of dark matter signatures, hitherto unconfirmed, contribute to a feeling of growing excitement in the particle astrophysics community. We should revisit the situation in a decade. If by then we have not identified a dark matter particle candidate, I certainly will be more enthusiastic about exploring alternative gravity theories. Perhaps we will identify a theory that simultaneously accounts for dark matter and dark energy.

References

1. F. Zwicky, Helv. Phys. Acta **6**, 110–127 (1933)
2. Y. Sofue, Publ. Astron. Soc. Jpn. **64**(2) (2012)
3. M.S. Roberts, A.H. Rots, Astron. Astrophys. **26**, 483 (1973)
4. V.C. Rubin, W.K. Ford, S.E. Thonnard, Astrophys. J. **225**, L107 (1978)
5. V.C. Rubin, W.K. Ford, Astrophys. J. **159**, 379 (1970)
6. Y. Sofu, V.C. Rubin, Annu. Rev. Astron. Astrophys. **39**, 137 (2001)
7. D.H. Rogstad, G.S. Shostak, Astrophys. J. **176**, 315 (1972)
8. K. Freeman, Astrophys. J. **160**, 811 (1970)
9. J.P.E. Ostriker, P.J.E. Peebles, A. Yahil, Astrophys. J. Lett. **193**, L1 (1974)
10. J. Einasto, A. Kaasik, E. Saar, Nature **250**, 309 (1974)
11. J. Sellwood, arXiv:1006.4855 (2010)
12. C. Carignan, K. Freeman, Astrophys. J. **294**, 494
13. P. Sackett, Astrophys. J. **483**, 103 (1997)
14. J. Kormendy, N. Drory, R. Bender, M. Cornell, Astrophys. J. **723**, 54 (2010)
15. J.D. Bekenstein, arXiv:1201.2759 (2012)
16. S.J.M. Koposov, H. Yoo, H.-W. Rix, D. Weinberg, A. Macciò, J. Escudé, Astrophys. J. **696**, 2179 (2009)
17. M. Boylan-Kolchin, J.S. Bullock, M. Kaplinghat, Mon. Not. R. Astron. Soc. **415**, L40 (2011)
18. S. Mashchenko, H.M.P. Couchman, J. Wadsley, Nature **442**, 539 (2006)
19. S.-H. Oh, C. Brook, F. Governato, E. Brinks, L. Mayer, W.J.G. de Blok, A. Brooks, F. Walter, Astron. J. **142**, 24. arXiv:1011.2777
20. J.I. Read, L. Mayer, A.M. Brooks, F. Governato, G. Lake, Mon. Not. R. Astron. Soc. arXiv:0902.0009
21. S. Garbari, J.I. Read, G. Lake, Mon. Not. R. Astron. Soc. arXiv:1105.6339
22. A. Zitrin, T. Broadhurst, D. Coe et al., Astrophys. J. **742**, 117 (2011)
23. K. Umetsu, T. Broadhurst, A. Zitrin, E. Medezinski, D. Coe, M. Postman, Astrophys. J. **738**, 41 (2011)
24. D.A. Buote, F. Gastaldello, P. Humphrey et al., Astrophys. J. **664**, 123 (2007)
25. M.B. Gralla, K. Sharon, M.D. Gladders et al., Astrophys. J. **737**, 74 (2011)
26. S.W. Allen, A.E. Evrard, A.B. Mantz, Annu. Rev. Astron. Astrophys. **49**, 409–470 (2011)
27. N. Kaiser, Mon. Not. R. Astron. Soc. **227**, 1–27 (1987)
28. C. Blake et al., Mon. Not. R. Astron. Soc. **415**, 2892 (2011). arXiv:1104.2948
29. C. Blake et al., Mon. Not. R. Astron. Soc. **418** 1707. arXiv:1108.2635

30. S. Ho et al., arXiv:1201.2137
31. R. Hlozek et al., Astrophys. J. arXiv:1105.4887
32. D.H. Weinberg, M.J. Mortonson, D.J. Eisenstein, C. Hirata, A.G. Riess, E. Rozo, Phys. Rep. arXiv:1201.2434
33. C.B. Brook, G. Stinson, B.K. Gibson, R. Roškar, J. Wadsley, T. Quinn, Mon. Not. R. Astron. Soc. **419**, 771. arXiv:1105.2562
34. A.V. Maccio, Astrophys. J. Lett. **744**, L9. arXiv:1111.5620
35. F. Governato et al., Nature **463**, 203–206 (2010)
36. J. Silk, A. Nusser, Astrophys. J. **725**, 556 (2011)
37. M.S. Bovill, M. Ricotti, Astrophys. J. **741**, 18 (2011). arXiv:1010.2233
38. A.V. Maccio, X. Kang, F. Fontanot, R.S. Somerville, S.E. Koposov, P. Monaco, Mon. Not. R. Astron. Soc. **402**, 1995 (2010)
39. A.V. Kravtsov, Adv. Astron. 281913 (2010). arXiv:0906.3295
40. M. Boylan-Kolchin, J.S. Bullock, M. Kaplinghat, Mon. Not. R. Astron. Soc. arXiv:1111.2048
41. I. Ferrero, M.G. Abadi, J.F. Navarro, L.V. Sales, S. Gurovich, Mon. Not. R. Astron. Soc. **425**, 2817 (2012). arXiv:1111.6609
42. S. Nickerson, G. Stinson, H.M.P. Couchman, J. Bailin, J. Wadsley, Mon. Not. R. Astron. Soc. **415**, 257 (2011). arXiv:1103.3285
43. F. Fontanot, G. De Lucia, P. Monaco, R.S. Somerville, P. Santini, Mon. Not. R. Astron. Soc. **397**, 1776 (2009)
44. B. Henriques et al., Mon. Not. R. Astron. Soc. **415**, 3571 (2011). arXiv:1009.1392
45. D. Elbaz et al., Astron. Astrophys. **533**, 119 (2011). arXiv:1105.2537
46. R.A. Swaters, R. Sancisi, T.S. van Albada, J.M. van der Hulst, Astrophys. J. **729**, 118 (2011). arXiv:1101.3120
47. P. van Dokkum, C. Conroy, Astrophys. J. **735**, L13 (2011)
48. G. Bertin, N.C. Amorisco, Astron. Astrophys. arXiv:0912.3178
49. M. Milgrom, Astrophys. J. **270**, 365 (1983)
50. S. McGaugh, Phys. Rev. Lett. **106**, 121303. arXiv:1102.3913
51. M.E. Anderson, J.N. Bregman, Astrophys. J. **714**, 320 (2010)
52. X. Dai, J.N. Bregman, C.S. Kochanek, E. Rasia, Astrophys. J. **719**, 119–125 (2010)
53. S. Andreon, Mon. Not. R. Astron. Soc. **407**, 263 (2010). arXiv:1004.2785
54. J.M. Shull, B.D. Smith, C.W. Danforth, Astrophys. J. arXiv:1112.2706
55. H.A. Feldman, R. Watkins, M.J. Hudson, Mon. Not. R. Astron. Soc. **407**, 2328–2338 (2010)
56. A. Kashlinsky, F. Atrio-Barandela, H. Ebeling, A. Edge, D. Kocevski, Astrophys. J. **712**, L81–L85 (2010)
57. A. Natarajan, C. Savage, K. Freese, Phys. Rev. D. arXiv:1109.0014
58. H. Iminniyaz, M. Drees, X. Chen, J. Cosmol. Astropart. Phys. arXiv:1104.5548
59. T. Lin, H. Yu, K. Zurek, arXiv:1111.0293
60. M. Ackermann (The Fermi LAT Collaboration), arXiv:1109.0521
61. D. Borla Tridon, P. Colin, L. Cossio, M. Doro, V. Scalzotto, in *32nd ICRC* (2011). arXiv:1110.4008
62. D. Grasso, D. Gaggero, in *Contribution to the 2011 Fermi Symposium—eConf Proceedings C110509* (2011). arXiv:1110.2591
63. Astropart. Phys. **34**, 528–538 (2011)
64. T.R. Slatyer, N. Toro, N. Weiner, Phys. Rev. D. arXiv:1107.3546
65. A. Geringer-Sameth, S.M. Koushiappas, arXiv:1108.2914
66. S. Galli, F. Iocco, G. Bertone, A. Melchiorri, Phys. Rev. D **84**, 027302 (2011)
67. D. Hooper, D.P. Finkbeiner, G. Dobler, Phys. Rev. D **76**, 083012 (2007)
68. D. Hooper, G. Zaharias, Phys. Rev. D **77**, 043511 (2008)
69. M. Su, T.R. Slatyer, D.P. Finkbeiner, Astrophys. J. **724**, 1044–1082 (2010)
70. D. Hooper, T. Linden, arXiv:1110.0006 (2011)
71. A. Morselli et al. (Fermi-LAT Collaboration), arXiv:1012.2292 [astro-ph.HE]
72. D. Pietrobon et al., Astrophys. J. arXiv:1110.5418
73. D. Hooper, T. Linden, Phys. Rev. D **83**, 083517 (2011)

74. A. Ibarra, D. Tran, C. Weniger, J. Cosmol. Astropart. Phys. **1**, 9 (2010)
75. T. Delahaye, J. Silk, Phys. Rev. Lett. **105**, 221301 (2010)
76. L. Dugger, T.E. Jeltema, S. Profumo, J. Cosmol. Astropart. Phys. **12**, 15 (2010)
77. S. Turck-Chieze, R.A. Garcia, I. Lopes, J. Ballot, S. Couvidat, S. Mathur, D. Salabert, J. Silk, Astrophys. J. Lett. **746**, L12 (2012)
78. I. Lopes, J. Silk, Science **330**, 462 (2010)
79. Y. Akrami et al., J. Cosmol. Astropart. Phys. **04**, 012 (2011)
80. L. Bergstrom, T. Bringmann, J. Edsjo, Phys. Rev. D **83**, 045024 (2010)
81. A. Perez-Garcia, J. Silk, J. Stone, Phys. Rev. Lett. **105**, 1101 (2010)
82. F. Aharonian et al., Astron. Astrophys. **503**, 817 (2009)
83. M. Chernyakova, D. Malyshev, F.A. Aharonian, R.M. Crocker, D.I. Jones, Astrophys. J. **726** (2011). arXiv:1009.2630
84. G. Bertone, D.G. Cerdeno, M. Fornasa, L. Pieri, R. Ruiz de Austri, R. Trotta, arXiv:1111.2607

Chapter 11
Dark Energy: Observational Status and Theoretical Models

Shinji Tsujikawa

Abstract About 70 % of the energy density of the Universe today consists of dark energy responsible for cosmic acceleration. We present observational bounds on dark energy constrained by the type Ia supernovæ, cosmic microwave background, and baryon acoustic oscillations. We also review theoretical attempts to explain the origin of dark energy. This includes the cosmological constant, modified matter models (such as quintessence, k-essence, coupled dark energy, unified models of dark energy and dark matter), and modified gravity models (such as $f(R)$ gravity, scalar-tensor theories, braneworlds).

11.1 Introduction

The observational discovery of the late-time cosmic acceleration reported in 1998 [1, 2] based on the type Ia supernovæ (SN Ia) opened up a new field of research in cosmology.[1] The source for this acceleration, dubbed dark energy [3], is unknown, in spite of tremendous efforts to understand its origin over the last decade [4–9]. Dark energy is distinguished from ordinary matter in that it has a negative pressure whose equation of state w_{DE} is close to -1. Independent observational data such as SN Ia [10–13], cosmic microwave background (CMB) [14, 15], and baryon acoustic oscillations (BAO) [16–18] have continued to confirm that about 70 % of the energy density of the present Universe consists of dark energy.

The simplest candidate for dark energy is the so-called cosmological constant Λ whose equation of state is $w_{DE} = -1$. If the cosmological constant originates from a vacuum energy of particle physics, its energy scale is significantly larger than the dark energy density today [19] ($\rho_{DE}^{(0)} \approx 10^{-47}$ GeV^4). Hence we need to find a mechanism to obtain the tiny value of Λ consistent with observations. A lot of efforts have been made in this direction under the framework of particle physics. For

---

[1] Saul Perlmutter, Adam Riess, and Brian Schmidt won the Nobel prize of physics in 2012.

S. Tsujikawa (✉)
Department of Physics, Faculty of Science, Tokyo University of Science, 1–3 Kagurazaka, Shinjuku-ku, Tokyo, 162-8601, Japan
e-mail: shinji@rs.kagu.tus.ac.jp

G. Calcagni et al. (eds.), *Quantum Gravity and Quantum Cosmology*,
Lecture Notes in Physics 863, DOI 10.1007/978-3-642-33036-0_11,
© Springer-Verlag Berlin Heidelberg 2013

example, the recent development of string theory shows that it is possible to construct de Sitter vacua by compactifying extra dimensions in the presence of fluxes with an account of non-perturbative corrections [20].

The first step toward understanding the property of dark energy is to clarify whether it is a simple cosmological constant or it originates from other sources that dynamically change in time. Dynamical dark energy models can be distinguished from the cosmological constant by studying the evolution of w_{DE}. The scalar field models of DE such as quintessence [21–30] and k-essence [31–33] predict a wide variety of variations of w_{DE}, but still the current observational data are not sufficient to provide some preference of such models over the Λ-Cold-Dark-Matter (ΛCDM) model. Moreover, the field potentials need to be sufficiently flat such that the field evolves slowly enough to drive the present cosmic acceleration. This demands that the field mass be extremely small ($m_\phi \approx 10^{-33}$ eV) relative to typical mass scales appearing in particle physics [34, 35]. However it is not entirely hopeless to construct viable scalar-field dark energy models in the framework of particle physics. We note that there is another class of modified matter models based on perfect fluids—the so-called (generalized) Chaplygin gas model [36, 37]. If these models are responsible for explaining the origin of dark matter as well as dark energy, then they are severely constrained from the matter power spectrum in galaxy clustering [38].

There exists another class of dynamical dark energy models that modify general relativity. The models that belong to this class are $f(R)$ gravity [39–42] (f is a function of the Ricci scalar R), scalar-tensor theories [43–47], and the Dvali, Gabadadze and Porrati (DGP) braneworld model [48]. The attractive feature of these models is that the cosmic acceleration can be realized without recourse to a dark energy component. If we modify gravity from general relativity, however, there are stringent constraints coming from local gravity tests as well as a number of observational constraints such as large-scale structure (LSS) and CMB. Hence the restriction on modified gravity models is in general very tight compared to modified matter models. We shall construct viable modified gravity models and discuss their observational and experimental signatures.

This review is organized as follows. In Sect. 11.2 we provide recent observational constraints on dark energy obtained by SN Ia, CMB, and BAO data. In Sect. 11.3 we review theoretical attempts to explain the origin of the cosmological constant consistent with the low-energy scale of dark energy. In Sect. 11.4 we discuss modified gravity models of dark energy—including quintessence, k-essence, coupled dark energy, and unified models of dark energy and dark matter. In Sect. 11.5 we review modified gravity models and provide a number of ways to distinguish those models observationally from the ΛCDM model. We conclude in Sect. 11.6.

We use units such that $c = \hbar = 1$, where c is the speed of light and $\hbar$ is reduced Planck's constant. The gravitational constant G is related to the Planck mass $m_{pl} = 1.2211 \times 10^{19}$ GeV via $G = 1/m_{pl}^2$ and the reduced Planck mass $M_{pl} = 2.4357 \times 10^{18}$ GeV via $\kappa^2 \equiv 8\pi G = 1/M_{pl}^2$, respectively. We write the Hubble constant today as $H_0 = 100h$ km s^{-1} Mpc^{-1}, where h describes the uncertainty on the value H_0. We use the metric signature $(-, +, +, +)$.

11.2 Observational Constraints on Dark Energy

The late-time cosmic acceleration is supported by a number of independent observations—such as (i) supernovæ observations, (ii) Cosmic Microwave Background (CMB), and (iii) Baryon acoustic oscillations (BAO). In this section we discuss observational constraints on the property of dark energy.

11.2.1 Supernovæ Ia Observations

In 1998 Riess et al. [1] and Perlmutter et al. [2] independently reported the late-time cosmic acceleration by observing distant supernovæ of type Ia (SN Ia). The line-element describing a 4-dimensional homogeneous and isotropic Universe is given by

$$ds^2 = g_{\mu\nu}dx^\mu dx^\nu = -dt^2 + a^2(t)\left[\frac{dr^2}{1-Kr^2} + r^2\left(d\theta^2 + \sin^2\theta d\phi^2\right)\right], \quad (11.1)$$

which is called the Friedmann–Lemaître–Robertson–Walker (FLRW) metric. Here $a(t)$ is the scale factor with cosmic time t, and $K = +1, -1, 0$ correspond to closed, open and flat geometries, respectively. The redshift z is defined by $z = a_0/a - 1$, where $a_0 = 1$ is the scale factor today.

In order to discuss the cosmological evolution in the low-redshift regime ($z < \mathscr{O}(1)$), let us consider non-relativistic matter with energy density ρ_m and dark energy with energy density ρ_{DE} and pressure P_{DE}, satisfying the continuity equations

$$\dot{\rho}_m + 3H\rho_m = 0, \quad (11.2)$$

$$\dot{\rho}_{DE} + 3H(\rho_{DE} + P_{DE}) = 0, \quad (11.3)$$

which correspond to the conservation of the energy-momentum tensor $T_{\mu\nu}$ for each component ($\nabla_\mu T^{\mu\nu} = 0$, where ∇ represents a covariant derivative). Note that a dot represents a derivative with respect to t. The cosmological dynamics is known by solving the Einstein equations

$$G_{\mu\nu} = 8\pi G T_{\mu\nu}, \quad (11.4)$$

where $G_{\mu\nu}$ is the Einstein tensor. For the metric (11.1) the 00 component of the Einstein equations gives

$$H^2 = \frac{8\pi G}{3}(\rho_m + \rho_{DE}) - \frac{K}{a^2}, \quad (11.5)$$

where $H \equiv \dot{a}/a$ is the Hubble parameter. We define the density parameters

$$\Omega_m \equiv \frac{8\pi G\rho_m}{3H^2}, \qquad \Omega_{DE} \equiv \frac{8\pi G\rho_{DE}}{3H^2}, \qquad \Omega_K \equiv -\frac{K}{(aH)^2}, \quad (11.6)$$

which satisfy the relation $\Omega_m + \Omega_{DE} + \Omega_K = 1$ from Eq. (11.5). Integrating Eqs. (11.2) and (11.3), we obtain

$$\rho_m = \rho_m^{(0)}(1+z)^3, \qquad \rho_{DE} = \rho_{DE}^{(0)} \exp\left[\int_0^z \frac{3(1+w_{DE})}{1+\tilde{z}} d\tilde{z}\right], \qquad (11.7)$$

where '0' represents the values today and $w_{DE} = P_{DE}/\rho_{DE}$ is the equation of state of dark energy. Plugging these relations into Eq. (11.5), it follows that

$$H^2(z) = H_0^2\left[\Omega_m^{(0)}(1+z)^3 + \Omega_{DE}^{(0)} \exp\left\{\int_0^z \frac{3(1+w_{DE})}{1+\tilde{z}} d\tilde{z}\right\} + \Omega_K^{(0)}(1+z)^2\right].$$

$$(11.8)$$

The expansion rate $H(z)$ can be known observationally by measuring the luminosity distance $d_L(z)$ of SN Ia. The luminosity distance is defined by $d_L^2 \equiv L_s/(4\pi\mathscr{F})$, where L_s is the absolute luminosity of a source and $\mathscr{F}$ is an observed flux. It is a textbook exercise [8, 9] to derive $d_L(z)$ for the FLRW metric (11.1):

$$d_L(z) = \frac{1+z}{H_0\sqrt{\Omega_K^{(0)}}} \sinh\left(\sqrt{\Omega_K^{(0)}} \int_0^z \frac{d\tilde{z}}{E(\tilde{z})}\right), \qquad (11.9)$$

where $E(z) \equiv H(z)/H_0$. The function $f_K(\chi) \equiv 1/\sqrt{\Omega_K^{(0)}} \sinh(\sqrt{\Omega_K^{(0)}}\chi)$ can be understood as $f_K(\chi) = \sin\chi$ (for $K = +1$), $f_K(\chi) = \chi$ (for $K = 0$), and $f_K(\chi) = \sinh\chi$ (for $K = -1$). For the flat case ($K = 0$), Eq. (11.9) reduces to $d_L(z) = (1+z)\int_0^z d\tilde{z}/H(\tilde{z})$, i.e.,

$$H(z) = \left[\frac{d}{dz}\left(\frac{d_L(z)}{1+z}\right)\right]^{-1}. \qquad (11.10)$$

Hence the measurement of the luminosity distance $d_L(z)$ of SN Ia allows us to find the expansion history of the Universe for $z < \mathcal{O}(1)$.

The luminosity distance d_L is expressed in terms of an apparent magnitude m and an absolute magnitude M of an object, as

$$m - M = 5\log_{10}\left(\frac{d_L}{10\,\text{pc}}\right). \qquad (11.11)$$

The absolute magnitude M at the peak of brightness is the same for any SN Ia under the assumption of standard candles, which is around $M \approx -19$ [1, 2]. The luminosity distance $d_L(z)$ is known from Eq. (11.11) by observing the apparent magnitude m. The redshift z of an object is known by measuring the wavelength λ_0 of light relative to its wavelength λ in the rest frame, i.e., $z = \lambda_0/\lambda - 1$. The observations of many SN Ia provide the dependence of the luminosity distance d_L in terms of z.

Expanding the function (11.9) around $z = 0$, it follows that

$$d_L(z) = \frac{1}{H_0}\left[z + \left\{1 - \frac{E'(0)}{2}\right\}z^2 + \mathcal{O}(z^3)\right]$$

$$= \frac{1}{H_0}\left[z + \frac{1}{4}(1 - 3w_{DE}\Omega_{DE}^{(0)} + \Omega_K^{(0)})z^2 + \mathcal{O}(z^3)\right], \qquad (11.12)$$

Fig. 11.1 68.3 %, 95.4 %,
and 99.7 % confidence level
contours on w_{DE} and $\Omega_m^{(0)}$
(denoted as w and Ω_m in the
figure) constrained by the
Union08 SN Ia datasets. The
equation of state w_{DE} is
assumed to be constant. From
Ref. [13]

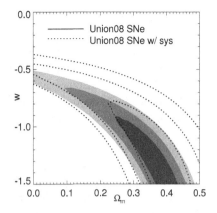

where a prime represents a derivative with respect to z. Note that, in the second line, we have used Eq. (11.8). In the presence of dark energy ($w_{DE} < 0$ and $\Omega_{DE}^{(0)} > 0$) the luminosity distance gets larger than that in the flat Universe without dark energy. For smaller (negative) w_{DE} and for larger $\Omega_{DE}^{(0)}$ this tendency becomes more significant. The open Universe without dark energy can also give rise to a larger value of $d_L(z)$, but the density parameter $\Omega_K^{(0)}$ is constrained to be close to 0 from the WMAP data (more precisely, $-0.0133 < \Omega_K^{(0)} < 0.0084$ [15]). Hence, in the low-redshift regime ($z < 1$), the luminosity distance in the open Universe is not very different from that in the flat Universe without dark energy.

As we see from Eq. (11.12), the observational data in the high-redshift regime ($z > 0.5$) allows us to confirm the presence of dark energy. The SN Ia data released by Riess et al. [1] and Perlmutter et al. [2] in 1998 in the redshift regime $0.2 < z < 0.8$ showed that the luminosity distances of observed SN Ia tend to be larger than those predicted in the flat Universe without dark energy. Assuming a flat Universe with a dark-energy equation of state $w_{DE} = -1$ (i.e., the cosmological constant), Perlmutter et al. [2] found that the cosmological constant is present at the 99 % confidence level. According to their analysis the density parameter of non-relativistic matter today was constrained to be $\Omega_m^{(0)} = 0.28^{+0.09}_{-0.08}$ (68 % confidence level) in the flat universe with the cosmological constant.

Over the past decade, more SN Ia data have been collected by a number of high-redshift surveys—such as SuperNova Legacy Survey (SNLS) [10], Hubble Space Telescope (HST) [11], and 'Equation of State: SupErNovæ trace Cosmic Expansion' (ESSENCE) [12] survey. These data also confirmed that the Universe entered the epoch of cosmic acceleration after the matter-dominated epoch. If we allow the case in which dark energy is different from the cosmological constant (i.e., $w_{DE} \neq -1$), then observational constraints on w_{DE} and $\Omega_{DE}^{(0)}$ (or $\Omega_m^{(0)}$) are not so stringent. In Fig. 11.1 we show the observational contours on ($w_{DE}, \Omega_m^{(0)}$) for constant w_{DE} obtained from the 'Union08' SN Ia data by Kowalski et al. [13]. Clearly the SN Ia data alone are not yet sufficient to place tight bounds on w_{DE}.

In the flat Universe dominated by dark energy with constant $w_{\rm DE}$, it follows from Eq. (11.8) that $H^2 \approx H_0^2 \Omega_{\rm DE}^{(0)} (1 + z)^{3(1+w_{\rm DE})} \propto a^{-3(1+w_{\rm DE})}$. Integrating this equation, we find that the scale factor evolves as $a \propto t^{2/[3(1+w_{\rm DE})]}$ for $w_{\rm DE} > -1$ and $a \propto e^{Ht}$ for $w_{\rm DE} = -1$. The cosmic acceleration occurs for $-1 \le w_{\rm DE} < -1/3$. In fact, Fig. 11.1 shows that $w_{\rm DE}$ is constrained to be smaller than $-1/3$. If $w_{\rm DE} < -1$, which is called phantom or ghost dark energy [49], the solution corresponding to the expanding Universe is given by $a \propto (t_s - t)^{2/[3(1+w_{\rm DE})]}$, where t_s is a constant. In this case the Universe ends at $t = t_s$ with a so-called big rip singularity [50, 51] at which the curvature grows toward infinity. The current observations allow the possibility of the phantom equation of state. We note, however, that a dark-energy equation-of-state index smaller than -1 does not necessarily imply the appearance of the big rip singularity. In fact, in some modified gravity models such as $f(R)$ gravity it is possible to realize $w_{\rm DE} < -1$ without having a future big rip singularity [52].

11.2.2 CMB

The temperature anisotropies in CMB are affected by the presence of dark energy. The position of the acoustic peaks in CMB anisotropies depends on the expansion history from the decoupling epoch to the present. Hence the presence of dark energy leads to a shift in the positions of acoustic peaks. There is also another effect called the integrated Sachs–Wolfe (ISW) effect [53] induced by the variation of the gravitational potential during the epoch of the cosmic acceleration. Since the ISW effect is limited to large-scale perturbations, the former effect is typically more important.

The cosmic inflation in the early Universe [54–57] predicts nearly scale-invariant spectra of density perturbations through the quantum fluctuation of a scalar field. This is consistent with the CMB temperature anisotropies observed by COBE [58] and WMAP [14]. The perturbations are 'frozen' after the scale $\lambda = (2\pi/k)a$ (k is a comoving wavenumber) leaves the Hubble radius H^{-1} during inflation ($\lambda > H^{-1}$) [59, 60]. After inflation, perturbations cross inside the Hubble radius again ($\lambda < H^{-1}$) and they start to oscillate as sound waves. This second horizon crossing occurs earlier for larger k (i.e., for smaller-scale perturbations).

We define the sound horizon as $r_s(\eta) = \int_0^\eta d\tilde{\eta} c_s(\tilde{\eta})$, where c_s is the sound speed and $d\eta = a^{-1}dt$. The squared sound speed is given by

$$c_s^2 = \frac{1}{3(1 + R_s)}, \qquad R_s = \frac{3\rho_b}{4\rho_\gamma}, \tag{11.13}$$

where ρ_b and ρ_γ are the energy densities of baryons and photons, respectively. The characteristic angle for the location of CMB acoustic peaks is [61]

$$\theta_A \equiv \frac{r_s(z_{\rm dec})}{d_A^{(c)}(z_{\rm dec})}, \tag{11.14}$$

where $d_A^{(c)}$ is the comoving angular diameter distance related to the luminosity distance d_L via the duality relation $d_A^{(c)} = d_L/(1+z)$ [9], and $z_{dec} \approx 1090$ is the redshift at the decoupling epoch. The CMB multipole ℓ_A that corresponds to the angle (11.14) is

$$\ell_A = \frac{\pi}{\theta_A} = \pi \frac{d_A^{(c)}(z_{dec})}{r_s(z_{dec})}. \tag{11.15}$$

Using Eq. (11.9) and the background equation $3H^2 = 8\pi G(\rho_m + \rho_r)$ for the redshift $z > z_{dec}$ (where ρ_m and ρ_r are the energy density of non-relativistic matter and radiation, respectively), we obtain [62, 63]

$$\ell_A = \frac{3\pi}{4}\sqrt{\frac{\omega_b}{\omega_\gamma}}\left[\ln\left(\frac{\sqrt{R_s(a_{dec}) + R_s(a_{eq})} + \sqrt{1 + R_s(a_{dec})}}{1 + \sqrt{R_s(a_{eq})}}\right)\right]^{-1}\mathcal{R}, \tag{11.16}$$

where $\omega_b \equiv \Omega_b^{(0)}h^2$ and $\omega_\gamma \equiv \Omega_\gamma^{(0)}h^2$, and $\mathcal{R}$ is the so-called CMB shift parameter defined by [64]

$$\mathcal{R} \equiv \sqrt{\frac{\Omega_m^{(0)}}{\Omega_K^{(0)}}}\sinh\left(\sqrt{\Omega_K^{(0)}}\int_0^{z_{dec}}\frac{dz}{E(z)}\right). \tag{11.17}$$

The quantity $R_s = 3\rho_b/(4\rho_\gamma)$ can be expressed as

$$R_s(a) = \frac{3\omega_b}{4\omega_\gamma}a. \tag{11.18}$$

In Eq. (11.16), a_{dec} and a_{eq} correspond to the scale factor at the decoupling epoch and at the radiation-matter equality, respectively.

The change of cosmic expansion history from the decoupling epoch to the present affects the CMB shift parameter, which gives rise to the shift for the multipole ℓ_A. The general relation for all peaks and troughs of observed CMB anisotropies is given by [65]

$$\ell_m = \ell_A(m - \phi_m), \tag{11.19}$$

where m represents peak numbers ($m = 1$ for the first peak, $m = 1.5$ for the first trough, ...) and ϕ_m is the shift of multipoles. For a given cosmic curvature $\Omega_K^{(0)}$, the quantity ϕ_m depends weakly on ω_b and $\omega_m \equiv \Omega_m^{(0)}h^2$. The shift of the first peak can be fitted as $\phi_1 = 0.265$ [65]. The WMAP 7-year bound on the CMB shift parameter is given by [15]

$$\mathcal{R} = 1.725 \pm 0.018, \tag{11.20}$$

at the 68 % confidence level. Taking $\mathcal{R} = 1.72$ together with other values $\omega_b = 0.02265$, $\omega_m = 0.1369$, and $\omega_\gamma = 2.469 \times 10^{-5}$ constrained by the WMAP 5-year data, we obtain $\ell_A \approx 300$ from Eq. (11.16). Using the relation (11.19) with $\phi_1 = 0.265$ we find that the first acoustic peak corresponds to $\ell_1 \approx 220$, as observed in CMB anisotropies.

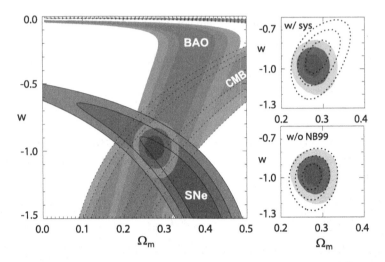

Fig. 11.2 68.3 %, 95.4 % and 99.7 % confidence level contours on w_{DE} and $\Omega_m^{(0)}$ (denoted as w and Ω_m in the figure, respectively) for a flat Universe. The *left panel* illustrates the individual constraints from SN Ia, CMB, and BAO, as well as the combined constraints (*filled gray contours*, statistical errors only). The *upper right panel* shows the effect of including systematic errors. The *lower right panel* illustrates the impact of the Supernova Cosmology Project (SCP) Nearby 1999 data. From Ref. [13]

In the flat Universe ($K = 0$) the CMB shift parameter is simply given by $\mathscr{R} = \sqrt{\Omega_m^{(0)}} \int_0^{z_{dec}} dz/E(z)$. For smaller $\Omega_m^{(0)}$ (i.e., for larger $\Omega_{DE}^{(0)}$), $\mathscr{R}$ tends to be smaller. For the cosmological constant ($w_{DE} = -1$) the normalized Hubble expansion rate is given by $E(z) = [\Omega_m^{(0)}(1 + z)^3 + \Omega_{DE}^{(0)}]^{1/2}$. Under the bound (11.20) the density parameter is constrained to be $0.72 < \Omega_{DE}^{(0)} < 0.77$. This is consistent with the bound coming from the SN Ia data. One can also show that, for increasing w_{DE}, the observationally allowed values of $\Omega_m^{(0)}$ gets larger. However, $\mathscr{R}$ depends weakly on the w_{DE}. Hence the CMB data alone do not provide a tight constraint on w_{DE}. In Fig. 11.2 we show the joint observational constraints on w_{DE} and $\Omega_m^{(0)}$ (for constant w_{DE}) obtained from the WMAP 5-year data and the Union08 SN Ia data [13]. The joint observational constraints provide much tighter bounds compared to the individual constraint from CMB and SN Ia. For the flat Universe Kowalski et al. [13] obtained the bounds $w_{DE} = -0.955^{+0.060+0.059}_{-0.066-0.060}$ and $\Omega_m^{(0)} = 0.265^{+0.022+0.018}_{-0.021-0.016}$ (with statistical and systematic errors) from the combined data analysis of CMB and SN Ia.

11.2.3 BAO

The detection of baryon acoustic oscillations first reported in 2005 by Eisenstein et al. [16] in a spectroscopic sample of 46,748 luminous red galaxies observed by the Sloan Digital Sky Survey (SDSS) has provided another test for probing the property of dark energy. Since baryons are strongly coupled to photons prior to the decoupling epoch, the oscillation of sound waves is imprinted in baryon perturbations as well as CMB anisotropies.

The sound horizon at which baryons were released from the Compton drag of photons determines the location of baryon acoustic oscillations. This epoch, called the drag epoch, occurs at the redshift z_d. The sound horizon at $z = z_d$ is given by $r_s(z_d) = \int_0^{\eta_d} d\eta c_s(\eta)$, where c_s is the sound speed. According to the fitting formula of z_d by Eisenstein and Hu [66], z_d and $r_s(z_d)$ are constrained to be around $z_d \approx 1020$ and $r_s(z_d) \approx 150$ Mpc.

We observe the angular and redshift distributions of galaxies as a power spectrum $P(k_\perp, k_\parallel)$ in the redshift space, where $k_\perp$ and $k_\parallel$ are the wavenumbers perpendicular and parallel to the direction of light, respectively. In principle, we can measure the following two ratios [67]

$$\theta_s(z) = \frac{r_s(z_d)}{d_A^{(c)}(z)}, \qquad \delta z_s(z) = \frac{r_s(z_d)H(z)}{c}, \qquad (11.21)$$

where the speed of light c is recovered for clarity. In the first equation $d_A^{(c)}$ is the comoving angular diameter distance related to the proper angular diameter distance d_A via the relation $d_A^{(c)} = d_A/a = d_A(1 + z)$. The quantity $\theta_s(z)$ characterizes the angle orthogonal to the line of sight, whereas the quantity δz_s corresponds to the oscillations along the line of sight.

The current BAO observations are not sufficient to measure both $\theta_s(z)$ and $\delta z_s(z)$ independently. From the spherically averaged spectrum one can find a combined distance scale ratio given by [67]

$$\left[\theta_s(z)^2 \delta z_s(z)\right]^{1/3} \equiv \frac{r_s(z_d)}{[(1+z)^2 d_A^2(z)c/H(z)]^{1/3}}, \qquad (11.22)$$

or, alternatively, the effective distance ratio [16]

$$D_V(z) \equiv \left[(1+z)^2 d_A^2(z)cz/H(z)\right]^{1/3}. \qquad (11.23)$$

In 2005 Eisenstein et al. [16] obtained the constraint $D_V(z) = 1370 \pm 64$ Mpc at the redshift $z = 0.35$. In 2007 Percival et al. [17] measured the effective distance ratio defined by

$$r_{BAO}(z) \equiv \frac{r_s(z_d)}{D_V(z)}, \qquad (11.24)$$

at the two redshifts: $r_{BAO}(z = 0.2) = 0.1980 \pm 0.0058$ and $r_{BAO}(z = 0.35) = 0.1094 \pm 0.0033$. This is based on the data from the 2-degree Field (2dF) Galaxy Redshift Survey. These data provide the observational contour of BAO plotted in Fig. 11.2. From the joint data analysis of SN Ia, WMAP 5-year, and BAO data, Kowalski et al. [13] placed the constraints $w_{DE} = -0.969^{+0.059}_{-0.063}(\text{stat})^{+0.063}_{-0.066}(\text{sys})$ and $\Omega_m^{(0)} = 0.274^{+0.016}_{-0.016}(\text{stat})^{+0.013}_{-0.012}(\text{sys})$ for the constant equation of state of dark energy.

The recent measurement of the 2dF as well as the SDSS data provided the effective distance ratio to be $r_{BAO}(z = 0.2) = 0.1905 \pm 0.0061$ and $r_{BAO}(z = 0.35) = 0.1097 \pm 0.0036$ [18]. Using these data together with the WMAP 7-year data [15] and the Gaussian prior on the Hubble constant $H_0 = 74.2 \pm 3.6$ km s^{-1} Mpc^{-1}, Komatsu et al. [15] derived the constraint $w_{DE} = -1.10 \pm 0.14$ (68 % confidence level) for the constant equation of state in the flat Universe. Adding the high-z SN Ia in their analysis they found the most stringent bound: $w_{DE} = -0.980 \pm 0.053$ (68 % confidence level). Hence the ΛCDM model is well consistent with a number of independent observational data.

Finally we should mention that there are other constraints coming from the cosmic age [68], large-scale clustering [69–71], gamma ray bursts [72], and weak lensing [73]. So far we have not found strong evidence for supporting dynamical dark energy models over the ΛCDM model, but future high-precision observations may break this degeneracy.

11.3 Cosmological Constant

The cosmological constant Λ is one of the simplest candidates of dark energy, and as we have seen in the previous section, it is favored by a number of observations. However, if the origin of the cosmological constant is a vacuum energy, it suffers from a serious problem of its energy scale relative to the dark energy density today [19]. The zero-point energy of some field of mass m with momentum k and frequency ω is given by $E = \omega/2 = \sqrt{k^2 + m^2}/2$. Summing over the zero-point energies of this field up to a cut-off scale k_{max} ($\gg m$), we obtain the vacuum energy density

$$\rho_{vac} = \int_0^{k_{max}} \frac{d^3k}{(2\pi)^3} \frac{1}{2} \sqrt{k^2 + m^2}. \qquad (11.25)$$

Since the integral is dominated by the mode with large k ($\gg m$), we find that

$$\rho_{vac} \approx \int_0^{k_{max}} \frac{4\pi k^2 dk}{(2\pi)^3} \frac{1}{2} k = \frac{k_{max}^4}{16\pi^2}. \qquad (11.26)$$

Taking the cut-off scale k_{max} to be the Planck mass m_{pl}, the vacuum energy density can be estimated as $\rho_{vac} \approx 10^{74}$ GeV^4. This is about 10^{121} times larger than the observed value $\rho_{DE}^{(0)} \approx 10^{-47}$ GeV^4.

Before the observational discovery of dark energy in 1998, most people believed that the cosmological constant is exactly zero and tried to explain why it is so. The vanishing of a constant may imply the existence of some symmetry. In supersymmetric theories the bosonic degree of freedom has its Fermi counterpart which contributes to the zero-point energy with an opposite sign.[2] If supersymmetry is unbroken, an equal number of bosonic and fermionic degrees of freedom is present

---

[2]The readers who are not familiar with supersymmetric theories may consult the books [74, 75].

such that the total vacuum energy vanishes. However, it is known that supersymmetry is broken at sufficient high energies (for the typical scale $M_{SUSY} \approx 10^3$ GeV). Therefore, the vacuum energy is generally non-zero in the world of broken supersymmetry.

Even if supersymmetry is broken, there is a hope to obtain a vanishing Λ or a tiny amount of Λ. In supergravity theory the effective cosmological constant is given by an expectation value of the potential V for chiral scalar fields φ^i [74]:

$$V(\varphi, \varphi^*) = e^{\kappa^2 K}\left[D_i W \left(K^{ij^*} \right)(D_j W)^* - 3\kappa^2 |W|^2 \right], \tag{11.27}$$

where K and W are the so-called Kähler potential and the superpotential, respectively, which are functions of φ^i and its complex conjugate φ^{i*}. The quantity K^{ij^*} is an inverse of the derivative $K_{ij^*} \equiv \partial^2 K / \partial \varphi^i \partial \varphi^{j^*}$, whereas the derivative $D_i W$ is defined by $D_i W \equiv \partial W / \partial \varphi^i + \kappa^2 W (\partial K / \partial \varphi^i)$.

The condition $D_i W \neq 0$ corresponds to the breaking of supersymmetry. In this case, it is possible to find scalar-field values leading to a vanishing potential ($V = 0$), but this is not in general an equilibrium point of the potential V. Nevertheless, there is a class of Kähler potentials and superpotentials giving a stationary scalar-field configuration at $V = 0$. The gluino condensation model in $E_8 \times E_8$ superstring theory proposed by Dine [76] belongs to this class. The reduction of the 10-dimensional action to the 4-dimensional action gives rise to a so-called modulus field T. This field characterizes the scale of the compactified 6-dimensional manifold. Generally one has another complex scalar field S corresponding to 4-dimensional dilaton/axion fields. The fields T and S are governed by the Kähler potential

$$K(T, S) = -\frac{3}{\kappa^2} \ln \left(T + T^* \right) - \frac{1}{\kappa^2} \ln \left(S + S^* \right), \tag{11.28}$$

where $(T + T^*)$ and $(S + S^*)$ are positive definite. The field S couples to the gauge fields, while T does not. An effective superpotential for S can be obtained by integrating out the gauge fields under the use of the R-invariance [77]:

$$W(S) = M_{pl}^3 \left[c_1 + c_2 \exp\left(-\frac{3S}{2c_3} \right) \right], \tag{11.29}$$

where c_1, c_2, and c_3 are constants.

Substituting Eqs. (11.28) and (11.29) into Eq. (11.27), we obtain the field potential

$$
\begin{aligned}
V &= \frac{1}{(T + T^*)^3 (S + S^*)} (D_S W) K^{SS^*} (D_S W)^* \\
&= \frac{M_{pl}^4}{(T + T^*)^3 (S + S^*)} \left| c_1 + c_2 \exp\left(-\frac{3S}{2c_3} \right) \left[1 + \frac{3}{2c_3}(S + S^*) \right] \right|^2,
\end{aligned} \tag{11.30}
$$

where, in the first line, we have used the property $(D_T W) K^{TT^*} (D_T W)^* = 3\kappa^2 |W|^2$ for the modulus term. This potential is positive because of the cancellation of the last term in Eq. (11.27). The stationary field configuration with $V = 0$ is realized under the condition $D_S W = \partial W / \partial S - W/(S + S^*) = 0$. The derivative, $D_T W = \kappa^2 W \partial K / \partial T = -3W/(T + T^*)$, does not necessarily vanish. When $D_T W \neq 0$ the supersymmetry is broken with a vanishing potential energy. Therefore it is possible to obtain a stationary field configuration with $V = 0$ even if supersymmetry is broken.

The discussion above is based on the lowest-order perturbation theory. This picture is not necessarily valid to all finite orders of perturbation theory because the non-supersymmetric field configuration is not protected by any symmetry. Moreover, some non-perturbative effect can provide a large contribution to the effective cosmological constant [35]. The so-called flux compactification in type IIB string theory allows us to realize a metastable de Sitter (dS) vacuum by taking into account a non-perturbative correction to the superpotential (coming from brane instantons) as well as a number of anti $D3$-branes in a warped geometry [20]. Hence it is not hopeless to obtain a small value of Λ or a vanishing Λ even in the presence of some non-perturbative corrections.

Kachru, Kallosh, Linde and Trivedi (KKLT) [20] constructed dS solutions in type II string theory compactified on a Calabi–Yau manifold in the presence of flux. The construction of the dS vacua in the KKLT scenario consists of two steps. The first step is to freeze all moduli fields in the flux compactification at a supersymmetric anti de Sitter (AdS) vacuum. Then a small number of anti $D3$-branes is added in a warped geometry with a throat, so that the AdS minimum is uplifted to yield a dS vacuum with broken supersymmetry. If we want to use the KKLT dS minimum derived above for the present cosmic acceleration, we require that the potential energy V_{dS} at the minimum is of the order of $V_{dS} \approx 10^{-47}$ GeV^4. Depending on the number of fluxes there is a wealth of dS vacua, which introduced the notion of string landscape [78].

The question why the vacuum we live in has a very small energy density among many possible vacua has been sometimes answered with the anthropic principle [79, 80]. Using the anthropic arguments, Weinberg put the bound on the vacuum energy density [81]

$$-10^{-123} m_{pl}^4 \lesssim \rho_\Lambda \lesssim 3 \times 10^{-121} m_{pl}^4. \tag{11.31}$$

The upper bound comes from the requirement that the vacuum energy does not dominate over the matter density for redshift $z \gtrsim 1$. Meanwhile, the lower bound comes from the condition that ρ_Λ does not cancel the present cosmological density. Some people have studied landscape statistics by considering the relative abundance of long-lived low-energy vacua satisfying the bound (11.31) [82–85]. These statistical approaches are still under study, but it will be interesting to pursue the possibility to obtain high probabilities for the appearance of low-energy vacua.

11.4 Modified Matter Models

In this section we discuss 'modified matter models' in which the energy-momentum tensor $T_{\mu\nu}$ on the right-hand side of the Einstein equations contains an exotic matter source with a negative pressure. The models that belong to this class are quintessence, k-essence, coupled dark energy, and generalized Chaplygin gas.

11.4.1 Quintessence

A canonical scalar field ϕ responsible for dark energy is dubbed quintessence [28, 29] (see also Refs. [21–25] for earlier works). The action of quintessence is described by

$$S = \int d^4x \sqrt{-g} \left[\frac{1}{2\kappa^2} R - \frac{1}{2} g^{\mu\nu} \partial_\mu \phi \partial_\nu \phi - V(\phi) \right] + S_M, \quad (11.32)$$

where R is a Ricci scalar and ϕ is a scalar field with a potential $V(\phi)$. As a matter action S_M, we consider perfect fluids of radiation (energy density ρ_r, equation of state $w_r = 1/3$) and non-relativistic matter (energy density ρ_m, equation of state $w_m = 0$).

In the flat FLRW background radiation and non-relativistic matter satisfy the continuity equations $\dot{\rho}_r + 4H\rho_r = 0$ and $\dot{\rho}_m + 3H\rho_m = 0$, respectively. The energy density ρ_ϕ and the pressure P_ϕ of the field are $\rho_\phi = \dot{\phi}^2/2 + V(\phi)$ and $P_\phi = \dot{\phi}^2/2 - V(\phi)$, respectively. The continuity equation, $\dot{\rho}_\phi + 3H(\rho_\phi + P_\phi) = 0$, translates to

$$\ddot{\phi} + 3H\dot{\phi} + V_{,\phi} = 0, \quad (11.33)$$

where $V_{,\phi} \equiv dV/d\phi$. The field equation of state is given by

$$w_\phi \equiv \frac{P_\phi}{\rho_\phi} = \frac{\dot{\phi}^2 - 2V(\phi)}{\dot{\phi}^2 + 2V(\phi)}. \quad (11.34)$$

From the Einstein equations (11.4) we obtain the following equations:

$$H^2 = \frac{\kappa^2}{3} \left[\frac{1}{2}\dot{\phi}^2 + V(\phi) + \rho_m + \rho_r \right], \quad (11.35)$$

$$\dot{H} = -\frac{\kappa^2}{2} \left(\dot{\phi}^2 + \rho_m + \frac{4}{3}\rho_r \right). \quad (11.36)$$

Although $\{\rho_r, \rho_m\} \gg \rho_\phi$ during radiation and matter eras, the field energy density needs to dominate at late times to be responsible for dark energy. The condition to realize the late-time cosmic acceleration corresponds to $w_\phi < -1/3$, i.e., $\dot{\phi}^2 < V(\phi)$ from Eq. (11.34). This means that the scalar potential needs to be flat enough for the field to evolve slowly. If the dominant contribution to the energy density of the Universe is the slowly rolling scalar field satisfying the condition $\dot{\phi}^2 \ll V(\phi)$, we obtain the approximate relations $3H\dot{\phi} + V_{,\phi} \approx 0$ and $3H^2 \approx \kappa^2 V(\phi)$ from Eqs. (11.33) and (11.35), respectively. Hence the field equation of state in Eq. (11.34) is approximately given by

$$w_\phi \approx -1 + \frac{2}{3}\varepsilon_s, \quad (11.37)$$

where $\varepsilon_s \equiv (V_{,\phi}/V)^2/(2\kappa^2)$ is the so-called slow-roll parameter [59]. During the accelerated expansion of the Universe, ε_s is much smaller than 1 because the potential is sufficiently flat. Unlike the cosmological constant, the field equation of state deviates from -1 ($w_\phi > -1$).

Introducing the dimensionless variables $x_1 \equiv \kappa\dot\phi/(\sqrt6 H)$, $x_2 \equiv \kappa\sqrt{V}/(\sqrt3 H)$, and $x_3 \equiv \kappa\sqrt{\rho_r}/(\sqrt3 H)$, we obtain the following equations from Eqs. (11.33), (11.35), and (11.36) [8, 27, 86, 87]:

$$x_1' = -3x_1 + \frac{\sqrt6}{2}\lambda x_2^2 + \frac{1}{2}x_1\left(3 + 3x_1^2 - 3x_2^2 + x_3^2\right), \tag{11.38}$$

$$x_2' = -\frac{\sqrt6}{2}\lambda x_1 x_2 + \frac{1}{2}x_2\left(3 + 3x_1^2 - 3x_2^2 + x_3^2\right), \tag{11.39}$$

$$x_3' = -2x_3 + \frac{1}{2}x_3\left(3 + 3x_1^2 - 3x_2^2 + x_3^2\right), \tag{11.40}$$

where a prime represents a derivative with respect to $N = \ln a$, and λ is defined by $\lambda \equiv -V_{,\phi}/(\kappa V)$. The density parameters of the field, radiation, and non-relativistic matter are given by $\Omega_\phi = x_1^2 + x_2^2$, $\Omega_r = x_3^2$, and $\Omega_m = 1 - x_1^2 - x_2^2 - x_3^2$, respectively. One has constant λ for the exponential potential [27]

$$V(\phi) = V_0 e^{-\kappa\lambda\phi}, \tag{11.41}$$

in which case the fixed points of the system (11.38)–(11.40) can be derived by setting $x_i' = 0$ ($i = 1, 2, 3$). The fixed point that can be used for dark energy is given by

$$(x_1, x_2, x_3) = \left(\frac{\lambda}{\sqrt6}, \sqrt{1 - \frac{\lambda^2}{6}}, 0\right), \qquad w_\phi = -1 + \frac{\lambda^2}{3}, \qquad \Omega_\phi = 1. \tag{11.42}$$

Cosmic acceleration can be realized for $w_\phi < -1/3$, i.e., $\lambda^2 < 2$. One can show that, in this case, the accelerated fixed point is a stable attractor [27]. Hence the solutions finally approach the fixed point (11.42) after the matter era (characterized by the fixed point $(x_1, x_2, x_3) = (0, 0, 0)$).

If λ varies with time, we have the following relation

$$\lambda' = -\sqrt6\lambda^2(\Gamma - 1)x_1, \tag{11.43}$$

where $\Gamma \equiv VV_{,\phi\phi}/V_{,\phi}^2$. For monotonically decreasing potentials one has $\lambda > 0$ and $x_1 > 0$ for $V_{,\phi} < 0$ and $\lambda < 0$ and $x_1 < 0$ for $V_{,\phi} > 0$. If the condition

$$\Gamma = \frac{VV_{,\phi\phi}}{V_{,\phi}^2} > 1, \tag{11.44}$$

is satisfied, the absolute value of λ decreases toward 0 irrespective of the signs of $V_{,\phi}$ [30]. Then the solutions finally approach the accelerated 'instantaneous' fixed point (11.42) even if λ^2 is larger than 2 during radiation and matter eras [86, 87]. In this case the field equation of state gradually decreases to -1, so the models showing this behavior are called 'freezing' models [88]. The condition (11.44) is the so-called tracking condition under which the field density eventually catches up that of the background fluid.

A representative potential of the freezing model is the inverse power-law potential $V(\phi) = M^{4+n}\phi^{-n}$ ($n > 0$) [24, 30], which can appear in the fermion condensate model as a dynamical supersymmetry breaking [89]. In this case one has

Fig. 11.3 The allowed region in the (w_ϕ, w'_ϕ) plane for thawing and freezing models of quintessence (w_ϕ is denoted as w in the figure). The thawing models correspond to the region between the curves (a) $w'_\phi = 3(1 + w_\phi)$ and (b) $w'_\phi = 1 + w_\phi$, whereas the freezing models are characterized by the region between the curves (c) $w'_\phi = 0.2w_\phi(1 + w_\phi)$ and (d) $w'_\phi = 3w_\phi(1 + w_\phi)$. The *dotted line* shows the border between the acceleration and deceleration of the field ($\ddot\phi = 0$), which corresponds to $w'_\phi = 3(1 + w_\phi)^2$. From Ref. [88]

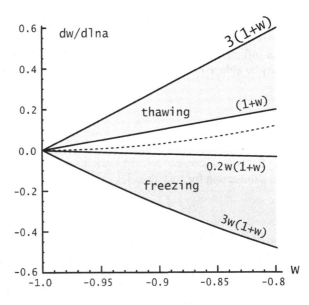

$\Gamma = (n + 1)/n > 1$ and hence the tracking condition is satisfied. Unlike the cosmological constant, even if the field energy density is not negligible relative to the background fluid density around the beginning of the radiation era, the field eventually enters the tracking regime to lead to the late-time cosmic acceleration [30]. Another example of freezing models is $V(\phi) = M^{4+n}\phi^{-n} \exp(\alpha\phi^2/m_{\rm pl}^2)$, which has a minimum with a positive energy density at which the field is eventually trapped. This potential is motivated in the framework of supergravity [90].

There is another class of quintessence potentials called 'thawing' models [88]. In thawing models the field with mass m_ϕ has been frozen by the Hubble friction (i.e., the term $H\dot\phi$) until recently and then it begins to evolve after H drops below m_ϕ. At early times the equation of state of dark energy is $w_\phi \approx -1$, but it begins to grow for $H < m_\phi$. The representative potentials that belong to this class are (a) $V(\phi) = V_0 + M^{4-n}\phi^n$ ($n > 0$) and (b) $V(\phi) = M^4 \cos^2(\phi/f)$. The potential (a) with $n = 1$ was originally proposed by Linde [91] to replace the cosmological constant by a slowly evolving scalar field. In Ref. [92] this was revised to allow for negative values of $V(\phi)$. The universe will collapse in the future if the system enters the region with $V(\phi) < 0$. The potential (b) is motivated by the pseudo-Nambu–Goldstone boson (PNGB), which was introduced in Ref. [93] in response to the first tentative suggestions for the existence of the cosmological constant. The small mass of the PNGB model required for dark energy is protected against radiative corrections, so this model is favored theoretically. In fact there are a number of interesting works to explain the small energy scale $M \approx 10^{-3}$ eV required for the PNGB quintessence in supersymmetric theories [94–97].

The freezing models and the thawing models are characterized by the conditions $w'_\phi \equiv dw_\phi/dN < 0$ and $w'_\phi > 0$, respectively. More precisely, the allowed regions for the freezing and thawing models are given by $3w_\phi(1 + w_\phi) \lesssim w'_\phi \lesssim 0.2w_\phi(1 +$

w_ϕ) and $1 + w_\phi \lesssim w'_\phi \lesssim 3(1 + w_\phi)$, respectively [88] (see Ref. [98] for details). These regions are illustrated in Fig. 11.3. While the observational data up to now are not sufficient to distinguish freezing and thawing models by the variation of w_ϕ, we may be able to do so with the next-decade high-precision observations.

11.4.2 k-Essence

Scalar fields with non-canonical kinetic terms often appear in particle physics. In general, the action for such theories can be expressed as

$$S = \int d^4 x \sqrt{-g} \left[\frac{1}{2\kappa^2} R + P(\phi, X) \right] + S_M, \tag{11.45}$$

where $P(\phi, X)$ is a function in terms of a scalar field ϕ and its kinetic energy $X = -(1/2) g^{\mu\nu} \partial_\mu \phi \partial_\nu \phi$, and S_M is a matter action. Even in the absence of the field potential $V(\phi)$ it is possible to realize the cosmic acceleration due to the kinetic energy X [99]. The application of these theories to dark energy was first carried out by Chiba et al. [31]. In Ref. [32] this was extended to more general cases and the models based on the action (11.45) were named 'k-essence'. The action (11.45) includes a wide variety of theories listed below.

- (A) Low-energy effective string theory.

  The action of low-energy effective string theory in the presence of a higher-order derivative term $(\tilde{\nabla}\phi)^4$ is given by [100, 101]

  $$S = \frac{1}{2\kappa^2} \int d^4 \tilde{x} \sqrt{-\tilde{g}} \left[F(\phi) \tilde{R} + \omega(\phi)(\tilde{\nabla}\phi)^2 + \alpha' B(\phi)(\tilde{\nabla}\phi)^4 + \mathcal{O}(\alpha'^2) \right], \tag{11.46}$$

  which is derived by the expansion in terms of the Regge slope parameter α' (this is related to the string mass scale M_s via the relation $M_s = \sqrt{2/\alpha'}$). The scalar field ϕ, dubbed dilaton field, is coupled to the Ricci scalar R with the strength $F(\phi)$. This frame is called the Jordan frame, in which the tilde is used in the action (11.46). Under a conformal transformation, $g_{\mu\nu} = F(\phi)\tilde{g}_{\mu\nu}$, we obtain the action in the Einstein frame [99]

  $$S_E = \int d^4 x \sqrt{-g} \left[\frac{1}{2\kappa^2} R + K(\phi)X + L(\phi)X^2 + \cdots \right], \tag{11.47}$$

  where $K(\phi) = 3(F_{,\phi}/F)^2 - 2\omega/F$ and $L(\phi) = 2\alpha' B(\phi)/\kappa^2$.

- (B) Ghost condensate.

  The theories with a negative kinetic energy $-X$ generally suffers from a vacuum instability [102, 103], but the presence of the quadratic term X^2 can evade this problem. The model constructed in this vein is the ghost condensate model characterized by the Lagrangian [104]

  $$P = -X + \frac{X^2}{M^4}, \tag{11.48}$$

where M is a constant. A more general version of this model, called the dilatonic ghost condensate [105], is

$$P = -X + e^{\kappa\lambda\phi} \frac{X^2}{M^4}, \tag{11.49}$$

which is motivated by a dilatonic higher-order correction to the tree-level action (as we have discussed in the case (A)).

- (C) Tachyon.
  A tachyon field appears as an unstable mode of D-branes (non-Bogomol'nyi–Prasad–Sommerfield, non-BPS, branes). The effective 4-dimensional Lagrangian is given by [106–108]

$$P = -V(\phi)\sqrt{1 - 2X}, \tag{11.50}$$

where $V(\phi)$ is a potential of the tachyon field ϕ. While it is difficult for the tachyon model to be compatible with inflation in the early Universe because of the problem for ending inflation, one can use it for dark energy provided that the potential is shallower than $V(\phi) = V_0\phi^{-2}$ [109–113].

- (D) Dirac–Born–Infeld (DBI) theory.
  In the so-called 'D-cceleration' mechanism in which a scalar field ϕ parametrizes a direction on the approximate Coulomb branch of the system in $\mathcal{N} = 4$ supersymmetric Yang–Mills theory, the field dynamics can be described by the DBI action for a probe $D3$-brane moving in a radial direction of the anti de Sitter space-time [114, 115]. The Lagrangian density with the field potential $V(\phi)$ is given by

$$P = -f(\phi)^{-1}\sqrt{1 - 2f(\phi)X} + f(\phi)^{-1} - V(\phi), \tag{11.51}$$

where $f(\phi)$ is a warped factor of the AdS throat. In this theory one can realize the acceleration of the Universe even in the regime where $2f(\phi)X$ is close to 1. The application of this theory to dark energy has been carried out in Refs. [116, 117].

For the theories with the action (11.45), the pressure P_ϕ and the energy density ρ_ϕ of the field are $P_\phi = P$ and $\rho_\phi = 2XP_{,X} - P$, respectively. The equation of state of k-essence is given by

$$w_\phi = \frac{P_\phi}{\rho_\phi} = \frac{P}{2XP_{,X} - P}. \tag{11.52}$$

As long as the condition $|2XP_{,X}| \ll |P|$ is satisfied, w_ϕ can be close to -1. In the ghost condensate model (11.48) we have

$$w_\phi = \frac{1 - X/M^4}{1 - 3X/M^4}, \tag{11.53}$$

which gives $-1 < w_\phi < -1/3$ for $1/2 < X/M^4 < 2/3$. In particular, the de Sitter solution ($w_\phi = -1$) is realized at $X/M^4 = 1/2$. Since the field energy density is $\rho_\phi = M^4/4$ at the de Sitter point, it is possible to explain the cosmic acceleration today for $M \approx 10^{-3}$ eV.

In order to discuss stability conditions of k-essence in the ultraviolet (UV) regime, we decompose the field into the homogenous and perturbed parts as $\phi(t, \mathbf{x}) = \phi_0(t) + \delta\phi(t, \mathbf{x})$ in the Minkowski background and derive the Lagrangian and the Hamiltonian for perturbations. The resulting second-order Hamiltonian reads [105]

$$\delta H = (P_{,X} + 2X P_{,XX})\frac{(\dot{\delta\phi})^2}{2} + P_{,X}\frac{(\nabla\delta\phi)^2}{2} - P_{,\phi\phi}\frac{(\delta\phi)^2}{2}. \quad (11.54)$$

The term $P_{,\phi\phi}$ is related to the effective mass of the field, which is unimportant in the UV regime as long as the field is responsible for dark energy. The positivity of the first two terms in Eq. (11.54) leads to the following stability conditions

$$P_{,X} + 2X P_{,XX} \geq 0, \qquad P_{,X} \geq 0. \quad (11.55)$$

The phantom model with a negative kinetic energy $-X$ with a potential $V(\phi)$, i.e., $P = -X - V(\phi)$, do not satisfy the above conditions. Although the phantom model with $P = -X - V(\phi)$ can lead to the background cosmological dynamics allowed by SN Ia observations ($w_\phi < -1$) [102, 118, 119], it suffers from a catastrophic particle production of ghosts and normal fields because of the instability of the vacuum [102, 103]. This problem is overcome in the ghost condensate model (11.48) in which the conditions (11.55) are satisfied for $X/M^4 > 1/2$. Thus, successful k-essence models need to be constructed to be consistent with the conditions (11.55), while the field is responsible for dark energy under the condition $|2X P_{,X}| \ll |P|$.

The propagation speed c_s of the field is given by [120]

$$c_s^2 = \frac{P_{\phi,X}}{\rho_{\phi,X}} = \frac{P_{,X}}{P_{,X} + 2X P_{,XX}}, \quad (11.56)$$

which is positive under the conditions (11.55). The speed c_s remains subluminal provided that

$$P_{,XX} > 0. \quad (11.57)$$

This condition is ensured for the models (11.48), (11.49), (11.50), and (11.51).

There are some k-essence models proposed to solve the coincidence problem of dark energy. One example is [32, 33]

$$P = \frac{1}{\phi^2}\left(-2.01 + 2\sqrt{1 + X} + 3 \cdot 10^{-17} X^3 - 10^{-24} X^4\right). \quad (11.58)$$

In these models the solutions can finally approach the accelerating phase even if they start from relatively large values of the k-essence energy density Ω_ϕ in the radiation era. In such cases, there is a period in which the sound speed becomes superluminal before reaching the accelerated attractor [121]. Moreover, it was shown that the basins of attraction of a radiation scaling solution in such models are restricted to be very small [122]. We stress that these properties arise only for the k-essence models constructed to solve the coincidence problem.

11.4.3 Coupled Dark Energy

Since the energy density of dark energy is of the same order as that of dark matter in
the present Universe, this implies that dark energy may have some relation with dark
matter. In this section we discuss the cosmological viability of coupled dark energy
models and related topics such as scaling solutions, the chameleon mechanism, and
varying α.

11.4.3.1 The Coupling Between Dark Energy and Dark Matter

In the flat FLRW cosmological background, a general coupling between dark energy
(with energy density ρ_{DE} and equation of state w_{DE}) and dark matter (with energy
density ρ_m) may be described by the following equations

$$\dot{\rho}_{DE} + 3H(1 + w_{DE})\rho_{DE} = -\beta, \tag{11.59}$$

$$\dot{\rho}_m + 3H\rho_m = +\beta, \tag{11.60}$$

where β is the rate of the energy exchange in the dark sector.

There are several forms of couplings proposed so far. Two simple examples are
given by

$$(A) \quad \beta = \kappa Q\rho_m\dot{\phi}, \tag{11.61}$$

$$(B) \quad \beta = \alpha H\rho_m, \tag{11.62}$$

where Q and α are dimensionless constants. The coupling (A) arises in scalar-tensor
theories after the conformal transformation to the Einstein frame [123–126]. In general the coupling Q is field-dependent [127, 128], but Brans–Dicke theory [129]
(including $f(R)$ gravity) gives rise to a constant coupling [130]. The coupling (B)
is more phenomenological, but this form is useful to place observational bounds
from the cosmic expansion history.

Let us consider the coupling (A) in the presence of a coupled quintessence field
with the exponential potential (11.41). We assume that the coupling Q is constant.
Taking into account radiation uncoupled to dark energy ($\rho_r \propto a^{-4}$), the Friedmann
equation is given by $3H^2 = \kappa^2(\rho_{DE} + \rho_m + \rho_r)$, where $\rho_{DE} = \dot{\phi}^2/2 + V(\phi)$. Introducing the dimensionless variables $x_1 = \kappa\dot{\phi}/(\sqrt{6}H)$, $x_2 = \kappa\sqrt{V}/(\sqrt{3}H)$, and
$x_3 = \kappa\sqrt{\rho_r}/(\sqrt{3}H)$ as in Sect. 11.4.1, we obtain

$$x_1' = -3x_1 + \frac{\sqrt{6}}{2}\lambda x_2^2 + \frac{1}{2}x_1\left(3 + 3x_1^2 - 3x_2^2 + x_3^2\right) - \frac{\sqrt{6}}{2}Q\left(1 - x_1^2 - x_2^2 - x_3^2\right), \tag{11.63}$$

and the same differential equations for x_2 and x_3 as given in Eqs. (11.39) and
(11.40). For this dynamical system there is a scalar-field dominated fixed point given
in Eq. (11.42) as well as the radiation point $(x_1, x_2, x_3) = (0, 0, 1)$. In the presence

of the coupling Q, the standard matter era is replaced by a 'ϕ-matter-dominated epoch (ϕMDE)' [125] characterized by

$$(x_1, x_2, x_3) = \left(-\frac{\sqrt{6}}{3}Q, 0, 0\right), \qquad \Omega_\phi = \frac{2}{3}Q^2, \qquad w_\phi = 1. \qquad (11.64)$$

Defining the effective equation of state

$$w_{\text{eff}} = -1 - \frac{2\dot{H}}{3H^2}, \qquad (11.65)$$

one has $w_{\text{eff}} = 2Q^2/3$ for the ϕMDE, which is different from 0 in the uncoupled case. Provided that $2Q^2/3 < 1$, the ϕMDE is a saddle followed by the accelerated point (11.42) [125].

The evolution of the scale factor during the ϕMDE is given by $a \propto t^{2/(3+2Q^2)}$, which is different from that in the uncoupled quintessence. This leads to a change to the CMB shift parameter defined in Eq. (11.17). From the CMB likelihood analysis, the strength of the coupling is constrained to be $|Q| < 0.1$ [125]. The evolution of matter density perturbations is also subject to change by the effect of the coupling. Under a quasi-static approximation on sub-horizon scales the matter perturbation δ_m obeys the following equation [131, 132]

$$\ddot{\delta}_m + (2H + Q\dot{\phi})\dot{\delta}_m - 4\pi G_{\text{eff}}\rho_m\delta_m \approx 0, \qquad (11.66)$$

where the effective gravitational coupling is given by $G_{\text{eff}} = (1 + 2Q^2)G$. During the ϕMDE one can obtain the analytic solution to Eq. (11.66), as $\delta_m \propto a^{1+2Q^2}$. Hence the presence of the coupling Q leads to a larger growth rate relative to the uncoupled quintessence. We can parameterize the growth rate of matter perturbations, as [133]

$$f \equiv \frac{\dot{\delta}_m}{H\delta_m} = (\Omega_m)^\gamma, \qquad (11.67)$$

where $\Omega_m \equiv \kappa^2\rho_m/(3H^2)$ is the density parameter of non-relativistic matter. In the ΛCDM model the growth index γ can be approximately given by $\gamma \approx 0.55$ [134, 135]. In the coupled quintessence the growth rate can be fitted to the numerical solution by the formula $f = (\Omega_m)^\gamma(1 + cQ^2)$, where $c = 2.1$ and $\gamma = 0.56$ are the best-fit values [136]. Using the galaxy and Lyman-α power spectra, the growth index γ and the coupling Q are constrained to be $\gamma = 0.6^{+0.4}_{-0.3}$ and $|Q| < 0.52$ (95 % confidence level), respectively. This is weaker than the bound coming from the CMB constraint [136]. We also note that the equation for matter perturbations has been derived for the coupled k-essence scenario with a field-dependent coupling $Q(\phi)$ [137, 138]. In principle, it is possible to reconstruct the coupling from observations if the evolution of matter perturbations is known accurately [139].

11.4.3.2 Chameleon Mechanism

If a scalar field ϕ is coupled to baryons as well as dark matter, this gives rise to a fifth-force interaction that can be constrained experimentally. A large coupling of the order of unity arises in modified gravity theories as well as superstring theories. In such cases, we need to suppress such a strong interaction with baryons for the compatibility with local gravity experiments. There is a way to screen the fifth force under the so-called chameleon mechanism [140, 141], in which the field mass is different depending on the matter density in the surrounding environment. If the field is sufficiently heavy in the regions of high density, a spherically-symmetric body can have a 'thin-shell' around its surface such that the effective coupling between the field and matter is suppressed outside the body.

The action of a chameleon scalar field ϕ with a potential $V(\phi)$ is given by

$$S = \int d^4x \sqrt{-g} \left[\frac{1}{2\kappa^2} R - \frac{1}{2} g^{\mu\nu} \partial_\mu \phi \partial_\nu \phi - V(\phi) \right] + \int d^4x \mathcal{L}_M \left(g^{(i)}_{\mu\nu}, \Psi^{(i)}_M \right),$$

(11.68)

where g is the determinant of the metric $g_{\mu\nu}$ (in the Einstein frame) and $\mathcal{L}_M$ is a matter Lagrangian with matter fields $\Psi^{(i)}_M$ coupled to a metric $g^{(i)}_{\mu\nu}$. The metric $g^{(i)}_{\mu\nu}$ is related to the Einstein frame metric $g_{\mu\nu}$ via $g^{(i)}_{\mu\nu} = e^{2\kappa Q_i \phi} g_{\mu\nu}$, where Q_i are the strengths of the couplings for each matter component with the field ϕ. The typical field potential is chosen to be of the runaway type (such as $V(\phi) = M^{4+n} \phi^{-n}$). We also restrict the form of the potential such that $|V_{,\phi}| \to \infty$ as $\phi \to 0$.

Varying the action (11.68) with respect to ϕ, we obtain the field equation

$$\Box \phi - V_{,\phi} = -\sum_i \kappa Q_i e^{4\kappa Q_i \phi} g^{\mu\nu}_{(i)} T^{(i)}_{\mu\nu},$$

(11.69)

where $T^{(i)}_{\mu\nu} = -(2/\sqrt{-g^{(i)}}) \delta \mathcal{L}_M / \delta g^{\mu\nu}_{(i)}$ is the stress-energy tensor for the i-th form of matter. For non-relativistic matter we have $g^{\mu\nu}_{(i)} T^{(i)}_{\mu\nu} = -\tilde{\rho}_i$, where $\tilde{\rho}_i$ is an energy density. It is convenient to introduce the energy density $\rho_i \equiv \tilde{\rho}_i e^{3\kappa Q_i \phi}$, which is conserved in the Einstein frame. In the following, let us consider the case in which the couplings Q_i are the same for all species, i.e., $Q_i = Q$. In a spherically symmetric space-time under the weak gravitational background (i.e., neglecting the backreaction of gravitational potentials), Eq. (11.69) reads

$$\frac{d^2\phi}{dr^2} + \frac{2}{r} \frac{d\phi}{dr} = \frac{dV_{\rm eff}}{d\phi},$$

(11.70)

where r is a distance from the center of symmetry, and $V_{\rm eff}$ is the effective potential given by

$$V_{\rm eff}(\phi) = V(\phi) + e^{\kappa Q\phi} \rho,$$

(11.71)

and $\rho \equiv \sum_i \rho_i$. For the runaway potential with $V_{,\phi} < 0$ the positive coupling Q leads to a minimum of the effective potential. In $f(R)$ gravity the negative coupling ($Q = -1/\sqrt{6}$) gives rise to a minimum for the potential with $V_{,\phi} > 0$ (as we will see in Sect. 11.5.1.3).

We assume that a spherically-symmetric body has a constant density $\rho = \rho_A$ inside the body ($r < r_c$) and that the energy density outside the body ($r > r_c$) is $\rho = \rho_B$. The mass M_c of the body and the gravitational potential Φ_c at the radius r_c

are given by $M_c = (4\pi/3)r_c^3 \rho_A$ and $\Phi_c = G M_c/r_c$, respectively. The effective potential $V_{\mathrm{eff}}(\phi)$ has two minima at the field values ϕ_A and ϕ_B satisfying $V'_{\mathrm{eff}}(\phi_A) = 0$ and $V'_{\mathrm{eff}}(\phi_B) = 0$, respectively. The former corresponds to the region with a high density that gives rise to a heavy mass squared $m_A^2 \equiv V''_{\mathrm{eff}}(\phi_A)$, whereas the latter to the lower density region with a lighter mass squared $m_B^2 \equiv V''_{\mathrm{eff}}(\phi_B)$. When we consider the 'dynamics' of the field ϕ according to Eq. (11.70) we need to consider the inverted effective potential $(-V_{\mathrm{eff}})$ having two *maxima* at $\phi = \phi_A$ and $\phi = \phi_B$.

The boundary conditions for the field are given by $\frac{d\phi}{dr}(r = 0) = 0$ and $\phi(r \to \infty) = \phi_B$. The field ϕ is at rest at $r = 0$ and begins to roll down the potential when the matter-coupling term $\kappa Q \rho_A e^{\kappa Q\phi}$ becomes important at a radius r_1 in Eq. (11.70). As long as r_1 is close to r_c so that $\Delta r_c \equiv r_c - r_1 \ll r_c$, the body has a thin-shell inside the body. Since the field acquires a sufficient kinetic energy in the thin-shell regime ($r_1 < r < r_c$), it climbs up the potential hill outside the body ($r > r_c$). The field profile can be obtained by matching the solutions of Eq. (11.70) at the radius $r = r_1$ and $r = r_c$. Neglecting the mass term m_B, we obtain the thin-shell field profile outside the body [140–142]

$$\phi(r) \approx \phi_B - \frac{2 Q_{\mathrm{eff}}}{\kappa} \frac{G M_c}{r}, \tag{11.72}$$

where

$$Q_{\mathrm{eff}} = 3 Q \varepsilon_{\mathrm{th}}, \qquad \varepsilon_{\mathrm{th}} \equiv \frac{\kappa(\phi_B - \phi_A)}{6 Q \Phi_c}. \tag{11.73}$$

Here $\varepsilon_{\mathrm{th}}$ is called the thin-shell parameter. Under the conditions $\Delta r_c/r_c \ll 1$ and $1/(m_A r_c) \ll 1$, the thin-shell parameter is approximately given by $\varepsilon_{\mathrm{th}} \approx \Delta r_c/r_c + 1/(m_A r_c)$ [142]. As long as $\varepsilon_{\mathrm{th}} \ll 1$, the amplitude of the effective coupling Q_{eff} can be much smaller than 1. Hence it is possible for the large coupling models ($|Q| = \mathcal{O}(1)$) to be consistent with local gravity experiments if the body has a thin-shell.

Let us study the constraint on the thin-shell parameter from the possible violation of the equivalence principle (EP). The tightest bound comes from the solar system tests of weak EP using the free-fall acceleration of Moon (a_{Moon}) and Earth ($a_\oplus$) toward Sun [141]. The experimental bound on the difference of two accelerations is given by [143, 144]

$$\frac{|a_{\mathrm{Moon}} - a_\oplus|}{(a_{\mathrm{Moon}} + a_\oplus)/2} < 10^{-13}. \tag{11.74}$$

If Earth, Sun, and Moon have thin-shells, the field profiles outside the bodies are given by Eq. (11.72) with the replacement of corresponding quantities. The acceleration induced by a fifth force with the field profile $\phi(r)$ and the effective coupling Q_{eff} is $a^{\mathrm{fifth}} = |Q_{\mathrm{eff}} \nabla \phi(r)|$. Using the thin-shell parameter $\varepsilon_{\mathrm{th},\oplus}$ for Earth, the accelerations $a_\oplus$ and a_{Moon} toward Sun (mass $M_\odot$) are [141]

$$a_\oplus \approx \frac{G M_\odot}{r^2} \left[1 + 18 Q^2 \varepsilon_{\mathrm{th},\oplus}^2 \frac{\Phi_\oplus}{\Phi_\odot} \right], \tag{11.75}$$

$$a_{\mathrm{Moon}} \approx \frac{G M_\odot}{r^2} \left[1 + 18 Q^2 \varepsilon_{\mathrm{th},\oplus}^2 \frac{\Phi_\oplus^2}{\Phi_\odot \Phi_{\mathrm{Moon}}} \right], \tag{11.76}$$

where $\Phi_\odot \approx 2.1 \times 10^{-6}$, $\Phi_\oplus \approx 7.0 \times 10^{-10}$, and $\Phi_{\text{Moon}} \approx 3.1 \times 10^{-11}$ are the gravitational potentials of Sun, Earth and Moon, respectively. Hence the condition (11.74) translates into

$$\varepsilon_{\text{th},\oplus} < 8.8 \times 10^{-7}/|Q|. \tag{11.77}$$

Since the condition $|\phi_B| \gg |\phi_A|$ is satisfied for the field potentials under consideration, one has $\varepsilon_{\text{th},\oplus} \approx \kappa\phi_B/(6Q\Phi_\oplus)$ from Eq. (11.73). Then the condition (11.77) translates into

$$|\kappa\phi_{B,\oplus}| < 3.7 \times 10^{-15}. \tag{11.78}$$

For example, let us consider the inverse power-law potential $V(\phi) = M^{4+n}\phi^{-n}$. In this case we have $\phi_{B,\oplus} = [(n/Q)(M_{\text{pl}}^4/\rho_B)(M/M_{\text{pl}})^{n+4}]^{1/(n+1)}M_{\text{pl}}$, where we recovered the reduced Planck mass $M_{\text{pl}} = 1/\kappa$. For n and Q of the order of unity, the constraint (11.78) gives $M < 10^{-(15n+130)/(n+4)}M_{\text{pl}}$. When $n = 1$, for example, one has $M < 10^{-2}$ eV. If the same potential is responsible for dark energy, the mass M is constrained to be larger than this value [145]. For the potential $V(\phi) = M^4 \exp(M^n/\phi^n)$, however, we have that $V(\phi) \approx M^4 + M^{4+n}\phi^{-n}$ for $\phi > M$, which is responsible for dark energy for $M \approx 10^{-3}$ eV. This can be compatible with the mass scale M constrained by (11.78) [145].

11.4.4 Unified Models of Dark Energy and Dark Matter

There are a number of works to explain the origin of dark energy and dark matter using a single fluid or a single scalar field. Let us first discuss the generalized Chaplygin gas (GCG) model as an example of a single fluid model [36, 37]. In this model the pressure P of the perfect fluid is related to its energy density ρ via

$$P = -A\rho^{-\alpha}, \tag{11.79}$$

where A is a positive constant. The original Chaplygin gas model corresponds to $\alpha = 1$ [36].

Plugging the relation (11.79) into the continuity equation $\dot{\rho} + 3H(\rho + P) = 0$, we obtain the integrated solution

$$\rho(t) = \left[A + \frac{B}{a^{3(1+\alpha)}}\right]^{1/(1+\alpha)}, \tag{11.80}$$

where B is an integration constant. In the early epoch ($a \ll 1$) the energy density evolves as $\rho \propto a^{-3}$, which means that the fluid behaves as dark matter. In the late epoch ($a \gg 1$) the energy density approaches a constant value $A^{1/(a+\alpha)}$ and hence the fluid behaves as dark energy. A fluid with the generalized Chaplygin gas therefore interpolates between dark matter and dark energy.

Although this model is attractive to provide unified description of two dark components, it is severely constrained by the matter power spectrum in large-scale structure. The gauge-invariant matter perturbation δ_m with a comoving wavenumber k obeys the following equation of motion [38]

$$\ddot{\delta}_m + \left(2 + 3c_s^2 - 6w\right)H\dot{\delta}_m - \left[\frac{3}{2}H^2\left(1 - 6c_s^2 - 3w^2 + 8w\right) - \left(\frac{c_s k}{a}\right)^2\right]\delta_m = 0,$$

(11.81)

where $w = P/\rho$ is the fluid equation of state and c_s is the sound speed given by

$$c_s^2 = \frac{dP}{d\rho} = -\alpha w.$$

(11.82)

Since $w \to 0$ and $c_s^2 \to 0$ in the limit $z \gg 1$, the sound speed is much smaller than unity in the deep matter era and starts to grow around the end of it. Since w is negative, c_s^2 is positive for $\alpha > 0$ and negative for $\alpha < 0$.

From Eq. (11.81) the perturbations satisfying the following condition grow via the gravitational instability

$$\left|c_s^2\right| < \frac{3}{2}\left(\frac{aH}{k}\right)^2.$$

(11.83)

When $|c_s^2| > (3/2)(aH/k)^2$, the perturbations exhibit either rapid growth or damped oscillations depending on the sign of c_s^2. The violation of the condition (11.83) mainly occurs around the present epoch in which $|w|$ is of the order of unity and hence $|c_s^2| \sim |\alpha|$. The smallest scale relevant to the galaxy matter power spectrum in the linear regime corresponds to a wavenumber around $k = 0.1h$ Mpc^{-1}. Then the constraint (11.83) gives the upper bound on the values of $|\alpha|$ [38]:

$$|\alpha| \lesssim 10^{-5}.$$

(11.84)

Hence the generalized Chaplygin gas model is hardly distinguishable from the ΛCDM model. In particular the original Chaplygin gas model ($\alpha = 1$) is excluded from the observations of large-scale structure. Although non-linear clustering may change the evolution of perturbations in this model [146, 147], it is unlikely that the constraint (11.84) is relaxed significantly.

The above conclusion comes from the fact that in the Chaplygin gas model the sound speed is too large to match with observations. There is a way to avoid this problem by adding a non-adiabatic contribution to Eq. (11.81) to make c_s vanish [148]. It is also possible to construct unified models of dark energy and dark matter using a purely kinetic scalar field [149]. Let us consider k-essence models in which the Lagrangian density $P(X)$ has an extremum at some value $X = X_0$, e.g. [149]

$$P = P_0 + P_2(X - X_0)^2.$$

(11.85)

The pressure $P_\phi = P$ and the energy density $\rho_\phi = 2XP_{,X} - P$ satisfy the continuity equation $\dot{\rho}_\phi + 3H(\rho_\phi + P_\phi) = 0$, i.e.,

$$(P_{,X} + 2XP_{,XX})\dot{X} + 6HP_{,X}X = 0.$$

(11.86)

The solution around $X = X_0$ can be derived by introducing a small parameter $\varepsilon = (X - X_0)/X_0$. Plugging Eq. (11.85) into Eq. (11.86), we find that ε satisfies the equation $\dot{\varepsilon} = -3H\varepsilon$ at linear order. Hence we obtain the solution $X = X_0[1 + \varepsilon_1(a/a_1)^{-3}]$, where ε_1 and a_1 are constants. The validity of the above approximation demands that $\varepsilon_1(a/a_1)^{-3} \ll 1$. Since $P_\phi \approx P_0$ and $\rho_\phi \approx -P_0 + 4P_2 X_0^2 \varepsilon_1 (a/a_1)^{-3}$ in the regime where X is close to X_0, the field equation of state is given by

$$w_\phi \approx -\left[1 - \frac{4P_2}{P_0}X_0^2\varepsilon_1\left(\frac{a}{a_1}\right)^{-3}\right]^{-1}. \tag{11.87}$$

Since $w_\phi \to -1$ at late times it is possible to give rise to the cosmic acceleration. One can also realize $w_\phi \approx 0$ during the matter era, provided that the condition $4P_2 X_0^2/|P_0| \gg 1$ is satisfied. The squared sound speed defined in Eq. (11.56) is approximately given by

$$c_s^2 \approx \frac{1}{2}\varepsilon_1\left(\frac{a}{a_1}\right)^{-3}, \tag{11.88}$$

which is much smaller than unity. Hence the large sound speed problem can be evaded in the model (11.85). In Ref. [150] it was shown that the above purely k-essence model is equivalent to a fluid with a closed-form barotropic equation of state plus a constant term that works as a cosmological constant to all orders in structure formation.

11.5 Modified Gravity Models

There is another class of dark energy models in which gravity is modified from general relativity (GR). We review a number of cosmological and gravitational aspects of $f(R)$ gravity, Gauss–Bonnet gravity, scalar-tensor theories, and a braneworld model. We also discuss observational signatures of those models to distinguish them from other dark energy models.

11.5.1 $f(R)$ Gravity

The simplest modification to GR is $f(R)$ gravity with the action

$$S = \frac{1}{2\kappa^2}\int d^4x\sqrt{-g}f(R) + \int d^4x\mathscr{L}_M(g_{\mu\nu}, \Psi_M), \tag{11.89}$$

where f is a function of the Ricci scalar R and $\mathscr{L}_M$ is a matter Lagrangian for perfect fluids. The Lagrangian $\mathscr{L}_M$ depends on the metric $g_{\mu\nu}$ and the matter fields Ψ_M. We do not consider a direct coupling between the Ricci scalar and matter.

11.5.1.1 Viable $f(R)$ Dark Energy Models

In the standard variational approach called metric formalism, the affine connections $\Gamma^\lambda_{\mu\nu}$ are related to the metric $g_{\mu\nu}$. In this formalism the field equation can be derived by varying the action (11.89) with respect to $g_{\mu\nu}$:

$$F(R)R_{\mu\nu}(g) - \frac{1}{2}f(R)g_{\mu\nu} - \nabla_\mu\nabla_\nu F(R) + g_{\mu\nu}\Box F(R) = \kappa^2 T_{\mu\nu}, \qquad (11.90)$$

where $F(R) \equiv \partial f/\partial R$, and $T_{\mu\nu} = -(2/\sqrt{-g})\delta\mathscr{L}_M/\delta g^{\mu\nu}$ is the energy-momentum tensor of matter. The trace of Eq. (11.90) is given by

$$3\Box F(R) + F(R)R - 2f(R) = \kappa^2 T, \qquad (11.91)$$

where $T = g^{\mu\nu}T_{\mu\nu} = -\rho_M + 3P_M$. Here ρ_M and P_M are the energy density and the pressure of matter, respectively.

The de Sitter point corresponds to a vacuum solution at which the Ricci scalar is constant. Since $\Box F(R) = 0$ at this point, we obtain

$$F(R)R - 2f(R) = 0. \qquad (11.92)$$

The model $f(R) = \alpha R^2$ satisfies this condition and hence it gives rise to an exact de Sitter solution. In fact the first model of inflation proposed by Starobinsky [54] corresponds to $f(R) = R + \alpha R^2$, in which the cosmic acceleration ends when the term αR^2 becomes smaller than R. Dark energy models based on $f(R)$ theories can be also constructed to realize the late-time de Sitter solution satisfying the condition (11.92).

The possibility of the late-time cosmic acceleration in $f(R)$ gravity was first suggested by Capozziello [39] in 2002. An $f(R)$ dark energy model of the form $f(R) = R - \mu^{2(n+1)}/R^n$ $(n > 0)$ was proposed in Refs. [40–42], but it became clear that this model suffers from a number of problems such as the matter instability [151], absence of the matter era [152, 153], and inability to satisfy local gravity constraints [154–159]. This problem arises from the fact that $f_{,RR} < 0$ in this model.

In order to see why the models with negative values of $f_{,RR}$ are excluded, let us consider local fluctuations on a background characterized by a curvature R_0 and a density ρ_0. We expand Eq. (11.91) in powers of fluctuations under a weak-field approximation. We decompose the quantities $F(R)$, $g_{\mu\nu}$, and $T_{\mu\nu}$ into the background part and the perturbed part: $R = R^{(0)} + \delta R$, $F = F^{(0)}(1 + \delta_F)$, $g_{\mu\nu} = \eta_{\mu\nu} + h_{\mu\nu}$, and $T_{\mu\nu} = T^{(0)}_{\mu\nu} + \delta T_{\mu\nu}$, where we have used the approximation that $g^{(0)}_{\mu\nu}$ corresponds the metric $\eta_{\mu\nu}$ in the Minkowski space-time. Then the trace Eq. (11.91) reads [156, 157]

$$\left(\frac{\partial^2}{\partial t^2} - \nabla^2\right)\delta_F + M^2\delta_F = -\frac{\kappa^2}{3F^{(0)}}\delta T, \qquad (11.93)$$

where $\delta T \equiv \eta^{\mu\nu}\delta T_{\mu\nu}$, and

$$M^2 \equiv \frac{1}{3}\left[\frac{f_{,R}(R^{(0)})}{f_{,RR}(R^{(0)})} - R^{(0)}\right] = \frac{R^{(0)}}{3}\left[\frac{1}{m(R^{(0)})} - 1\right]. \qquad (11.94)$$

Here the quantity $m = R f_{,RR}/f_{,R}$ characterizes the deviation from the ΛCDM model $(f(R) = R - 2\Lambda)$. In the homogeneous and isotropic cosmological background (without a Hubble friction), δ_F is a function of the cosmic time t only and Eq. (11.93) reduces to

$$\ddot{\delta}_F + M^2 \delta_F = \frac{\kappa^2}{3 F^{(0)}} \rho, \tag{11.95}$$

where $\rho \equiv -\delta T$. For the models where the deviation from the ΛCDM model is small, we have $m(R^{(0)}) \ll 1$ so that $|M^2|$ is much larger than $R^{(0)}$. If $M^2 < 0$, the perturbation δ_F exhibits a violent instability. Then the condition $M^2 \approx f_{,R}(R^{(0)})/[3 f_{,RR}(R^{(0)})] > 0$ is needed for the stability of cosmological perturbations. We also require that $f_{,R}(R^{(0)}) > 0$ to avoid anti-gravity (i.e., to avoid that the graviton becomes a ghost). Hence the condition $f_{,RR}(R^{(0)}) > 0$ needs to hold for avoiding a tachyonic instability associated with the negative mass squared [160–164].

For the consistency with local gravity constraints in solar system, the function $f(R)$ needs to be close to that in the ΛCDM model in the region of high density (in the region where the Ricci scalar R is much larger than the cosmological Ricci scalar R_0 today). We also require the existence of a stable late-time de Sitter point given in Eq. (11.92). From the stability analysis about the de Sitter point, one can show that it is stable for $0 < m = R f_{,RR}/f_{,R} < 1$ [165–167]. Then we can summarize the conditions for the viability of $f(R)$ dark energy models:

- (i) $f_{,R} > 0$ for $R \geq R_0$.
- (ii) $f_{,RR} > 0$ for $R \geq R_0$.
- (iii) $f(R) \to R - 2\Lambda$ for $R \gg R_0$.
- (iv) $0 < R f_{,RR}/f_{,R} < 1$ at the de Sitter point satisfying $R f_{,R} = 2f$.

The examples of viable models satisfying all these requirements are [168–170]

$$\text{(A)} \quad f(R) = R - \mu R_c \frac{(R/R_c)^{2n}}{(R/R_c)^{2n} + 1} \quad \text{with } n, \mu, R_c > 0, \tag{11.96}$$

$$\text{(B)} \quad f(R) = R - \mu R_c \left[1 - \left(1 + R^2/R_c^2\right)^{-n}\right] \quad \text{with } n, \mu, R_c > 0, \tag{11.97}$$

$$\text{(C)} \quad f(R) = R - \mu R_c \tanh\left(R/R_c\right) \quad \text{with } \mu, R_c > 0, \tag{11.98}$$

where μ, R_c, and n are constants. Models similar to (C) were proposed in Refs. [171, 172]. Note that R_c is roughly of the order of the present cosmological Ricci scalar R_0. If $R \gg R_c$ the models are close to the ΛCDM model $(f(R) \approx R - \mu R_c)$, so that GR is recovered in the region of high density. The models (A) and (B) have the following asymptotic behavior

$$f(R) \approx R - \mu R_c \left[1 - \left(\frac{R}{R_c}\right)^{-2n}\right] \quad (R \gg R_c), \tag{11.99}$$

which rapidly approaches the ΛCDM model for $n \gtrsim 1$. The model (C) shows an even faster decrease of m in the region $R \gg R_c$. The model $f(R) = R - \mu R_c (R/R_c)^n$ $(0 < n < 1)$ proposed in Refs. [167, 173] is also viable, but it does not allow the rapid decrease of m in the region of high density required for the consistency with local gravity tests.

For example, let us consider the model (B). The de Sitter point given by the condition (11.92) satisfies

$$\mu = \frac{x_1(1 + x_1^2)^{n+1}}{2[(1 + x_1^2)^{n+1} - 1 - (n+1)x_1^2]}, \tag{11.100}$$

where $x_1 \equiv R_1/R_c$ and R_1 is the Ricci scalar at the de Sitter point. The stability condition ($0 < m < 1$) at this point gives [169]

$$\left(1 + x_1^2\right)^{n+2} > 1 + (n+2)x_1^2 + (n+1)(2n+1)x_1^4. \tag{11.101}$$

The condition (11.101) gives the lower bound on the parameter μ. When $n = 1$ one has $x_1 > \sqrt{3}$ and $\mu > 8\sqrt{3}/9$. Under Eq. (11.101) one can show that the conditions $f_{,R} > 0$ and $f_{,RR} > 0$ are also satisfied for $R \geq R_1$.

11.5.1.2 Observational Signatures of $f(R)$ Dark Energy Models

In the flat FLRW space-time we obtain the following equations of motion from Eqs. (11.90) and (11.91):

$$3FH^2 = \kappa^2 \rho_m + \frac{1}{2}(FR - f) - 3H\dot{F}, \tag{11.102}$$

$$2F\dot{H} = -\kappa^2 \rho_m - \ddot{F} + H\dot{F}, \tag{11.103}$$

where, for the perfect fluid, we have taken into account only the non-relativistic matter with energy density ρ_m. In order to confront $f(R)$ dark energy models with SN Ia observations, we rewrite Eqs. (11.102) and (11.103) as follows:

$$3AH^2 = \kappa^2(\rho_m + \rho_{DE}), \tag{11.104}$$

$$-2A\dot{H} = \kappa^2(\rho_m + \rho_{DE} + P_{DE}), \tag{11.105}$$

where A is some constant and

$$\kappa^2 \rho_{DE} \equiv \frac{1}{2}(FR - f) - 3H\dot{F} + 3H^2(A - F), \tag{11.106}$$

$$\kappa^2 P_{DE} \equiv \ddot{F} + 2H\dot{F} - \frac{1}{2}(FR - f) - \left(3H^2 + 2\dot{H}\right)(A - F). \tag{11.107}$$

By defining ρ_{DE} and P_{DE} in this way, one can easily show that the following continuity equation holds:

$$\dot{\rho}_{DE} + 3H(\rho_{DE} + P_{DE}) = 0. \tag{11.108}$$

We define the dark energy equation of state $w_{DE} \equiv P_{DE}/\rho_{DE}$, which is directly related to the one used in SN Ia observations. From Eqs. (11.104) and (11.105), it is given by [52, 174]

$$w_{\mathrm{DE}} = -\frac{2A\dot{H} + 3AH^2}{3AH^2 - \kappa^2\rho_m} = \frac{w_{\mathrm{eff}}}{1 - (F/A)\tilde{\Omega}_m}, \qquad (11.109)$$

where $\tilde{\Omega}_m \equiv \kappa^2\rho_m/(3FH^2)$. Viable $f(R)$ models approach the ΛCDM model in the past, i.e., $F \to 1$ as $R \to \infty$. In order to reproduce the standard matter era in the high-redshift regime we can choose $A = 1$ in Eqs. (11.104) and (11.105). Another possible choice is $A = F_0$, where F_0 is the present value of F. This choice is suitable if the deviation of F_0 from 1 is small (as in scalar-tensor theory with a massless scalar field [175, 176]). In both cases the equation of state w_{DE} can be smaller than -1 before reaching the de Sitter attractor [52, 168, 170, 172, 177]. This originates from the fact that the presence of non-relativistic matter makes the denominator in Eq. (11.109) smaller than 1. Thus $f(R)$ dark energy models give rise to a phantom equation of state without violating stability conditions of the system. The models (A) and (B) are allowed by the SN Ia observations provided that n is larger than the order of unity [178, 179].

The modification of gravity manifests itself in the effective gravitational coupling that appears in the equation of cosmological perturbations. The full perturbation equations in $f(R)$ gravity are presented in Refs. [180–182]. When we confront $f(R)$ models with the observations of large-scale structure, the wavenumbers k of interest are sub-horizon modes with $k/a \gg H$. We can employ a so-called quasi-static approximation under which the dominant terms in perturbation equations correspond to those including k^2/a^2, $\delta\rho_m$, and M^2 [8, 176, 183]. Then the matter density perturbation δ_m approximately satisfies the following equation [183, 184]:

$$\ddot{\delta}_m + 2H\dot{\delta}_m - 4\pi G_{\mathrm{eff}}\rho_m\delta_m \approx 0, \qquad (11.110)$$

where ρ_m is the energy density of non-relativistic matter, and

$$G_{\mathrm{eff}} = \frac{G}{f_{,R}}\frac{1 + 4mk^2/(a^2 R)}{1 + 3mk^2/(a^2 R)}, \qquad (11.111)$$

where $m \equiv Rf_{,RR}/f_{,R}$. This approximation is accurate for viable $f(R)$ dark energy models as long as an oscillating mode of the scalar-field degree of freedom is suppressed relative to the matter-induced mode [169, 170, 185, 186].

In the regime where the deviation from the ΛCDM model is small such that $mk^2/(a^2 R) \ll 1$, the effective gravitational coupling G_{eff} is very close to the gravitational constant G. Then the matter perturbation evolves as $\delta_m \propto t^{2/3}$ during the matter dominance. Meanwhile in the regime $mk^2/(a^2 R) \gg 1$ one has $G_{\mathrm{eff}} \approx 4G/(3f_{,R})$, so that the evolution of δ_m during the matter era is given by $\delta_m \propto t^{(\sqrt{33}-1)/6}$ [169, 170]. The transition from the former regime to the latter regime occurs at the critical redshift [187]

$$z_k \approx \left[\left(\frac{k}{a_0 H_0}\right)^2 \frac{2n(2n+1)}{\mu^{2n}}\frac{(2(1 - \Omega_m^{(0)}))^{2n+1}}{(\Omega_m^{(0)})^{2(n+1)}}\right]^{1/(6n+4)} - 1, \qquad (11.112)$$

where '0' represents the values today. The time t_k at the transition has a scale-dependence $t_k \propto k^{-3/(6n+4)}$, which means that the transition occurs earlier for larger k. The matter power spectrum $P_{\delta_m} = |\delta_m|^2$ at the onset of cosmic acceleration (at time t_Λ) shows a difference compared to the case of the ΛCDM model [169]:

$$\frac{P_{\delta_m}}{P_{\delta_m}^{\Lambda\text{CDM}}} = \left(\frac{t_\Lambda}{t_k}\right)^{2(\frac{\sqrt{33}-1}{6} - \frac{2}{3})} \propto k^{\frac{\sqrt{33}-5}{6n+4}}. \tag{11.113}$$

The ratio of the two power spectra today, i.e., $P_{\delta_m}(t_0)/P_{\delta_m}^{\Lambda\text{CDM}}(t_0)$ is in general different from Eq. (11.113), but the difference is small for n of the order of unity [170].

The modified evolution of perturbations for the redshift $z < z_k$ gives rise to the integrated Sachs–Wolfe effect in CMB anisotropies [161, 173, 188, 189], but this is limited to very large scales (low multipoles). Since the CMB spectrum on the scales relevant to the large-scale structure ($k \gtrsim 0.01h$ Mpc^{-1}) is hardly affected by this modification, there is a difference between the spectral indices of the CMB spectrum and the galaxy power spectrum: $\Delta n_s = (\sqrt{33} - 5)/(6n + 4)$. Observationally we do not find any strong signature for the difference of slopes of the two spectra. If we take the mild bound $\Delta n_s < 0.05$, we obtain the constraint $n > 2$.

11.5.1.3 Local Gravity Constraints

Let us discuss local gravity constraints on $f(R)$ dark energy models. In the region of high density where gravitational experiments are carried out, the linear expansion of R in terms of the cosmological value $R^{(0)}$ and the perturbation δR is no longer valid because of the violation of the condition $\delta R \ll R^{(0)}$. In such a non-linear regime, the chameleon mechanism [140, 141] can be at work to suppress the effective coupling between dark energy and non-relativistic matter. In order to study how the chameleon mechanism works in $f(R)$ gravity, we transform the action (11.89) to the Einstein frame action under the conformal transformation $\tilde{g}_{\mu\nu} = F g_{\mu\nu}$ [190],

$$S_E = \int d^4x \sqrt{-\tilde{g}} \left[\frac{1}{2\kappa^2}\tilde{R} - \frac{1}{2}\tilde{g}^{\mu\nu}\partial_\mu\phi\partial_\nu\phi - V(\phi)\right] + \int d^4x \mathscr{L}_M(g_{\mu\nu}, \Psi_m), \tag{11.114}$$

where $\kappa\phi \equiv \sqrt{3/2}\ln F$, $V(\phi) = (RF - f)/(2\kappa^2 F^2)$, and a tilde represents quantities in the Einstein frame.

The action (11.114) is the same as (11.68) with the correspondence that $g_{\mu\nu}$ in the Jordan frame is equivalent to $g_{\mu\nu}^{(i)}$ in the action (11.68). Since the quantity F is given by $F = e^{-2\kappa Q\phi}$ with $Q = -1/\sqrt{6}$ in metric $f(R)$ gravity, the field ϕ is coupled to non-relativistic matter (including baryons and dark matter) with a universal coupling $Q = -1/\sqrt{6}$. Let us consider the models (11.96) and (11.97), which behave as Eq. (11.99) in the region of high density ($R \gg R_c$). For the functional form (11.99) the effective potential defined in Eq. (11.71) is

$$V_{\text{eff}}(\phi) \approx \frac{\mu R_c}{2\kappa^2}e^{-4\kappa\phi/\sqrt{6}}\left[1 - (2n+1)\left(\frac{-\kappa\phi}{\sqrt{6}n\mu}\right)^{2n/(2n+1)}\right] + \rho e^{-\kappa\phi/\sqrt{6}}, \tag{11.115}$$

where

$$F = e^{2\kappa\phi/\sqrt{6}} = 1 - 2n\mu \left(\frac{R}{R_c}\right)^{-(2n+1)}. \qquad (11.116)$$

Inside and outside a spherically-symmetric body, the effective potential (11.115) has the following minima given, respectively, by

$$\kappa\phi_A \approx -\sqrt{6}n\mu \left(\frac{R_c}{\kappa^2 \rho_A}\right)^{2n+1}, \qquad \kappa\phi_B \approx -\sqrt{6}n\mu \left(\frac{R_c}{\kappa^2 \rho_B}\right)^{2n+1}. \qquad (11.117)$$

One has $|\phi_B| \gg |\phi_A|$ provided that $\rho_A \gg \rho_B$.

The bound (11.78) translates into

$$\frac{n\mu}{x_1^{2n+1}} \left(\frac{R_1}{\rho_B}\right)^{2n+1} < 1.5 \times 10^{-15}, \qquad (11.118)$$

where $x_1 = R_1/R_c$ and R_1 is the Ricci scalar at the de Sitter point. Let us consider the case in which the Lagrangian density is given by (11.99) for $R \geq R_1$. In the original models of Hu and Sawicki [168] and Starobinsky [169] there are some modification to the estimation of R_1, but this change is not significant when we place constraints on model parameters. The de Sitter point for the model (11.99) corresponds to $\mu = x_1^{2n+1}/[2(x_1^{2n} - n - 1)]$. Substituting this relation into Eq. (11.118), we find

$$\frac{n}{2(x_1^{2n} - n - 1)} \left(\frac{R_1}{\rho_B}\right)^{2n+1} < 1.5 \times 10^{-15}. \qquad (11.119)$$

The stability of the de Sitter point requires that $m(R_1) < 1$, which translates into the condition $x_1^{2n} > 2n^2 + 3n + 1$. Then the term $n/[2(x_1^{2n} - n - 1)]$ is smaller than 0.25 for $n > 0$. Using the approximation that R_1 and ρ_B are of the orders of the present cosmological density 10^{-29} g/cm^3 and the baryonic/dark matter density 10^{-24} g/cm^3 in our galaxy, respectively, we obtain the following constraint from (11.119) [191]:

$$n > 0.9. \qquad (11.120)$$

Thus n does not need to be much larger than unity. Under the condition (11.120), the deviation from the ΛCDM becomes important as R decreases to the order of R_c.

11.5.2 Scalar-Tensor Theories

There is another class of modified gravity called scalar-tensor theories in which the Ricci scalar R is coupled to a scalar field φ. One of the simplest examples is Brans–Dicke (BD) theory [129] with the action

$$S = \int d^4x \sqrt{-g} \left[\frac{1}{2}\varphi R - \frac{\omega_{BD}}{2\varphi}(\nabla\varphi)^2 - U(\varphi)\right] + S_M(g_{\mu\nu}, \Psi_M), \qquad (11.121)$$

where ω_{BD} is the BD parameter, $U(\varphi)$ is the field potential, and S_M is a matter action that depends on the metric $g_{\mu\nu}$ and matter fields Ψ_m. The original BD theory [129] does not have the field potential $U(\varphi)$.

The general action for scalar-tensor theories can be written as

$$S = \int d^4x \sqrt{-g} \left[\frac{1}{2} f(\varphi, R) - \frac{1}{2} \omega(\varphi)(\nabla\varphi)^2 \right] + S_M(g_{\mu\nu}, \Psi_M), \qquad (11.122)$$

where f is a general function of the scalar field φ and the Ricci scalar R, ω is a function of φ. We choose the unit $\kappa^2 = 1$. We consider theories of the type

$$f(\varphi, R) = F(\varphi)R - 2U(\varphi). \qquad (11.123)$$

Under the conformal transformation $\tilde{g}_{\mu\nu} = F g_{\mu\nu}$, the action in the Einstein frame is given by [190]

$$S_E = \int d^4x \sqrt{-\tilde{g}} \left[\frac{1}{2} \tilde{R} - \frac{1}{2} (\tilde{\nabla}\phi)^2 - V(\phi) \right] + S_M(g_{\mu\nu}, \Psi_M), \qquad (11.124)$$

where $V = U/F^2$. We have introduced a new scalar field ϕ in order to make the field kinetic term canonical:

$$\phi \equiv \int d\varphi \sqrt{\frac{3}{2} \left(\frac{F_{,\varphi}}{F} \right)^2 + \frac{\omega}{F}}. \qquad (11.125)$$

We define the coupling between dark energy and non-relativistic matter, as

$$Q \equiv -\frac{F_{,\phi}}{2F} = -\frac{F_{,\varphi}}{F} \left[\frac{3}{2} \left(\frac{F_{,\varphi}}{F} \right)^2 + \frac{\omega}{F} \right]^{-1/2}. \qquad (11.126)$$

In $f(R)$ gravity we have $\omega = 0$ and hence $F = \exp(\sqrt{2/3}\phi)$ from Eq. (11.125). Then the coupling is given by $Q = -1/\sqrt{6}$ from Eq. (11.126). If Q is constant as in $f(R)$ gravity, the following relations hold from Eqs. (11.125) and (11.126):

$$F = e^{-2Q\phi}, \qquad \omega = \left(1 - 6Q^2 \right) F \left(\frac{d\phi}{d\varphi} \right)^2. \qquad (11.127)$$

In this case the action (11.122) in the Jordan frame reads [130]

$$S = \int d^4x \sqrt{-g} \left[\frac{1}{2} F(\phi)R - \frac{1}{2} (1 - 6Q^2) F(\phi)(\nabla\phi)^2 - U(\phi) \right] + S_M(g_{\mu\nu}, \Psi_m). \qquad (11.128)$$

In the limit that $Q \to 0$ the action (11.128) reduces to the one for a minimally coupled scalar field ϕ with the potential $U(\phi)$. The transformation of the Jordan frame action (11.128) under the conformal transformation $\tilde{g}_{\mu\nu} = e^{-2Q\phi} g_{\mu\nu}$ gives rise to the Einstein frame action (11.124) with a constant coupling Q.

One can compare (11.128) with the action (11.121) in BD theory. Setting $\varphi = F = e^{-2Q\phi}$, one finds that the two actions are equivalent if the parameter ω_{BD} is related to Q via the relation [130, 141]

$$3 + 2\omega_{BD} = \frac{1}{2Q^2}. \qquad (11.129)$$

Using this relation, we find that the general-relativistic limit ($\omega_{BD} \to \infty$) corresponds to the vanishing coupling ($Q \to 0$). Since $Q = -1/\sqrt{6}$ in $f(R)$ gravity, this corresponds to the Brans–Dicke parameter $\omega_{BD} = 0$ [154, 192, 193].

There are also other scalar-tensor theories that give rise to field-dependent couplings $Q(\phi)$. For a non-minimally coupled scalar field with $F(\varphi) = 1 - \xi\varphi^2$ and $\omega(\varphi) = 1$ in the action (11.122) with (11.123), the coupling is field-dependent, i.e., $Q(\varphi) = \xi\varphi/[1 - \xi\varphi^2(1 - 6\xi)]^{1/2}$. The cosmological dynamics of dark energy models based on such theories have been studied by a number of authors [44–47, 124, 194, 195].

Let us consider BD theory with the action (11.128). In the absence of the potential $U(\phi)$ the BD parameter ω_{BD} is constrained to be $\omega_{BD} > 4.0 \times 10^4$ from solar-system experiments [144]. This bound also applies to the case of a nearly massless field with the potential $U(\phi)$ in which the Yukawa correction e^{-Mr} is close to unity (where M is the scalar field mass and r is an interaction length). Using the bound $\omega_{BD} > 4.0 \times 10^4$ in Eq. (11.129), we find

$$|Q| < 2.5 \times 10^{-3}. \tag{11.130}$$

In this case the cosmological evolution for such theories is hardly distinguishable from the $Q = 0$ case. Even for scalar-tensor theories with such small couplings, it was shown that the phantom equation state of dark energy can be realized without the appearance of a ghost state [196–199].

In the presence of the field potential it is possible for large coupling models ($|Q| = \mathcal{O}(1)$) to satisfy local gravity constraints under the chameleon mechanism, provided that the mass M of the field ϕ is sufficiently large in the region of high density. In metric $f(R)$ gravity ($Q = -1/\sqrt{6}$) the field potential $U(\phi)$ in Eq. (11.128) corresponds to $U = (FR - f)/2$ with $\phi = \sqrt{3/2}\ln F$. The viable $f(R)$ dark energy models (11.96) and (11.97) have the asymptotic form (11.99), in which case the field potential is given by

$$U(\phi) = \frac{\mu R_c}{2}\left[1 - \frac{2n+1}{(2n\mu)^{2n/(2n+1)}}\left(1 - e^{2\phi/\sqrt{6}}\right)^{2n/(2n+1)}\right]. \tag{11.131}$$

For BD theories with the constant coupling Q, one can generalize the potential (11.131) to the form

$$U(\phi) = U_0\left[1 - C\left(1 - e^{-2Q\phi}\right)^p\right] \quad (U_0 > 0,\ C > 0,\ 0 < p < 1). \tag{11.132}$$

As $\phi \to 0$, the potential (11.132) approaches the finite value U_0 with a divergence of the field mass squared $M^2 = U_{,\phi\phi} \to \infty$. This model has a curvature singularity at $\phi = 0$ as in the case of the $f(R)$ models (11.96) and (11.97). The mass M decreases as the field evolves away from $\phi = 0$. The late-time cosmic acceleration can be realized by the potential (11.132) provided that U_0 is of the order of H_0^2.

Since the action (11.124) in the Einstein frame is equivalent to the action (11.68), the chameleon mechanism can be at work even for BD theories with large couplings ($|Q| = \mathcal{O}(1)$). Considering a spherically symmetric body with homogenous densities ρ_A and ρ_B inside and outside bodies respectively, the effective potential

$V_{\rm eff} = V(\phi) + e^{Q\phi}\rho$ in the Einstein frame (where $V(\phi) = U(\phi)/F^2$) has two minima characterized by

$$\phi_A \approx \frac{1}{2Q}\left(\frac{2U_0 pC}{\rho_A}\right)^{1/(1-p)}, \qquad \phi_B \approx \frac{1}{2Q}\left(\frac{2U_0 pC}{\rho_B}\right)^{1/(1-p)}. \qquad (11.133)$$

Using the experimental bound (11.78) coming from the violation of equivalence principle together with the condition for realizing the cosmic acceleration today, we obtain the constraint [130]

$$p > 1 - \frac{5}{13.8 - \log_{10}|Q|}. \qquad (11.134)$$

When $|Q| = 10^{-2}$ and $|Q| = 10^{-1}$ we have $p > 0.68$ and $p > 0.66$, respectively. In $f(R)$ gravity the above bound corresponds to $p > 0.65$, which translates into $n > 0.9$ for the model (11.99).

The evolution of cosmological perturbations in scalar-tensor theories has been discussed in Refs. [130, 176, 183, 200, 201]. Under the quasi-static approximation on sub-horizon scales, the matter perturbation δ_m for the theory (11.128) obeys the following equation of motion [130, 201]:

$$\ddot{\delta}_m + 2H\dot{\delta}_m - 4\pi G_{\rm eff}\rho_m\delta_m \approx 0, \qquad (11.135)$$

where the effective (cosmological) gravitational coupling is

$$G_{\rm eff} = \frac{G}{F}\frac{(k^2/a^2)(1+2Q^2)F + M^2}{(k^2/a^2)F + M^2}. \qquad (11.136)$$

Here $M^2 \equiv U_{,\phi\phi}$ is the field mass squared. In the 'general relativistic' regime characterized by $M^2/F \gg k^2/a^2$, one has $G_{\rm eff} \approx G/F$ and $\delta_m \propto t^{2/3}$. In the 'scalar-tensor' regime characterized by $M^2/F \ll k^2/a^2$, it follows that $G_{\rm eff} \approx (1+2Q^2)G/F$ and $\delta_m \propto t^{(\sqrt{25+48Q^2}-1)/6}$. If the transition from the former regime to the latter regime occurs during the matter era, this gives rise to a difference between the spectral indices of the matter power spectrum and of the CMB spectrum on the scales $0.01h$ Mpc$^{-1} \lesssim k \lesssim 0.2h$ Mpc^{-1} [130]:

$$\Delta n_s = \frac{(1-p)(\sqrt{25+48Q^2}-5)}{4-p}. \qquad (11.137)$$

Under the criterion $\Delta n_s < 0.05$, we obtain the bounds $p > 0.957$ for $Q = 1$ and $p > 0.855$ for $Q = 0.5$. As long as p is close to 1, the model can be consistent with both cosmological and local gravity constraints.

For the perturbed metric $ds^2 = -(1 + 2\Psi)dt^2 + a^2(t)(1 - 2\Phi)\delta_{ij}dx^i dx^j$, the gravitational potentials obey the following equations under a quasi-static approximation on sub-horizon scales [130]

$$\frac{k^2}{a^2}\Psi \approx -\frac{4\pi G}{F}\frac{(k^2/a^2)(1+2Q^2)F + M^2}{(k^2/a^2)F + M^2}\rho_m\delta_m, \qquad (11.138)$$

$$\frac{k^2}{a^2}\Phi \approx -\frac{4\pi G}{F}\frac{(k^2/a^2)(1-2Q^2)F + M^2}{(k^2/a^2)F + M^2}\rho_m\delta_m, \qquad (11.139)$$

where we have recovered the gravitational constant G. The results (11.138) and (11.139) include those in $f(R)$ gravity by setting $Q = -1/\sqrt{6}$. In the regime $M^2/F \ll k^2/a^2$, the evolution of Ψ and Φ is subject to change compared to that in the GR regime characterized by $M^2/F \gg k^2/a^2$. In general, the difference from GR may be quantified by the parameters q and ζ [202]:

$$\frac{k^2}{a^2}\Phi = -4\pi G q \rho_m \delta_m, \qquad \frac{\Phi - \Psi}{\Phi} = \zeta. \qquad (11.140)$$

In the regime $M^2/F \ll k^2/a^2$ of scalar-tensor theory (11.128), it follows that $q \approx (1 - 2Q^2)/F$ and $\zeta \approx -4Q^2/(1 - 2Q^2)$.

In order to confront dark energy models with the observations of weak lensing, it may be convenient to introduce the following quantity [202]:

$$\Sigma \equiv q\left(1 - \frac{1}{2}\zeta\right). \qquad (11.141)$$

From the definition (11.140) we find that the weak lensing potential $\psi = \Phi + \Psi$ can be expressed as

$$\psi = -8\pi G \frac{a^2}{k^2} \rho_m \delta_m \Sigma. \qquad (11.142)$$

In scalar-tensor theory (11.128) one has $\Sigma = 1/F$. The effect of modified gravity theories manifests itself in weak lensing observations in at least two ways. One is the multiplication of the term Σ on the right-hand side of Eq. (11.142). Another is the modification of the evolution of δ_m. The latter depends on two parameters q and ζ, or equivalently, Σ and ζ. Thus two parameters (Σ, ζ) will be useful to detect signatures of modified gravity theories from future surveys of weak lensing.

11.5.3 DGP Model

In the so-called Dvali–Gabadadze–Porrati (DGP) [48] braneworld it is possible to realize a 'self-accelerating Universe' even in the absence of dark energy. In braneworlds standard model particles are confined on a 3-dimensional (3D) brane embedded in the 5-dimensional bulk space-time with large extra dimensions. In the DGP braneworld model [48] the 3-brane is embedded in a Minkowski bulk space-time with infinitely large extra dimensions. Newton gravity can be recovered by adding a 4D Einstein–Hilbert action sourced by the brane curvature to the 5D action. Such a 4D term may be induced by quantum corrections coming from the bulk gravity and its coupling with matter on the brane. In the DGP model the standard 4D gravity is recovered for small distances, whereas the effect from the 5D gravity manifests itself for large distances. The late-time cosmic acceleration can be realized without introducing a dark energy component [203, 204].

The action for the DGP model is given by

$$S = \frac{1}{2\kappa_{(5)}^2} \int d^5X \sqrt{-\tilde{g}}\tilde{R} + \frac{1}{2\kappa_{(4)}^2} \int d^4X \sqrt{-g}R - \int d^5X \sqrt{-\tilde{g}}\mathcal{L}_M, \qquad (11.143)$$

where $\tilde{g}_{AB}$ is the metric in the $5D$ bulk and $g_{\mu\nu} = \partial_\mu X^A \partial_\nu X^B \tilde{g}_{AB}$ is the induced metric on the brane with $X^A(x^c)$ being the coordinates of an event on the brane labelled by x^c. The $5D$ and $4D$ (reduced) gravitational constants, $\kappa^2_{(5)}$ and $\kappa^2_{(4)}$, are related to the $5D$ and $4D$ Planck masses, $M_{(5)}$ and $M_{(4)}$, via $\kappa^2_{(5)} = 1/M^3_{(5)}$ and $\kappa^2_{(4)} = 1/M^2_{(4)}$. The first and second terms in Eq. (11.143) correspond to Einstein–Hilbert actions in the $5D$ bulk and on the brane, respectively. The matter action consists of a brane-localized matter whose action is given by $\int d^4x \sqrt{-g}(\sigma + \mathcal{L}_M^{\text{brane}})$, where σ is the 3-brane tension and $\mathcal{L}_M^{\text{brane}}$ is the Lagrangian density on the brane. Since the tension is not related to the Ricci scalar R, it can be adjusted to be zero.

The Einstein equation in the $5D$ bulk is given by $G^{(5)}_{AB} = 0$, where $G^{(5)}_{AB}$ is the $5D$ Einstein tensor. Imposing the Israel junction conditions on the brane with a $\mathbb{Z}_2$ symmetry, we obtain the $4D$ Einstein equation [205]

$$G_{\mu\nu} - \frac{1}{r_c}(K_{\mu\nu} - K g_{\mu\nu}) = \kappa^2_{(4)} T_{\mu\nu}, \qquad (11.144)$$

where $K_{\mu\nu}$ is the extrinsic curvature on the brane and $T_{\mu\nu}$ is the energy-momentum tensor of localized matter. The cross-over scale r_c is defined by $r_c \equiv \kappa^2_{(5)}/(2\kappa^2_{(4)})$. The Friedmann equation on the flat FLRW brane takes a simple form [203, 204]

$$H^2 - \frac{\varepsilon}{r_c}H = \frac{\kappa^2_{(4)}}{3}\rho_M, \qquad (11.145)$$

where $\varepsilon = \pm 1$, and ρ_M is the energy density of matter on the brane (with pressure P_M) satisfying the continuity equation

$$\dot{\rho}_M + 3H(\rho_M + P_M) = 0. \qquad (11.146)$$

If r_c is much larger than the Hubble radius H^{-1}, the first term in Eq. (11.145) dominates over the second one. In this case the standard Friedmann equation, $H^2 = \kappa^2_{(4)}\rho_M/3$, is recovered. Meanwhile, in the regime $r_c < H^{-1}$, the presence of the second term in Eq. (11.145) leads to a modification to the standard Friedmann equation. In the Universe dominated by non-relativistic matter ($\rho_M \propto a^{-3}$), the Universe approaches a de Sitter solution for $\varepsilon = +1$: $H \to H_{dS} = 1/r_c$. Hence it is possible to realize the present cosmic acceleration provided that r_c is of the order of the present Hubble radius H_0^{-1}.

Although the DGP braneworld is an attractive model allowing a self acceleration, the joint constraints from SNLS, BAO, and CMB data shows that this model is disfavored observationally [206–210]. Moreover the DGP model contains a ghost mode for the branch of the self acceleration [211, 212].

In the DGP model, a brane bending mode ϕ in the bulk corresponds to a scalar-field degree of freedom. In general, such a field can mediate a long-range fifth force incompatible with local gravity experiments, but the presence of a self-interaction of ϕ allows the so-called Vainshtein mechanism [213] to work within a radius $r_* = (r_g r_c^2)^{1/3}$ (r_g is the Schwarzschild radius of a source). The DGP model can be consistent with local gravity constraints under some range of conditions on the energy-momentum tensor [214–216].

The DGP model stimulated other approaches for constructing ghost-free theories in the presence of non-linear self-interactions of a scalar field ϕ. It is important to keep the field equations at second order in time derivatives to avoid that an extra degree of freedom gives rise to a ghost state. In particular Nicolis et al. [217] imposed a constant gradient-shift symmetry ('Galilean' symmetry), $\partial_\mu \phi \rightarrow \partial_\mu \phi + b_\mu$, to restrict the equations of motion at second order, while keeping a universal gravitational coupling with matter. In the 4-dimensional Minkowski space-time they found five terms $\mathcal{L}_i$ ($i = 1, \ldots 5$) giving rise to equations of motion satisfying the Galilean symmetry. The first three terms are given by $\mathcal{L}_1 = \phi$, $\mathcal{L}_2 = \nabla_\mu \phi \nabla^\mu \phi$, and $\mathcal{L}_3 = \Box \phi \nabla_\mu \phi \nabla^\mu \phi$. The term $\mathcal{L}_3$ is the non-linear field derivative that appears in the DGP model, which allows the possibility for the consistency with solar system experiments through the Vainshtein mechanism. Deffayet et al. [218, 219] derived the covariant expression of the terms $\mathcal{L}_i$ ($i = 1, \ldots 5$) by extending the analysis to the curved space-time. The cosmology based on such a covariant Galileon was studied in Ref. [220]. For the covariant Galileon there is a tracker solution that finally approaches a de Sitter fixed point responsible for cosmic acceleration today. Since the equation of state of the tracker is $w_{\mathrm{DE}} = -2$ during the matter era, the solutions approaching the tracker at late-times are favored observationally [221].

11.6 Conclusions

We summarize the results presented in this review.

- The cosmological constant ($w_{\mathrm{DE}} = -1$) is favored by a number of observations, but theoretically it is still challenging to explain why its energy scale is very small.
- Quintessence leads to the variation of the field equation of state in the region $w_\phi > -1$, but the current observations are not sufficient to distinguish between quintessence potentials.
- In k-essence it is possible to realize the cosmic acceleration by a field kinetic energy, while avoiding the instability problem associated with a phantom field. The k-essence models that aim to solve the coincidence problem inevitably leads to the superluminal propagation of the sound speed.
- In coupled dark energy models there is an upper bound on the strength of the coupling from the observations of CMB, large-scale structure and SN Ia.
- The generalized Chaplygin gas model allows the unified description of dark energy and dark matter, but it needs to be very close to the ΛCDM model to explain the observed matter power spectrum. There is a class of viable unified models of dark energy and dark matter using a purely k-essence field.
- In $f(R)$ gravity and scalar-tensor theories it is possible to construct viable models that satisfy both cosmological and local gravity constraints. These models leave several interesting observational signatures such as modifications to the matter power spectrum and to the weak-lensing spectrum.
- The DGP model allows the self-acceleration of the Universe, but it is effectively ruled out from observational constraints and the ghost problem. However, some of

the extension of works such as Galileon gravity allow the possibility for avoiding the ghost problem, while satisfying cosmological and local gravity constraints.

We hope that future high-precision observations will allow us to approach the origin of dark energy.

Acknowledgements The author thanks the organizers of the Sixth Aegean Summer School for inviting him to give a lecture on dark energy. This work was supported by Grant-in-Aid for Scientific Research Fund of the JSPS (No. 30318802) and Grant-in-Aid for Scientific Research on Innovative Areas (No. 21111006).

References

1. A.G. Riess et al. (Supernova Search Team Collaboration), Astron. J. **116**, 1009 (1998). astro-ph/9805201
2. S. Perlmutter et al. (Supernova Cosmology Project Collaboration), Astrophys. J. **517**, 565 (1999). astro-ph/9812133
3. D. Huterer, M.S. Turner, Phys. Rev. D **60**, 081301 (1999). astro-ph/9808133
4. V. Sahni, A.A. Starobinsky, Int. J. Mod. Phys. D **9**, 373 (2000). astro-ph/9904398
5. S.M. Carroll, Living Rev. Relativ. **4**, 1 (2001). http://www.livingreviews.org/lrr-2001-1. astro-ph/0004075
6. P.J.E. Peebles, B. Ratra, Rev. Mod. Phys. **75**, 559 (2003). astro-ph/0207347
7. T. Padmanabhan, Phys. Rep. **380**, 235 (2003). hep-th/0212290
8. E.J. Copeland, M. Sami, S. Tsujikawa, Int. J. Mod. Phys. D **15**, 1753 (2006). hep-th/0603057
9. L. Amendola, S. Tsujikawa, *Dark Energy: Theory and Observations* (Cambridge University Press, Cambridge, 2010)
10. P. Astier et al. (The SNLS Collaboration), Astron. Astrophys. **447**, 31 (2006). astro-ph/0510447
11. A.G. Riess et al., Astrophys. J. **659**, 98 (2007). astro-ph/0612666
12. W.M. Wood-Vasey et al. (ESSENCE Collaboration), Astrophys. J. **666**, 694 (2007). astro-ph/0701041
13. M. Kowalski et al. (Supernova Cosmology Project Collaboration), Astrophys. J. **686**, 749 (2008). arXiv:0804.4142
14. D.N. Spergel et al. (WMAP Collaboration), Astrophys. J. Suppl. **148**, 175 (2003). astro-ph/0302209
15. E. Komatsu et al., Astrophys. J. Suppl. **192**, 18 (2011). arXiv:1001.4538
16. D.J. Eisenstein et al. (SDSS Collaboration), Astrophys. J. **633**, 560 (2005). astro-ph/0501171
17. W.J. Percival, S. Cole, D.J. Eisenstein, R.C. Nichol, J.A. Peacock, A.C. Pope, A.S. Szalay, Mon. Not. R. Astron. Soc. **381**, 1053 (2007). arXiv:0705.3323
18. W.J. Percival et al., Mon. Not. R. Astron. Soc. **401**, 2148 (2010). arXiv:0907.1660
19. S. Weinberg, Rev. Mod. Phys. **61**, 1 (1989)
20. S. Kachru, R. Kallosh, A.D. Linde, S.P. Trivedi, Phys. Rev. D **68**, 046005 (2003). hep-th/0301240
21. Y. Fujii, Phys. Rev. D **26**, 2580 (1982)
22. L.H. Ford, Phys. Rev. D **35**, 2339 (1987)
23. C. Wetterich, Nucl. Phys. B **302**, 668 (1988)
24. B. Ratra, J. Peebles, Phys. Rev. D **37**, 321 (1988)
25. T. Chiba, N. Sugiyama, T. Nakamura, Mon. Not. R. Astron. Soc. **289**, L5 (1997). astro-ph/9704199
26. P.G. Ferreira, M. Joyce, Phys. Rev. Lett. **79**, 4740 (1997). astro-ph/9707286
27. E.J. Copeland, A.R. Liddle, D. Wands, Phys. Rev. D **57**, 4686 (1998). gr-qc/9711068

28. R.R. Caldwell, R. Dave, P.J. Steinhardt, Phys. Rev. Lett. **80**, 1582 (1998). astro-ph/9708069
29. I. Zlatev, L.M. Wang, P.J. Steinhardt, Phys. Rev. Lett. **82**, 896 (1999). astro-ph/9807002
30. P.J. Steinhardt, L.M. Wang, I. Zlatev, Phys. Rev. D **59**, 123504 (1999). astro-ph/9812313
31. T. Chiba, T. Okabe, M. Yamaguchi, Phys. Rev. D **62**, 023511 (2000). astro-ph/9912463
32. C. Armendariz-Picon, V.F. Mukhanov, P.J. Steinhardt, Phys. Rev. Lett. **85**, 4438 (2000). astro-ph/0004134
33. C. Armendariz-Picon, V.F. Mukhanov, P.J. Steinhardt, Phys. Rev. D **63**, 103510 (2001). astro-ph/0006373
34. S.M. Carroll, Phys. Rev. Lett. **81**, 3067 (1998). astro-ph/9806099
35. C.F. Kolda, D.H. Lyth, Phys. Lett. B **458**, 197 (1999). hep-ph/9811375
36. A.Yu. Kamenshchik, U. Moschella, V. Pasquier, Phys. Lett. B **511**, 265 (2001). gr-qc/0103004
37. M.C. Bento, O. Bertolami, A.A. Sen, Phys. Rev. D **66**, 043507 (2002). gr-qc/0202064
38. H. Sandvik, M. Tegmark, M. Zaldarriaga, I. Waga, Phys. Rev. D **69**, 123524 (2004). astro-ph/0212114
39. S. Capozziello, Int. J. Mod. Phys. D **11**, 483 (2002). gr-qc/0201033
40. S. Capozziello, S. Carloni, A. Troisi, Recent Res. Dev. Astron. Astrophys. **1**, 625 (2003). astro-ph/0303041
41. S. Capozziello, V.F. Cardone, S. Carloni, A. Troisi, Int. J. Mod. Phys. D **12**, 1969 (2003). astro-ph/0307018
42. S.M. Carroll, V. Duvvuri, M. Trodden, M.S. Turner, Phys. Rev. D **70**, 043528 (2004). astro-ph/0306438
43. L. Amendola, Phys. Rev. D **60**, 043501 (1999). astro-ph/9904120
44. J.P. Uzan, Phys. Rev. D **59**, 123510 (1999). gr-qc/9903004
45. T. Chiba, Phys. Rev. D **60**, 083508 (1999). gr-qc/9903094
46. N. Bartolo, M. Pietroni, Phys. Rev. D **61**, 023518 (2000). hep-ph/9908521
47. F. Perrotta, C. Baccigalupi, S. Matarrese, Phys. Rev. D **61**, 023507 (2000). astro-ph/9906066
48. G.R. Dvali, G. Gabadadze, M. Porrati, Phys. Lett. B **485**, 208 (2000). hep-th/0005016
49. R.R. Caldwell, Phys. Lett. B **545**, 23 (2002). astro-ph/9908168
50. A.A. Starobinsky, Gravit. Cosmol. **6**, 157 (2000). astro-ph/9912054
51. R.R. Caldwell, M. Kamionkowski, N.N. Weinberg, Phys. Rev. Lett. **91**, 071301 (2003). astro-ph/0302506
52. L. Amendola, S. Tsujikawa, Phys. Lett. B **660**, 125 (2008). arXiv:0705.0396
53. R.K. Sachs, A.M. Wolfe, Astrophys. J. **147**, 73 (1967)
54. A.A. Starobinsky, Phys. Lett. B **91**, 99 (1980)
55. D. Kazanas, Astrophys. J. **241**, L59 (1980)
56. K. Sato, Mon. Not. R. Astron. Soc. **195**, 467 (1981)
57. A.H. Guth, Phys. Rev. D **23**, 347 (1981)
58. G.F. Smoot et al., Astrophys. J. **396**, L1 (1992)
59. A.R. Liddle, D.H. Lyth, *Cosmological Inflation and Large-Scale Structure* (Cambridge University Press, Cambridge, 2000)
60. B.A. Bassett, S. Tsujikawa, D. Wands, Rev. Mod. Phys. **78**, 537 (2006). astro-ph/0507632
61. L. Page et al. (WMAP Collaboration), Astrophys. J. Suppl. **148**, 233 (2003). astro-ph/0302220
62. W. Hu, N. Sugiyama, Astrophys. J. **444**, 489 (1995). astro-ph/9407093
63. W. Hu, N. Sugiyama, Astrophys. J. **471**, 542 (1996). astro-ph/9510117
64. G. Efstathiou, J.R. Bond, Mon. Not. R. Astron. Soc. **304**, 75 (1999). astro-ph/9807103
65. M. Doran, M. Lilley, Mon. Not. R. Astron. Soc. **330**, 965 (2002). astro-ph/0104486
66. D.J. Eisenstein, W. Hu, Astrophys. J. **496**, 605 (1998). astro-ph/9709112
67. M. Shoji, D. Jeong, E. Komatsu, Astrophys. J. **693**, 1404 (2009). arXiv:0805.4238
68. B. Feng, X.L. Wang, X.M. Zhang, Phys. Lett. B **607**, 35 (2005). astro-ph/0404224
69. M. Tegmark et al. (SDSS Collaboration), Phys. Rev. D **69**, 103501 (2004). astro-ph/0310723
70. U. Seljak et al., Phys. Rev. D **71**, 103515 (2005). astro-ph/0407372
71. M. Tegmark et al. (SDSS Collaboration), Phys. Rev. D **74**, 123507 (2006). astro-ph/0608632

72. D. Hooper, S. Dodelson, Astropart. Phys. **27**, 113 (2007). astro-ph/0512232
73. B. Jain, A. Taylor, Phys. Rev. Lett. **91**, 141302 (2003). astro-ph/0306046
74. D. Bailin, A. Love, *Supersymmetric Gauge Field Theory and String Theory* (Institute of Physics Publishing, Bristol, 1994)
75. M. Green, J.H. Schwarz, E. Witten, *Superstring Theory* (Cambridge University Press, Cambridge, 1987)
76. M. Dine, R. Rohm, N. Seiberg, E. Witten, Phys. Lett. B **156**, 55 (1985)
77. I. Affleck, M. Dine, N. Seiberg, Nucl. Phys. B **256**, 557 (1985)
78. L. Susskind, in *Universe or Multiverse?* ed. by B. Carr (Cambridge University Press, Cambridge, 2007), pp. 247–266. hep-th/0302219
79. B. Carter, in *Confrontation of Cosmological Theories with Observational Data*, ed. by M.S. Longair (Kluwer, Dordrecht, 1974), pp. 291–298
80. J. Barrow, F. Tipler, *The Cosmological Anthropic Principle* (Oxford University Press, Oxford, 1988)
81. S. Weinberg, Phys. Rev. Lett. **59**, 2607 (1987)
82. J. Garriga, A. Vilenkin, Phys. Rev. D **61**, 083502 (2000). astro-ph/9908115
83. F. Denef, M.R. Douglas, J. High Energy Phys. **0405**, 072 (2004). hep-th/0404116
84. J. Garriga, A.D. Linde, A. Vilenkin, Phys. Rev. D **69**, 063521 (2004). hep-th/0310034
85. R. Blumenhagen, F. Gmeiner, G. Honecker, D. Lüst, T. Weigand, Nucl. Phys. B **713**, 83 (2005). hep-th/0411173
86. A. de la Macorra, G. Piccinelli, Phys. Rev. D **61**, 123503 (2000). hep-ph/9909459
87. S.C.C. Ng, N.J. Nunes, F. Rosati, Phys. Rev. D **64**, 083510 (2001). astro-ph/0107321
88. R.R. Caldwell, E.V. Linder, Phys. Rev. Lett. **95**, 141301 (2005). astro-ph/0505494
89. P. Binétruy, Phys. Rev. D **60**, 063502 (1999). hep-ph/9810553
90. P. Brax, J. Martin, Phys. Lett. B **468**, 40 (1999). astro-ph/9905040
91. A.D. Linde, in *Three Hundred Years of Gravitation*, ed. by S.W. Hawking, W. Israel (Cambridge University Press, Cambridge, 1987), pp. 604–630
92. R. Kallosh, J. Kratochvil, A. Linde, E.V. Linder, M. Shmakova, J. Cosmol. Astropart. Phys. **0310**, 015 (2003). astro-ph/0307185
93. J.A. Frieman, C.T. Hill, A. Stebbins, I. Waga, Phys. Rev. Lett. **75**, 2077 (1995). astro-ph/9505060
94. Y. Nomura, T. Watari, T. Yanagida, Phys. Lett. B **484**, 103 (2000). hep-ph/0004182
95. K. Choi, Phys. Rev. D **62**, 043509 (2000). hep-ph/9902292
96. J.E. Kim, H.P. Nilles, Phys. Lett. B **553**, 1 (2003). hep-ph/0210402
97. L.J. Hall, Y. Nomura, S.J. Oliver, Phys. Rev. Lett. **95**, 141302 (2005). astro-ph/0503706
98. E.V. Linder, Phys. Rev. D **73**, 063010 (2006). astro-ph/0601052
99. C. Armendariz-Picon, T. Damour, V.F. Mukhanov, Phys. Lett. B **458**, 209 (1999). hep-th/9904075
100. M. Gasperini, G. Veneziano, Astropart. Phys. **1**, 317 (1993). hep-th/9211021
101. M. Gasperini, G. Veneziano, Phys. Rep. **373**, 1 (2003). hep-th/0207130
102. S.M. Carroll, M. Hoffman, M. Trodden, Phys. Rev. D **68**, 023509 (2003). astro-ph/0301273
103. J.M. Cline, S. Jeon, G.D. Moore, Phys. Rev. D **70**, 043543 (2004). hep-ph/0311312
104. N. Arkani-Hamed, H.C. Cheng, M.A. Luty, S. Mukohyama, J. High Energy Phys. **0405**, 074 (2004). hep-th/0312099
105. F. Piazza, S. Tsujikawa, J. Cosmol. Astropart. Phys. **0407**, 004 (2004). hep-th/0405054
106. M.R. Garousi, Nucl. Phys. B **584**, 284 (2000). hep-th/0003122
107. A. Sen, J. High Energy Phys. **0204**, 048 (2002). hep-th/0203211
108. G.W. Gibbons, Phys. Lett. B **537**, 1 (2002). hep-th/0204008
109. T. Padmanabhan, Phys. Rev. D **66**, 021301 (2002). hep-th/0204150
110. L.R.W. Abramo, F. Finelli, Phys. Lett. B **575**, 165 (2003). astro-ph/0307208
111. J.M. Aguirregabiria, R. Lazkoz, Phys. Rev. D **69**, 123502 (2004). hep-th/0402190
112. M.R. Garousi, M. Sami, S. Tsujikawa, Phys. Rev. D **70**, 043536 (2004). hep-th/0402075
113. E.J. Copeland, M.R. Garousi, M. Sami, S. Tsujikawa, Phys. Rev. D **71**, 043003 (2005). hep-th/0411192

114. E. Silverstein, D. Tong, Phys. Rev. D **70**, 103505 (2004). hep-th/0310221
115. M. Alishahiha, E. Silverstein, D. Tong, Phys. Rev. D **70**, 123505 (2004). hep-th/0404084
116. J. Martin, M. Yamaguchi, Phys. Rev. D **77**, 123508 (2008). arXiv:0801.3375
117. Z.K. Guo, N. Ohta, J. Cosmol. Astropart. Phys. **0804**, 035 (2008). arXiv:0803.1013
118. P. Singh, M. Sami, N. Dadhich, Phys. Rev. D **68**, 023522 (2003). hep-th/0305110
119. M. Sami, A. Toporensky, Mod. Phys. Lett. A **19**, 1509 (2004). gr-qc/0312009
120. J. Garriga, V.F. Mukhanov, Phys. Lett. B **458**, 219 (1999). hep-th/9904176
121. C. Bonvin, C. Caprini, R. Durrer, Phys. Rev. Lett. **97**, 081303 (2006). astro-ph/0606584
122. M. Malquarti, E.J. Copeland, A.R. Liddle, M. Trodden, Phys. Rev. D **67**, 123503 (2003). astro-ph/0302279
123. C. Wetterich, Astron. Astrophys. **301**, 321 (1995). hep-th/9408025
124. L. Amendola, Phys. Rev. D **60**, 043501 (1999). astro-ph/9904120
125. L. Amendola, Phys. Rev. D **62**, 043511 (2000). astro-ph/9908023
126. D.J. Holden, D. Wands, Phys. Rev. D **61**, 043506 (2000). gr-qc/9908026
127. G. Huey, B.D. Wandelt, Phys. Rev. D **74**, 023519 (2006). astro-ph/0407196
128. S. Das, P.S. Corasaniti, J. Khoury, Phys. Rev. D **73**, 083509 (2006). astro-ph/0510628
129. C. Brans, R.H. Dicke, Phys. Rev. **124**, 925 (1961)
130. S. Tsujikawa, K. Uddin, S. Mizuno, R. Tavakol, J. Yokoyama, Phys. Rev. D **77**, 103009 (2008). arXiv:0803.1106
131. L. Amendola, D. Tocchini-Valentini, Phys. Rev. D **66**, 043528 (2002). astro-ph/0111535
132. L. Amendola, Phys. Rev. D **69**, 103524 (2004). astro-ph/0311175
133. J. Peebles, *The Large-Scale Structure of the Universe* (Princeton University Press, Princeton, 1980)
134. L.M. Wang, P.J. Steinhardt, Astrophys. J. **508**, 483 (1998). astro-ph/9804015
135. E.V. Linder, Phys. Rev. D **72**, 043529 (2005). astro-ph/0507263
136. C. Di Porto, L. Amendola, Phys. Rev. D **77**, 083508 (2008). arXiv:0707.2686
137. L. Amendola, Phys. Rev. Lett. **93**, 181102 (2004). hep-th/0409224
138. L. Amendola, S. Tsujikawa, M. Sami, Phys. Lett. B **632**, 155 (2006). astro-ph/0506222
139. S. Tsujikawa, Phys. Rev. D **72**, 083512 (2005). astro-ph/0508542
140. J. Khoury, A. Weltman, Phys. Rev. Lett. **93**, 171104 (2004). astro-ph/0309300
141. J. Khoury, A. Weltman, Phys. Rev. D **69**, 044026 (2004). astro-ph/0309411
142. T. Tamaki, S. Tsujikawa, Phys. Rev. D **78**, 084028 (2008). arXiv:0808.2284
143. C.M. Will, Living Rev. Relativ. **4**, 4 (2001). http://www.livingreviews.org/lrr-2001-4. gr-qc/0103036
144. C.M. Will, Living Rev. Relativ. **9**, 3 (2005). http://www.livingreviews.org/lrr-2006-3. gr-qc/0510072
145. P. Brax, C. van de Bruck, A.C. Davis, J. Khoury, A. Weltman, Phys. Rev. D **70**, 123518 (2004). astro-ph/0408415
146. P.P. Avelino, L.M.G. Beca, J.P.M. de Carvalho, C.J.A. Martins, E.J. Copeland, Phys. Rev. D **69**, 041301 (2004). astro-ph/0306493
147. L. Amendola, F. Finelli, C. Burigana, D. Carturan, J. Cosmol. Astropart. Phys. **0307**, 005 (2003). astro-ph/0304325
148. L. Amendola, I. Waga, F. Finelli, J. Cosmol. Astropart. Phys. **0511**, 009 (2005). astro-ph/0509099
149. R.J. Scherrer, Phys. Rev. Lett. **93**, 011301 (2004). astro-ph/0402316
150. D. Giannakis, W. Hu, Phys. Rev. D **72**, 063502 (2005). astro-ph/0501423
151. A.D. Dolgov, M. Kawasaki, Phys. Lett. B **573**, 1 (2003). astro-ph/0307285
152. L. Amendola, D. Polarski, S. Tsujikawa, Phys. Rev. Lett. **98**, 131302 (2007). astro-ph/0603703
153. L. Amendola, D. Polarski, S. Tsujikawa, Int. J. Mod. Phys. D **16**, 1555 (2007). astro-ph/0605384
154. T. Chiba, Phys. Lett. B **575**, 1 (2003). astro-ph/0307338
155. G.J. Olmo, Phys. Rev. Lett. **95**, 261102 (2005). gr-qc/0505101
156. G.J. Olmo, Phys. Rev. D **72**, 083505 (2005). gr-qc/0505135

157. I. Navarro, K. Van Acoleyen, J. Cosmol. Astropart. Phys. **0702**, 022 (2007). gr-qc/0611127
158. A.L. Erickcek, T.L. Smith, M. Kamionkowski, Phys. Rev. D **74**, 121501 (2006). astro-ph/0610483
159. T. Chiba, T.L. Smith, A.L. Erickcek, Phys. Rev. D **75**, 124014 (2007). astro-ph/0611867
160. S.M. Carroll, I. Sawicki, A. Silvestri, M. Trodden, New J. Phys. **8**, 323 (2006). astro-ph/0607458
161. Y.S. Song, W. Hu, I. Sawicki, Phys. Rev. D **75**, 044004 (2007). astro-ph/0610532
162. R. Bean, D. Bernat, L. Pogosian, A. Silvestri, M. Trodden, Phys. Rev. D **75**, 064020 (2007). astro-ph/0611321
163. T. Faulkner, M. Tegmark, E.F. Bunn, Y. Mao, Phys. Rev. D **76**, 063505 (2007). astro-ph/0612569
164. L. Pogosian, A. Silvestri, Phys. Rev. D **77**, 023503 (2008). arXiv:0709.0296
165. V. Muller, H.J. Schmidt, A.A. Starobinsky, Phys. Lett. B **202**, 198 (1988)
166. V. Faraoni, Phys. Rev. D **70**, 044037 (2004). gr-qc/0407021
167. L. Amendola, R. Gannouji, D. Polarski, S. Tsujikawa, Phys. Rev. D **75**, 083504 (2007). gr-qc/0612180
168. W. Hu, I. Sawicki, Phys. Rev. D **76**, 064004 (2007). arXiv:0705.1158
169. A.A. Starobinsky, JETP Lett. **86**, 157 (2007). arXiv:0706.2041
170. S. Tsujikawa, Phys. Rev. D **77**, 023507 (2008). arXiv:0709.1391
171. S.A. Appleby, R.A. Battye, Phys. Lett. B **654**, 7 (2007). arXiv:0705.3199
172. E.V. Linder, Phys. Rev. D **80**, 123528 (2009). arXiv:0905.2962
173. B. Li, J.D. Barrow, Phys. Rev. D **75**, 084010 (2007). gr-qc/0701111
174. R. Gannouji, B. Moraes, D. Polarski, J. Cosmol. Astropart. Phys. **0902**, 034 (2009). arXiv:0809.3374
175. D.F. Torres, Phys. Rev. D **66**, 043522 (2002). astro-ph/0204504
176. B. Boisseau, G. Esposito-Farese, D. Polarski, A.A. Starobinsky, Phys. Rev. Lett. **85**, 2236 (2000). gr-qc/0001066
177. H. Motohashi, A.A. Starobinsky, J. Yokoyama, Prog. Theor. Phys. **123**, 887 (2010). arXiv:1002.1141
178. M. Martinelli, A. Melchiorri, L. Amendola, Phys. Rev. D **79**, 123516 (2009). arXiv:0906.2350
179. A. Ali, R. Gannouji, M. Sami, A.A. Sen, Phys. Rev. D **81**, 104029 (2010). arXiv:1001.5384
180. J.c. Hwang, H.r. Noh, Phys. Rev. D **65**, 023512 (2002). astro-ph/0102005
181. J.c. Hwang, H. Noh, Phys. Rev. D **71**, 063536 (2005). gr-qc/0412126
182. A. De Felice, S. Tsujikawa, Living Rev. Relativ. **13**, 3 (2010). http://www.livingreviews.org/lrr-2010-3. arXiv:1002.4928
183. S. Tsujikawa, Phys. Rev. D **76**, 023514 (2007). arXiv:0705.1032
184. S. Tsujikawa, K. Uddin, R. Tavakol, Phys. Rev. D **77**, 043007 (2008). arXiv:0712.0082
185. S.A. Appleby, R.A. Battye, J. Cosmol. Astropart. Phys. **0805**, 019 (2008). arXiv:0803.1081
186. H. Motohashi, A.A. Starobinsky, J. Yokoyama, Int. J. Mod. Phys. D **18**, 1731 (2009). arXiv:0905.0730
187. S. Tsujikawa, R. Gannouji, B. Moraes, D. Polarski, Phys. Rev. D **80**, 084044 (2009). arXiv:0908.2669
188. P. Zhang, Phys. Rev. D **73**, 123504 (2006). astro-ph/0511218
189. Y.S. Song, H. Peiris, W. Hu, Phys. Rev. D **76**, 063517 (2007). arXiv:0706.2399
190. K.i. Maeda, Phys. Rev. D **39**, 3159 (1989)
191. S. Capozziello, S. Tsujikawa, Phys. Rev. D **77**, 107501 (2008). arXiv:0712.2268
192. J. O'Hanlon, Phys. Rev. Lett. **29**, 137 (1972)
193. P. Teyssandier, P. Tourrenc, J. Math. Phys. **24**, 2793 (1983)
194. C. Baccigalupi, S. Matarrese, F. Perrotta, Phys. Rev. D **62**, 123510 (2000). astro-ph/0005543
195. A. Riazuelo, J.P. Uzan, Phys. Rev. D **66**, 023525 (2002). astro-ph/0107386
196. L. Perivolaropoulos, J. Cosmol. Astropart. Phys. **0510**, 001 (2005). astro-ph/0504582
197. S. Nesseris, L. Perivolaropoulos, J. Cosmol. Astropart. Phys. **0701**, 018 (2007). astro-ph/0610092

198. J. Martin, C. Schimd, J.P. Uzan, Phys. Rev. Lett. **96**, 061303 (2006). astro-ph/0510208
199. R. Gannouji, D. Polarski, A. Ranquet, A.A. Starobinsky, J. Cosmol. Astropart. Phys. **0609**, 016 (2006). astro-ph/0606287
200. V. Acquaviva, L. Verde, J. Cosmol. Astropart. Phys. **0712**, 001 (2007). arXiv:0709.0082
201. Y.S. Song, L. Hollenstein, G. Caldera-Cabral, K. Koyama, J. Cosmol. Astropart. Phys. **1004**, 018 (2010). arXiv:1001.0969
202. L. Amendola, M. Kunz, D. Sapone, J. Cosmol. Astropart. Phys. **0804**, 013 (2008). arXiv: 0704.2421
203. C. Deffayet, Phys. Lett. B **502**, 199 (2001). hep-th/0010186
204. C. Deffayet, G.R. Dvali, G. Gabadadze, Phys. Rev. D **65**, 044023 (2002). astro-ph/0105068
205. K. Hinterbichler, A. Nicolis, M. Porrati, J. High Energy Phys. **0909**, 089 (2009). arXiv:0905.2359
206. I. Sawicki, S.M. Carroll, astro-ph/0510364
207. M. Fairbairn, A. Goobar, Phys. Lett. B **642**, 432 (2006). astro-ph/0511029
208. R. Maartens, E. Majerotto, Phys. Rev. D **74**, 023004 (2006). astro-ph/0603353
209. U. Alam, V. Sahni, Phys. Rev. D **73**, 084024 (2006). astro-ph/0511473
210. Y.S. Song, I. Sawicki, W. Hu, Phys. Rev. D **75**, 064003 (2007). astro-ph/0606286
211. A. Nicolis, R. Rattazzi, J. High Energy Phys. **0406**, 059 (2004). hep-th/0404159
212. M.A. Luty, M. Porrati, R. Rattazzi, J. High Energy Phys. **0309**, 029 (2003). hep-th/0303116
213. A.I. Vainshtein, Phys. Lett. B **39**, 393 (1972)
214. C. Deffayet, G.R. Dvali, G. Gabadadze, A.I. Vainshtein, Phys. Rev. D **65**, 044026 (2002). hep-th/0106001
215. A. Gruzinov, New Astron. **10**, 311 (2005). astro-ph/0112246
216. M. Porrati, Phys. Lett. B **534**, 209 (2002). hep-th/0203014
217. A. Nicolis, R. Rattazzi, E. Trincherini, Phys. Rev. D **79**, 064036 (2009). arXiv:0811.2197
218. C. Deffayet, G. Esposito-Farese, A. Vikman, Phys. Rev. D **79**, 084003 (2009). arXiv:0901. 1314
219. C. Deffayet, S. Deser, G. Esposito-Farese, Phys. Rev. D **80**, 064015 (2009). arXiv:0906.1967
220. A. De Felice, S. Tsujikawa, Phys. Rev. Lett. **105**, 111301 (2010). arXiv:1007.2700
221. S. Nesseris, A. De Felice, S. Tsujikawa, Phys. Rev. D **82**, 124054 (2010). arXiv:1010.0407

Chapter 12
Unconventional Cosmology

Robert H. Brandenberger

Abstract I review two cosmological paradigms which are alternative to the current inflationary scenario. The first alternative is the "matter bounce", a non-singular bouncing cosmology with a matter-dominated phase of contraction. The second is an "emergent" scenario, which can be implemented in the context of "string gas cosmology". I will compare these scenarios with the inflationary one and demonstrate that all three lead to an approximately scale-invariant spectrum of cosmological perturbations.

12.1 Introduction

12.1.1 Overview

"Unconventional cosmology" is the title which I was given for my lectures. Based on my interpretation of this title my job is to lecture on alternatives to the current paradigm of early universe cosmology, the "conventional theory". The fact that almost all cosmologists agree that there is a current paradigm speaks to the remarkable progress of cosmology over the past three decades. At the time when the current paradigm of early universe cosmology, the inflationary scenario [1] (see also [2–4]), was developed, we had very little observational information about the large-scale structure of the universe. The success of inflationary cosmology at that point in time is that it could explain some of the puzzles which the previous paradigm—Standard Big Bang cosmology—could not address. It was very soon realized [5] (see also [4, 6, 7]) that inflation was much more powerful than simply being able to explain puzzles of Standard Big Bang cosmology such as the flatness and horizon problems. In fact, inflationary cosmology gave rise to the first explanation for the origin of inhomogeneities in the universe based on causal physics: It yields a mechanism for generating an approximately scale-invariant spectrum of primordial density fluctuations, i.e. the kind of spectrum which had already been suggested as

R.H. Brandenberger (✉)
Physics Department, McGill University, 3600 University Street, Montreal, QC H3A 2T8, Canada
e-mail: rhb@physics.mcgill.ca

G. Calcagni et al. (eds.), *Quantum Gravity and Quantum Cosmology*,
Lecture Notes in Physics 863, DOI 10.1007/978-3-642-33036-0_12,
© Springer-Verlag Berlin Heidelberg 2013

a reasonable one to be consistent with the (at that time limited) information about the distribution of galaxies [8, 9]. As already realized earlier, such a primordial spectrum of density fluctuations leads to an angular power spectrum of anisotropies in the cosmic microwave background (CMB) radiation which is scale-invariant on large scales and characterized by acoustic oscillations on angular scales of a degree and lower [10, 11]. This prediction has now been confirmed observationally [12] with high accuracy. It is important, however, to keep in mind that any theory which yields an approximately scale-invariant spectrum of primordial fluctuations—and I will present a couple of such theories in these lectures—will agree with the recent high-precision observations of the large-scale structure and CMB anisotropies. Thus, current observations cannot be interpreted as a proof that inflation took place.

In spite of its phenomenological success, inflationary cosmology suffers from some important conceptual problems, which may imply that it is not so "conventional" after all. These problems motivate the search for alternative proposals for the evolution of the early universe and for the generation of structure. These alternatives must be consistent with the current observations, and they must make predictions with which they can be observationally distinguished from inflationary cosmology.

There are indeed paradigms alternative to inflation which generate an almost scale-invariant spectrum of primordial cosmological fluctuations. In these lectures I will present two examples—first the string gas realization [13, 14] (see [15–17] for reviews) of the emergent universe paradigm [18], and second the "matter bounce scenario" [19, 20] (see [21] for reviews). I should emphasize, however, that these are not the only alternatives to the inflationary scenario. The Pre-Big-Bang scenario [22], the Ekpyrotic scenario [23], and the pseudo-conformal construction [24, 25] are other promising models, and there are others.

The outline of this lecture series is as follows. The first lecture (Sects. 12.1–12.3) focuses on background (homogeneous and isotropic) cosmologies. I begin with a review of the inflationary scenario, the current paradigm of early universe cosmology. After discussing the phenomenological successes of the scenario, I will list a number of conceptual problems which in part motivate the search for alternative scenarios. In Sect. 12.2 I introduce the first alternative paradigm which will be discussed here, the "matter bounce" scenario. After presenting the basic idea of the scenario, I will discuss various ways to realize it. In Sect. 12.3 I turn to the "emergent Universe" scenario. Once again, I begin by presenting the basic ideas before turning to a discussion of "string gas cosmology", the specific realization which has provided some very interesting results.

The second lecture (Sects. 12.4–12.7) focuses on the question of how the inhomogeneities and anisotropies which are observed now in the distribution of galaxies on large scales and in temperature maps of the CMB, respectively, are generated. I will first (Sect. 12.4) briefly review the theory of cosmological perturbations. Then, I will emphasize that all three scenarios (inflation, the matter bounce and string gas cosmology), yield fluctuations in agreement with current data, but are distinguishable by future observations. Fluctuations in inflation are reviewed in Sect. 12.5, those in the matter bounce in Sect. 12.6, and those in string gas cosmology in

Fig. 12.1 A sketch showing the time line of inflationary cosmology. The period of accelerated expansion begins at time t_i and end at t_R. The time evolution after t_R corresponds to what happens in Standard Cosmology

Sect. 12.7. The final section focuses on outstanding problems of the various scenarios, and contains some general discussion.

These lectures are a modified version of lectures given previously [21] at various summer schools.

12.1.2 Review of Inflationary Cosmology

Inflationary cosmology [1] addresses several shortcomings of Standard Big Bang cosmology (the previous paradigm of early universe cosmology). It explains why the universe is to a good approximation homogeneous and isotropic on large scales (the "horizon problem"), it explains why it is to an excellent accuracy spatially flat (the "flatness problem"), and it can explain its large size and entropy from initial conditions where the universe is of microscopic size.

The idea of inflationary cosmology is to add a period to the evolution of the very early universe during which the scale factor undergoes accelerated expansion—most often nearly exponential growth. To obtain exponential expansion of space in the context of Einstein gravity, the energy density must be constant. Thus, during inflation the total energy and size of the universe both increase exponentially. In this way, inflation can solve the size and entropy problems of Standard Cosmology. Since the horizon expands exponentially during the period of inflation and all classical fluctuations redshift, inflation produces an approximately homogeneous and isotropic space. In addition, the relative contribution of spatial curvature decreases during the period of inflation. Thus, inflation can also address the "flatness problem" of Standard Big Bang cosmology. Any "unconventional cosmology" which claims to provide an alternative to inflation must also address the basic problems of Standard Cosmology mentioned above.

The time line of inflationary cosmology is sketched in Fig. 12.1. The time t_i is the beginning of the inflationary period, and t_R is its end (the meaning of the subscript R will become clear later). Although inflation is usually associated with physics at very high energy scales, e.g. $E \sim 10^{16}$ GeV, all that is required from the initial basic considerations is that inflation ends before the time of nucleosynthesis.

During the period of inflation, the density of any pre-existing particles is diluted exponentially. Hence, if inflation is to be viable, it must contain a mechanism to heat the universe at t_R, a "reheating" mechanism—hence the subscript R on t_R. This mechanism must involve dramatic entropy generation. It is this non-adiabatic evolution which leads to a solution of the flatness problem.

Inflationary cosmology, however, does more than simply solve some conceptual problems of the previous paradigm. It for the first time provided a causal theory of structure formation. Any proposed alternative to cosmological inflation must also match this success. Here we review the basic idea of why inflationary cosmology can provide a causal explanation of the observed inhomogeneities in the universe. The calculations will be reviewed in the second lecture.

In order to understand why inflation provides a causal structure formation mechanism, we start with a space-time sketch of inflationary cosmology as presented in Fig. 12.2. The vertical axis is time, the horizontal axis corresponds to physical distance. Three different distance scales are shown. The solid line labelled by k is the physical length corresponding to a fixed comoving perturbation. The second solid line (blue) is the Hubble radius

$$l_H(t) \equiv H^{-1}(t). \tag{12.1}$$

As will be shown in Lecture 2, he Hubble radius separates scales where microphysics dominates over gravity (sub-Hubble scales) from ones on which the effects of microphysics are negligible (super-Hubble scales). Hence, a necessary requirement for a causal theory of structure formation is that scales we observe today originate at sub-Hubble lengths in the early universe. The third length is the "horizon", the forward light cone of our position at the Big Bang. The horizon is the causality limit. Note that because of the exponential expansion of space during inflation, the horizon is exponentially larger than the Hubble radius. It is important not to confuse these two scales. Hubble radius and horizon are the same in Standard Cosmology, but in all three early universe scenarios which will be discussed in these lectures they are completely different (in inflationary cosmology the horizon is exponentially larger, in the matter bounce scenario it is in fact infinite, and in the emergent scenario it is infinite if the emergent phase extends to $t = -\infty$). In fact, in any structure formation scenario the two scales need to be different.

From Fig. 12.2 it is clear that provided the period of inflation is sufficiently long, all scales which are currently observed originate as sub-Hubble scales at the beginning of the inflationary phase. Thus, in inflationary cosmology it is possible to have a causal generation mechanism of fluctuations [4–6]. Since matter pre-existing at t_i is redshifted away, we are left with a matter vacuum. The inflationary universe scenario of structure formation is based on the hypothesis that all current structure originated as quantum vacuum fluctuations. From Fig. 12.2 it is also clear that the horizon problem of standard cosmology can be solved provided that the period of inflation lasts sufficiently long. For inflation to solve the horizon and flatness problem of Standard cosmology, the period of exponential expansion must be greater than about $50H^{-1}$ (this number depends very slightly on the energy scale at which inflation takes place).

In order to obtain exponential expansion of space in the context of Einstein gravity, matter with an equation of state $p = -\rho$ is required, where p and ρ are pressure

Fig. 12.2 Space-time sketch
of inflationary cosmology.
The *vertical axis* is time, the
horizontal axis corresponds
to physical distance. The
solid line labelled k is the
physical length of a fixed
comoving fluctuation scale.
The role of the Hubble radius
and the horizon are discussed
in the text

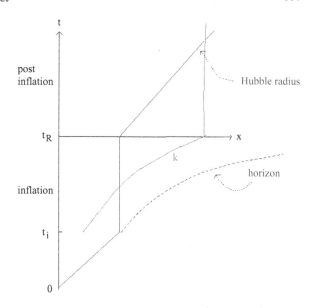

and energy density, respectively. In the context of renormalizable quantum field theory, a phase dominated by almost constant (both in space and time) potential energy of a scalar matter field is required.

12.1.3 Conceptual Problems of Inflationary Cosmology

In spite of the phenomenological success of the inflationary paradigm, conventional scalar field-driven inflation suffers from several important conceptual problems.

The first problem concern the nature of the inflaton, the scalar field which generates the inflationary expansion. No particle corresponding to the excitation of a scalar field has yet been observed in nature, and the Higgs field which is introduced to give elementary particles masses in the Standard Model of particle physics does not have the required flatness of the potential to yield inflation, unless it is non-minimally coupled to gravity [26]. In particle physics theories beyond the Standard Model there are often many scalar fields, but it is in general very hard to obtain the required flatness properties on the potential

The second problem (the *amplitude problem*) relates to the amplitude of the spectrum of cosmological perturbations. In a wide class of inflationary models, obtaining the correct amplitude requires the introduction of a hierarchy in scales, namely [27]

$$\frac{V(\varphi)}{\Delta\varphi^4} \leq 10^{-12}, \tag{12.2}$$

where $\Delta\varphi$ is the change in the inflaton field during the minimal length of the inflationary period, and $V(\varphi)$ is the potential energy during inflation.

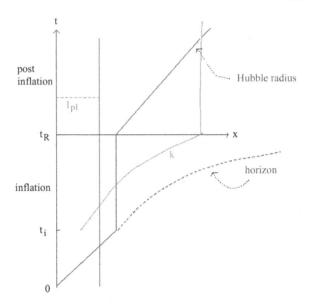

Fig. 12.3 Space-time diagram (sketch) of inflationary cosmology where we have added an extra length scale, namely the Planck length l_{pl} (*majenta vertical line*). The symbols have the same meaning as in Fig. 12.2. Note, specifically, that—as long as the period of inflation lasts a couple of e-foldings longer than the minimal value required for inflation to address the problems of Standard Big Bang cosmology—all wavelengths of cosmological interest to us today start out at the beginning of the period of inflation with a wavelength which is smaller than the Planck length

A more serious problem is the *trans-Planckian problem* [28]. Returning to the space-time diagram of inflation (see Fig. 12.3), we can immediately deduce that, provided that the period of inflation lasted sufficiently long (for GUT scale inflation the number is about 70 e-foldings), then all scales inside the Hubble radius today started out with a physical wavelength smaller than the Planck scale at the beginning of inflation. Now, the theory of cosmological perturbations is based on Einstein's theory of General Relativity coupled to a simple semi-classical description of matter. It is clear that these building blocks of the theory are inapplicable on scales comparable and smaller than the Planck scale. Thus, the key successful prediction of inflation (the theory of the origin of fluctuations) is based on suspect calculations since new physics *must* enter into a correct computation of the spectrum of cosmological perturbations. The key question is as to whether the predictions obtained using the current theory are sensitive to the specifics of the unknown theory which takes over on small scales. Simple toy models of new physics on super-Planck scales based on modified dispersion relations were used in [29, 30] (see also [31–33]) to show that the resulting spectrum of cosmological fluctuations indeed depends on what is assumed about physics on trans-Planckian scales.

A fourth problem is the *singularity problem*. It was known for a long time that Standard Big Bang cosmology cannot be the complete story of the early universe because of the initial singularity, a singularity which is unavoidable when basing cosmology on Einstein's field equations in the presence of a matter source obeying the weak energy conditions (see e.g. [34] for a textbook discussion). The singularity theorems have been generalized to apply to Einstein gravity coupled to scalar field matter, i.e. to scalar field-driven inflationary cosmology [35]. It was shown that, in this context, a past singularity at some point in space is unavoidable. Thus we know, from the outset, that scalar field-driven inflation cannot be the ultimate theory of the very early universe.

The Achilles heel of scalar field-driven inflationary cosmology may be the *cosmological constant problem*. We know from observations that the large quantum vacuum energy of field theories does not gravitate today. However, to obtain a period of inflation one is using the part of the energy-momentum tensor of the scalar field which looks like the vacuum energy. In the absence of a solution of the cosmological constant problem it is unclear whether scalar field-driven inflation is robust, i.e. whether the mechanism which renders the quantum vacuum energy gravitationally inert today will not also prevent the vacuum energy from gravitating during the period of slow-rolling of the inflaton field.

A final problem which we will mention here is the concern that the energy scale at which inflation takes place is too high to justify an effective field theory analysis based on Einstein gravity. In simple toy models of inflation, the energy scale during the period of inflation is about 10^{16} GeV, very close to the string scale in many string models, and not too far from the Planck scale. Thus, correction terms in the effective action for matter and gravity may already be important at the energy scale of inflation, and the cosmological dynamics may be rather different from what is obtained when neglecting the correction terms.

In Fig. 12.4 we show once again the space-time sketch of inflationary cosmology. In addition to the length scales which appear in the previous versions of this figure, we have now shaded the "zones of ignorance", zones where the Einstein gravity effective action is sure to break down. As described above, fluctuations emerge from the short distance zone of ignorance (except if the period of inflation is very short), and the energy scale of inflation might put the period of inflation too close to the high energy density zone of ignorance to trust the predictions based on using the Einstein action.

The arguments in this subsection provide a motivation for considering alternative scenarios of early universe cosmology. Below we will focus on two scenarios, the *matter bounce* and *string gas cosmology*, a realization of the *emergent universe* paradigm.

12.2 Matter Bounce

12.2.1 The Idea

The first alternative to cosmological inflation as a theory of structure formation is the "matter bounce" , an alternative which is not yet well appreciated (for an overview

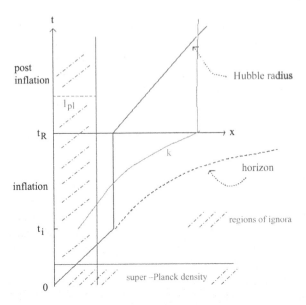

Fig. 12.4 Space-time diagram (sketch) of inflationary cosmology including the two zones of ignorance—sub-Planckian wavelengths and trans-Planckian densities. The symbols have the same meaning as in Fig. 12.2. Note, specifically, that—as long as the period of inflation lasts a couple of e-foldings longer than the minimal value required for inflation to address the problems of Standard Big Bang cosmology—all wavelengths of cosmological interest to us today start out at the beginning of the period of inflation with a wavelength which is in the zone of ignorance

the reader is referred to [21]). The scenario is based on a cosmological background in which the scale factor $a(t)$ bounces in a non-singular manner.

Figure 12.5 shows a space-time sketch of such a bouncing cosmology. Without loss of generality we can adjust the time axis such that the bounce point (minimal value of the scale factor) occurs at $t = 0$. There are three phases in such a non-singular bounce: the initial contracting phase during which the Hubble radius is decreasing linearly in $|t|$, a bounce phase during which a transition from contraction to expansion takes place, and thirdly the usual expanding phase of Standard Cosmology. There is no prolonged inflationary phase after the bounce, nor is there a time-symmetric deflationary contracting period before the bounce point. As is obvious from the Figure, scales which we observe today started out early in the contracting phase at sub-Hubble lengths. The matter bounce scenario assumes that the contracting phase is matter-dominated at the times when scales we observe today exit the Hubble radius. A model in which the contracting phase is the time reverse of our current expanding phase would obey this condition. The assumption of an initial matter-dominated phase will be seen later in Lecture 2 to be important if we want to obtain a scale-invariant spectrum of cosmological perturbations from initial vacuum fluctuations [19, 20].

Fig. 12.5 Space-time sketch in the matter bounce scenario. The *vertical axis* is conformal time η, the *horizontal axis* denotes a co-moving space coordinate. The *vertical line* indicates the wavelength of a fluctuation mode. Also, $\mathscr{H}^{-1}$ denotes the co-moving Hubble radius

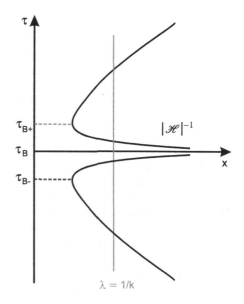

Let us make a first comparison with inflation. A non-deflationary contracting phase replaces the accelerated expanding phase as a mechanism to bring fixed co-moving scales within the Hubble radius as we go back in time, allowing us to consider the possibility of a causal generation mechanism of fluctuations. Starting with vacuum fluctuations, a matter-dominated contracting phase is required in order to obtain a scale-invariant spectrum. This corresponds to the requirement in inflationary cosmology that the accelerated expansion be nearly exponential.

How are the problems of Standard Big Bang cosmology addressed in the matter bounce scenario? First of all, note that since the universe begins cold and large, the size and entropy problems of Standard Cosmology do not arise. As is obvious from Fig. 12.5, there is no horizon problem for the matter bounce scenario as long as the contracting period is long (to be specific, of similar duration as the post-bounce expanding phase until the present time). By the same argument, it is possible to have a causal mechanism for generating the primordial cosmological perturbations which evolve into the structures we observe today. Specifically, as will be discussed in Sect. 12.6, if the fluctuations originate as vacuum perturbations on sub-Hubble scales in the contracting phase, then the resulting spectrum at late times for scales exiting the Hubble radius in the matter-dominated phase of contraction is scale-invariant [19, 20].

The flatness problem is the one which is only partially addressed in the matter bounce setup. The contribution of the spatial curvature decreases in the contracting phase at the same rate as it increases in the expanding phase. Thus, to explain the observed spatial flatness, comparable spatial flatness at early times in the contracting phase is required. This is an improved situation compared to the situation in Standard Big Bang cosmology where spatial flatness is overall an unstable fixed point and hence extreme fine tuning of the initial conditions is required to explain the observed degree of flatness. But the situation is not as good as it is in a model with a long period of inflation where spatial flatness is a local attractor in initial condition space (it is not a global attractor, though!).

How does the matter bounce scenario address the conceptual problem of inflation? First of all, the length scale of fluctuations of interest for current observations on cosmological scales is many orders of magnitude larger than the Planck length throughout the evolution. If the energy scale at the bounce point is comparable to the particle physics GUT scale, then typical wavelengths at the bounce point are not too different from 1 mm. Hence, the fluctuations never get close to the small wavelength zone of ignorance in Figs. 12.3 and 12.4, and thus a description of the evolution of fluctuations using Einstein gravity should be well justified modulo possible difficulties at the bounce point which we will return to in Sect. 12.6. Thus, there is no trans-Planckian problem for fluctuations in the matter bounce scenario.

As will be discussed below, new physics is required in order to provide a non-singular bounce. Thus, the "solution" of the singularity problem is put in by hand and cannot be counted as a success, except in realizations of the matter bounce in the context of a string theory background in which the non-singular evolution follows from general principles. Such a theory has recently been presented in [36, 37] (see [38] for an analysis of fluctuations in these models). Existing matter bounce models do not address the cosmological constant problem. However, I would like to emphasize that the mechanism which drives the evolution in the matter bounce scenario is robust against our ignorance of what solves the cosmological constant problem, an improvement of the situation compared to the situation in inflationary cosmology.

With Einstein gravity and matter satisfying the usual energy conditions it is not possible to obtain a non-singular bounce. Thus, new physics is required in order to obtain a non-singular bouncing cosmology. Such new physics can arise by modifying the gravitational sector, or by modifying the matter sector. The study of bouncing cosmologies has a long history (see [39] for an in-depth review of a lot of these past approaches). We will now turn to a brief overview of some more recent work on non-singular bouncing cosmology with the matter bounce in mind.

12.2.2 Realizing a Matter Bounce with Modified Matter

In order to obtain a cosmological bounce in the context of Einstein gravity, it is necessary to introduce a new form of matter which violates the Null Energy Condition (NEC). A simple way to do this is by introducing quintom matter [40–43]. Resulting nonsingular quintom bouncing models have been discussed in [44, 45]. Quintom matter is a set of two matter fields, one of them regular matter (obeying the NEC), the second a "phantom" field with opposite sign kinetic term which violates the NEC. Even though this model is plagued by ghost instabilities [46], we will use it to illustrate the basic idea of how a bouncing cosmology can be obtained.

We [44, 45] model both matter components with scalar fields, the mass of the regular one (φ) being m, and M being that of the field $\tilde{\varphi}$ with wrong sign kinetic term. We consider a contracting universe and assume that early on both fields are oscillating, but that the amplitude $\mathscr{A}$ of φ greatly exceeds the corresponding amplitude $\tilde{\mathscr{A}}$ of $\tilde{\varphi}$ such that the energy density is dominated by φ. During the initial period of contraction, both amplitudes grow at the same rate. At some point, $\mathscr{A}$ will become so large that the oscillations of φ freeze out.[1] Then, $\mathscr{A}$ will grow only slowly, whereas $\tilde{\mathscr{A}}$ will continue to increase. Thus, the (negative) energy density in $\tilde{\varphi}$ will grow in absolute value relative to that of φ. The total energy density will decrease towards 0. At that point, $H = 0$ by the Friedmann equations. Since it is only the phantom field which has large kinetic energy, it follows that $\dot{H} > 0$ when $H = 0$. Hence, a non-singular bounce occurs.

The Higgs sector of the Lee-Wick model [47] provides a concrete realization of the quintom bounce model, as studied in [48]. Quintom models like all other models with negative sign kinetic terms suffer from an instability problem [46] in the matter sector and are hence problematic. In addition, they are unstable against the addition of radiation (see e.g. [49, 50]) and anisotropic stress (the BKL instability [51]).

An improved way of obtaining a non-singular bouncing cosmology using modified matter is by using a ghost condensate field [52] (see also [53, 54] where ghost condensates have been used to produce non-singular bounces in different contexts). The ghost condensation mechanism is the analog of the Higgs mechanism in the kinetic sector of the theory. In the Higgs mechanism we take a field ϕ whose mass when evaluated at $\phi = 0$ is tachyonic, add higher powers of ϕ^2 to the potential term in the Lagrangian such that there is a stable fixed point $\phi = v \neq 0$, and thus when expanded about $\phi = v$ the mass term has the "safe" non-tachyonic sign. In the ghost condensate construction we take a field ϕ whose kinetic term

$$X \equiv -g^{\mu\nu}\partial_\mu\phi\partial_\nu\phi \qquad (12.3)$$

appears with the wrong sign in the Lagrangian. Then, we add higher powers of X to the kinetic Lagrangian such that there is a stable fixed point $X = c^2$ and such that when expanded about $X = c^2$ the fluctuations have the regular sign of the kinetic term:

$$\mathscr{L} = \frac{1}{8}M^4(X - c^2)^2 - V(\phi), \qquad (12.4)$$

where $V(\phi)$ is a usual potential function, M is a characteristic mass scale and the dimensions of ϕ are chosen such that X is dimensionless.

In the context of cosmology, the ghost condensate is

$$\phi = ct \qquad (12.5)$$

and breaks local Lorentz invariance. Now let us expand the homogeneous component of ϕ about the ghost condensate:

$$\phi(t) = ct + \pi(t). \qquad (12.6)$$

If $\dot{\pi} < 0$ then the gravitational energy density is negative, and a non-singular bounce is possible. Thus, in [52] we constructed a model in which the ghost condensate field

---

[1] This corresponds to the time reverse of entering a region of large-field inflation.

starts at negative values and the potential $V(\phi)$ is negligible. As ϕ approaches $\phi = 0$ it encounters a positive potential which slows it down, leading to $\dot{\pi} < 0$ and hence to negative gravitational energy density. Thus, a non-singular bounce can occur. We take the potential to be of the form

$$V(\phi) \sim \phi^{-\alpha} \tag{12.7}$$

for $|\phi| \gg M$, where M is the mass scale above which the higher derivative kinetic terms are important. For sufficiently large values of α, namely

$$\alpha \geq 6, \tag{12.8}$$

the energy density in the ghost condensate increases faster than that of radiation and anisotropic stress at the universe contracts. Hence, this bouncing cosmology is locally stable against the addition of radiation and anisotropic stress (there is still an instability to the development of anisotropic stress in the contracting phase prior to the time when the ghost condensate starts to dominate).

Non-singular bouncing cosmologies can also be obtained making use of Galileon models [55, 56]. However, these models also suffer from an instability against the development of anisotropic stress.

The Ekpyrotic contracting universe (contracting phase with an equation of state $w \gg 1$ is stable against the growth of anisotropies, as shown in [57]). Thus, one way of obtaining a matter bounce which is stable against the development of anisotropic stress is to have a phase of Ekpyrotic contraction set in shortly after the time t_{eq}^- of equal matter and radiation in the contracting phase. A model in which this is realized and in which the non-singular bounce is generated by a ghost condensate and Galileon construction has recently been worked out in [58].

12.2.3 Realizing a Matter Bounce with Modified Gravity

It is unreasonable to expect that Einstein gravity will provide a good description of the physics at very high energy densities. In particular, all approaches to quantum gravity lead to correction terms in the gravitational action (compared to the pure Einstein term) which become dominant at the Planck scale. It is possible (and in some approaches to quantum gravity such as string theory even likely) that the new terms will tend to prevent cosmological singularities from appearing, and hence might allow a bouncing cosmology even in the presence of matter which obeys the NEC.

One early study is based on a higher derivative Lagrangian resulting from the "nonsingular universe construction" of [59] which is based on a Lagrange multiplier construction which forces all space-time curvature invariants to stay bounded as the energy density increases. This Lagrangian admits bouncing solutions in the presence of regular matter. Another model is the non-local higher derivative action of [60] which is constructed to be ghost-free about Minkowski space-time and which admits bouncing solutions. Mirage cosmology [61] (induced gravity on a brane which is moving into and out of a high-curvature throat of a higher-dimensional bulk space-time also admits bouncing cosmologies [62].

A few years ago there was a lot of interest in Horava-Lifshitz gravity [63], an approach to quantum gravity in four space-time dimensions which is based on a gravitational Lagrangian which is power-counting renormalizable with respect to the reduced symmetry group of spatial diffeomorphisms only (we drop the invariance requirement under space-dependent time reparametrizations), the lost symmetry is replaced by an anisotropic scaling symmetry between space and time. The Lagrangian contains higher space derivative terms. As was realized in [64], in the presence of spatial curvature these higher space derivative terms act as ghost radiation and ghost anisotropic stress and lead to the possibility of a non-singular bouncing cosmology.

Loop quantum cosmology is an approach to quantum cosmology which also leads to bouncing solutions (see e.g. [65, 66] for a review). What is responsible here for singularity avoidance is the fundamental discreteness of the area which comes from quantization. Other lecturers at this school have discussed loop quantum cosmology in depth.

Superstring theory as a quantum theory which includes gravity will likely also resolve cosmological singularities. As will be discussed in detail in the section on string gas cosmology, the new degrees of freedom which string theory admits compared to point particle theories lead to duality symmetries which relate large and small spaces. Physical quantities such as the temperature remain bounded, and it is hence likely to obtain bouncing cosmological solutions. Our understanding of string cosmology is hampered by the lack of a fully non-perturbative formulation of string theory in a cosmological space-time. Most analyses of string cosmology are performed using string-motivated field theory. A specific theory in which the field theory approximations are under good control is the Type II string cosmology of [36, 37].

12.3 Emergent Universe

12.3.1 The Idea

The "emergent universe" scenario [18] is another non-singular cosmological scenario in which time runs from $-\infty$ to $+\infty$. The idea is that if we follow the evolution of our homogeneous and isotropic space-time into the past, the expansion rate H ceases to increase as we approach a certain limiting scale (most likely related to the Planck energy). Instead of further increasing, H decreases to zero, and the scale factor approaches a constant value as we tend to past infinity. The time evolution of the scale factor is sketched in Fig. 12.6.

In Fig. 12.7 we sketch the space-time diagram in emergent cosmology. Since the early emergent phase is quasi-static, the Hubble radius is infinite. For the same reason, the physical wavelength of fluctuations remains constant in this phase. At

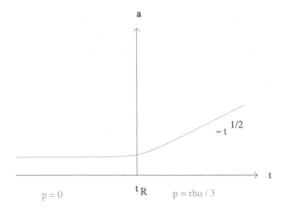

Fig. 12.6 The dynamics of emergent universe cosmology. The *vertical axis* represents the scale factor of the universe, the *horizontal axis* is time

the end of the emergent phase, the Hubble radius decreases to a microscopic value and makes a transition to its evolution in Standard Cosmology.

Once again, we see that fluctuations originate on sub-Hubble scales. In emergent cosmology, it is the existence of a quasi-static phase which leads to this result. What sources fluctuations depends on the realization of the emergent scenario. String gas cosmology is the example which I will consider later on. In this case, the source of perturbations is thermal: string thermodynamical fluctuations in a compact space with stable winding modes, and this in fact leads to a scale-invariant spectrum [14].

How does emergent cosmology address the problems of Standard Cosmology? As in the case of a bouncing cosmology, the horizon is infinite and hence there is no horizon problem. Since there is likely thermal equilibrium in the emergent phase, a mechanism to homogenize the universe exists, and hence spatial flatness is not a mystery. As discussed in the previous paragraph, there is no causality obstacle against producing cosmological fluctuations. The scenario is non-singular, but this cannot in general be weighted as a success unless the emergent phase can be shown to arise from some well controlled ultraviolet physics.

Like in the case of a bouncing cosmology, there is no trans-Planckian problem for fluctuations—their wavelength never gets close to the Planck scale. And like in the case of a bouncing cosmology, the physics driving the background dynamics is robust against our ignorance of what solves the cosmological constant problem. These are two advantages of the emergent scenario compared to inflation.

On the negative side, the origin of the large size and entropy of our universe remains a mystery in emergent cosmology. Also, the physics yielding the emergent phase is not well understood in terms of an effective field theory setting, in contrast to the physics yielding inflation.

Whereas there are a lot of toy models for a bouncing cosmology, there are not many models that realize an emergent universe. The "String Gas Cosmology" model discussed below is a concrete proposal. Another recent proposal is in the context of Galileon cosmology [67] (see [68] for a discussion of the termination of the emergent phase in the context of the model of [67]). There is also a relationship with the work of [24, 25]. The small number of concrete models, however, does not mean

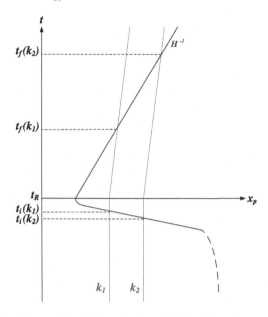

Fig. 12.7 Space-time diagram (sketch) showing the evolution of fixed co-moving scales in emergent cosmology. The *vertical axis* is time, the *horizontal axis* is physical distance. The *solid curve* represents the Einstein frame Hubble radius H^{-1} which shrinks abruptly to a micro-physical scale at t_R and then increases linearly in time for $t > t_R$. Fixed co-moving scales (the *dotted lines* labeled by k_1 and k_2) which are currently probed in cosmological observations have wavelengths which were smaller than the Hubble radius long before t_R. They exit the Hubble radius at times $t_i(k)$ just prior to t_R, and propagate with a wavelength larger than the Hubble radius until they re-enter the Hubble radius at times $t_f(k)$

that this approach is not promising. I suspect that any non-perturbative approach to quantum gravity which leads to an emergence of space after some phase transition will lead to a convincing realization of emergent cosmology.

12.3.2 String Gas Cosmology

String gas cosmology [13] (see also [69], and see [15–17] for a review) is a realization of the emergent cosmology paradigm which results from coupling a gas of fundamental strings to a background space-time metric. It is assumed that the spatial sections are compact. For simplicity, we take all spatial directions to be toroidal and denote the radius of the torus by R.

The guiding principle of string gas cosmology is to focus on symmetries and degrees of freedom which are new to string theory (compared to point particle theories) and which will be part of any non-perturbative string theory, and to use them to develop a new cosmology. The symmetry we make use of is *T-duality*, and the new degrees of freedom are the *string oscillatory modes* (corresponding to fluctuations in the shape of a string) and the *string winding modes* (strings winding the background space). Strings also have *momentum modes* which correspond to the center of mass motion of the strings. Point particles only have momentum modes.

The first key feature of string theory is that there is a limiting temperature for a gas of strings in thermal equilibrium, the *Hagedorn temperature* [70] T_H. This stems from the fact that the number of string oscillatory states increases exponentially with energy. Thus, if we take a box of strings and adiabatically decrease the box size, the temperature will never diverge. This is the first indication that string theory has the potential to resolve the cosmological singularity problem.

The second key feature of string theory upon which string gas cosmology is based is *T-duality*. To introduce this symmetry, let us discuss the radius dependence of the energy of the basic string states: The energy of an oscillatory mode is independent of R, momentum mode energies are quantized in units of $1/R$, i.e.

$$E_n = n\mu \frac{l_s^2}{R}, \qquad (12.9)$$

where l_s is the string length and μ is the mass per unit length of a string. The winding mode energies are quantized in units of R, i.e.

$$E_m = m\mu R, \qquad (12.10)$$

where both n and m are integers. Thus, a new symmetry of the spectrum of string states emerges: Under the change

$$R \to 1/R \qquad (12.11)$$

in the radius of the torus (in units of l_s) the energy spectrum of string states is invariant if winding and momentum quantum numbers are interchanged

$$(n, m) \to (m, n). \qquad (12.12)$$

The above symmetry is the simplest element of a larger symmetry group, the T-duality symmetry group which in general also mixes fluxes and geometry. The string vertex operators are consistent with this symmetry, and thus T-duality is a symmetry of perturbative string theory. Postulating that T-duality extends to non-perturbative string theory leads [71] to the need of adding D-branes to the list of fundamental objects in string theory. With this addition, T-duality is expected to be a symmetry of non-perturbative string theory. Specifically, T-duality will take a spectrum of stable Type IIA branes and map it into a corresponding spectrum of stable Type IIB branes with identical masses [72].

As discussed in [13], the above T-duality symmetry leads to an equivalence between small and large spaces, an equivalence elaborated on further in [73, 74].

Let us now turn to the background cosmology which emerges from string gas cosmology. First consider the adiabatic evolution of a box of strings as the box radius R decreases. If the initial radius is much larger than the string length, then in thermal equilibrium most of the energy is initially in the momentum modes since they are the lightest ones. As R decreases, the temperature first rises as in standard cosmology since the string states which are occupied (the momentum modes) get

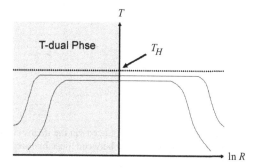

Fig. 12.8 The temperature (*vertical axis*) as a function of radius (*horizontal axis*) of a gas of closed strings in thermal equilibrium. Note the absence of a temperature singularity. The range of values of R for which the temperature is close to the Hagedorn temperature T_H depends on the total entropy of the universe. The upper of the two curves corresponds to a universe with larger entropy

heavier. However, as the temperature approaches the Hagedorn temperature, the energy begins to flow into the oscillatory modes and the temperature levels off. As the radius R decreases below the string scale, the temperature begins to decrease as the energy begins to flow into the winding modes whose energy decreases as R decreases (see Fig. 12.8). Thus, as argued in [13], the temperature singularity of early universe cosmology is resolved in string gas cosmology.

The equations that govern the background of string gas cosmology are not known. The Einstein equations are not the correct equations since they do not obey the T-duality symmetry of string theory. Many early studies of string gas cosmology were based on using the dilaton gravity equations [75–77]. However, these equations are not satisfactory, either. Firstly, we expect that string theory correction terms to the low energy effective action of string theory become dominant in the Hagedorn phase. Secondly, the dilaton gravity equations yield a rapidly changing dilaton during the Hagedorn phase (in the string frame). Once the dilaton becomes large, it becomes inconsistent to focus on fundamental string states rather than brane states. In other words, using dilaton gravity as a background for string gas cosmology does not correctly reflect the S-duality symmetry of string theory. A background for string gas cosmology including a rolling tachyon was proposed [78] which allows a background in the Hagedorn phase with constant scale factor and constant dilaton; but this construction is rather ad hoc. Another study of this problem was given in [79].

Some conclusions about the time-temperature relation in string gas cosmology can be derived based on thermodynamical considerations alone. One possibility is that R starts out much smaller than the self-dual value and increases monotonically. From Fig. 12.8 it then follows that the time-temperature curve will correspond to that of a bouncing cosmology. A specific realization of this possibility in the context of a string theory background in which the effective background equations of motion are well justified is given in [36, 37].

Alternatively, it is possible that the universe starts out in a meta-stable state near the Hagedorn temperature, the *Hagedorn phase*, and then smoothly evolves into an

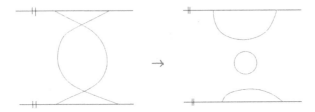

Fig. 12.9 The process by which string loops are produced via the intersection of winding strings. The top and bottom lines are identified and the space between these lines represents space with one toroidal dimension un-wrapped

expanding phase dominated by radiation like in Standard Cosmology. Note that we are assuming that not only the scale factor but also the dilaton is constant in time. This is the setup which is assumed in the string gas realization of the emergent universe scenario.

Note that it is the annihilation (see Fig. 12.9) of winding strings into string loops (which acts as stringy radiation) which leads to the transition from the early quasi-static phase to the radiation phase of Standard Cosmology.

The evolution of the scale factor in string gas cosmology is as in any general emergent universe scenario (see Fig. 12.6). In this figure, along the horizontal axis, the approximate equation of state for the string gas cosmology realization of the emergent scenario is also indicated. During the Hagedorn phase the pressure is negligible due to the cancellation between the positive pressure of the momentum modes and the negative pressure of the winding modes, after time t_R the equation of state is that of a radiation-dominated universe.

As pointed out in [13], the annihilation process which allows for the expansion of spatial radii is only possible in at most three large spatial dimensions. This is a simple dimension counting argument: string world sheets have measure zero intersection probability in more than four large space-time dimensions. Hence, string gas cosmology may provide a natural mechanism for explaining why there are exactly three large spatial dimensions. This argument was supported by numerical studies of string evolution in three and four spatial dimensions [80] (see also [81]). The flow of energy from winding modes to string loops can be modelled by effective Boltzmann equations [82] analogous to those used to describe the flow of energy between infinite cosmic strings and cosmic string loops (see e.g. [83–85] for reviews).

There is a caveat regarding the above mechanism. In the analysis of [82] it was assumed that the string interaction rates were time-independent. If the dynamics of the Hagedorn phase is modelled by dilaton gravity, the dilaton is rapidly changing during the phase in which the string frame scale factor is static. As discussed in [86, 87] (see also [88]), in this case the mechanism which tells us that exactly three spatial dimensions become macroscopic does not work.

An important question which has to be addressed in any model of string cosmology is what stabilizes the moduli, in particular the sizes and shapes of the extra spatial dimensions. In this respect string gas cosmology in the context of heterotic string theory has some major advantages over other approaches to string cosmology, at least in the context of toroidal compactifications, the ones which have been studied to date. This issue is reviewed in detail in [89]. The basic idea [90] is that winding modes about the extra spatial dimensions create an energy barrier against expansion, whereas momentum modes cause an energy barrier against contraction. There is hence an energetically favored value for the radius R of an extra spatial dimension (which is typically the string length). This is the self-dual radius. This mechanism is a special case of the general principle of moduli trapping at enhanced symmetry states [91, 92].

In order to avoid a cosmological constant problem, it is important that the induced potential energy of the four-dimensional effective field theory vanishes at the self-dual radius. This issue has been studied in detail [93, 94] in the case of heterotic superstring theory, and it was shown that the special massless enhanced symmetry states which appear at the self-dual radius and dominate the potential at that point have vanishing potential energy. Thus, in heterotic string gas cosmology the radion moduli are dynamically stabilized. By studying the off-diagonal Einstein equations in the presence of a string gas with both momentum and winding modes it can also be shown [95] that the shape moduli are stabilized at points of extra symmetry.

The only modulus which is not stabilized by string winding and momentum modes is the dilaton. One can [96] introduce gaugino condensation, the same mechanism used in string inflation model building (see e.g. [97] for a recent review) and show that this generates a stabilizing potential for the dilaton without interfering with the radion stabilization force provided by the string winding and momentum modes. Gaugino condensation also leads [98] to supersymmetry breaking (typically at a high energy scale).

A final comment concerns the isotropy of the three large dimensions. In contrast to the situation in Standard cosmology, in string gas cosmology the anisotropy decreases in the expanding phase [99]. Thus, there is a natural isotropization mechanism for the three large spatial dimensions.

12.4 Cosmological Perturbations

12.4.1 Overview

The topic of the second lecture is the theory of cosmologonal perturbations and its applications to both inflationary cosmology and the "un-conventional" cosmological alternatives discussed in the first lecture.

The theory of cosmological perturbations is the main tool of modern cosmology. It allows us to follow the evolution of small inhomogeneities generated in the very early universe and propagate their evolution to the present time, which then allows us to work out predictions of models of the early universe. For an extensive overview of the subject the reader is referred to [100], and to [101] for an overview.

As we have seen in the first lecture, in many models of the very early Universe, in particular in inflationary cosmology, in the emergent universe paradigm and in the matter bounce scenario, primordial inhomogeneities are generated in an initial phase on sub-Hubble scales. The wavelength is then stretched relative to the Hubble radius $H^{-1}(t)$, where H is the cosmological expansion rate, becomes larger than the Hubble radius at some time and then propagates on super-Hubble scales until re-entering at late cosmological times. In a majority of the current structure formation scenarios (string gas cosmology is an exception in this respect), fluctuations are assumed to emerge as quantum vacuum perturbations. Hence, to describe the generation and evolution of the inhomogeneities, both General Relativity and quantum mechanics are required. What makes the theory of cosmological perturbations tractable is that the amplitude of the fractional fluctuations is small today and hence (since gravity is a purely attractive force) that it was even smaller in the early universe. This justifies the linear analysis of the generation and evolution of fluctuations.

In the context of a Universe with an inflationary period, the quantum origin of cosmological fluctuations was first discussed in [5] and also [4, 6] for earlier ideas. In particular, Mukhanov [5] and Press [6] realized that in an exponentially expanding background, the curvature fluctuations would be scale-invariant, and Mukhanov provided a quantitative calculation which also yielded the logarithmic deviation from exact scale-invariance.

Here we give a very abbreviated overview of the quantum theory of cosmological perturbations. The reader is referred to [101] for a description which is closer to what was presented at the Naxos school.

The basic idea of the theory of cosmological perturbations is simple. In order to obtain the action for linearized cosmological perturbations, we expand the action for gravity and matter to quadratic order in the fluctuating degrees of freedom. The linear terms cancel because the background is taken to satisfy the background equations of motion.

At first sight, it appears that there are ten degrees of freedom for the metric fluctuations, in addition to the matter perturbations. However, four of these degrees of freedom are equivalent to space-time diffeomorphisms. To study the remaining six degrees of freedom for metric fluctuations it proves very useful to classify them according to how they transform under spatial rotations. There are two scalar modes, two vector modes and two tensor modes. At linear order in cosmological perturbation theory, scalar, vector and tensor modes decouple. For simple forms of matter such as scalar fields or perfect fluids, the matter fluctuations couple only to the scalar metric modes. These are the so-called "cosmological perturbations" which we study below.

If matter has no anisotropic stress, then one of the scalar metric degrees of freedom disappears. In addition, one of the Einstein constraint equations couples the remaining metric degree of freedom to matter. Thus, if there is only one matter component (e.g. one scalar matter field), there is only one independent scalar cosmological fluctuation mode.

To obtain the action and equation of motion for this mode, we begin with the Einstein-Hilbert action for gravity and the action for matter (which we take for simplicity to be a scalar field φ—for the more complicated case of general hydrodynamical fluctuations the reader is referred to [100])

$$S = \int d^4x \sqrt{-g} \left[-\frac{1}{16\pi G} R + \frac{1}{2} \partial_\mu \varphi \partial^\mu \varphi - V(\varphi) \right], \tag{12.13}$$

where R is the Ricci curvature scalar.

The simplest way to proceed is to work in longitudinal gauge, in which the metric and matter take the form (assuming no anisotropic stress)

$$ds^2 = a^2(\eta) \left[(1 + 2\phi(\eta, \mathbf{x})) d\eta^2 - (1 - 2\phi(t, \mathbf{x})) d\mathbf{x}^2 \right]$$

$$\varphi(\eta, \mathbf{x}) = \varphi_0(\eta) + \delta\varphi(\eta, \mathbf{x}), \tag{12.14}$$

where η in conformal time. The two fluctuation variables ϕ and $\delta\varphi$ must be linked by the Einstein constraint equations since there cannot be matter fluctuations without induced metric fluctuations.

The two nontrivial tasks of the lengthy [100] computation of the quadratic piece of the action is to find out what combination of $\delta\varphi$ and ϕ gives the variable v in terms of which the action has canonical kinetic term, and what the form of the time-dependent mass is. This calculation involves inserting the ansatz (12.14) into the action (12.13), expanding the result to second order in the fluctuating fields, making use of the background and of the constraint equations, and dropping total derivative terms from the action. In the context of scalar field matter, the quantum theory of cosmological fluctuations was developed by Mukhanov [102, 103] and Sasaki [104]. The result is the following contribution $S^{(2)}$ to the action quadratic in the perturbations:

$$S^{(2)} = \frac{1}{2} \int d^4x \left[v'^2 - v_{,i} v_{,i} + \frac{z''}{z} v^2 \right], \tag{12.15}$$

where the canonical variable v (the "Sasaki-Mukhanov variable" introduced in [103]—see also [105]) is given by

$$v = a \left[\delta\varphi + \frac{\varphi_0'}{\mathcal{H}} \phi \right], \tag{12.16}$$

with $\mathcal{H} = a'/a$, and where

$$z = \frac{a\varphi_0'}{\mathcal{H}}. \tag{12.17}$$

As long as the equation of state does not change over time

$$z(\eta) \sim a(\eta). \tag{12.18}$$

Note that the variable v is related to the curvature perturbation $\mathcal{R}$ in comoving coordinates introduced in [108] and closely related to the variable ζ used in [106, 107]:

$$v = z\mathcal{R}. \tag{12.19}$$

The equation of motion which follows from the action (12.15) is (in momentum space)

$$v_k'' + k^2 v_k - \frac{z''}{z} v_k = 0, \tag{12.20}$$

where v_k is the k'th Fourier mode of v. The mass term in the above equation is in general given by the Hubble scale (the scale whose wave-number will be denoted k_H). Thus, it immediately follows that on small length scales, i.e. for $k > k_H$, the solutions for v_k are constant amplitude oscillations. These oscillations freeze out at Hubble radius crossing, i.e. when $k = k_H$. On longer scales ($k \ll k_H$), there is a mode of v_k which scales as z. This mode is the dominant one in an expanding universe, but not in a contracting one.

Given the action (12.15), the cosmological perturbations can be quantized by canonical quantization (in the same way that a scalar matter field on a fixed cosmological background is quantized [109]).

The final step in the quantum theory of cosmological perturbations is to specify an initial state. Since in inflationary cosmology all pre-existing classical fluctuations are red-shifted by the accelerated expansion of space, one usually assumes that the field v starts out at the initial time t_i mode by mode in its vacuum state. This prescription, however, can be criticized in light of the trans-Planckian problem for cosmological fluctuations. It assumes that ultraviolet modes which are continuously crossing the Planck scale cutoff $k = m_{pl}$ are in their vacuum state, which is a strong constraint on physics on trans-Planckian scales.

There are two other questions which immediately emerge: what is the initial time t_i, and which of the many possible vacuum states should be chosen. It is usually assumed that since the fluctuations only oscillate on sub-Hubble scales, the choice of the initial time is not important, as long as it is earlier than the time when scales of cosmological interest today cross the Hubble radius during the inflationary phase. The state is usually taken to be the Bunch-Davies vacuum (see e.g. [109]), since this state is empty of particles at t_i in the coordinate frame determined by the FRW coordinates Thus, we choose the initial conditions

$$v_k(\eta_i) = \frac{1}{\sqrt{2k}}$$
$$v'_k(\eta_i) = \frac{\sqrt{k}}{\sqrt{2}}$$

(12.21)

where η_i is the conformal time corresponding to the physical time t_i.

Returning to the case of an expanding universe, the scaling

$$v_k \sim z \sim a$$

(12.22)

implies that the wave function of the quantum variable v_k which performs quantum vacuum fluctuations on sub-Hubble scales, stops oscillating on super-Hubble scales and instead is squeezed (the amplitude increases in configuration space but decreases in momentum space). This squeezing corresponds to quantum particle production. It is also one of the two conditions which are required for the classicalization of the fluctuations. The second condition is decoherence which is induced by the non-linearities in the dynamical system which are inevitable since the Einstein action leads to highly nonlinear equations (see [110] for an in-depth discussion of this point, and [111] for related work).

Note that the squeezing of cosmological fluctuations on super-Hubble scales occurs in all models, in particular in string gas cosmology and in the bouncing universe scenario since also in these scenarios perturbations propagate on super-Hubble scales for a long period of time. In a contracting phase, the dominant mode of v_k on super-Hubble scales is not the one given in (12.22) (which in this case is a decaying mode), but rather the second mode which scales as z^{-p} with an exponent p which is positive and whose exact value depends on the background equation of state.

Applications of this theory in inflationary cosmology, in the matter bounce scenario and in string gas cosmology will be considered in the following sections.

12.5 Fluctuations in Inflationary Cosmology

We will now use the quantum theory of cosmological perturbations developed in the previous section to calculate the spectrum of curvature fluctuations in inflationary cosmology. The starting point are quantum vacuum initial conditions for the canonical fluctuation variable v_k:

$$v_k(\eta_i) = \frac{1}{\sqrt{2k}} \tag{12.23}$$

for all k for which the wavelength is smaller than the Hubble radius at the initial time t_i.

The amplitude remains unchanged until the modes exit the Hubble radius at the respective times $t_H(k)$ given by

$$a^{-1}\big(t_H(k)\big)k = H. \tag{12.24}$$

We need to compute the power spectrum $\mathscr{P}_{\mathscr{R}}(k)$ of the curvature fluctuation $\mathscr{R}$ defined in (12.19) at some late time t when the modes are super-Hubble. We first relate the power spectrum via the growth rate (12.22) of v on super-Hubble scales to the power spectrum at the time $t_H(k)$ and then use the constancy of the amplitude of v on sub-Hubble scales to relate it to the initial conditions (12.23). Thus

$$
\begin{aligned}
\mathscr{P}_{\mathscr{R}}(k,t) \equiv k^3 \mathscr{R}_k^2(t) &= k^3 z^{-2}(t) \big|v_k(t)\big|^2 \\
&= k^3 z^{-2}(t) \left(\frac{a(t)}{a(t_H(k))}\right)^2 \big|v_k\big(t_H(k)\big)\big|^2 \\
&= k^3 z^{-2}\big(t_H(k)\big)\big|v_k\big(t_H(k)\big)\big|^2 \\
&\sim k^3 \left(\frac{a(t)}{z(t)}\right)^2 a^{-2}\big(t_H(k)\big)\big|v_k(t_i)\big|^2,
\end{aligned}
\tag{12.25}
$$

where in the final step we have used (12.18) and the constancy of the amplitude of v on sub-Hubble scales. Making use of the condition (12.24) for Hubble radius crossing, and of the initial conditions (12.23), we immediately see that

$$\mathscr{P}_{\mathscr{R}}(k,t) \sim \left(\frac{a(t)}{z(t)}\right)^2 k^3 k^{-2} k^{-1} H^2, \tag{12.26}$$

and that thus a scale invariant power spectrum with amplitude proportional to H^2 results, in agreement with what was argued on heuristic grounds in the overview of inflation in the first section. To obtain the precise amplitude, we need to make use of the relation between z and a. We obtain

$$\mathcal{P}_{\mathcal{R}}(k, t) \sim \frac{H^4}{\dot{\varphi}_0^2} \qquad (12.27)$$

which for any given value of k is to be evaluated at the time $t_H(k)$ (before the end of inflation). For a scalar field potential (see following subsection)

$$V(\varphi) = \lambda \varphi^4 \qquad (12.28)$$

the resulting amplitude in (12.27) is λ. Thus, in order to obtain the observed value of the power spectrum of the order of 10^{-10}, the coupling constant λ must be tuned to a very small value.

12.6 Matter Bounce and Structure Formation

As we already discussed in Sect. 12.2 of these notes, in a non-singular bouncing cosmology fluctuations on scales relevant to current cosmological observations have a physical wavelength which at early times during the contracting phase was smaller than the Hubble radius. Hence, a causal generation mechanism for fluctuations is possible. In fact, in [19, 20] it was realized that fluctuations which originate on sub-Hubble scales in their quantum vacuum state and exit the Hubble radius during a matter-dominated contracting phase acquire a scale-invariant spectrum. As we review below, this is due to the particular growth rate of the dominant fluctuation mode in the contracting phase which is exactly right to convert a vacuum spectrum into a scale-invariant one. During any non-matter phase of contraction which might follow the initial matter-dominated phase the slope of the spectrum remains unchanged on super-Hubble scales since all corresponding mode functions grow by the same factor. Thus, the spectrum of fluctuations right before the bounce is scale-invariant. Provided that the spectrum does not change its slope during the bounce phase, a model falling into the matter bounce category will provide an alternative to inflation for generating s scale-invariant spectrum of curvature perturbations.

The propagation of infrared (IR) fluctuations through the non-singular bounce was analyzed in the case of the higher derivative gravity model of [60] in [112], in mirage cosmology in [62], in the case of the quintom bounce in [44, 45, 48], for a ghost condensate bounce in [52], for a Horava-Lifshitz bounce in [113, 114], and more recently [38] in the string theory bounce model of [36, 37]. The result of these studies is that the scale-invariance of the spectrum before the bounce persists after the bounce as long as the time period of the bounce phase is short compared to the wavelengths of the modes being considered. Note that if the fluctuations have a thermal origin, then the condition on the background cosmology to yield scale-invariance of the spectrum of fluctuations is different [115].

12.6.1 Basics

First we will consider fluctuations in a matter bounce without extra degrees of freedom. In this case, we need only focus on the usual fluctuation variable v. The equation of motion its Fourier mode v_k is

$$v_k'' + \left(k^2 - \frac{z''}{z}\right)v_k = 0. \tag{12.29}$$

If the equation of state of the background is time-independent, then $z \sim a$ and hence the negative square mass term in (12.29) is H^2. Thus, on length scales smaller than the Hubble radius, the solutions of (12.29) are oscillating, whereas on larger scales they are frozen in, and their amplitude depends on the time evolution of z.

In the case of an expanding universe the dominant mode of v scales as z. However, in a contracting universe it is the second of the two modes which dominates. If the contracting phase is matter-dominated, i.e. $a(t) \sim t^{2/3}$ and $\eta(t) \sim t^{1/3}$ the dominant mode of v scales as η^{-1} and hence

$$v_k(\eta) = c_1 \eta^2 + c_2 \eta^{-1}, \tag{12.30}$$

where c_1 and c_2 are constants. The c_1 mode is the mode for which ζ is constant on super-Hubble scales. However, in a contracting universe it is the c_2 mode which dominates and leads to a scale-invariant spectrum [19, 20]:

$$P_\zeta(k, \eta) \sim k^3 |v_k(\eta)|^2 a^{-2}(\eta)$$

$$\sim k^3 |v_k(\eta_H(k))|^2 \left(\frac{\eta_H(k)}{\eta}\right)^2 \sim k^{3-1-2}$$

$$\sim \text{const}, \tag{12.31}$$

where we have made use of the scaling of the dominant mode of v_k, the Hubble radius crossing condition $\eta_H(k) \sim k^{-1}$, and the assumption that we have a vacuum spectrum at Hubble radius crossing.

At this point we have shown that the spectrum of fluctuations is scale-invariant on super-Hubble scales before the bounce phase. The evolution during the bounce depends in principle on the specific realization of the non-singular bounce. In any concrete model, the equations of motion can be solved numerically without approximation during the bounce. Alternatively, we can solve them approximately using analytical techniques. Key to the analytical analysis are the General Relativistic matching conditions for fluctuations across a phase transition in the background [116, 117]. These conditions imply that both Φ and $\tilde{\zeta}$ are conserved at the bounce, where

$$\tilde{\zeta} = \zeta + \frac{1}{3}\frac{k^2\Phi}{\mathcal{H}^2 - \mathcal{H}'}. \tag{12.32}$$

However, as stressed in [118], these matching conditions can only be used at a transition when the background metric obeys the matching conditions. This is not the case if we were to match directly between the contracting matter phase and the expanding matter phase, as was done in early studies [119–122] of fluctuations in the Ekpyrotic scenario.

In the case of a non-singular bounce we have three phases: the initial contracting phase with a fixed equation of state (e.g. $w = 0$), a bounce phase during which the universe smoothly transits between contraction and expansion, and finally the expanding phase with constant w. We need to apply the matching conditions twice: first at the transition between the contracting phase and the bounce phase (on both sides of the matching surface the universe is contracting), and then between the bouncing phase and the expanding phase. The bottom line of the studies of [38, 44, 45, 48, 52, 62, 112–114] is that on length scales large compared to the time of the bounce, the spectrum of curvature fluctuations is not changed during the bounce phase. Since typically the bounce time is set by a microphysical scale whereas the wavelength of fluctuations which we observe today is macroscopic (about 1 mm if the bounce scale is set by the particle physics GUT scale), we conclude that for scales relevant to current observations the spectrum is unchanged during the bounce. This completes the demonstration that a non-singular matter bounce leads to a scale-invariant spectrum of cosmological perturbations after the bounce provided that the initial spectrum on sub-Hubble scales is vacuum.

The fact that fluctuations grow both in the contracting and expanding phase has implications for cyclic cosmologies in four space-time dimensions: In the presence of fluctuations, no such cyclic models are possible—the growth of fluctuations breaks this cyclicity. As we showed above, the spectral index of the power spectrum of the fluctuations changes during the bounce. Hence, four space-time-dimensional cyclic background cosmologies are not predictive—the index of the power spectrum changes from cycle to cycle [123]. Note that the cyclic version of the Ekpyrotic scenario [124] avoids these problems because it is not cyclic in the above sense: it is a higher space-time-dimensional model in which the radius of an extra dimensions evolves cyclically, but the four-dimensional scale factor does not.

The above analysis is applicable only as long as no new degrees of freedom become relevant at high energy densities, in particular during the bounce phase. In non-singular bounce models obtained by modifying the matter sector, new degrees of freedom arise from the extra matter fields. They can thus give entropy fluctuations which may compete with the adiabatic mode studied above. In the quintom bounce model this issue has recently been studied in [125]. It was found that fluctuations in the ghost field which yields the bounce are unimportant on large scales since they have a blue spectrum. However, entropy fluctuations due to extra low-mass fields can be important. Their spectrum is also scale-invariant, and this yields the "matter bounce curvaton" mechanism.

In non-singular bouncing models obtained by modifying the gravitational sector of the theory the identification of potential extra degrees of freedom is more difficult. As an example, let us mention the situation in the case of the Horava-Lifshitz bounce. The theory has the same number of geometric degrees of freedom as General Relativity, but less symmetries. Thus, more of the degrees of freedom are physical. Recall from the discussion of the theory of cosmological perturbations in Sect. 12.2 that there are ten total geometrical degrees of freedom for linear cosmological perturbations, four of them being scalar, four vector and two tensor. In Einstein gravity the symmetry group of space-time diffeomorphisms is generated at the level of linear fluctuations by four functions, leaving six of the ten geometrical variables as physical—two scalar, two vector and two tensor modes. In the absence of anisotropic stress the number of scalar variables is reduced by one, and the Hamiltonian constraint relates the remaining scalar metric fluctuation to matter.

In Horava-Lifshitz gravity one loses one scalar gauge degree of freedom, namely that of space-dependent time reparametrizations. Thus, one expects an extra physical degree of freedom. It has been recently been shown [126] that in the projectable version of the theory (in which the lapse function $N(t)$ is constrainted to be a function of time only) the extra degree of scalar cosmological perturbations is either ghost-like or tachyonic, depending on parameters in the Lagrangian. Thus, the theory appears to be ill-behaved in the context of cosmology. However, in the full nonprojectable version (in which the lapse $N(t, \mathbf{x})$ is a function of both space and time, the extra degree of freedom is well behaved. It is important on ultraviolet scales but decouples in the infrared [127].

12.6.2 Specific Predictions

Canonical single field inflation models predict very small non-Gaussianities in the spectrum of fluctuations. One way to characterize the non-Gaussianities is via the three point function of the curvature fluctuation, also called the "bispectrum". As realized in [128], the bispectrum induced in the minimal matter bounce scenario (no entropy modes considered) has an amplitude which is at the borderline of what the Planck satellite experiment will be able to detect, and it has a special form. These are specific predictions of the matter bounce scenario with which the matter bounce scenario can be distinguished from those of standard inflationary models (see [129] for a recent detailed review of non-Gaussianities in inflationary cosmology and a list of references). In the following we give a very brief summary of the analysis of non-Gaussianities in the matter bounce scenario.

Non-Gaussianities are induced in any cosmological model simply because the Einstein equations are non-linear. In momentum space, the bispectrum contains amplitude and shape information. The bispectrum is a function of the three momenta. Momentum conservation implies that the three momenta have to add up to zero. However, this still leaves a rich shape information in the bispectrum in addition to the information about the overall amplitude.

A formalism to compute the non-Gaussianities for the curvature fluctuation variable ζ was developed in [130]. Working in the interaction representation, the three-point function of ζ is given to leading order by

$$\langle \zeta(t, \mathbf{k}_1)\zeta(t, \mathbf{k}_2)\zeta(t, \mathbf{k}_3)\rangle$$
$$= i \int_{t_i}^{t} dt' \langle [\zeta(t, \mathbf{k}_1)\zeta(t, \mathbf{k}_2)\zeta(t, \mathbf{k}_3), L_{int}(t')]\rangle, \qquad (12.33)$$

where t_i corresponds to the initial time before which there are any non-Gaussianities. The square parentheses indicate the commutator, and L_{int} is the interaction Lagrangian

The interaction Lagrangian contains many terms. In particular, there are terms containing the time derivative of ζ. Each term leads to a particular shape of the bispectrum. In an expanding universe such as in inflationary cosmology $\dot{\zeta} = 0$. However, in a contracting phase the time derivative of ζ does not vanish since the dominant mode is growing in time. Hence, there are new contributions to the shape which have a very different form from the shape of the terms which appear in inflationary cosmology. The larger value of the amplitude of the bispectrum follows again from the fact that there is a mode function which grows in time in the contracting phase.

The three-point function can be expressed in the following general form:

$$\langle \zeta(\mathbf{k}_1)\zeta(\mathbf{k}_2)\zeta(\mathbf{k}_3)\rangle = (2\pi)^7 \delta\left(\sum \mathbf{k}_i\right) \frac{P_\zeta^2}{\prod k_i^3}$$
$$\times \mathscr{A}(\mathbf{k}_1, \mathbf{k}_2, \mathbf{k}_3), \tag{12.34}$$

where $k_i = |\mathbf{k}_i|$ and $\mathscr{A}$ is the shape function. In this expression we have factored out the dependence on the power spectrum P_ζ. In inflationary cosmology it has become usual to express the bispectrum in terms of a non-Gaussianity parameter f_{NL}. However, this is only useful if the shape of the three point function is known. As a generalization, we here use [128]

$$|\mathscr{B}|_{NL}(\mathbf{k}_1, \mathbf{k}_2, \mathbf{k}_3) = \frac{10}{3} \frac{\mathscr{A}(\mathbf{k}_1, \mathbf{k}_2, \mathbf{k}_3)}{\sum_i k_i^3}. \tag{12.35}$$

The computation of the bispectrum is tedious. In the case of the matter bounce (no entropy fluctuations) the result is

$$\mathscr{A} = \frac{3}{256 \prod k_i^2}\left\{3\sum k_i^9 + \sum_{i\neq j} k_i^7 k_j^2 \right.$$
$$- 9\sum_{i\neq j} k_i^6 k_j^3 + 5\sum_{i\neq j} k_i^5 k_j^4$$
$$\left. - 66\sum_{i\neq j\neq k} k_i^5 k_j^2 k_k^2 + 9\sum_{i\neq j\neq k} k_i^4 k_j^3 k_k^2\right\}. \tag{12.36}$$

This equation describes the shape which is predicted. Some of the terms (such as the last two) are the same as those which occur in single field slow-roll inflation, but the others are new. Note, in particular, that the new terms are not negligible.

If we project the resulting shape function $\mathscr{A}$ onto some popular shape masks we

$$|\mathscr{B}|_{NL}^{local} = -\frac{35}{8}, \tag{12.37}$$

for the local shape ($k_1 \ll k_2 = k_3$). This is negative and of order $O(1)$. For the equilateral form ($k_1 = k_2 = k_3$) the result is

$$|\mathscr{B}|_{NL}^{\text{equil}} = -\frac{255}{64}, \tag{12.38}$$

and for the folded form ($k_1 = 2k_2 = 2k_3$) one obtains the value

$$|\mathscr{B}|_{NL}^{\text{folded}} = -\frac{9}{4}. \tag{12.39}$$

These amplitudes are close to what the Planck CMB satellite experiment will be able to detect.

The matter bounce scenario also predicts a change in the slope of the primordial power spectrum on small scales [131]: scales which exit the Hubble radius in the radiation phase retain a blue spectrum since the squeezing rate on scales larger than the Hubble radius is insufficient to give longer wavelength modes a sufficient boost relative to the shorter wavelength ones.

12.7 String Gas Cosmology and Structure Formation

In this section we discuss cosmological fluctuations in one particular realization of the emergent universe scenario, namely string gas cosmology. In contrast to the case of exponential inflation and the matter bounce, where a scale-invariant spectrum emerges from initial quantum vacuum fluctuations independent of the specific realization of the background cosmology, in the case of the emergent universe scenario a scale-invariant spectrum is generated only in the string gas cosmology realization, and in other realizations which share some general properties which will be mentioned at the end of this section.

12.7.1 Overview

The analysis of cosmological perturbations in string gas cosmology (pioneered in [14]) is based on the cosmological background of string gas cosmology represented in Fig. 12.6. In turn, this background yields the space-time diagram sketched in Fig. 12.10. As in Fig. 12.2, the vertical axis is time and the horizontal axis denotes the physical distance. For times $t < t_R$, we are in the static Hagedorn phase and the Hubble radius is infinite. For $t > t_R$, the Einstein frame Hubble radius is expanding as in standard cosmology. The time t_R is when the string winding modes begin to decay into string loops, and the scale factor starts to increase, leading to the transition to the radiation phase of standard cosmology.

Let us now compare the evolution of the physical wavelength corresponding to a fixed co-moving scale with that of the Einstein frame Hubble radius $H^{-1}(t)$. The evolution of scales in string gas cosmology is identical to the evolution in standard and in inflationary cosmology for $t > t_R$. If we follow the physical wavelength of the co-moving scale which corresponds to the current Hubble radius back to the time t_R, then—taking the Hagedorn temperature to be of the order 10^{16} GeV—we

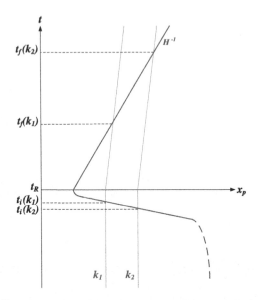

Fig. 12.10 Space-time diagram (sketch) showing the evolution of fixed co-moving scales in string gas cosmology. The *vertical axis* is time, the *horizontal axis* is physical distance. The *solid curve* represents the Einstein frame Hubble radius H^{-1} which shrinks abruptly to a micro-physical scale at t_R and then increases linearly in time for $t > t_R$. Fixed co-moving scales (the *dotted lines* labeled by k_1 and k_2) which are currently probed in cosmological observations have wavelengths which are smaller than the Hubble radius long before t_R. They exit the Hubble radius at times $t_i(k)$ just prior to t_R, and propagate with a wavelength larger than the Hubble radius until they re-enter the Hubble radius at times $t_f(k)$

obtain a length of about 1 mm. Compared to the string scale and the Planck scale, this is in the far infrared.

The physical wavelength is constant in the Hagedorn phase since space is static. But, as we enter the Hagedorn phase going back in time, the Hubble radius diverges to infinity. Hence, as in the case of inflationary cosmology, fluctuation modes begin sub-Hubble during the Hagedorn phase, and thus a causal generation mechanism for fluctuations is possible.

However, the physics of the generation mechanism is very different. In the case of inflationary cosmology, fluctuations are assumed to start as quantum vacuum perturbations because classical inhomogeneities are red-shifting. In contrast, in the Hagedorn phase of string gas cosmology there is no red-shifting of classical matter. Hence, it is the fluctuations in the classical matter which dominate. Since classical matter is a string gas, the dominant fluctuations are string thermodynamic fluctuations.

Our proposal for string gas structure formation is the following [14] (see [132] for a more detailed description). For a fixed co-moving scale with wavenumber k we compute the matter fluctuations while the scale in sub-Hubble (and therefore gravitational effects are sub-dominant). When the scale exits the Hubble radius at time $t_i(k)$ we use the gravitational constraint equations to determine the induced metric fluctuations, which are then propagated to late times using the usual equations of gravitational perturbation theory. Since the scales we are interested in are in the far infrared, we use the Einstein constraint equations for fluctuations.

12.7.2 Spectrum of Cosmological Fluctuations

We write the metric including cosmological perturbations (scalar metric fluctuations) and gravitational waves in the following form (using conformal time η)

$$ds^2 = a^2(\eta)\{(1+2\Phi)d\eta^2 - [(1-2\Phi)\delta_{ij} + h_{ij}]dx^i dx^j\}. \qquad (12.40)$$

As in previous sections, we are working in the longitudinal gauge for the scalar metric perturbations and we have taken matter to be free of anisotropic stress. The spatial tensor $h_{ij}(\mathbf{x}, t)$ is transverse and traceless and represents the gravitational waves.

Note that in contrast to the case of slow-roll inflation, scalar metric fluctuations and gravitational waves are generated by matter at the same order in cosmological perturbation theory. This could lead to the expectation that the amplitude of gravitational waves in string gas cosmology should be generically larger than in inflationary cosmology. This expectation, however, is not realized [133] since there is a different mechanism which suppresses the gravitational waves relative to the density perturbations (namely the fact that the gravitational wave amplitude is set by the amplitude of the pressure, and the pressure is suppressed relative to the energy density in the Hagedorn phase).

Assuming that the fluctuations are described by the perturbed Einstein equations (they are *not* if the dilaton is not fixed [134, 135]), then the spectra of cosmological perturbations Φ and gravitational waves h are given by the energy-momentum fluctuations in the following way [132]

$$\langle |\Phi(k)|^2 \rangle = 16\pi^2 G^2 k^{-4} \langle \delta T^0{}_0(k)\delta T^0{}_0(k) \rangle, \qquad (12.41)$$

where the pointed brackets indicate expectation values, and

$$\langle |h(k)|^2 \rangle = 16\pi^2 G^2 k^{-4} \langle \delta T^i{}_j(k)\delta T^i{}_j(k) \rangle, \qquad (12.42)$$

where on the right hand side of (12.42) we mean the average over the correlation functions with $i \neq j$, and h is the amplitude of the gravitational waves.[2]

Let us now use (12.41) to determine the spectrum of scalar metric fluctuations. We first calculate the root mean square energy density fluctuations in a sphere of radius $R = k^{-1}$. For a system in thermal equilibrium they are given by the specific heat capacity C_V via

$$\langle \delta\rho^2 \rangle = \frac{T^2}{R^6} C_V. \qquad (12.43)$$

---

[2]The gravitational wave tensor h_{ij} can be written as the amplitude h multiplied by a constant polarization tensor.

The specific heat of a gas of closed strings on a torus of radius R can be derived
from the partition function of a gas of closed strings. This computation was carried
out in [136] (see also [137]) with the result

$$C_V \approx 2 \frac{R^2/\ell^3}{T(1 - T/T_H)}. \tag{12.44}$$

The specific heat capacity scales holographically with the size of the box. This result
follows rigorously from evaluating the string partition function in the Hagedorn
phase. The result, however, can also be understood heuristically: in the Hagedorn
phase the string winding modes are crucial. These modes look like point particles in
one less spatial dimension. Hence, we expect the specific heat capacity to scale like
in the case of point particles in one less dimension of space.[3]

With these results, the power spectrum $P(k)$ of scalar metric fluctuations can be
evaluated as follows

$$\begin{aligned}
P_\Phi(k) &\equiv \frac{1}{2\pi^2} k^3 |\Phi(k)|^2 \\
&= 8G^2 k^{-1} \langle |\delta\rho(k)|^2 \rangle \\
&= 8G^2 k^2 \langle (\delta M)^2 \rangle_R \\
&= 8G^2 k^{-4} \langle (\delta\rho)^2 \rangle_R \\
&= 8G^2 \frac{T}{\ell_s^3} \frac{1}{1 - T/T_H},
\end{aligned} \tag{12.45}$$

where in the first step we have used (12.41) to replace the expectation value of
$|\Phi(k)|^2$ in terms of the correlation function of the energy density, and in the second
step we have made the transition to position space.

The first conclusion from the result (12.45) is that the spectrum is approximately scale-invariant ($P(k)$ is independent of k). It is the 'holographic' scaling
$C_V(R) \sim R^2$ which is responsible for the overall scale-invariance of the spectrum
of cosmological perturbations. However, there is a small k dependence which comes
from the fact that in the above equation for a scale k the temperature T is to be evaluated at the time $t_i(k)$. Thus, the factor $(1 - T/T_H)$ in the denominator is responsible
for giving the spectrum a slight dependence on k. Since the temperature slightly decreases as time increases at the end of the Hagedorn phase, shorter wavelengths for
which $t_i(k)$ occurs later obtain a smaller amplitude. Thus, the spectrum has a slight
red tilt.

---

[3]We emphasize that it was important for us to have compact spatial dimensions in order to obtain the winding modes which are crucial to get the holographic scaling of the thermodynamic
quantities.

12.7.3 Key Prediction of String Gas Cosmology

As discovered in [133], the spectrum of gravitational waves is also nearly scale invariant. However, in the expression for the spectrum of gravitational waves the factor $(1 - T/T_H)$ appears in the numerator, thus leading to a slight blue tilt in the spectrum. This is a prediction with which the cosmological effects of string gas cosmology can be distinguished from those of inflationary cosmology, where quite generically a slight red tilt for gravitational waves results. The physical reason for the blue tilt in string gas cosmology is that large scales exit the Hubble radius earlier when the pressure and hence also the off-diagonal spatial components of $T_{\mu\nu}$ are closer to zero.

Let us analyze this issue in a bit more detail and estimate the dimensionless power spectrum of gravitational waves. First, we make some general comments. In slow-roll inflation, to leading order in perturbation theory matter fluctuations do not couple to tensor modes. This is due to the fact that the matter background field is slowly evolving in time and the leading order gravitational fluctuations are linear in the matter fluctuations. In our case, the background is not evolving (at least at the level of our computations), and hence the dominant metric fluctuations are quadratic in the matter field fluctuations. At this level, matter fluctuations induce both scalar and tensor metric fluctuations. Based on this consideration we might expect that in our string gas cosmology scenario, the ratio of tensor to scalar metric fluctuations will be larger than in simple slow-roll inflationary models. However, since the amplitude h of the gravitational waves is proportional to the pressure, and the pressure is suppressed in the Hagedorn phase, the amplitude of the gravitational waves will also be small in string gas cosmology.

The method for calculating the spectrum of gravitational waves is similar to the procedure outlined in the last section for scalar metric fluctuations. For a mode with fixed co-moving wavenumber k, we compute the correlation function of the off-diagonal spatial elements of the string gas energy-momentum tensor at the time $t_i(k)$ when the mode exits the Hubble radius and use (12.42) to infer the amplitude of the power spectrum of gravitational waves at that time. We then follow the evolution of the gravitational wave power spectrum on super-Hubble scales for $t > t_i(k)$ using the equations of general relativistic perturbation theory.

The power spectrum of the tensor modes is given by (12.42):

$$P_h(k) = 16\pi^2 G^2 k^{-4} k^3 \langle \delta T^i{}_j(k) \delta T^i{}_j(k) \rangle \tag{12.46}$$

for $i \neq j$. Note that the k^3 factor multiplying the momentum space correlation function of $T^i{}_j$ gives the position space correlation function $C^i{}_j{}^i{}_j(R)$, namely the root mean square fluctuation of $T^i{}_j$ in a region of radius $R = k^{-1}$. Thus,

$$P_h(k) = 16\pi^2 G^2 k^{-4} C^i{}_j{}^i{}_j(R). \tag{12.47}$$

The correlation function $C^i{}_j{}^i{}_j$ on the right hand side of the above equation follows from the thermal closed string partition function and was computed explicitly in [137] (see also [132]). We obtain

$$P_h(k) \sim 16\pi^2 G^2 \frac{T}{l_s^3}(1 - T/T_H)\ln^2\left[\frac{1}{l_s^2 k^2}(1 - T/T_H)\right], \qquad (12.48)$$

which, for temperatures close to the Hagedorn value reduces to

$$P_h(k) \sim \left(\frac{l_{Pl}}{l_s}\right)^4 (1 - T/T_H)\ln^2\left[\frac{1}{l_s^2 k^2}(1 - T/T_H)\right]. \qquad (12.49)$$

This shows that the spectrum of tensor modes is—to a first approximation, namely neglecting the logarithmic factor and neglecting the k-dependence of $T(t_i(k))$— scale-invariant.

On super-Hubble scales, the amplitude h of the gravitational waves is constant. The wave oscillations freeze out when the wavelength of the mode crosses the Hubble radius. As in the case of scalar metric fluctuations, the waves are squeezed. Whereas the wave amplitude remains constant, its time derivative decreases. Another way to see this squeezing is to change variables to

$$\psi(\eta, \mathbf{x}) = a(\eta)h(\eta, \mathbf{x}) \qquad (12.50)$$

in terms of which the action has canonical kinetic term. The action in terms of ψ becomes

$$S = \frac{1}{2}\int d^4x\left(\psi'^2 - \psi_{,i}\psi_{,i} + \frac{a''}{a}\psi^2\right) \qquad (12.51)$$

from which it immediately follows that on super-Hubble scales $\psi \sim a$. This is the squeezing of gravitational waves [138].

Since there is no k-dependence in the squeezing factor, the scale-invariance of the spectrum at the end of the Hagedorn phase will lead to a scale-invariance of the spectrum at late times.

Note that in the case of string gas cosmology, the squeezing factor $z(\eta)$ for scalar metric fluctuations does not differ substantially from the squeezing factor $a(\eta)$ for gravitational waves. In the case of inflationary cosmology, $z(\eta)$ and $a(\eta)$ differ greatly during reheating, leading to a much larger squeezing for scalar metric fluctuations, and hence to a suppressed tensor to scalar ratio of fluctuations. In the case of string gas cosmology, there is no difference in squeezing between the scalar and the tensor modes.

Let us return to the discussion of the spectrum of gravitational waves. The result for the power spectrum is given in (12.49), and we mentioned that to a first approximation this corresponds to a scale-invariant spectrum. As realized in [133], the logarithmic term and the k-dependence of $T(t_i(k))$ both lead to a small blue-tilt of the spectrum. This feature is characteristic of our scenario and cannot be reproduced in inflationary models. In inflationary models, the amplitude of the gravitational waves is set by the Hubble constant H. The Hubble constant cannot increase during inflation, and hence no blue tilt of the gravitational wave spectrum is possible.

A heuristic way of understanding the origin of the slight blue tilt in the spectrum of tensor modes is as follows. The closer we get to the Hagedorn temperature, the more the thermal bath is dominated by long string states, and thus the smaller the pressure will be compared to the pressure of a pure radiation bath. Since the pressure terms (strictly speaking the anisotropic pressure terms) in the energy-momentum tensor are responsible for the tensor modes, we conclude that the smaller the value of the wavenumber k (and thus the higher the temperature $T(t_i(k))$ when the mode exits the Hubble radius, the lower the amplitude of the tensor modes. In contrast, the scalar modes are determined by the energy density, which increases at $T(t_i(k))$ as k decreases, leading to a slight red tilt.

The reader may ask about the predictions of string gas cosmology for non-Gaussianities. The answer is [139] that the non-Gaussianities from the thermal string gas perturbations are Poisson-suppressed on scales larger than the thermal wavelength in the Hagedorn phase. However, if the spatial sections are initially large, then it is possible that a network of cosmic superstrings [140, 141] will be left behind. These strings—if stable—would achieve a scaling solution (constant number of strings crossing each Hubble volume at each time [83–85]). Such strings give rise to linear discontinuities in the CMB temperature maps [142], lines which can be searched for using edge detection algorithms such as the Canny algorithm (see [143–146] for recent feasibility studies).

12.7.4 Comments

At the outset of this section we mentioned that not all emergent universe scenarios will produce a scale-invariant spectrum. For example, string gas cosmology in a non-compact three-dimensional space will not have the holographic scaling of the specific heat capacity and hence will not yield a scale-invariant spectrum.

Under which conditions does our above analysis generalize? Three conditions appear to be necessary in order to obtain scale-invariant cosmological fluctuations from an emergent background. Firstly, the background cosmology should have a quasi-static early phase followed after a short transition period by the radiation phase of Standard Big Bang cosmology. Secondly, the evolution of cosmological fluctuations on the infrared scales relevant to current cosmological observations should be describable in terms of perturbed Einstein gravity, i.e. using the formalism discussed in Sect. 12.4, even if the background cosmology cannot. Finally, the specific heat capacity $C_V(R)$ in a region of radius R should scale holographically, i.e.

$$C_V(R) \sim R^2. \tag{12.52}$$

12.8 Conclusions

In these lectures I have given an overview of the matter bounce and emergent universe scenarios of primordial cosmology. Both yield causal mechanisms for the generation of a scale-invariant spectrum of cosmological perturbations, the same kind of spectrum which is predicted by inflationary cosmology. In all three scenarios, the fluctuations are to a good approximation Gaussian. Thus, current cosmological observations cannot tell these models apart.

I have discussed specific predictions for future observations with which the three early universe can be teased apart observationally. The string gas cosmology realization of the emergent Universe predicts a small blue tilt in the spectrum of gravitational waves. Since inflationary models generically predict a red tilt, the tilt in the gravitational wave spectrum is a very promising characteristic. The simplest realization of the matter bounce scenario produces a distinguished shape of the cosmological bispectrum—and this appears to be an interesting distinctive signal to explore.

Part of the motivation for looking for alternatives to inflation comes from the realization that (at least current versions of) inflationary cosmology suffers from a number of conceptual problems, in particular a trans-Planckian problem for the fluctuations, and the singularity problem for the background cosmology. As I hope to have convinced the reader, these problems are resolved both in string gas cosmology and in the matter bounce scenario (in the latter, the singularity problem is "solved" by construction).

The matter bounce and emergent universe scenarios successfully address many of the problems of Standard Big Bang cosmology which inflationary cosmology addresses. In particular, neither scenario has a horizon problem. However, they do not solve all of the problems. The biggest challenge for the matter bounce scenario appears to be the instability to anisotropic stress. The biggest problem of string gas cosmology is our lack of an effective field theory which is consistent with the cosmological background evolution which is required. String gas cosmology also does not explain the size and entropy of the universe.

I have focused on two alternative cosmological scenarios. As mentioned earlier, there are more, e.g. the Ekpyrotic universe. A goal of future research should be to find improved realizations of all three cosmological scenarios considered here, and also to develop new paradigms which hopefully will have fewer conceptual problems that current ones.

Acknowledgements I wish to thank the organizers of this school for inviting me to give these lectures on the beautiful island of Naxos, and for their hospitality. I wish to thank all of my collaborators on whose work I have drawn. This work has been supported in part by funds from an NSERC Discovery Grant and from the Canada Research Chair program. I also acknowledge support from the Killam foundation for the period 9/09–8/11.

References

1. A.H. Guth, The inflationary universe: a possible solution to the horizon and flatness problems. Phys. Rev. D **23**, 347 (1981)
2. R. Brout, F. Englert, E. Gunzig, The creation of the universe as a quantum phenomenon. Ann. Phys. **115**, 78 (1978)

3. A.A. Starobinsky, A new type of isotropic cosmological models without singularity. Phys. Lett. B **91**, 99 (1980)
4. K. Sato, First order phase transition of a vacuum and expansion of the universe. Mon. Not. R. Astron. Soc. **195**, 467 (1981)
5. V. Mukhanov, G. Chibisov, Quantum fluctuation and nonsingular universe. JETP Lett. **33**, 532 (1981) (In Russian) [Pisma Zh. Eksp. Teor. Fiz. **33**, 549 (1981)]
6. W. Press, Spontaneous production of the Zel'dovich spectrum of cosmological fluctuations. Phys. Scr. **21**, 702 (1980)
7. A.A. Starobinsky, Spectrum of relict gravitational radiation and the early state of the universe. JETP Lett. **30**, 682 (1979) [Pisma Zh. Eksp. Teor. Fiz. **30** 719 (1979)]
8. Y.B. Zeldovich, A hypothesis, unifying the structure and the entropy of the universe. Mon. Not. R. Astron. Soc. **160**, 1P (1972)
9. E.R. Harrison, Fluctuations at the threshold of classical cosmology. Phys. Rev. D **1**, 2726 (1970)
10. R.A. Sunyaev, Y.B. Zeldovich, Small scale fluctuations of relic radiation. Astrophys. Space Sci. **7**, 3–19 (1970)
11. P.J.E. Peebles, J.T. Yu, Primeval adiabatic perturbation in an expanding universe. Astrophys. J. **162**, 815–836 (1970)
12. C.L. Bennett et al., First year Wilkinson Microwave Anisotropy Probe (WMAP) observations: preliminary maps and basic results. Astrophys. J. Suppl. **148**, 1 (2003). arXiv:astro-ph/0302207
13. R.H. Brandenberger, C. Vafa, Superstrings in the early universe. Nucl. Phys. B **316**, 391 (1989)
14. A. Nayeri, R.H. Brandenberger, C. Vafa, Producing a scale-invariant spectrum of perturbations in a Hagedorn phase of string cosmology. Phys. Rev. Lett. **97**, 021302 (2006). arXiv:hep-th/0511140
15. R.H. Brandenberger, String gas cosmology, in *String Cosmology*, ed. by J. Erdmenger (Wiley, New York, 2009). arXiv:0808.0746 [hep-th]
16. T. Battefeld, S. Watson, String gas cosmology. Rev. Mod. Phys. **78**, 435 (2006). arXiv:hep-th/0510022
17. R.H. Brandenberger, String gas cosmology: progress and problems. Class. Quantum Gravity **28**, 204005 (2011). arXiv:1105.3247 [hep-th]
18. G.F.R. Ellis, R. Maartens, The emergent universe: inflationary cosmology with no singularity. Class. Quantum Gravity **21**, 223 (2004). gr-qc/0211082
19. D. Wands, Duality invariance of cosmological perturbation spectra. Phys. Rev. D **60**, 023507 (1999). arXiv:gr-qc/9809062
20. F. Finelli, R. Brandenberger, On the generation of a scale-invariant spectrum of adiabatic fluctuations in cosmological models with a contracting phase. Phys. Rev. D **65**, 103522 (2002). arXiv:hep-th/0112249
21. R.H. Brandenberger, Cosmology of the very early universe. AIP Conf. Proc. **1268**, 3–70 (2010). arXiv:1003.1745 [hep-th]
22. M. Gasperini, G. Veneziano, Pre-big bang in string cosmology. Astropart. Phys. **1**, 317 (1993). arXiv:hep-th/9211021
23. J. Khoury, B.A. Ovrut, P.J. Steinhardt, N. Turok, The ekpyrotic universe: colliding branes and the origin of the hot big bang. Phys. Rev. D **64**, 123522 (2001). arXiv:hep-th/0103239
24. V.A. Rubakov, Harrison-Zeldovich spectrum from conformal invariance. J. Cosmol. Astropart. Phys. **0909**, 030 (2009). arXiv:0906.3693 [hep-th]
25. K. Hinterbichler, J. Khoury, The pseudo-conformal universe: scale invariance from spontaneous breaking of conformal symmetry. arXiv:1106.1428 [hep-th]
26. F.L. Bezrukov, M. Shaposhnikov, The standard model Higgs boson as the inflaton. Phys. Lett. B **659**, 703 (2008). arXiv:0710.3755 [hep-th]
27. F.C. Adams, K. Freese, A.H. Guth, Constraints on the scalar field potential in inflationary models. Phys. Rev. D **43**, 965 (1991)
28. R.H. Brandenberger, Inflationary cosmology: progress and problems. arXiv:hep-ph/9910410

29. R.H. Brandenberger, J. Martin, The robustness of inflation to changes in superPlanck scale physics. Mod. Phys. Lett. A **16**, 999 (2001). arXiv:astro-ph/0005432

30. J. Martin, R.H. Brandenberger, The transPlanckian problem of inflationary cosmology. Phys. Rev. D **63**, 123501 (2001). arXiv:hep-th/0005209

31. J.C. Niemeyer, Inflation with a high frequency cutoff. Phys. Rev. D **63**, 123502 (2001). arXiv: astro-ph/0005533

32. J.C. Niemeyer, R. Parentani, Minimal modifications of the primordial power spectrum from an adiabatic short distance cutoff. Phys. Rev. D **64**, 101301 (2001). arXiv:astro-ph/0101451

33. S. Shankaranarayanan, Is there an imprint of Planck scale physics on inflationary cosmology? Class. Quantum Gravity **20**, 75 (2003). arXiv:gr-qc/0203060

34. S.W. Hawking, G.F.R. Ellis, *The Large Scale Structure of Space-Time* (Cambridge University Press, Cambridge, 1973)

35. A. Borde, A. Vilenkin, Eternal inflation and the initial singularity. Phys. Rev. Lett. **72**, 3305 (1994). arXiv:gr-qc/9312022

36. C. Kounnas, H. Partouche, N. Toumbas, Thermal duality and non-singular cosmology in d-dimensional superstrings. Nucl. Phys. B **855**, 280 (2012). arXiv:1106.0946 [hep-th]

37. C. Kounnas, H. Partouche, N. Toumbas, S-brane to thermal non-singular string cosmology. arXiv:1111.5816 [hep-th]

38. R. Brandenberger, C. Kounnas, H. Partouche, S. Patil, N. Toumbas, Fluctuations in non-singular bouncing cosmologies from Type II superstrings, to be submitted

39. M. Novello, S.E.P. Bergliaffa, Bouncing cosmologies. Phys. Rep. **463**, 127 (2008). arXiv:0802.1634 [astro-ph]

40. B. Feng, X.L. Wang, X.M. Zhang, Dark energy constraints from the cosmic age and supernova. Phys. Lett. B **607**, 35 (2005). arXiv:astro-ph/0404224

41. B. Feng, M. Li, Y.S. Piao, X. Zhang, Oscillating quintom and the recurrent universe. Phys. Lett. B **634**, 101 (2006). arXiv:astro-ph/0407432

42. Y.F. Cai, T. Qiu, Y.S. Piao, M. Li, X. Zhang, Bouncing universe with quintom matter. J. High Energy Phys. **0710**, 071 (2007). arXiv:0704.1090 [gr-qc]

43. Y.F. Cai, T.T. Qiu, J.Q. Xia, X. Zhang, A model of inflationary cosmology without singularity. Phys. Rev. D **79**, 021303 (2009). arXiv:0808.0819 [astro-ph]

44. Y.F. Cai, T. Qiu, R. Brandenberger, Y.S. Piao, X. Zhang, On perturbations of quintom bounce. J. Cosmol. Astropart. Phys. **0803**, 013 (2008). arXiv:0711.2187 [hep-th]

45. Y.F. Cai, X. Zhang, Evolution of metric perturbations in quintom bounce model. J. Cosmol. Astropart. Phys. **0906**, 003 (2009). arXiv:0808.2551 [astro-ph]

46. J.M. Cline, S. Jeon, G.D. Moore, The phantom menaced: constraints on low-energy effective ghosts. Phys. Rev. D **70**, 043543 (2004). arXiv:hep-ph/0311312

47. B. Grinstein, D. O'Connell, M.B. Wise, The Lee-Wick standard model. Phys. Rev. D **77**, 025012 (2008). arXiv:0704.1845 [hep-ph]

48. Y.F. Cai, T. Qiu, R. Brandenberger, X. Zhang, A nonsingular cosmology with a scale-invariant spectrum of cosmological perturbations from Lee-Wick theory. Phys. Rev. D **80**, 023511 (2009). arXiv:0810.4677 [hep-th]

49. J. Karouby, R. Brandenberger, A radiation bounce from the Lee-Wick construction? Phys. Rev. D **82**, 063532 (2010). arXiv:1004.4947 [hep-th]

50. J. Karouby, T. Qiu, R. Brandenberger, On the instability of the Lee-Wick bounce. Phys. Rev. D **84**, 043505 (2011). arXiv:1104.3193 [hep-th]

51. V.A. Belinsky, I.M. Khalatnikov, E.M. Lifshitz, Oscillatory approach to a singular point in the relativistic cosmology. Adv. Phys. **19**, 525 (1970)

52. C. Lin, R.H. Brandenberger, L.P. Levasseur, A matter bounce by means of Ghost condensation. J. Cosmol. Astropart. Phys. **1104**, 019 (2011). arXiv:1007.2654 [hep-th]

53. P. Creminelli, L. Senatore, A smooth bouncing cosmology with scale invariant spectrum. J. Cosmol. Astropart. Phys. **0711**, 010 (2007). arXiv:hep-th/0702165

54. E.I. Buchbinder, J. Khoury, B.A. Ovrut, New ekpyrotic cosmology. Phys. Rev. D **76**, 123503 (2007). arXiv:hep-th/0702154

55. T. Qiu, J. Evslin, Y.F. Cai, M. Li, X. Zhang, Bouncing Galileon cosmologies. J. Cosmol. Astropart. Phys. **1110**, 036 (2011). arXiv:1108.0593 [hep-th]
56. D.A. Easson, I. Sawicki, A. Vikman, G-bounce. J. Cosmol. Astropart. Phys. **1111**, 021 (2011). arXiv:1109.1047 [hep-th]
57. J.K. Erickson, D.H. Wesley, P.J. Steinhardt, N. Turok, Kasner and mixmaster behavior in universes with equation of state $w \geq 1$. Phys. Rev. D **69**, 063514 (2004). arXiv:hep-th/0312009
58. Y. Cai, D. Easson, R. Brandenberger, Towards a nonsingular bouncing cosmology. J. Cosmol. Astropart. Phys. **1208**, 020 (2012). arXiv:1206.2382
59. R.H. Brandenberger, V.F. Mukhanov, A. Sornborger, A cosmological theory without singularities. Phys. Rev. D **48**, 1629 (1993). arXiv:gr-qc/9303001
60. T. Biswas, A. Mazumdar, W. Siegel, Bouncing universes in string-inspired gravity. J. Cosmol. Astropart. Phys. **0603**, 009 (2006). arXiv:hep-th/0508194
61. A. Kehagias, E. Kiritsis, Mirage cosmology. J. High Energy Phys. **9911**, 022 (1999). arXiv:hep-th/9910174
62. R. Brandenberger, H. Firouzjahi, O. Saremi, Cosmological perturbations on a bouncing brane. J. Cosmol. Astropart. Phys. **0711**, 028 (2007). arXiv:0707.4181 [hep-th]
63. P. Horava, Quantum gravity at a Lifshitz point. Phys. Rev. D **79**, 084008 (2009). arXiv:0901.3775 [hep-th]
64. R. Brandenberger, Matter bounce in Horava-Lifshitz cosmology. Phys. Rev. D **80**, 043516 (2009). arXiv:0904.2835 [hep-th]
65. M. Bojowald, Quantum cosmology. Lect. Notes Phys. **835**, 1 (2011)
66. A. Ashtekar, P. Singh, Loop quantum cosmology: a status report. Class. Quantum Gravity **28**, 213001 (2011). arXiv:1108.0893 [gr-qc]
67. P. Creminelli, A. Nicolis, E. Trincherini, Galilean genesis: an alternative to inflation. J. Cosmol. Astropart. Phys. **1011**, 021 (2010). arXiv:1007.0027 [hep-th]
68. L. Levasseur Perreault, R. Brandenberger, A.-C. Davis, Defrosting in an emergent Galileon cosmology. Phys. Rev. D **84**, 103512 (2011). arXiv:1105.5649 [astro-ph.CO]
69. J. Kripfganz, H. Perlt, Cosmological impact of winding strings. Class. Quantum Gravity **5**, 453 (1988)
70. R. Hagedorn, Statistical thermodynamics of strong interactions at high-energies. Nuovo Cimento Suppl. **3**, 147 (1965)
71. J. Polchinski, *String Theory, vols. 1 and 2* (Cambridge Univ. Press, Cambridge, 1998)
72. T. Boehm, R. Brandenberger, On T-duality in brane gas cosmology. J. Cosmol. Astropart. Phys. **0306**, 008 (2003). arXiv:hep-th/0208188
73. K. Hotta, K. Kikkawa, H. Kunitomo, Correlation between momentum modes and winding modes in Brandenberger-Vafa's string cosmological model. Prog. Theor. Phys. **98**, 687 (1997). arXiv:hep-th/9705099
74. M.A.R. Osorio, M.A. Vazquez-Mozo, A cosmological interpretation of duality. Phys. Lett. B **320**, 259 (1994). arXiv:hep-th/9311080
75. A.A. Tseytlin, C. Vafa, Elements of string cosmology. Nucl. Phys. B **372**, 443 (1992). arXiv:hep-th/9109048
76. G. Veneziano, Scale factor duality for classical and quantum strings. Phys. Lett. B **265**, 287 (1991)
77. A.A. Tseytlin, Dilaton, winding modes and cosmological solutions. Class. Quantum Gravity **9**, 979 (1992). arXiv:hep-th/9112004
78. R.H. Brandenberger, A.R. Frey, S. Kanno, Towards a nonsingular tachyonic big crunch. Phys. Rev. D **76**, 063502 (2007). arXiv:0705.3265 [hep-th]
79. S. Arapoglu, A. Karakci, A. Kaya, S-duality in string gas cosmology. Phys. Lett. B **645**, 255 (2007). arXiv:hep-th/0611193
80. M. Sakellariadou, Numerical experiments in string cosmology. Nucl. Phys. B **468**, 319 (1996). arXiv:hep-th/9511075
81. G.B. Cleaver, P.J. Rosenthal, String cosmology and the dimension of space-time. Nucl. Phys. B **457**, 621 (1995). arXiv:hep-th/9402088

82. R. Brandenberger, D.A. Easson, D. Kimberly, Loitering phase in brane gas cosmology. Nucl. Phys. B **623**, 421 (2002). arXiv:hep-th/0109165
83. A. Vilenkin, E.P.S. Shellard, *Cosmic Strings and Other Topological Defects* (Cambridge Univ. Press, Cambridge, 1994)
84. M.B. Hindmarsh, T.W.B. Kibble, Cosmic strings. Rep. Prog. Phys. **58**, 477 (1995). arXiv: hep-ph/9411342
85. R.H. Brandenberger, Topological defects and structure formation. Int. J. Mod. Phys. A **9**, 2117 (1994). arXiv:astro-ph/9310041
86. R. Easther, B.R. Greene, M.G. Jackson, D. Kabat, String windings in the early universe. J. Cosmol. Astropart. Phys. **0502**, 009 (2005). arXiv:hep-th/0409121
87. R. Danos, A.R. Frey, A. Mazumdar, Interaction rates in string gas cosmology. Phys. Rev. D **70**, 106010 (2004). arXiv:hep-th/0409162
88. B. Greene, D. Kabat, S. Marnerides, Dynamical decompactification and three large dimensions. Phys. Rev. D **82**, 043528 (2010). arXiv:0908.0955 [hep-th]
89. R.H. Brandenberger, Moduli stabilization in string gas cosmology. Prog. Theor. Phys. Suppl. **163**, 358 (2006). arXiv:hep-th/0509159
90. S. Watson, R. Brandenberger, Stabilization of extra dimensions at tree level. J. Cosmol. Astropart. Phys. **0311**, 008 (2003). arXiv:hep-th/0307044
91. L. Kofman, A. Linde, X. Liu, A. Maloney, L. McAllister, E. Silverstein, Beauty is attractive: moduli trapping at enhanced symmetry points. J. High Energy Phys. **0405**, 030 (2004). arXiv:hep-th/0403001
92. S. Watson, Moduli stabilization with the string Higgs effect. Phys. Rev. D **70**, 066005 (2004). arXiv:hep-th/0404177
93. S.P. Patil, R. Brandenberger, Radion stabilization by stringy effects in general relativity and dilaton gravity. Phys. Rev. D **71**, 103522 (2005). arXiv:hep-th/0401037
94. S.P. Patil, R.H. Brandenberger, The cosmology of massless string modes. J. Cosmol. Astropart. Phys. **0601**, 005 (2006). arXiv:hep-th/0502069
95. R. Brandenberger, Y.K. Cheung, S. Watson, Moduli stabilization with string gases and fluxes. J. High Energy Phys. **0605**, 025 (2006). arXiv:hep-th/0501032
96. R.J. Danos, A.R. Frey, R.H. Brandenberger, Stabilizing moduli with thermal matter and nonperturbative effects. Phys. Rev. D **77**, 126009 (2008). arXiv:0802.1557 [hep-th]
97. C.P. Burgess, L. McAllister, Challenges for string cosmology. Class. Quantum Gravity **28**, 204002 (2011). arXiv:1108.2660 [hep-th]
98. S. Mishra, W. Xue, R. Brandenberger, U. Yajnik, Supersymmetry breaking and dilaton stabilization in string gas cosmology. arXiv:1103.1389 [hep-th]
99. S. Watson, R.H. Brandenberger, Isotropization in brane gas cosmology. Phys. Rev. D **67**, 043510 (2003). arXiv:hep-th/0207168
100. V.F. Mukhanov, H.A. Feldman, R.H. Brandenberger, Theory of cosmological perturbations. Part 1. Classical perturbations. Part 2. Quantum theory of perturbations. Part 3. Extensions. Phys. Rep. **215**, 203 (1992)
101. R.H. Brandenberger, Lectures on the theory of cosmological perturbations. Lect. Notes Phys. **646**, 127 (2004). arXiv:hep-th/0306071
102. V.F. Mukhanov, Quantum theory of gauge invariant cosmological perturbations. Sov. Phys. JETP **67**(7), 1297 (1988) [Zh. Eksp. Teor. Fiz. **94**, 1 (1988)]
103. V.F. Mukhanov, Gravitational instability of the universe filled with a scalar field. JETP Lett. **41**, 493 (1985) [Pisma Zh. Eksp. Teor. Fiz. **41**, 402 (1985)]
104. M. Sasaki, Large scale quantum fluctuations in the inflationary universe. Prog. Theor. Phys. **76**, 1036 (1986)
105. V.N. Lukash, Production of phonons in an isotropic universe. Sov. Phys. JETP **52**, 807 (1980) [Zh. Eksp. Teor. Fiz. **79**]
106. J.M. Bardeen, P.J. Steinhardt, M.S. Turner, Spontaneous creation of almost scale—free density perturbations in an inflationary universe. Phys. Rev. D **28**, 679 (1983)
107. R.H. Brandenberger, R. Kahn, Cosmological perturbations in inflationary universe models. Phys. Rev. D **29**, 2172 (1984)

108. D.H. Lyth, Large scale energy density perturbations and inflation. Phys. Rev. D **31**, 1792 (1985)
109. N.D. Birrell, P.C.W. Davies, *Quantum Fields in Curved Space* (Cambridge University Press, Cambridge, 1982), 340p
110. C. Kiefer, I. Lohmar, D. Polarski, A.A. Starobinsky, Pointer states for primordial fluctuations in inflationary cosmology. Class. Quantum Gravity **24**, 1699 (2007). arXiv:astro-ph/0610700
111. P. Martineau, On the decoherence of primordial fluctuations during inflation. Class. Quantum Gravity **24**, 5817 (2007). arXiv:astro-ph/0601134
112. S. Alexander, T. Biswas, R.H. Brandenberger, On the transfer of adiabatic fluctuations through a nonsingular cosmological bounce. arXiv:0707.4679 [hep-th]
113. X. Gao, Y. Wang, W. Xue, R. Brandenberger, Fluctuations in a Hořava-Lifshitz bouncing cosmology. J. Cosmol. Astropart. Phys. **1002**, 020 (2010). arXiv:0911.3196 [hep-th]
114. X. Gao, Y. Wang, R. Brandenberger, A. Riotto, Cosmological perturbations in Hořava-Lifshitz gravity. Phys. Rev. D **81**, 083508 (2010). arXiv:0905.3821 [hep-th]
115. Y.F. Cai, W. Xue, R. Brandenberger, X.m. Zhang, Thermal fluctuations and bouncing cosmologies. J. Cosmol. Astropart. Phys. **0906**, 037 (2009). arXiv:0903.4938 [hep-th]
116. J.C. Hwang, E.T. Vishniac, Gauge-invariant joining conditions for cosmological perturbations. Astrophys. J. **382**, 363 (1991)
117. N. Deruelle, V.F. Mukhanov, On matching conditions for cosmological perturbations. Phys. Rev. D **52**, 5549 (1995). arXiv:gr-qc/9503050
118. R. Durrer, F. Vernizzi, Adiabatic perturbations in pre big bang models: matching conditions and scale invariance. Phys. Rev. D **66**, 083503 (2002). arXiv:hep-ph/0203275
119. D.H. Lyth, The failure of cosmological perturbation theory in the new ekpyrotic scenario. Phys. Lett. B **526**, 173 (2002). hep-ph/0110007
120. D.H. Lyth, The primordial curvature perturbation in the ekpyrotic universe. Phys. Lett. B **524**, 1 (2002). hep-ph/0106153
121. R. Brandenberger, F. Finelli, On the spectrum of fluctuations in an effective field theory of the ekpyrotic universe. J. High Energy Phys. **0111**, 056 (2001). hep-th/0109004
122. J. Khoury, B.A. Ovrut, P.J. Steinhardt, N. Turok, Density perturbations in the ekpyrotic scenario. Phys. Rev. D **66**, 046005 (2002). hep-th/0109050
123. R.H. Brandenberger, Processing of cosmological perturbations in a cyclic cosmology. Phys. Rev. D **80**, 023535 (2009). arXiv:0905.1514 [hep-th]
124. P.J. Steinhardt, N. Turok, Cosmic evolution in a cyclic universe. Phys. Rev. D **65**, 126003 (2002). hep-th/0111098
125. Y. Cai, R. Brandenberger, X. Zhang, The matter bounce curvaton scenario. J. Cosmol. Astropart. Phys. **1103**, 003 (2011). arXiv:1101.0822 [hep-th]
126. A. Cerioni, R.H. Brandenberger, Cosmological perturbations in the projectable version of Horava-Lifshitz gravity. J. Cosmol. Astropart. Phys. **1108**, 015 (2011). arXiv:1007.1006 [hep-th]
127. A. Cerioni, R.H. Brandenberger, Cosmological perturbations in the "healthy extension" of Horava-Lifshitz gravity. arXiv:1008.3589 [hep-th]
128. Y.F. Cai, W. Xue, R. Brandenberger, X. Zhang, Non-Gaussianity in a matter bounce. J. Cosmol. Astropart. Phys. **0905**, 011 (2009). arXiv:0903.0631 [astro-ph.CO]
129. X. Chen, Primordial non-Gaussianities from inflation models. Adv. Astron. **2010**, 638979 (2010). arXiv:1002.1416 [astro-ph.CO]
130. J.M. Maldacena, Non-Gaussian features of primordial fluctuations in single field inflationary models. J. High Energy Phys. **0305**, 013 (2003). arXiv:astro-ph/0210603
131. H. Li, J.Q. Xia, R. Brandenberger, X. Zhang, Constraints on models with a break in the primordial power spectrum. Phys. Lett. B **690**, 451 (2010). arXiv:0903.3725 [astro-ph.CO]
132. R.H. Brandenberger, A. Nayeri, S.P. Patil, C. Vafa, String gas cosmology and structure formation. Int. J. Mod. Phys. A **22**, 3621 (2007). arXiv:hep-th/0608121
133. R.H. Brandenberger, A. Nayeri, S.P. Patil, C. Vafa, Tensor modes from a primordial Hagedorn phase of string cosmology. Phys. Rev. Lett. **98**, 231302 (2007). arXiv:hep-th/0604126

134. R.H. Brandenberger et al., More on the spectrum of perturbations in string gas cosmology. J. Cosmol. Astropart. Phys. **0611**, 009 (2006). arXiv:hep-th/0608186

135. N. Kaloper, L. Kofman, A. Linde, V. Mukhanov, On the new string theory inspired mechanism of generation of cosmological perturbations. J. Cosmol. Astropart. Phys. **0610**, 006 (2006). arXiv:hep-th/0608200

136. N. Deo, S. Jain, O. Narayan, C.I. Tan, The effect of topology on the thermodynamic limit for a string gas. Phys. Rev. D **45**, 3641 (1992)

137. A. Nayeri, Inflation free, stringy generation of scale-invariant cosmological fluctuations in $D = 3 + 1$ dimensions. arXiv:hep-th/0607073

138. L.P. Grishchuk, Amplification of gravitational waves in an isotropic universe. Sov. Phys. JETP **40**, 409 (1975) [Zh. Eksp. Teor. Fiz. **67**, 825 (1974)]

139. B. Chen, Y. Wang, W. Xue, R. Brandenberger, String gas cosmology and non-Gaussianities. arXiv:0712.2477 [hep-th]

140. E. Witten, Cosmic superstrings. Phys. Lett. B **153**, 243 (1985)

141. E.J. Copeland, R.C. Myers, J. Polchinski, Cosmic F and D strings. J. High Energy Phys. **0406**, 013 (2004). hep-th/0312067

142. N. Kaiser, A. Stebbins, Microwave anisotropy due to cosmic strings. Nature **310**, 391 (1984)

143. S. Amsel, J. Berger, R.H. Brandenberger, Detecting cosmic strings in the CMB with the Canny algorithm. J. Cosmol. Astropart. Phys. **0804**, 015 (2008). arXiv:0709.0982 [astro-ph]

144. A. Stewart, R. Brandenberger, Edge detection, cosmic strings and the south pole telescope. J. Cosmol. Astropart. Phys. **0902**, 009 (2009). arXiv:0809.0865 [astro-ph]

145. R.J. Danos, R.H. Brandenberger, Canny algorithm, cosmic strings and the cosmic microwave background. Int. J. Mod. Phys. D **19**, 183 (2010). arXiv:0811.2004 [astro-ph]

146. R.J. Danos, R.H. Brandenberger, Searching for signatures of cosmic superstrings in the CMB. J. Cosmol. Astropart. Phys. **1002**, 033 (2010). arXiv:0910.5722 [astro-ph.CO]

Chapter 13
Quantum Gravity and Inflation

M.G. Romania, N.C. Tsamis, and R.P. Woodard

Abstract We review some perturbative results obtained in quantum gravity in an accelerating cosmological background. We then describe a class of non-local, purely gravitational models which have the correct structure to reproduce the leading infrared logarithms of quantum gravitational back-reaction during the inflationary regime. These models end inflation in a distinctive phase of oscillations with slight and short violations of the weak energy condition and should, when coupled to matter, lead to rapid reheating. By elaborating this class of models we exhibit one that has the same behaviour during inflation, goes quiescent until the onset of matter domination, and induces a small, positive cosmological constant of about the right size thereafter. We also briefly comment on the primordial density perturbations that this class of models predict.

13.1 Introduction

FRW Cosmology and Inflation   On scales larger than about 100 Mpc the universe is well described by the FRW geometry:

$$ds^2 = -dt^2 + a^2(t)d\mathbf{x} \cdot d\mathbf{x}. \tag{13.1}$$

The time variation of the scale factor $a(t)$ gives the instantaneous values of the *Hubble parameter* $H(t)$ and the *deceleration parameter* $q(t)$:

$$H(t) \equiv \frac{\dot{a}(t)}{a(t)} = \frac{d}{dt}\ln a(t), \tag{13.2}$$

M.G. Romania (✉) · N.C. Tsamis
Department of Physics, University of Crete, 710 03 Heraklion, Greece
e-mail: romania@physics.uoc.gr

N.C. Tsamis
e-mail: tsamis@physics.uoc.gr

R.P. Woodard
Department of Physics, University of Florida, Gainesville, FL 32611, USA
e-mail: woodard@phys.ufl.edu

G. Calcagni et al. (eds.), *Quantum Gravity and Quantum Cosmology*,
Lecture Notes in Physics 863, DOI 10.1007/978-3-642-33036-0_13,
© Springer-Verlag Berlin Heidelberg 2013

$$q(t) \equiv -\frac{\dot{a}(t)\ddot{a}(t)}{\dot{a}^2(t)} = -1 - \frac{\dot{H}(t)}{H^2(t)} \equiv -1 + \varepsilon(t). \tag{13.3}$$

Their current values are: $H_{\text{now}} \simeq (73.8 \pm 2.4)$ km/sec Mpc $\simeq 2.4 \times 10^{-18}$ Hz [1] and $\varepsilon_{\text{now}} \simeq 0.33 \pm 0.13$ [2, 3].

There is overwhelming evidence that the history of the universe included a period of accelerated expansion known as inflation and defined by $H > 0$ with $\varepsilon < 1$ [4, 5]. This expansion occurred very early—$t \sim 10^{-33}$ sec—and the latest data [3, 6], plus the assumption of single scalar inflation imply: $H_I \lesssim 1.7 \times 10^{38}$ Hz with $\varepsilon_I \lesssim 0.011$ [7].

The Horizon Problem   The strongest evidence in favor of primordial inflation is the fact that we can detect epochs of cosmological history during which the observable universe was in thermal equilibrium. Without an early phase of primordial acceleration there is no way such distant regions can even have exchanged a single photon, much less interacted strongly enough to have equilibrated. To understand why, let us use the fact that photons travel on paths with zero invariant interval to calculate the size of our horizon:

$$ds^2 = -dt^2 + a^2(t)dr^2 = 0 \quad \Longrightarrow \quad dr = \frac{dt}{a(t)}. \tag{13.4}$$

Now consider some past time t_{past}, and compare the coordinate distance R_{past} we can observe at t_{now} with the coordinate radius of light which propagated from the beginning of the universe at t_{initial} to t_{past}:

$$R_{\text{past}} = \int \frac{dt}{a(t)}, \quad \text{for } t_{\text{past}} < t < t_{\text{now}}, \tag{13.5}$$

$$R_{\text{future}} = \int \frac{dt}{a(t)}, \quad \text{for } t_{\text{initial}} < t < t_{\text{past}}. \tag{13.6}$$

For the universe at t_{past} to have reached thermal equilibrium by causal processes requires:

$$\left(\frac{R_{\text{past}}}{R_{\text{future}}} \right)^2 \leq 1. \tag{13.7}$$

Suppose the universe expanded with constant $\varepsilon \equiv -\dot{H}H^{-2}$:

$$\text{constant } \varepsilon \quad \Longrightarrow \quad H = \frac{1}{\varepsilon t} \quad \Longrightarrow \quad a \sim t^{\frac{1}{\varepsilon}} \tag{13.8}$$

$$\Longrightarrow \quad \int \frac{dt}{a(t)} = \frac{1}{(\varepsilon - 1)Ha}. \tag{13.9}$$

If the universe was decelerating throughout its existence the upper limit of (13.9) dominates over the lower one:

$$\varepsilon > 1 \quad \Longrightarrow \quad Ha \sim t^{1-\frac{1}{\varepsilon}} \text{ falls} \tag{13.10}$$

$$\Longrightarrow \quad R_{\text{past}} \sim \frac{1}{(\varepsilon - 1)Ha}\bigg|_{\text{now}}, \quad R_{\text{future}} \sim \frac{1}{(\varepsilon - 1)Ha}\bigg|_{\text{past}} \tag{13.11}$$

$$\Longrightarrow \quad R_{\text{future}} \ll R_{\text{past}}. \tag{13.12}$$

For instance, at recombination—when the universe is observed to be in thermal equilibrium to one part in 10^5!—and at nucleosynthesis, respectively:

$$\left(\frac{R_{\text{past}}}{R_{\text{future}}}\right)^2 \sim 2000 \text{ and } 10^9. \tag{13.13}$$

Hence the observed equilibrium during these epochs could not have come about by causal processes; it would have had to be an accidental feature of the way the universe began. No one knows how the universe began, but assuming it began in a very high degree of thermal equilibrium seems problematic. This sort of unsatisfactory conclusion can be avoided if we assume the universe went through a phase of acceleration before t_{past}. In that case the integral (13.9) is dominated by its lower limit and we can make the past light-cone arbitrarily large by assuming t_{initial} is close to zero:

$$\varepsilon < 1 \quad \Longrightarrow \quad Ha \sim t^{1-\frac{1}{\varepsilon}} \text{ grows} \tag{13.14}$$

$$\Longrightarrow \quad R_{\text{future}} \sim \frac{1}{(\varepsilon - 1)Ha}\bigg|_{\text{initial}} \tag{13.15}$$

$$\Longrightarrow \quad \text{for } t_{\text{initial}} \to 0 : R_{\text{future}} \gg R_{\text{past}}. \tag{13.16}$$

Single-Scalar Inflation   Although the evidence for a phase of primordial inflation is very strong [8], there is no compelling mechanism for making it happen [9]. The simplest model consists of gravity plus a minimally coupled scalar field (called the *inflaton*) whose Lagrangian is [10]:

$$\mathcal{L} = \sqrt{-g}\left(-\frac{1}{2}g^{\mu\nu}\partial_\mu\varphi\partial_\nu\varphi - V(\varphi) + \frac{R}{16\pi G}\right). \tag{13.17}$$

Note that this model is general enough to support *any* expansion history $a(t)$, provided only that $\dot{H}(t) \leq 0$ throughout. To see this, note that the nontrivial Einstein equations are:

$$3H^2 = 8\pi G\left[\frac{1}{2}\left(\frac{d\varphi}{dt}\right)^2 + V(\varphi)\right], \tag{13.18}$$

$$-2\dot{H} + 3H^2 = 8\pi G\left[\frac{1}{2}\left(\frac{d\varphi}{dt}\right)^2 - V(\varphi)\right]. \tag{13.19}$$

One would usually take the scalar potential $V(\varphi)$ as given and use these equations to determine the expansion history, but let us adopt the opposite perspective. That is, we will assume $a(t)$ is known and we then use the equations to reconstruct the potential $V(\varphi)$ which supports that geometry. By adding (13.19) to (13.18) we get the inflaton as a function of time:

$$-2\dot{H} = 8\pi G\left(\frac{d\varphi}{dt}\right)^2 \quad \Longrightarrow \quad \varphi(t) = \varphi_I + \int^t dt'\left(-\frac{\dot{H}(t')}{4\pi G}\right)^{\frac{1}{2}}. \tag{13.20}$$

By inverting this relation we get the time as a function of the inflaton: $t = t(\varphi)$. Now subtract (13.19) from (13.18) to find the potential which gives the desired expansion history:

$$6H^2 = 16\pi G V(\varphi) \implies V(\varphi) = \frac{3}{8\pi G} H^2\big[t(\varphi)\big]. \tag{13.21}$$

This construction seems to have first appeared in [11], and independently in [12] and [13].

Scalar Inflation Problems   As we have seen, the potential energy of a minimally coupled scalar field can cause inflation, but this mechanism involves assumptions which seem unlikely and are sometimes contradictory:

- That the universe began with the scalar field approximately spatially homogeneous over more than a Hubble volume $V(\varphi) > H^{-3}$ [14].
- That the scalar field potential must be flat enough make inflation last a long time [9, 10].
- That the minimum of the scalar field potential has just the right value $V_{min} \simeq 0$ to leave the post-inflationary universe with only the small amount of vacuum energy we detect today [15–18].
- That the scalar field couples enough to ordinary matter so that its kinetic energy can create a hot, dense universe at the end of inflation, but not so much that loop corrections from ordinary matter compromise the flatness of the inflaton potential [19].

Gravity-Driven Inflation   A more natural mechanism for inflation can be found within gravitation—which, after all, plays the dominant role in shaping cosmological evolution—by supposing that the bare cosmological constant Λ is not unnaturally small but rather large and positive. Here "large" means a Λ induced by some matter scale which might be as high as 10^{18} GeV. Then, the value of the dimensionless coupling constant would be $G\Lambda \sim 10^{-4}$, rather than the putative value of 10^{-122} [15–18].

Because Λ is constant in *space*, no special initial condition is needed to start inflation. We also dispense with the need to employ a new, otherwise undetected scalar field. However, Λ is constant in *time* as well, and classical physics can offer no natural mechanism for stopping inflation once it has begun [20, 21]. Quantum physics can: accelerated expansion continually rips virtual infrared gravitons out of the vacuum [22, 23] and these gravitons attract one another, thereby slowing inflation [24, 25]. This is a very weak effect for $G\Lambda \ll 1$, but a cumulative one, so inflation would last a long time for no other reason than that gravity is a weak interaction [24, 25].

Graviton Physical Modes   In terms of the full metric field $g_{ij}(x)$, the fluctuating graviton field $h_{ij}^{TT}(x)$ is defined as:

$$g_{ij}(t, \mathbf{x}) = a^2(t)\big[\delta_{ij} + \sqrt{32\pi G} h_{ij}^{TT}(t, \mathbf{x})\big]. \tag{13.22}$$

The free field expansion of the graviton field is:

$$h_{ij}^{TT}(t, \mathbf{x}) = \int \frac{d^3k}{(2\pi)^3} \sum_{\lambda} \left\{ u(t, k)e^{i\mathbf{k}\cdot\mathbf{x}}\varepsilon_{ij}(\mathbf{k}, \lambda)\alpha(\mathbf{k}, \lambda) + (\text{c.c.}) \right\}, \tag{13.23}$$

where (c.c.) denotes complex conjugation, $\varepsilon_{ij}(\mathbf{k}, \lambda)$ are the same transverse and traceless polarization tensors as in flat space, $\alpha(\mathbf{k}, \lambda)$ is the annihilation operator, and $u(t, k)$ are the mode functions which obey:

$$\ddot{u}(t, k) + 3H(t)\dot{u}(t, k) + \frac{k^2}{a^2(t)}u(t, k) = 0. \tag{13.24}$$

The mechanism we have sketched is that inflation rips gravitons out of the vacuum, and then the self-gravitation of these particles slows inflation. Let us first estimate the energy $E(t, k)$ which is present in a single polarization of a single wave vector $\mathbf{k}$ at time t. Because the precise definition of energy is subtle for gravitons we base this estimate on a massless, minimally coupled scalar field $\varphi(x)$, whose mode equation is the same as (13.24). The scalar field Lagrangian density is:

$$\mathscr{L}(x) = -\frac{1}{2}\sqrt{-g}g^{\mu\nu}\partial_\mu\varphi\partial_\nu\varphi = \frac{1}{2}a^3(t)\dot{\varphi}^2 - \frac{1}{2}\nabla\varphi\cdot\nabla\varphi. \tag{13.25}$$

The Lagrangian diagonalizes in momentum space:

$$L(t) = \int d^3x\mathscr{L}(x) = \int \frac{d^3k}{(2\pi)^3}\left\{ \frac{1}{2}a^3(t)\left|\dot{\tilde{\varphi}}(t, \mathbf{k})\right|^2 - \frac{1}{2}a(t)k^2\left|\tilde{\varphi}(t, \mathbf{k})\right|^2 \right\} \tag{13.26}$$

so that any mode with wavenumber $\mathbf{k}$ evolves independently as a harmonic oscillator $q(t)$ with a time-dependent mass $m(t) \sim a^3(t)$ and angular frequency $\omega(t) \equiv ka^{-1}(t)$:

$$q(t) = u(t, k)A + u^*(t, k)A^\dagger, \quad [A, A^\dagger] = 1, \tag{13.27}$$

$$E(t, k) = \frac{1}{2}a^3(t)\dot{q}^2(t) + \frac{1}{2}a(t)k^2q^2(t). \tag{13.28}$$

For the special case of de Sitter the mode functions are given by:

$$u(t, k) = \frac{H}{\sqrt{2k^3}}\left[1 - \frac{ik}{Ha(t)} \right]\exp\left(\frac{ik}{Ha(t)} \right). \tag{13.29}$$

Although our conclusions are quite generic, it will simplify the subsequent analysis if we make this assumption of de Sitter.

At any instant t the minimum energy is $E_{\min}(t, k) = \frac{1}{2}ka^{-1}(t)$. However because both the mass and angular frequency are time-dependent, the state with minimum energy at one instant is not the minimum energy state at later times; there is particle production. Bunch-Davies vacuum $|\Omega\rangle$ is the minimum energy state in the distant past, and the expectation value of the energy operator (13.28) in this state is:

$$\langle\Omega|E(t, k)|\Omega\rangle = \frac{a^3(t)}{2}\left|\dot{u}(t, k)\right|^2 + \frac{k^2a(t)}{2}\left|u(t, k)\right|^2 \tag{13.30}$$

$$= \frac{k}{a(t)}\left(\frac{1}{2} + \left[\frac{Ha(t)}{2k} \right]^2 \right). \tag{13.31}$$

By setting this equal to $(\frac{1}{2} + N)\hbar\omega$, one can read off the instantaneous occupation number $N(t, k)$:

$$N(t, k) = \left[\frac{Ha(t)}{2k} \right]^2. \tag{13.32}$$

We can consider $N(t, k)$ to be the number of gravitons with one polarization and wave vector **k** that have been created by time t.

At this point a short digression is useful on the significance of the co-moving wave number k in an expanding universe. Because $k = 2\pi/\lambda$ is the inverse of a coordinate length, the physical wave number is $ka^{-1}(t)$. This falls exponentially during inflation. Horizon crossing is when the physical wave number equals the Hubble parameter:

$$Horizon\ Crossing \implies k_{phys} = ka^{-1}(t) = H. \tag{13.33}$$

It is natural to separate modes into "infrared" and "ultraviolet" depending upon whether or not they have experienced horizon crossing:

$$Infrared \implies H < k < Ha(t), \tag{13.34}$$

$$Ultraviolet \implies k > Ha(t). \tag{13.35}$$

From (13.32) we see that there is negligible production of ultraviolet gravitons, whereas the number of infrared gravitons in even a single wave vector grows exponentially. This is a crucial observation because it means that the physics of this effect is controlled by the known, low energy theory of gravity, without regard to its still unknown ultraviolet completion.

The energy density induced by both polarizations of these infrared gravitons equals:

$$\rho_{IR} = \frac{2}{a^3(t)} \int^{Ha} \frac{d^3k}{(2\pi)^3} N(t, k) \frac{k}{a(t)} = \frac{H^4}{8\pi^2}. \tag{13.36}$$

This is much less than the energy density of the cosmological constant:

$$\rho_\Lambda = \frac{3H^2}{8\pi G} \implies \frac{\rho_{IR}}{\rho_\Lambda} = \frac{GH^2}{3\pi} \lesssim 10^{-11}. \tag{13.37}$$

One may wonder if the gravitational self-interaction of ρ_{IR} can even screen itself, much less ρ_Λ. To see that it can, note that even a small energy density can induce significant screening if it interacts over a *sufficiently large volume*. A simple way to see this is to consider the total energy density ρ_{tot} produced by a static energy density ρ_{bare} distributed throughout a sphere of radius R. For simplicity, we follow ADM [26] in using the Newtonian formula assuming it is the total mass $\frac{4}{3}\pi\rho_{tot}c^{-2}R^3$ that gravitates:

$$\rho_{tot} \approx \rho_{bare} - \frac{4\pi G\rho_{tot}^2 R^2}{5c^4} \implies \rho_{tot} \approx \frac{5c^4}{8\pi GR^2} \left[\sqrt{1 + \frac{16\pi G\rho_{bare}R^2}{5c^4}} - 1 \right]. \tag{13.38}$$

As R goes to infinity the screening becomes total—i.e., ρ_{tot} goes to zero—independent of how small ρ_{bare} is.

Equation (13.38) means the gravitational self-interaction of infrared gravitons can screen ρ_{IR}, but what about the vastly larger energy density ρ_Λ of the cosmological constant? The key observation for realizing that even ρ_Λ can be screened is that *the gravitational self-interaction hasn't had time to reach a static limit.* Indeed, most of the universe is not even now in causal contact, and never will be if the current phase of accelerated expansion persists. The lower bound of $\rho_{\text{tot}} = 0$ implicit in the static result (13.38) arises because it is the instantaneous value of ρ_{tot} which gravitates, so making it smaller by screening also cuts off the effect. But that cutoff disappears when one takes account of the causal nature of the interaction. The source for the gravitational field at time t is not the instantaneous energy density of infrared gravitons but rather its value far back in the past light-cone. That is not reduced by the instantaneous energy density becoming small; indeed, the effect of screening is to make the past light-cone open outwards, which exposes more of the early times when the energy density of infrared gravitons was high.

This discussion does not prove the viability of gravity-driven inflation. Because inflationary particle production is itself a 1-loop effect, the gravitational response to it cannot occur at less than 2-loop order. Two-loop computations in quantum gravity are not simple around flat space background, and they are considerably tougher around de Sitter. Then there is the delicate gauge issue of how to invariantly quantify screening [27]. Good physicists on both sides of the question have debated whether or not there is a significant screening effect from the mechanism we have described [28, 29], or from any of the related relaxation mechanisms which have been proposed [30–50]. There is even disagreement about the basic formalism of perturbative quantum gravity on de Sitter background [51–58]. The aim of this introduction has been merely to establish the *plausibility* of the mechanism. Having hopefully done that, we will henceforth explore a simple class of effective field equations that might describe it.

13.2 Model Building

Perturbative Results Let use first review some perturbative results on de Sitter:

$$\text{de Sitter Inflation} \implies a(t) = e^{Ht}. \tag{13.39}$$

The gravitational Lagrangian is:

$$\mathscr{L}_{\text{gr}} = \frac{1}{16\pi G}(R - 2\Lambda)\sqrt{-g}. \tag{13.40}$$

It turns out that quantum corrections cannot grow faster than powers of $\ln(a) = Ht$ [59, 60]. We are interested in the regime of $\ln(a) \gg 1$, in which case one needs only the *leading logarithm* contributions at any loop order L which contain the most factors of $\ln(a)$. Explicit computations [24, 25, 61], and general counting rules

[59, 60], give the following behaviour for the leading logarithm contributions to the energy density induced by quantum gravitational effects:

$$\rho_1 \sim +\Lambda^2, \tag{13.41}$$

$$\rho_2 \sim -G\Lambda^3 \ln[a(t)], \tag{13.42}$$

$$\rho_L \sim -\Lambda^2 (G\Lambda \ln[a(t)])^{L-1}. \tag{13.43}$$

Because stress-energy is separately conserved at each loop order, the quantum gravitationally induced pressure must be that of negative vacuum energy, up to small subleading logarithm corrections:

$$\dot{\rho}_L = -3H(\rho_L + p_L) \implies p_L(t) \sim -\rho_L(t). \tag{13.44}$$

Hence the general form of the pressure is:

$$p(t) \sim \Lambda^2 f[G\Lambda \ln(a)]. \tag{13.45}$$

Perturbation theory is valid only if the effective dimensionless coupling constant $G\Lambda \ln(a)$ of the theory is small. Thus, perturbation theory breaks down after a large number of e-foldings—$N \equiv Ht = \ln(a) \sim (G\Lambda)^{-1}$. However, if we had the effective field equations, at least for a general FRW geometry, it would be possible to evolve arbitrarily far in the future. So we shall try to guess these equations based on some general principles, and on what we know from perturbation theory.

Guessing the Effective Field Equations The classical gravitational equations of motion coming from (13.40) are:

$$G_{\mu\nu} = -\Lambda g_{\mu\nu}. \tag{13.46}$$

The equations of motion in the presence of the quantum induced stress-energy tensor $T_{\mu\nu}[g]$ are:

$$G_{\mu\nu} = -\Lambda g_{\mu\nu} + 8\pi G T_{\mu\nu}[g]. \tag{13.47}$$

The full quantum induced stress-energy encodes all information about quantum gravity. For example, variations of it about flat space—with $\Lambda = 0$—give all scattering amplitudes to all orders in perturbation theory. There is absolutely no chance we can guess this, nor is there any need to do so. We require only the most cosmologically significant part of the full effective quantum gravitational equations; that is, a functional of the FRW scale factor $a(t)$.

A few basic principles can be used to guide us [62]:

(i) *Correspondence*: The form of $T_{\mu\nu}[g]$ must of course reproduce the known results from perturbation theory about de Sitter space.

(ii) *Non-locality*: It is easy to show that a purely local $T_{\mu\nu}[g](x)$ can only lead to a constant change in the cosmological constant. Note first that such a local $T_{\mu\nu}[g](x)$ must be composed of the Riemann tensor and its derivatives. Now consider the de Sitter geometry for an arbitrary Hubble parameter H', not necessarily equal to the one associated with $\Lambda = 3H^2$. The Riemann tensor for this geometry reduces to a constant times sums of products of the metric, and any covariant derivative of it therefore vanishes:

$$R_{\rho\sigma\mu\nu} = H'^2[g_{\rho\mu}g_{\sigma\nu} - g_{\rho\nu}g_{\sigma\mu}] \implies D_\alpha R_{\rho\sigma\mu\nu} = 0. \tag{13.48}$$

Hence any local stress-energy must reduce, for this geometry, to $\#H'^4 g_{\mu\nu}$, and the effective field equation would become:

$$G_{\mu\nu} = -3H^2 g_{\mu\nu} + \#8\pi G H'^4 g_{\mu\nu} = -3H'^2\left(\frac{H^2}{H'^2} - \frac{8}{3}\pi G H'^2\right) g_{\mu\nu}. \tag{13.49}$$

This amounts to merely a renormalization of Λ:

$$\Lambda' = \frac{9}{16\pi G}\left[\sqrt{1 + \frac{32}{9}\pi G\Lambda} - 1\right]. \tag{13.50}$$

If one began in this geometry—which our actual renormalization condition would require—then there would never be any deviation form it. We conclude that screening requires a non-local $T_{\mu\nu}[g]$.

(iii) *Causality*: The quantum induced stress-energy must be both conserved and causal, in the sense that $T_{\mu\nu}[g](x)$ depends only upon metrics on or within the past light-cone of the point x^μ. We would normally ensure conservation by defining the stress-energy from the variation of an invariant effective action:

$$T_{\mu\nu}[g](x) = -\frac{2}{\sqrt{-g}} \cdot \frac{\delta\Gamma[g]}{\delta g^{\mu\nu}(x)} \implies D^\nu T_{\mu\nu} = 0. \tag{13.51}$$

However, this procedure conflicts with causality for the sort of non-local contributions of greatest interest to us.

To understand the problem, consider the action of a point particle $q(t)$. Suppose the action contains a non-local term of the form $q(s) \times q(s - \Delta t)$. One might think that its non-locality is safely confined to the past of $q(s)$, but any variation must also affect the term $q(s - \Delta t)$. This gives rise to an equation which depends on the future as well as the past:

$$\Gamma[q] = \int ds\, q(s)q(s - \Delta t) \implies \tag{13.52}$$

$$\frac{\delta\Gamma}{\delta q(t)} = \int ds\left[\delta(s - t)q(s - \Delta t) + q(s)\delta(s - \Delta t - t)\right] \tag{13.53}$$

$$= q(t - \Delta t) + q(t + \Delta t). \tag{13.54}$$

This same problem must afflict any variation such as (13.51) which is based on a non-local effective action that contains only a single field.

The proper way to derive non-local effective field equations which are both causal and conserved is by varying the Schwinger-Keldysh effective action [63–70]. This avoids the single field conundrum by employing two fields $g_{\mu\nu}^\pm$; with the + sign corresponding to the background metric during forward evolution and the − sign to backwards evolution. The stress-energy tensor of the Schwinger-Keldysh formalism is the variation with respect to either field, after which the two fields are set equal:

$$T_{\mu\nu}[g](x) = -\frac{2}{\sqrt{-g}} \cdot \frac{\delta \Gamma[g^+, g^-]}{\delta g^+_{\mu\nu}(x)}\bigg|_{g^\pm = g}. \tag{13.55}$$

One can show that the $+$ and $-$ contributions from fields at any point x'^μ exactly cancel unless x'^μ is on or within the past light-cone of x^μ.

The Schwinger-Keldysh effective action is what one should use to *derive* the correct effective field equations. However, deriving anything is tough in quantum gravity. The point of this exercise was to try *guessing* the most cosmologically significant part of the effective field equations. Because it is those equations we seek, not the effective action, we shall adopt the shortcut of simply making an appropriately non-local and causal ansatz for them, and then enforce conservation directly.

Perfect Fluid Ansatz   The ansatz must apply to all *FRW* cosmologies. The "perfect fluid" form of $T_{\mu\nu}$ can represent any cosmology and in addition provides enough free parameters to enforce conservation and correspondence with perturbative results:

$$T_{\mu\nu}[g] = (\rho + p)u_\mu u_\nu + p g_{\mu\nu}. \tag{13.56}$$

Our stress-energy is defined by specifying three things:

(i)   the energy density ρ as a functional of the metric tensor $\rho[g](x)$,
(ii) the pressure p as a functional of the metric tensor $p[g](x)$, and
(iii) the 4-velocity field u_μ as a functional of the metric tensor $u_\mu[g](x)$, chosen to be timelike and normalized:

$$g^{\mu\nu}u_\mu u_\nu = -1 \implies u^\mu u_{\mu;\nu} = 0. \tag{13.57}$$

Because of the normalization (13.57), only three of the components of u_μ are algebraically independent. Hence our ansatz consists of five independent functionals in total. Stress-energy conservation:

$$D^\mu T_{\mu\nu} = 0, \tag{13.58}$$

provides four equations and allows us to determine any four of these functionals in terms of the fifth. It turns out to be most convenient to specify the induced pressure functional $p[g]$ and then use conservation to obtain the form of the induced energy density $\rho[g]$ and the 4-velocity $u_\mu[g]$, up to their initial value data.

Building $p[g]$ We want the pressure $p[g](x)$ to be a causal, non-local functional of the metric which reduces to the form (13.45) in the de Sitter limit. A very simple ansatz along these lines is:

$$p[g](x) = \Lambda^2 f\left(-G\Lambda X[g](x)\right), \tag{13.59}$$

where $-X[g](x)$ is a dimensionless, non-local functional of the metric that grows like $\ln(a)$ when the metric is de Sitter. A natural way of incorporating causal non-locality is through the inverse of some differential operator. The simplest choice for this operator is the covariant scalar d'Alembertian:

$$\Box \equiv \frac{1}{\sqrt{-g}} \partial_\mu \left(g^{\mu\nu} \sqrt{-g} \partial_\nu \right). \tag{13.60}$$

To make $X[g](x)$ dimensionless, we need to act the inverse of $\Box$ on a curvature scalar, the simplest choice for which is the Ricci scalar R. We are therefore led to consider $X[g] = \Box^{-1} R$, with the inverse defined using retarded boundary conditions.

To see that this simple ansatz has the right properties, we specialize $\Box$ and R to a general FRW geometry:

$$\Box = -\left(\partial_t^2 + 3H \partial_t \right), \qquad R(t) = 12H^2(t) + 6\dot{H}(t). \tag{13.61}$$

Hence the specialization of $X[g](x)$ to FRW is:

$$X = \frac{1}{\Box} R = -\int_0^t dt' a^{-3} \int_0^{t'} dt'' a^3 \left[12H^2 + 6\dot{H} \right]. \tag{13.62}$$

For de Sitter spacetime—$a(t) = e^{Ht}$ with constant H—we get the correct correspondence limit:

$$\frac{1}{\Box} R = -4 \ln(a) + \frac{4}{3} \left[1 - e^{-3Ht} \right]. \tag{13.63}$$

More generally, expression (13.62) implies that $-X[g](x)$ will grow during the inflationary regime of large Ricci curvature, and then freeze in to a constant during the radiation dominated era of $R(t) = 0$. As long as the function $f(x)$ in (13.59) grows monotonically and without bound, this ansatz for the pressure is bound to produce enough screening to end inflation in roughly the right way.

Numerical Results There is no hope of deriving an analytic solution for $a(t)$ when the pressure is as complicated as (13.59) with (13.62), but this is a simple problem to solve numerically. Figures 13.1, 13.2 give the evolution of the non-local source $X(t)$, Figs. 13.3, 13.4 present the Ricci scalar $R(t)$, and Figs. 13.5, 13.6 show the Hubble parameter $H(t)$. These results were generated for the choice $f(x) = \exp(x) - 1$—the "exponential model"—although any function $f(x)$ which grows monotonically and without bound gives the same qualitative behaviour, including even $f(x) = x$. To avoid a long preliminary evolution with negligible effect, we set the unrealistically high value of $G\Lambda = 1/200$. Again, the behaviour is qualitatively the same for any choice of $G\Lambda$.

The following basic features emerge from our numerical work [62]:

- During the era of inflation, the source $-X(t)$ grows while the curvature scalar $R(t)$ and the Hubble parameter $H(t)$ decrease.
- Inflationary evolution dominates roughly until we reach the critical point X_{cr} defined by:

$$1 - 8\pi G \Lambda f[-G\Lambda X_{cr}] \equiv 0. \tag{13.64}$$

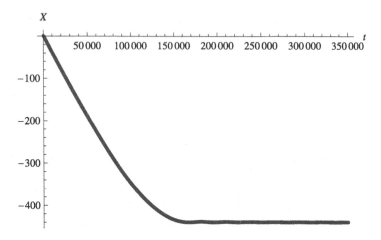

Fig. 13.1 The evolution of the source $X(t)$ over the full range for the exponential model

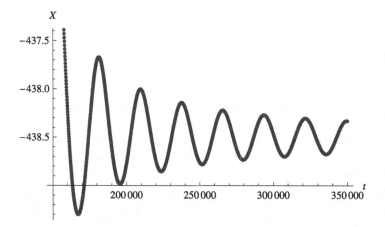

Fig. 13.2 The evolution of the source $X(t)$ during the oscillatory regime for the exponential model

- The epoch of inflation ($q < 0$) ends slightly before $X(t)$ reaches X_{cr}. This is most directly seen from the deceleration parameter because initially $q(t = 0) = -1$, while at criticality $q(t = t_{\mathrm{cr}}) = +\frac{1}{2}$.
- The source $X(t)$ oscillates with constant period and decreasing amplitude.
- Oscillations in $R(t)$ become significant as we approach the end of inflation; they are centered around $R = 0$, their frequency is given by:

$$\omega = G\Lambda H_0 \sqrt{72\pi f'_{\mathrm{cr}}},\tag{13.65}$$

and their amplitude decreases like the inverse of the number of oscillations.

- While there is net expansion during the era of oscillations, the Hubble parameter $H(t)$ attains small negative values for short time intervals. Of course negative

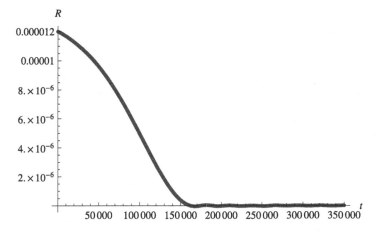

Fig. 13.3 The evolution of the curvature scalar $R(t)$ over the full range for the exponential model

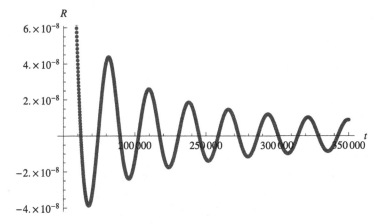

Fig. 13.4 The evolution of the curvature scalar $R(t)$ during the oscillatory regime for the exponential model

$H(t)$ corresponds to a compressing universe, which should lead to rapid reheating when matter couplings are included.

Analytic Results   Although one cannot obtain analytic results for the full evolution of $a(t)$, it is possible to give an approximate treatment for the period of oscillations. We use the evolution equation:

$$2\dot{H} + 3H^2 = \Lambda\{1 - 8\pi G\Lambda f[-G\Lambda X]\}, \quad X \equiv \frac{1}{\Box}R. \qquad (13.66)$$

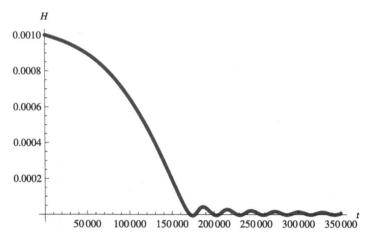

Fig. 13.5 The evolution of the Hubble parameter $H(t)$ over the full range for the exponential model

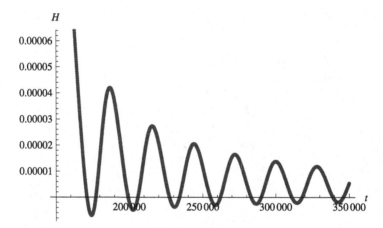

Fig. 13.6 The evolution of the Hubble parameter $H(t)$ during the oscillatory regime for the exponential model

Recall that we assume only that the function $f(x)$ grows monotonically and without bound. Hence there must exist a critical point X_{cr} such that:

$$1 - 8\pi G\Lambda f[-G\Lambda X_{cr}] = 0 \quad \Longrightarrow \quad X_{cr} = -\frac{1}{G\Lambda} f^{-1}\left(\frac{1}{8\pi G\Lambda}\right). \qquad (13.67)$$

Inflationary evolution dominates roughly until we reach the critical point. Close to the critical point the induced pressure p is nearly constant and, thus, it makes sense to expand f around its critical point:

$$f \simeq f_{cr} - G\Lambda\Delta X(t)f'_{cr}, \quad \Delta X(t) \equiv X(t) - X_{cr}. \qquad (13.68)$$

Now consider the linearized evolution equation:

$$2\dot{H} + 3H^2 \simeq 8\pi(G\Lambda)^2\Lambda(X - X_{cr})f'[-G\Lambda X_{cr}].$$ (13.69)

Using (13.61) we can express the co-moving time derivative of the Hubble parameter as:

$$\dot{H} = \frac{1}{6}R - 2H^2.$$ (13.70)

Because the amplitudes of both $R(t)$ and $H(t)$ fall like t^{-1} during the era of oscillations, the second term in (13.70) is irrelevant. Consequently, the evolution equation (13.69) becomes:

$$-R + 3H^2 \simeq -24\pi(G\Lambda)^2\Lambda(X - X_{cr})f'_{cr},$$ (13.71)

where we have defined:

$$f'_{cr} \equiv f'[-G\Lambda X_{cr}] \equiv -\frac{1}{G\Lambda}\frac{d}{dX}f[-G\Lambda X]\Big|_{X=X_{cr}}.$$ (13.72)

Action of the d'Alembertian operator (13.61) on (13.71) gives:

$$\ddot{R} + 2H\dot{R} + (\omega^2 - \dot{H})R + [3H^2R - 36H^4] \simeq 0,$$ (13.73)

where we define:

$$\omega^2 \equiv 24\pi(G\Lambda)^2\Lambda f'_{cr}.$$ (13.74)

We can again neglect the various "small" terms in (13.73) to infer:

$$\ddot{R} + 2H\dot{R} + \omega^2R \simeq 0 \implies R(t) \simeq \frac{\sin(\omega t)}{a(t)}.$$ (13.75)

This reveals the presence of oscillations. Note also that the frequency (13.74) agrees with numerical results.

Generic Results It is also possible to derive approximate analytic results for the period of inflation. If N is the number of e-foldings before criticality, the various geometrical parameters are [71–73]:

$$a(t) = a_{cr}e^{-N},$$ (13.76)

$$H(t) \simeq \frac{1}{3}\omega\sqrt{4N + \frac{4}{3}},$$ (13.77)

$$\varepsilon(t) \simeq \frac{2}{4N + \frac{4}{3}}.$$ (13.78)

During the oscillatory era it is best to describe these same parameters using the time $\Delta t \equiv t - t_{cr}$ since criticality. The following approximate relations hold [71–73]:

$$a(t) = a_{cr}C_2[C_1 + \omega\Delta t + \sqrt{2}\cos(\omega\Delta t + \phi)], \tag{13.79}$$

$$H(t) = \frac{\omega[1 - \sqrt{2}\sin(\omega\Delta t + \phi)]}{C_1 + \omega\Delta t + \sqrt{2}\cos(\omega\Delta t + \phi)}, \tag{13.80}$$

$$\varepsilon(t) = 1 + \frac{\sqrt{2}\cos(\omega\Delta t + \phi)[C_1 + \omega\Delta t + \sqrt{2}\cos(\omega\Delta t + \phi)]}{[1 - \sqrt{2}\sin(\omega\Delta t + \phi)]^2}. \tag{13.81}$$

The constants ϕ, C_1 and C_2 in relations (13.79)–(13.81) are chosen to match the two epochs at criticality ($N = 0$ and $\Delta t = 0$):

$$\phi = \arcsin\left(\frac{\sqrt{2} - \sqrt{2970}}{56}\right) \approx -\frac{\pi}{2}, \tag{13.82}$$

$$C_1 = \frac{\sqrt{27}}{2} - \frac{\sqrt{27}}{2}\sin\phi - \sqrt{2}\cos\phi \approx 3, \tag{13.83}$$

$$C_2 = \frac{1}{C_1 + \sqrt{2}\cos\phi} \approx \frac{1}{6}. \tag{13.84}$$

Primordial Density Perturbations

(i) *Scalar perturbations.* Initially ultraviolet modes in scalar driven inflation oscillate and redshift, and then become approximately constant around the time of horizon crossing [5]. The behaviour of scalar perturbations in this model differs in two significant ways. First, initially ultraviolet modes merely redshift, they do not oscillate. Like scalar driven inflation, the modes of this model become approximately constant around the time of horizon crossing. However, all super-horizon modes in this model begin oscillating with the same frequency ω at the end of inflation [74]. Because there are so many of these super-horizon modes after a long period of inflation, the fact that all of them start to oscillate at the end of inflation should lead to very rapid reheating, without the need to invoke anything other than the usual gravitational couplings to matter. After the universe reaches radiation domination one can show that the oscillations stop [24, 25], which is consistent with an approximately scale invariant power spectrum. What we cannot do is to evaluate the normalization. That is fixed by canonical quantization in scalar driven inflation, but we only have the effective field equations for this model. Recall that the combination of causality and non-locality means our effective field equations cannot derive from a conventional action principle.

(ii) *Tensor perturbations.* The analysis of tensor perturbations in this class of models is much simpler than that of scalar perturbations [72, 73]. The reason is that our perfect fluid stress-energy has no effect on the tensor perturbations h_{ij}^{TT}. Therefore the resulting power spectrum Δ_h^2 has the usual form:

$$\Delta_h^2(k) \simeq \frac{16GH^2(t_k)}{\pi}, \tag{13.85}$$

but with the expansion history peculiar to our model. There is nothing unconventional about our expansion history (13.76)–(13.78) during the epoch of inflation, so our prediction for the B mode of polarization in the cosmic microwave background is not distinct from that of scalar driven inflation. The period for which our model differs is the phase of oscillations (13.79)–(13.81), during which the usual Hubble "friction" term actually changes sign. Because the end of inflation comes about 50 e-foldings after the horizon crossing of the observable part of the cosmic microwave background, the corresponding enhancement in the stochastic background of gravitational radiation will be at the uncomfortably high frequency of $f \sim 10^9$ Hz [72, 73]. No current gravity wave detector has sensitivity at this frequency but one has been proposed [75].

13.3 Post-Inflationary Evolution

We assume that energy flows from the gravitational to the matter sector, leading to a radiation dominated universe at $t = t_r$. Because our model is sourced by the Ricci scalar, which vanishes during radiation domination, the quantum induced stress-energy simply cancels the bare cosmological constant. There is no deviation from conventional cosmology until the onset of matter domination at $t = t_m$. By that time the scales are so much below those of primordial inflation that only very small changes occur in $X(t)$, and we can use first order perturbation theory to compute the total pressure:

$$p_{\text{tot}} \equiv -\frac{\Lambda}{8\pi G} + p[g](x) \tag{13.86}$$

$$= -\frac{\Lambda}{8\pi G}\left\{1 - 8\pi G \Lambda f\left[-G\Lambda(X_{\text{cr}} + \Delta X)\right]\right\} \tag{13.87}$$

$$\simeq -\frac{\Lambda}{G} \times (G\Lambda)^2 f'_{\text{cr}} \Delta X. \tag{13.88}$$

The simple source (13.59) grows according to the formula:

$$\Delta X(t) \equiv X(t) - X_{\text{cr}} = -\frac{4}{3} \ln\left[1 + \frac{3}{2} H_m(t - t_m)\right] + O(1). \tag{13.89}$$

These facts give rise to two fatal problems for the model:

- *The Sign problem*: Because f is monotonically increasing and unbounded:

$$p_{\text{tot}} > 0 \quad \text{when } X(t) < X_{\text{cr}} \ll 0. \tag{13.90}$$

The observation of late time acceleration [15–18] implies the opposite.
- *The Magnitude problem*: The magnitude of the total pressure produced is unacceptably large:

$$\frac{p_{\text{tot}}}{p_{\text{now}}} \simeq \left(\frac{G\Lambda H_I}{H_{\text{now}}}\right)^2 f'_{\text{cr}} \Delta X \simeq 10^{86} \times f'_{\text{cr}} \times \Delta X, \tag{13.91}$$

where we have used:

$$p_{\text{now}} \simeq -\frac{3}{8\pi G} H_{\text{now}}^2, \quad H_I \sim 10^{13} \text{ GeV}, \quad H_{\text{now}} \sim 10^{-33} \text{ eV}. \quad (13.92)$$

Improved Ansatz   Both problems can be addressed by changing the source (13.62). What we need to do is add an extra curvature S inside the inverse d'Alembertian, divided by Λ to keep things dimensionless [76]:

$$p[g](x) = \Lambda^2 f[-G\Lambda X](x), \quad (13.93)$$

$$-G\Lambda X = -G\Lambda \frac{1}{\Box} R \quad \longrightarrow \quad -G\frac{1}{\Box}(R \times S) = -\frac{G\Lambda}{\Box}\left(R \times \frac{S}{\Lambda}\right). \quad (13.94)$$

In this way the magnitude falls with cosmological evolution so that $\Delta X(t)$ experiences only an acceptably small change at the onset of matter domination. To keep inflation ending successfully it is necessary to evaluate this curvature S far back in the past of the Ricci scalar R. We obtained acceptable results with a factor of ten.

That suffices for the magnitude problem. To solve the sign problem we note that the curvature scalar is positive during both inflation—$R = +12H^2$—and matter domination—$R = +3H^2$. A simple choice for S that changes its sign is R_{00} which equals $-3H^2$ during inflation and $+\frac{3}{2}H^2$ during matter domination [76]. Note that we can invariantly select the 00 component of $R_{\mu\nu}$ using the timelike 4-velocity field u^μ, which is just δ_0^μ for FRW. Hence the specialization of the improved ansatz to FRW is:

$$p[g](x) = \Lambda^2 f[-G\Lambda Y](x), \quad (13.95)$$

$$Y[g](t) = -\frac{1}{\Lambda}\frac{1}{\Box}\left[R(t) \times R_{00}\left(\frac{1}{10}t\right)\right] \equiv X_{\text{cr}} + \Delta Y. \quad (13.96)$$

Late Time Acceleration   Finally, we compute the total pressure in the improved ansatz [76]:

$$p_{\text{tot}} \simeq -G\Lambda^3 f_{\text{cr}}'\Delta Y \simeq -200 G\Lambda^2 f_{\text{cr}}' H_m^2. \quad (13.97)$$

For the exponential model:

$$f(x) = e^x - 1 \quad \Longrightarrow \quad f_{\text{cr}}' = \frac{1}{8\pi G\Lambda}, \quad (13.98)$$

the pressure ratio is:

$$t \gg t_m \quad \Rightarrow \quad \frac{p_{\text{tot}}}{p_{\text{now}}} \simeq \frac{200}{3} 8\pi (G\Lambda)^2 \times f_{\text{cr}}' \times \left(\frac{H_m}{H_{\text{now}}}\right)^2 \quad (13.99)$$

$$\simeq \frac{200}{3} 8\pi (G\Lambda)^2 \times f_{\text{cr}}' \times 10^{10} \quad (13.100)$$

$$\simeq \frac{2}{3} \times 10^{12} \times G\Lambda. \quad (13.101)$$

It is evident that for physically reasonable values of $G\Lambda = M^4 M_{Pl}^{-4}$ we can achieve the desired equality of p_{tot} with p_{now} whose ratio is given by (13.101).

13.4 Conclusions

There is very strong evidence that the universe underwent a very early phase of accelerated expansion known as primordial inflation. One can devise a scalar inflaton (13.17) to support this geometry but this entails positing a new and otherwise undetected degree of freedom, as well as making some unrealistic and sometimes contradictory assumptions about the inflaton's potential and its initial condition. On the other hand, there is no question that inflation results in the production of a vast sea of infrared gravitons, nor is there any question that these gravitons attract one another to some extent. Explicit results from perturbation theory indicate that this attraction grows stronger with time, until perturbation theory eventually breaks down.

Great controversy surrounds this final claim but, if it can be established, the phenomenological payoff is enormous. For then it becomes possible to dispense with the scalar inflaton and to make a virtue out of what is usually regarded as a terrible problem: namely, the fact that the observed cosmological constant is more than 120 orders of magnitude below its natural scale. We propose that the bare cosmological constant is not unnaturally small but instead only a few orders of magnitude below the Planck scale. What is being measured today is not this bare cosmological constant but rather the expansion rate, and we propose that the effect of the bare cosmological constant on the current expansion rate is subject to almost perfect screening by the self-gravitation of gravitons produced during a very long period of Λ-driven inflation.

We believe it is possible to use perturbation theory to establish the reality of quantum gravitational screening. We also feel one can resum the series of leading infrared logarithms to *derive* what happens at late times. However, neither thing will be easy, nor will they be quickly attained. In the meantime, we have devised a class of non-local effective field equations which might describe the eventual result of such a derivation. At this stage, one is free to dismiss our motivation from quantum gravitational inflation and simply regard these effective field equations in the same light as another classical model of inflation. They are at least no worse than scalar inflaton models, and they do have some remarkable and quite generic features. Chief of these are that inflation ends in a phase of oscillations which violate the weak energy condition, and for which there is participation from every super-horizon mode, not just the zero mode. The former feature may have left an observable signature in the stochastic background of gravitational radiation [72, 73]. And the last feature should lead to almost instantaneous reheating using only the universal gravitational coupling to matter [74].

Although the simplest of our models breaks down after the onset of matter domination, it can be easily fixed. Indeed, this can be done in such a way as to explain the current phase of cosmic acceleration. It will be interesting to see if any of these models can be derived from fundamental theory.

Acknowledgements We are grateful to L. Papantonopoulos for the invitation to deliver these lectures and for his extraordinary patience while they were written up. This work was partially supported by European Union grant FP-7-REGPOT-2008-1-CreteHEPCosmo-228644, by EU program "Thalis" ESF/NSRF 2007-2013, by NSF grant PHY-0855021, and by the Institute for Fundamental Theory at the University of Florida.

References

1. A.G. Riess et al., Astrophys. J. **730**, 119 (2011). arXiv:1103.2976
2. M. Hicken et al., Astrophys. J. **700**, 1097 (2009). arXiv:0901.4804
3. E. Komatsu et al., Astrophys. J. Suppl. **192**, 18 (2011). arXiv:1001.4538
4. A. Linde, *Particle Physics and Inflationary Cosmology* (Harwood, Amsterdam, 1990)
5. V.F. Mukhanov, *Physical Foundations of Comology* (Cambridge University Press, Cambridge, 2005)
6. R. Keisler et al., Astrophys. J. **743**, 28 (2011). arXiv:1105.3182
7. E.O. Kahya, V.K. Onemli, R.P. Woodard, Phys. Lett. B **694**, 101 (2010). arXiv:1006.3999
8. A.H. Guth, Phys. Rev. D **23**, 347 (1981)
9. A. Albrecht, P.J. Steinhardt, Phys. Rev. Lett. **48**, 1220 (1982)
10. A. Linde, Phys. Lett. B **108**, 389 (1982)
11. N.C. Tsamis, R.P. Woodard, Ann. Phys. **267**, 145 (1998). arXiv:hep-th/9712331
12. T.D. Saini, S. Raychaudhury, V. Saini, A.A. Starobinsky, Phys. Rev. Lett. **85**, 1162 (2000). arXiv:astro-ph/9910231
13. S. Capozziello, S. Nojiri, S.D. Odintsov, Phys. Lett. B **634**, 93 (2006). arXiv:hep-th/0512118
14. T. Vachaspati, M. Trodden, Phys. Rev. D **61**, 023502 (1999). arXiv:gr-qc/9811037
15. A.G. Riess et al., Astron. J. **116**, 1009–1038 (1998). arXiv:astro-ph/9805201
16. S. Perlmutter et al., Astrophys. J. **517**, 565–586 (1999). arXiv:astro-ph/9812133
17. Y. Wang, P. Mukherjee, Astrophys. J. **650**, 1 (2006). arXiv:astro-ph/0604051
18. U. Alam, V. Sahni, A.A. Starobinsky, J. Cosmol. Astropart. Phys. **0702**, 011 (2007). arXiv:astro-ph/0612381
19. R. Allahverdi, R. Brandenberger, F.Y. Cyr-Racine, A. Mazumdar, Annu. Rev. Nucl. Part. Sci. **60**, 27 (2010). arXiv:1001.2600
20. L.F. Abbott, S. Deser, Nucl. Phys. B **195**, 76 (1982)
21. P.H. Ginsparg, M.J. Perry, Nucl. Phys. B **222**, 245 (1983)
22. L.P. Grishchuck, Sov. Phys. JETP **40**, 409 (1975)
23. L.H. Ford, L. Parker, Phys. Rev. D **16**, 1601 (1977)
24. N.C. Tsamis, R.P. Woodard, Nucl. Phys. B **474**, 235 (1996). arXiv:hep-ph/9602315
25. N.C. Tsamis, R.P. Woodard, Ann. Phys. **253**, 1 (1997). arXiv:hep-ph/9602316
26. R. Arnowitt, S. Deser, C.W. Misner, Phys. Rev. Lett. **4**, 375 (1960)
27. N.C. Tsamis, R.P. Woodard, Class. Quantum Gravity **22**, 4171 (2005). arXiv:gr-qc/0506089
28. J. Garriga, T. Tanaka, Phys. Rev. D **77**, 024021 (2008). arXiv:0706.0295
29. N.C. Tsamis, R.P. Woodard, Phys. Rev. D **78**, 028501 (2008). arXiv:0708.2004
30. A.M. Polyakov, Sov. Phys. Usp. **25**, 187 (1982)
31. N.P. Myhrvold, Phys. Rev. D **28**, 2439 (1983)
32. E. Mottola, Phys. Rev. D **31**, 754 (1985)
33. E. Mottola, Phys. Rev. **33**, 2136 (1986)
34. P.O. Mazur, E. Mottola, Nucl. Phys. B **278**, 694 (1986)
35. I. Antoniadis, J. Iliopoulos, T.N. Tomaras, Phys. Rev. Lett. **56**, 1319 (1986)
36. N.C. Tsamis, R.P. Woodard, Phys. Lett. B **301**, 351 (1993)
37. A.D. Dolgov, M.B. Einhorn, V.I. Zakharov, Phys. Rev. D **52**, 717 (1995). arXiv:gr-qc/9403056
38. A.M. Polyakov, Nucl. Phys. B **834**, 316 (2010). arXiv:0912.5503
39. D. Krotov, A.M. Polyakov, Nucl. Phys. B **849**, 410 (2011). arXiv:1012.2107
40. A. Higuchi, Class. Quantum Gravity **26**, 072001 (2009). arXiv:0809.1255
41. E.T. Akhmedov, Mod. Phys. Lett. A **25**, 2815 (2010). arXiv:0909.3722

42. E.T. Akhmedov, P. Burda, Phys. Lett. B **687**, 267 (2010). arXiv:0912.3435
43. E.T. Akhmedov, A. Roura, A. Sadofyev, Phys. Rev. D **82**, 044035 (2010). arXiv:1006.3274
44. D. Marolf, I.A. Morrison, Phys. Rev. D **82**, 105032 (2010). arXiv:1006.0035
45. D. Marolf, I.A. Morrison, Phys. Rev. D **84**, 044040 (2011). arXiv:1010.5327
46. D. Marolf, I.A. Morrison, Gen. Relativ. Gravit. **43**, 3497 (2011). arXiv:1104.4343
47. H. Kitamoto, Y. Kitazawa, Nucl. Phys. B **839**, 552 (2010). arXiv:1004.2451
48. A. Higuchi, D. Marolf, I.A. Morrison, Phys. Rev. D **83**, 084029 (2011). arXiv:1012.3415
49. C.P. Burgess, R. Holman, L. Lelond, S. Shandera, J. Cosmol. Astropart. Phys. **1010**, 017 (2010). arXiv:1005.3551
50. D. Boyanovsky, R. Holman, J. High Energy Phys. **1105**, 047 (2011). arXiv:1103.4648
51. S.P. Miao, N.C. Tsamis, R.P. Woodard, J. Math. Phys. **50**, 122502 (2009). arXiv:0907.4930
52. S.P. Miao, N.C. Tsamis, R.P. Woodard, J. Math. Phys. **51**, 072503 (2010). arXiv:1002.4037
53. S.P. Miao, N.C. Tsamis, R.P. Woodard, J. Math. Phys. **52**, 122301 (2011). arXiv:1106.0925
54. E.O. Kahya, S.P. Miao, R.P. Woodard, J. Math. Phys. **53**, 022304 (2012). arXiv:1112.4420
55. A. Higuchi, D. Marolf, I.A. Morrison, Class. Quantum Gravity **28**, 245012 (2011). arXiv:1107.2712
56. S.P. Miao, N.C. Tsamis, R.P. Woodard, Class. Quantum Gravity **28**, 245013 (2011). arXiv:1107.4733
57. S.S. Kouros, Class. Quantum Gravity **18**, 4961 (2001). arXiv:gr-qc/0107064
58. P.J. Mora, R.P. Woodard, arXiv:1202.0999
59. N.C. Tsamis, R.P. Woodard, Nucl. Phys. B **724**, 295 (2005). arXiv:gr-qc/0505115
60. T. Prokopec, N.C. Tsamis, R.P. Woodard, Ann. Phys. **323**, 1324 (2008). arXiv:0707.0847
61. N.C. Tsamis, R.P. Woodard, Ann. Phys. **321**, 875 (2006). arXiv:gr-qc/0506056
62. N.C. Tsamis, R.P. Woodard, Phys. Rev. D **80**, 083512 (2009). arXiv:0904.2368
63. J. Schwinger, J. Math. Phys. **2**, 407 (1961)
64. K.T. Mahanthappa, Phys. Rev. **126**, 329 (1962)
65. P.M. Bakshi, K.T. Mahanthappa, J. Math. Phys. **4**, 1 (1963)
66. P.M. Bakshi, K.T. Mahanthappa, J. Math. Phys. **4**, 12 (1963)
67. L.V. Keldysh, Sov. Phys. JETP **20**, 1018 (1965)
68. K.C. Chou, Z.B. Su, B.L. Hao, L. Yu, Phys. Rep. **118**, 1 (1985)
69. R.D. Jordan, Phys. Rev. D **33**, 444 (1986)
70. E. Calzetta, B.L. Hu, Phys. Rev. D **35**, 495 (1987)
71. N.C. Tsamis, R.P. Woodard, Phys. Rev. D **80**, 083512 (2009). arXiv:0904.2368
72. M.G. Romania, N.C. Tsamis, R.P. Woodard, Class. Quantum Gravity **28**, 075013 (2011). arXiv:1006.5150
73. M.G. Romania, N.C. Tsamis, R.P. Woodard, arXiv:1108.1696
74. N.C. Tsamis, R.P. Woodard, Phys. Rev. D **82**, 063502 (2010). arXiv:1006.4834
75. A. Nishizawa et al., Phys. Rev. D **77**, 022002 (2008). arXiv:0710.1944
76. N.C. Tsamis, R.P. Woodard, Phys. Rev. D **81**, 103509 (2010). arXiv:gr-qc/1001492

Index

G. Calcagni et al. (eds.), *Quantum Gravity and Quantum Cosmology*,
Lecture Notes in Physics 863, DOI 10.1007/978-3-642-33036-0,
© Springer-Verlag Berlin Heidelberg 2013